Drong
206 Rosenberg Hall

Statistics for Environmental Biology and Toxicology

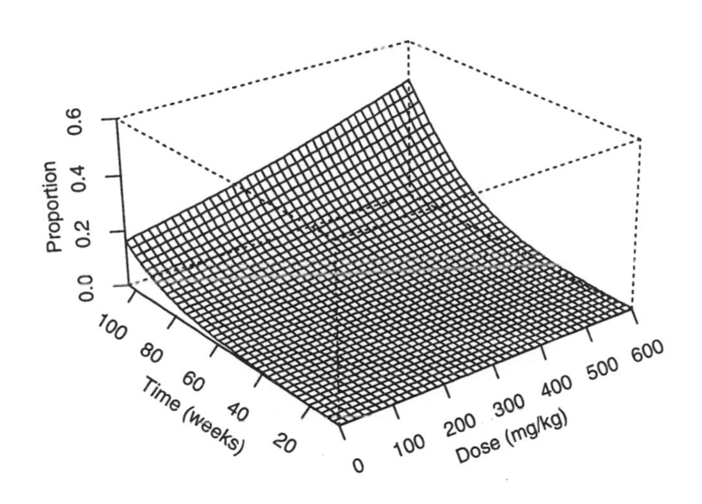

Statistics for Environmental Biology and Toxicology

Walter W. Piegorsch
Professor of Statistics
University of South Carolina
USA

and

A. John Bailer
Professor of Mathematics and Statistics
Miami University
USA

CHAPMAN & HALL
London · Weinheim · New York · Tokyo · Melbourne · Madras

Published by Chapman & Hall, 2–6 Boundary Row, London SE1 8HN, UK

Chapman & Hall, 2–6 Boundary Row, London SE1 8HN, UK

Chapman & Hall GmbH, Pappelallee 3, 69469 Weinheim, Germany

Chapman & Hall USA., 115 Fifth Avenue, New York, NY 10003, USA

Chapman & Hall Japan, ITP-Japan, Kyowa Building, 3F, 2-2-1 Hirakawacho, Chiyoda-ku, Tokyo 102, Japan

Chapman & Hall Australia, 102 Dodds Street, South Melbourne, Victoria 3205, Australia

Chapman & Hall India, R. Seshadri, 32 Second Main Road, CIT East, Madras 600 035, India

---

First edition 1997

Printed in Great Britain by T.J. Press Ltd, Padstow, Cornwall

ISBN 0 412 04731 4

A catalogue record for this book is available from the British Library

∞ Printed on permanent acid-free text paper, manufactured in accordance with ANSI/NISO Z39.48 - 1992 and ANSI/NISO Z39.48 - 1984 (Permanence of Paper).

To Karen
W.W.P.

To Jenny, Sara, Jacob, Christopher and Emily
A.J.B.

Contents

Preface

The study of toxins in the environment grew out of the post-World War II industrial age and the associated increase in use of synthetic pesticides, fertilizers, detergents and their additives, fossil fuels, derived industrial products, fissionable materials, and their generated waste products. As these uses expanded during the 1950s and 1960s, their potential for degradation of environmental resources became readily apparent. The resulting public recognition of a need to protect the environment and its inhabitants from toxic insult reached a peak in the 1970s. Many innovations in biological and ecological experimentation accompanied this increased awareness; these developments dramatically improved our ability to screen for detrimental toxins in the environment, estimate toxic impacts on biological systems, and assess the risks of toxic exposures to ecosystems and the organisms that inhabit them.

With the advent of these advanced experimental methods has come the need for sophisticated statistical procedures to analyze the data they generate. Data-analytic problems arising from innovative environmental research often require statistical analyses more complicated than the simple methods presented in standard textbooks and monographs, however. Methods learned in common first-year courses – such as two-sample t-tests, simple linear regression, analysis of variance (ANOVA), and other basic methods appropriate with normally distributed data – no longer provide investigators with sufficient design, analytic, and interpretive capabilities for complex environmental data. Indeed, in some cases a lack of awareness of appropriate or available statistical methods can unnecessarily constrain environmental investigators to simpler or less-ambitious research questions than they might otherwise wish to explore.

Development of quantitative methods necessary for modern environmental research often is classified under the broad umbrella of *environmetrics*. By its very nature, environmetric research involves close interdisciplinary collaboration among biological, environmental, and statistical scientists. Within this context, our goal in this text is to work from problems in environmental toxicology that have motivated statistical/environmetric advances, and present a selection of corresponding statistical techniques. We have targeted an audience that includes both environmental scientists and statisticians. These include: advanced graduate students in applied environmental fields, intermediate graduate students and advanced undergraduates in statistics, and also postgraduate applied researchers in statistics, toxicology, and biology requiring a source-book on quantitative models and methods in modern environmental data analysis.

For this audience, we illustrate statistical methods appropriate for use with data generated in environmental toxicology or associated ecological/biological studies. We extend beyond the basic approaches mentioned above, to nonnormal data and/or complex or nonstandard models (for example, overdispersed data, nonlinear models, time-to-event data). The methodology represents an intermediate step between basic statistical methods and advanced, research-level statistics.

We assume a basic cognizance of and familiarity with introductory statistics for understanding of the material. Background knowledge must include: significance testing, confidence intervals, one- and two-way analysis of variance, multiple regression and correlation, and associated notation. Also required is at least a passing familiarity with standard notation for probabilities of events, families of distributions, expectation, and variance. Readers with a full year (two semesters or three quarters) of introductory statistics and acquainted with fundamentals of calculus will benefit most from the material. The calculus background is not strictly necessary, but it will make many of the more advanced concepts and notation easier to digest. Where appropriate, we have highlighted with asterisks ($*$) those sub-sections and exercises that address more advanced material, or that require greater knowledge and understanding of probability distributions and/or calculus.

To review these requisite topics for the reader, and to establish concepts and notation, we begin with material on probability and random variables (Chapter 1), including statistical distributions that are common in the environmental sciences but not often discussed in introductory statistics courses. Chapter 2 continues the introductory material with a review of statistical inference, including methods of estimation and introductory likelihood theory, construction of confidence intervals, and principles of hypothesis testing. In these two chapters, environmental/environmetric applications are referred to only in passing, in order to introduce the material as efficiently as possible. Readers comfortable with these topics may wish move through the two chapters quickly, read only selected sections in detail (as appropriate), and/or refer back to certain sections in these chapters where needed for better comprehension of later material.

The next chapter is intended also as a form of review, covering introductory issues in experiment design (Chapter 3). Here, however, we begin to highlight specific examples in environmental biology and toxicology, in order to help illustrate the concepts. Thereafter follows material on analysis of treatment-versus-control differences (Chapter 4), and this chapter forms a bridge between prerequisite material and more advanced methods for environmental data analysis. Depending on the reader's statistical preparation and training, this chapter may serve as a form of review, but may also introduce new and modern methods that extend beyond basic statistical methods. In particular, the discussions of two-group comparisons for proportions or counts may prove novel to some readers.

More advanced, ostensibly distinct material begins in earnest in Chapter 5 on treatment-versus-control multiple comparisons, and continues through the rest of the text. The material is ordered so as to increase in difficulty as the text progresses, with common threads highlighted as often as possible. Nonetheless, the various chapters are designed to touch on a variety of data-analytic settings encountered in environmental toxicology and biology. Illustrative examples using data from (wherever possible) actual environmental studies are integral parts of the presentation, and examples often continue throughout the chapter to illustrate expanding or developing concepts. Short summary sections complete each chapter's presentation, reviewing the material and in some cases including extensions or additional reading. Also, all chapters (including the introductory ones) end with a set of exercises. These are available for instructors to assign as additional material, or for the general reader to practice the skills learned earlier in each chapter. Although the exercises are not formally subdivided (as are the chapters), they do follow roughly the order of each chapter's material. In doing so, they present a mix of applied and theoretical problems (where, as noted above, more advanced exercises are marked with asterisks). Here again, a rich collection of data from actual environmental studies is included.

An integral component of our presentation is the appeal to computer analysis with the more advanced methods. This represents a critical and necessary feature of modern statistics: advanced computer technology has made available quantitative methods that could not otherwise be applied to complex environmental data. In this vein, we highlight computer use wherever possible. A wealth of computer packages and programming languages are available for this use; we have chosen to highlight two: SAS® (SAS Institute Inc., 1989) and S-PLUS® (StatSci Division of MathSoft Inc., 1995). SAS's ubiquity and extent make it a natural choice, and we expect a majority of readers will already be familiar with at least basic SAS mechanics. In addition, the S-PLUS programming environment includes many functions similar to the SAS procedures, but provides the added benefits of easier customization, and pliant, yet advanced, graphical capabilities. We assume readers are already acquainted with the basics of each package, or can acquire such skills separately. For example, useful tutorials are available in Jaffe (1994), Spector (1994), or Venables and Ripley (1997). Figures containing computer code and output from these two packages are displayed throughout the text, starting in Chapter 2. Although these are not intended to be the most efficient way to program the desired operations, they will help illustrate use of the packages and (perhaps more importantly) interpretation of the outputs. Output from SAS procedures (version 6.11) is printed with the permission of SAS Institute Inc., Cary, NC, USA. Copyright © 1989–1995. S-PLUS programming and associated output (version 3.3) is printed with the permission of S-PLUS and MathSoft Inc., Seattle, WA, USA. Copyright © 1995.

For readers wishing to experiment further with SAS and S-PLUS, we have made available on an Internet site all the programming code presented in the text. The Uniform Resource Locator (URL) is

`http://www.muohio.edu/~ajbailer/book/codetable.htm`

Users may access this free site at their leisure. For more adventuresome programmers, we recommend the many statistical routines (including some in advanced languages such as FORTRAN or C) available at the *StatLib* Internet site, URL: `http://lib.stat.cmu.edu/`, or, for example, in the books by Griffiths and Hill (1985), Press *et al.* (1992), or Marazzi (1993). We gratefully acknowledge use of selected functions obtained from *StatLib* (`mixed.mtext` and `mixed.text` from

`http://lib.stat.cmu.edu/S/postscriptfonts`).

These functions, originally contributed by Alan Zaslavsky, were used to enhance many of the figures presented below.

For instructors wishing to use the material for an intermediate-level, one-semester course in environmetrics, a number of different syllabi are possible. For advanced graduate students *in applied fields*, consider

Chapter	Sections (omit sub-sections with asterisks)
1	1.1, 1.2, 1.3, 1.4
2	2.1, 2.2, 2.3, 2.4, 2.5
3-4	All
5	5.1, 5.2, 5.3
6	6.1, 6.3, 6.4
7	7.1, 7.2
8	8.1, 8.2, 8.4
9	9.1, 9.2, 9.3
11	11.1, 11.2

Time permitting, selected material from Chapter 10, or from Sections 7.3, 9.4, 9.5, 11.3, and/or 11.4 can be included as well. For intermediate/advanced students *in statistics*, a possible syllabus is

Chapter	Sections
1–4	review as necessary
5	5.1, 5.2, 5.3, 5.4, 5.5
6	6.1, 6.3, 6.4, 6.5, 6.6
7	7.1, 7.2, 7.3
8	All
9	9.1, 9.2, 9.3. 9.4
10	10.1, 10.2, 10.3
11	11.1, 11.2, 11.3

Here, material from Sections 6.7, 7.4, 9.5, 10.4, and/or 11.4 can be included, at the discretion of the instructor.

Although we focus on environmental biology and toxicology, the statistical methods we present are suitable for many other application areas, such as human biology or biomedicine. Different researchers or students may find certain sections and chapters to be of greater interest than others, depending on their own areas of research interest and training. This eclectic format is unavoidable, even intentional, in a single volume of intermediate and advanced environmetric methodology. We have based the contents on our own experiences in these areas, and hope our selection has produced a useful, unified presentation. Of course, certain important topic areas in environmetrics have been omitted or noted only briefly, in order to make the final product manageable. These include methods for data taken over space and/or time, population ecology including fertility and survival models employing Leslie matrices (Leslie, 1945), quantitative risk assessment, combining environmental information/meta-analysis, sampling methods, and descriptive and inferential multivariate methods. Useful texts that consider these various topics (often with environmental motivation) include: (a) for spatial statistics, Haining (1990), Christensen (1991), and Cressie (1993); (b) for time series analyses, Chatfield (1989), Box *et al*. (1994), and Christensen (1991); (c) for methods in population ecology, Green (1979) and Caswell (1989); (d) for risk analysis, Suter (1993), and Hallenbeck (1993); (e) for combining information and meta-analysis, Hedges and Olkin (1985) and Gaver *et al*. (1992); (f) for sampling methods, Green (1979), Scheaffer *et al*. (1996), and Thompson (1992); and (g) for multivariate methods, Davis (1986) and Christensen (1991). Many of these topics are also presented in selected chapters of Krewski and Franklin (1991), Hewitt (1992), Scheiner and Gurevitch (1993), Patil and Rao (1994), Cothern and

Ross (1994), and Morgan (1996). Interested readers are encouraged to peruse these various sources where appropriate.

Lastly, we wish to thank numerous colleagues at the University of South Carolina, Miami University, and elsewhere, including George Casella, Lawrence H. Cox, Philip M. Dixon, Don Edwards, Andrew S. Green, Joseph K. Haseman, Peter A. Lachenbruch, Brian G. Leroux, Barry H. Margolin, James T. Oris, Anthony J. Rossini, Terra L. Slaton, John D. Spurrier, Eric P. Smith, Wanzhu Tu, and David M. Umbach for their many helpful suggestions during the preparation of this material. Their comments helped make the presentation much more accessible and useful, although, of course, the material presented below is wholly our own responsibility. We thank also Michael D. Hogan for his early encouragement and recommendations on the project, and the editorial staff at Chapman & Hall, London, for their patience and guidance as this project developed and evolved into its final form.

Columbia, SC and Oxford, OH
February 1997

1

Basic probability and statistical distributions

Modern environmental science requires complex, multi-disciplinary collaborations among its practitioners, many of whom hail from diverse fields. At the core of this effort is the data-rich endeavor of observation and experimentation to study the relationships among environmental variables. This is no better exemplified than in environmental biology, where intricate response characterizations exist within biological systems, and where the addition of a toxic environmental agent can have adverse consequences. When data on environmental effects in biological systems are produced from biological sources, it is not uncommon to observe unusual interactions and other unexpected repercussions. The analysis and interpretation of such data often transcend basic statistical methodology, and require more advanced methods. In what follows, we will describe and illustrate analyses appropriate for nonstandard environmental data, centering on biological and toxicological outcomes used to indicate the effects of one or more environmental stimuli. As is true for most statistical approaches, however, the methods will have application in many nonbiological environmental scenarios as well.

We begin with a review of the basic vocabulary and concepts of probability and statistics. Readers familiar with these concepts may wish to skip forward to Section 1.2 on special statistical distributions, or farther on to Chapter 2, where we review basic principles in statistical inference.

1.1 INTRODUCTORY CONCEPTS IN PROBABILITY

1.1.1 Events and their probabilities

At a basic level, an experiment can be viewed as any random process that generates an outcome. We can collect all possible outcomes from a particular random process together into a set, $\mathscr{X}$, called the **sample space**. For any experimental process, an **event**, $\mathcal{E}$, is a particular outcome or collection of outcomes. **Probabilities** of observing events are defined in a long-term sense, by how frequent the events (or combinations of events) occur relative to all other elements of the sample space. That is, if we conduct an experiment time and time again, counting the number of occurrences of an event of interest, then the ratio of this count to the total number of times the experiment is conducted (a **relative frequency**) is the probability of the event of interest. We denote the probability of an event $\mathcal{E}$ as $P(\mathcal{E})$ for any $\mathcal{E}$ in the sample space, $\mathscr{X}$.

A number of basic rules, or **axioms**, of probability are employed in the interpretation of the quantity $P(\mathcal{E})$. Two of the most obvious are: (i) $0 \le P(\mathcal{E}) \le 1$, and (ii) $P(\mathscr{X}) = 1$. In addition, we have the following simple rules for combinations of events:

(a) **Addition rule**: $P(\mathcal{E}_1 \text{ or } \mathcal{E}_2) = P(\mathcal{E}_1) + P(\mathcal{E}_2) - P(\mathcal{E}_1 \text{ and } \mathcal{E}_2)$.

(b) **Conditionality rule**: $P(\mathcal{E}_1 \text{ given } \mathcal{E}_2) = P(\mathcal{E}_1 \text{ and } \mathcal{E}_2)/P(\mathcal{E}_2)$ for any event $\mathcal{E}_2$ such that $P(\mathcal{E}_2) > 0$. Conditional probabilities are denoted with the symbol '$|$', e.g. $P(\mathcal{E}_1 \mid \mathcal{E}_2) = P(\mathcal{E}_1 \text{ given } \mathcal{E}_2)$.

(c) **Multiplication rule**: $P(\mathcal{E}_1 \text{ and } \mathcal{E}_2) = P(\mathcal{E}_1 \mid \mathcal{E}_2)P(\mathcal{E}_2) = P(\mathcal{E}_2 \mid \mathcal{E}_1)P(\mathcal{E}_1)$.

When dealing with these rules, there are some special simplifying cases. For example, if two events, $\mathcal{E}_1$ and $\mathcal{E}_2$, never occur together, it is clear that $P(\mathcal{E}_1 \text{ and } \mathcal{E}_2) = 0$. In this case, we say $\mathcal{E}_1$ and $\mathcal{E}_2$ are **disjoint**. For disjoint events, the addition rule simplifies to $P(\mathcal{E}_1 \text{ or } \mathcal{E}_2) = P(\mathcal{E}_1) + P(\mathcal{E}_2)$. Two disjoint events, $\mathcal{E}_1$ and $\mathcal{E}_2$, are **complementary** events if the joint event $\{\mathcal{E}_1 \text{ or } \mathcal{E}_2\}$ equals $\mathscr{X}$, or equivalently, if $P(\mathcal{E}_1 \text{ or } \mathcal{E}_2) = 1$. If two events, $\mathcal{E}_1$ and $\mathcal{E}_2$, are complementary, then $P(\mathcal{E}_1) = 1 - P(\mathcal{E}_2)$ and $P(\mathcal{E}_2) = 1 - P(\mathcal{E}_1)$.

If two events, $\mathcal{E}_1$ and $\mathcal{E}_2$, occur in such a way that one has absolutely no impact on the other's occurrence, then the conditional probability that $\mathcal{E}_1$ occurs is unaffected by whether $\mathcal{E}_2$ has occurred, and vice versa. As such, $P(\mathcal{E}_1 \mid \mathcal{E}_2) = P(\mathcal{E}_1)$ and $P(\mathcal{E}_2 \mid \mathcal{E}_1) = P(\mathcal{E}_2)$, and we say the events are **independent**. Thus, knowledge that event $\mathcal{E}_2$ has occurred does not influence the probability of $\mathcal{E}_1$'s occurrence and, again, vice versa. (Notice that we assume implicitly here that both events occur with some nonzero probability: $P(\mathcal{E}_1) > 0$ and $P(\mathcal{E}_2) > 0$.) For two independent events, the multiplication rule simplifies to $P(\mathcal{E}_1 \text{ and } \mathcal{E}_2) = P(\mathcal{E}_1)P(\mathcal{E}_2)$.

1.1.2 Random variables

Probabilities allow us to describe the random nature of measured experimental outcomes. We call such outcomes **random variables**, and denote them by upper-case Roman letters, X or Z. We will assume that a random variable is a real number. Many other outcomes, such as categorical values, may be quantified into real numbers by coding them in some unambiguous fashion. For example, if an outcome is dichotomous ('yes' vs. 'no', 'dead' vs. 'alive', etc.) it may be quantified by defining the first outcome as $X = 1$ and the other as $X = 0$. Realized values of a random variable are denoted in general by lower-case Roman letters: $X = x$ or $Y = y$, etc.

Random variables may be characterized in one of two forms: either **discrete** or **continuous**. A discrete random variable takes on only discrete values, and is often associated with some sort of counting process. Examples include the dichotomous illustration above (say, 'dead' = 1 vs. 'alive' = 0 in a toxicity study), or counts of occurrences (numbers of tumors in a carcinogenesis study, numbers of observed species in an ecosystem appraisal, etc.). Conversely, a continuous random variable takes on values over a continuum. Continuous random variables are common when some form of measurement or level of response is under study (time to some event, blood concentrations of a toxin, weights of animals, etc.).

1.1.3 Probability functions

For discrete random variables, probabilities are fairly simple to describe. Suppose X is a discrete random variable, and consider the 'event' that X takes on some particular value, k. Then, the values of $P(X = k)$ over all possible values of k describe the **probability distribution** of X. We write $p_X(k) = P(X = k)$, and call this the **probability mass function** (or **p.m.f.**) of X. Notice that the p.m.f. is defined over a discrete set of values, hence the use of the term discrete random variable.

Since $p_X(k)$ is a probability, it must satisfy all of the basic axioms and rules described above. In particular, we have: (i) $0 \le p_X(k) \le 1$ for all arguments k, and (ii) $\sum p_X(k) = 1$, where the sum is taken over all points, k, of the sample space, $\mathcal{X}$. Summing the p.m.f. values as k increases produces the **cumulative distribution function** (or **c.d.f.**) of X:

$$F_X(k) = P(X \le k) = \sum_{i \le k} p_X(i). \tag{1.1}$$

The c.d.f. is important: it characterizes uniquely any random variable, so that two random variables with the same c.d.f. must have the same probability distribution.

For continuous random variables, the definition of probability functions is a bit more complex, and requires concepts from real-variable calculus. Continuous random variables have continuous probability functions, $f(x)$, that describe the density of probability over interval subsets of the real numbers: $a \le X \le b$. In a graphical sense, these probabilities are the area under $f(x)$ over $a \le x \le b$, and they are expressed as definite integrals:

$$P(a \le X \le b) = \int_a^b f(x)\,dx.$$

(Readers unfamiliar with the definite integral may wish to refer to introductory texts in calculus, such as Edwards and Penney (1990); readers requiring only a refresher may find targeted texts such as Khuri (1993) helpful.) Thus the c.d.f. of a continuous random variable is just the area under $f(x)$ integrated from $-\infty$ to the argument of the function:

$$F_X(x) = P(X \le x) = \int_{-\infty}^{x} f(y)\,dy. \tag{1.2}$$

From (1.2), we can see that $P(a \le X \le b) = F_X(b) - F_X(a)$ and so $P(X = a) = P(a \le X \le a) = F_X(a) - F_X(a) = 0$ for any a. Thus, while nonzero probability can be assigned to events that correspond to particular values for a discrete random variable, nonzero probability can be assigned only to events that correspond to *intervals* of values for a continuous random variable. In addition, the probability density $f(x)$ approaches a sort of incremental probability mass (over a constantly narrowing interval near x):

$$f(x) = \lim_{h \to 0} \frac{F_X(x + h) - F_X(x)}{h}.$$

We call $f(x)$ the **probability density function** (or **p.d.f.**) of X. Notice that if F_X is a differentiable function, then its derivative at the point x is the p.d.f.: $\partial F_X(x)/\partial x = f_X(x)$.

As with discrete random variables and their p.m.f.s, the p.d.f. from a continuous random variable must satisfy two basic axioms: (i) $f_X(x) \ge 0$ for all arguments x, and (ii) $\int_{-\infty}^{\infty} f_X(x)\,dx = 1$.

If more than one random variable is being considered, bivariate and multivariate extensions of these probability and distribution functions may be developed. For instance, if X and Y are discrete random variables, the

joint bivariate p.m.f. is $p_{X,Y}(j,k) = P(X = j \text{ and } Y = k)$, and the **joint bivariate c.d.f.** is $F_{X,Y}(j,k) = P(X \leq j \text{ and } Y \leq k)$. We can also describe the conditional and the marginal behavior of X and Y based on these quantities. For example the **conditional p.m.f.** of X given Y is defined as $p_{X|Y}(j \mid Y = k) = P[X = j \mid Y = k] = p_{X,Y}(j,k)/p_Y(k)$, where $p_Y(k) > 0$ for any k of interest. Marginally, X is itself a random variable; its p.m.f. is derived from the joint p.m.f. by averaging over all possible values of $Y = k$:

$$p_X(j) = \sum_k p_{X,Y}(j,k).$$

We call this the **marginal p.m.f.** of X.

For continuous random variables, the results are similar:

joint p.d.f.: $f_{X,Y}(x,y)$;

joint c.d.f.: $F_{X,Y}(x,y) = P(X \leq x \text{ and } Y \leq y)$;

conditional p.d.f.: $f_{X|Y}(x \mid y) = \dfrac{f_{X,Y}(x,y)}{f_Y(y)}$, with $f_Y(y) > 0$;

marginal p.d.f.: $f_X(x) = \displaystyle\int_{-\infty}^{\infty} f_{X,Y}(x,y)\, dy$.

For greater detail, interested readers may consult texts on statistics and probability, such as Wackerly *et al.* (1996) or Casella and Berger (1990).

One important characteristic of joint probability functions is that they factor under independence (as defined in Section 1.1.1); that is to say, if two random variables, X and Y, are statistically independent, their joint p.m.f. or joint p.d.f. factors into its marginal components: $p_{X,Y}(j,k) = p_X(j)p_Y(k)$, or $f_{X,Y}(j,k) = f_X(j)f_Y(k)$. Extended to multiple variables, say $X_1, \ldots, X_n$, this is

$$p_{X_1,\ldots,X_n}(k_1, \ldots, k_n) = \prod_{i=1}^{n} p_{X_i}(k_i) \tag{1.3}$$

for the discrete case (the continuous case is similar). This result is tremendously important for constructing likelihoods, as we will see in Chapter 2.

1.1.4* Mixture distributions

In many environmental applications, it is useful to derive a probability distribution for an observed random variable that is based on a mixture of other distributions. The mixing process in effect combines the probability functions of two or more random variables, and this can produce distributions that are more complex than often seen at an introductory level. To ease the transition to these complex forms, we present here only basic concepts of mixture distributions, including a relatively simple example. Although we will make some use of mixture distributions in the chapters below, readers wishing only a basic introduction to probability and statistical distributions may wish to move forward to Section 1.1.5, and return to the details in this section at a later time.

The simplest form of mixture distribution is a mixture of two random variables, say X and W. In the discrete case, this produces the mixture p.m.f.

$$p_Z(k) = \omega p_X(k) + (1 - \omega)p_W(k).$$

The **mixing parameter** ω is a constant between 0 and 1 that controls the amount of mixing, and the two component p.m.f.s $p_X(k)$ and $p_W(k)$ are being mixed together to produce a p.m.f. for the new random variable Z. The continuous case is similar, with p.d.f.s $f(\cdot)$ replacing p.m.f.s $p(\cdot)$. Such a combination of probability functions is said to produce a **mixture distribution** or a **compound distribution** of the original random variables.

The simple mixture can be extended in a number of ways, starting with any finite mixture of p.m.f.s

$$p_Z(k) = \sum_{i=1}^{n} \omega_i p_{X_i}(k)$$

where the mixing parameters must satisfy $\omega_1 + \omega_2 + \cdots + \omega_n = 1$. Or, the finite sum can be extended to an infinite sum

$$p_Z(k) = \sum_{i=1}^{\infty} \omega_i p_{X_i}(k),$$

where now the mixing parameters satisfy

$$\sum_{i=1}^{\infty} \omega_i = 1. \quad (1.4)$$

Notice that the mixing parameters ω_i satisfy the basic requirements of a discrete probability function: they are nonnegative and they sum to 1. Indeed, these are often viewed as probability functions in their own right: write ω_i as $p_Y(y_i)$ so that ω_i is the probability that some discrete random variable Y takes on the value y_i. The notation emphasizes that $p_Y(y_i)$

comprises a p.m.f. It can even be extended to allow the original random variable, X, to be a function of the mixing variable Y:

$$p_Z(k) = \sum_{i=1}^{\infty} p_Y(y_i)\, p_{X|y_i}(k \mid y_i). \tag{1.5}$$

As such, we often say that a random variable, X, is being mixed over another variable, Y, to achieve Z in (1.5).

We can even define mixture distributions over a continuous mixing set, using integral calculus:

$$p_Z(k) = \int_{-\infty}^{\infty} f(\psi)\, p_{X|\psi}(k \mid \psi)\, d\psi, \tag{1.6}$$

where the notation $p_{X|\psi}(k \mid \psi)$ emphasizes that the original p.m.f. must be a function of ψ in order to mix over it in a continuous manner. The mixture function $f(\psi)$ is assumed nonnegative for all values of ψ, and it must integrate to 1: $\int_{-\infty}^{\infty} f(\psi)\, d\psi = 1$. Extensions to the continuous case for X are straightforward: replace the p.m.f. $p_{X|\psi}(k \mid \psi)$ with a p.d.f. $f_{X|\psi}(x \mid \psi)$.

In this sense, a mixture distribution describes a **hierarchical model**, where first X and then ψ are assumed random. If we view the original variable X and the mixing variable ψ as jointly distributed, then the characterization in (1.6) indicates how ψ and X relate to each other: ψ is some parameter of the conditional distribution of X.

Under this joint distribution formulation, (1.6) corresponds to the marginal distribution of X after integrating out ψ (or, in (1.5), after summing over all y_i). That is, Z may be viewed as a marginal form of the original variable X, after removing the effect of the mixing parameter. For greater detail, interested readers are referred to the discussions in Casella and Berger (1990, Section 4.4) or Johnson *et al.* (1992, Chapter 8).

Example 1.1 Hierarchical count data for insect offspring survival
Casella and Berger (1990, Section 4.4) describe a simple field experiment where a hierarchical model is appropriate. Suppose we are observing the survival capability of an insect's offspring after some disturbance to their ecosystem (hurricane, pesticide application, etc.). The random variable of interest is $X = \{$number of surviving insect offspring$\}$, which when divided by $Y = \{$total number of eggs laid by the insect$\}$, estimates the probability of offspring survival.

Obviously, both Y and X are random variables. The way we have described it for this scenario, however, X is a function of the random outcome of Y, since, for example, it is bounded above by the realized value of $Y = y$. Notice that both variables are discrete counts.

Technically, this is a hierarchical model of the form represented in (1.5): X has a p.m.f., $p_{X|y}(k \mid y)$, that depends on the hierarchical variable $Y = y$, and Y is assumed random with its own p.m.f. $p_Y(y)$. (Trivially, if $Y = 0$ is 'observed,' then $p_{X|y}(k \mid 0)$ is defined as identically 0 also.) The marginal p.m.f. of X is found by mixing over all possible values of Y:

$$p_X(k) = \sum_{i=1}^{\infty} p_Y(y_i)\, p_{X|y_i}(k \mid y_i).$$

We will return to this example below, when we introduce specific parametric forms of p.m.f.s. ❑

1.1.5 Expected values: Means, moments, and variance

From the p.m.f. or p.d.f. of a random variable, X, we can calculate summary measures of the nature of the probability distribution. Perhaps the most important is a summary measure of the average value of X. This is called the **expected value** or **mean** of X, and is defined as the sum

$$\mathrm{E}[X] = \sum_{\mathscr{X}} x\, p_X(x) \tag{1.7a}$$

in the discrete case, or the integral

$$\mathrm{E}[X] = \int_{-\infty}^{\infty} x f_X(x)\, dx \tag{1.7b}$$

in the continuous case. The sum in (1.7a) is taken over all possible discrete values of x in $\mathscr{X}$, while the integral in (1.7b) is taken over the entire real line. In either case, the traditional notation for E[X] is μ, or for clarity μ_X. This quantity measures the **central tendency** of the probability distribution of X. In a sense, it is our best guess at a value for X before actually observing the random outcome. If we examine (1.7a) in more detail, we see that this quantity is a weighted sum in which each possible value of the random variable X is multiplied by the probability of observing that value. Thus, for discrete random variables, the expected value is a form of **weighted average**. In (1.7b), however, this weighted average notion does not carry over, since the probability of observing any particular value of a continuous random variable is zero. Expectation as calculated in (1.7b) for continuous random variables may be viewed as finding the point under the density function that would in fact balance it evenly. Imagine a playground see-saw with the probability density resting on it, and with the fulcrum of the see-saw moving back and forth to make the see-saw level. The leveling point of the fulcrum is the expected value of the density; cf. Moore and McCabe (1993, Fig. 1.14).

Notice, by the way, that applying the expectation operator to a constant always recovers the value of the constant: $\mathrm{E}[a] = a$.

Extending (1.7), we define the ***N*th moment of *X*** as the probability-weighted average value of X^N; for example, in the discrete case we have $\mathrm{E}[X^N] = \sum_X x^N p_X(x)$. A related value is the Nth central moment about the mean: $\mathrm{E}[(X-\mu)^N]$. Certain central moments have important interpretations. Among the most useful is the second central moment, also known as the **variance** of X: $\mathrm{Var}[X] = \mathrm{E}[(X-\mu)^2]$. Common notation for $\mathrm{Var}[X]$ is σ^2, or for clarity σ_X^2. This quantity measures the spread of the probability distribution of X, and is a useful measure of variability. Notice, however, that σ^2 is reported in squared units of measurement. To recover a measure on the same scale as the original observations, we take the square root of $\mathrm{Var}[X]$, called the **standard deviation** of X: $\sigma_X = \sqrt{\mathrm{Var}[X]}$.

We can also take the expected value of a function of X, say $g(X)$. In the discrete case, this is $\mathrm{E}[g(X)] = \sum_X g(x) p_X(x)$. (The continuous case is similar, using integrals instead of sums.) The mean and central moments noted above are special cases, with, for example, μ formed from the identity function $g(X) = X$. For linear functions of X, i.e. $g(X) = a + bX$, the expectation is linear: $\mathrm{E}[a + bX] = a + b\,\mathrm{E}[X]$. That is, shifting X (by a) and scaling X (by b) shifts and scales μ_X in a similar manner. This is not true of the variance, however: $\mathrm{Var}[a + bX] = b^2\mathrm{Var}[X]$. That is, shifting X by a has no effect on $\mathrm{Var}[X]$, while scaling X by b scales the variance by b^2. (A consequence of this is that the variance of a constant is zero: $\mathrm{Var}[a] = 0$.)

Bivariate and multivariate expectations are also possible. For instance, with two discrete random variables, X and Y, a general expression for the expected value of some bivariate function $g(X, Y)$ is

$$\mathrm{E}[g(X,Y)] = \sum_X \sum_Y g(x,y)\, p_{X,Y}(x,y).$$

(Again, the continuous case is similar.) In particular, a summary measure of the joint variability between X and Y is known as the **covariance**: $\mathrm{Cov}[X, Y] = \mathrm{E}[(X - \mu_X)(Y - \mu_Y)]$. (For a short-hand notation, we often use σ_{XY} for $\mathrm{Cov}[X,Y]$.) If large values of X tend to be observed with large values of Y, and small values of X with small values of Y, σ_{XY} becomes larger and more positive. Conversely, if large values of X are associated with small values of Y, and vice versa, then σ_{XY} will be negative. The greater this inverse association, the smaller (more negative) σ_{XY} becomes. Note that σ_{XY} is not a unitless measure of association. In fact, the units of σ_{XY} are X-units times Y-units. To obtain a unitless measure of association that is also bounded between –1 and 1, and hence somewhat easier to interpret, we divide by the standard deviations of X and Y, producing the **correlation coefficient** $\rho_{XY} = \sigma_{XY}/\sigma_X\sigma_Y$. When X and Y are independent, it can be shown that $\sigma_{XY} = 0$, making $\rho_{XY} = 0$. The reader should be warned, however, that the reverse is not true: there are cases of joint random variables where $\rho_{XY} = 0$, but the two variables are not independent; see Casella and Berger (1990, Section 4.5). Lastly, it is important to note that ρ_{XY} is a measure of linear association. One can show that ρ_{XY} is a multiple of the slope when X and Y are linearly related (Neter *et al.*, 1996, Section 2.9; Hogg and Tanis, 1997, Section 11.2). Thus, $\rho_{XY} = 0$ does not necessarily mean that X and Y are unrelated; it could be simply that the relationship between the two variables is not well described by a straight line.

We close this section with some general results on expected values of linear functions of random variables. A **linear combination** of a set of random variables, $X_1, \ldots, X_n$, is the sum

$$L = \sum_{i=1}^{n} c_i X_i$$

where the constants c_i are pre-specified. (If the c_i satisfy the special constraint that $\sum_{i=1}^{n} c_i = 0$, we call L a **contrast** among the X_is.) The expected value of any linear combination of random variables may be recovered through the linear features of the expectation operator:

$$\mu_L = \mathrm{E}[L] = \sum_{i=1}^{n} c_i \mathrm{E}[X_i].$$

For instance, if X has mean μ_X and Y has mean μ_Y, then the mean of their difference is $\mathrm{E}[X - Y] = \mu_X - \mu_Y$.

Variances of linear combinations are not quite as simple as their means, however, due to the squared nature of the variance operation. In the simplest case of two random variables, X and Y, we have $\mathrm{Var}[c_1X + c_2Y] = c_1^2\sigma_X^2 + c_2^2\sigma_Y^2 + 2c_1c_2\sigma_{XY}$, where σ_{XY} is the covariance of X and Y. Notice that if X and Y are independent random variables then we know $\sigma_{XY} = 0$, so the variance term simplifies to a sum of individual variances, each multiplied by the square of the combination coefficient. More generally, we have

$$\sigma_L^2 = \mathrm{Var}[L] = \sum_{i=1}^{n} c_i^2 \mathrm{Var}[X_i] + 2\sum_{i=1}^{n-1}\sum_{j=i+1}^{n} c_i c_j \mathrm{Cov}[X_i, X_j], \quad (1.8)$$

where the values $\mathrm{Cov}[X_i, X_j]$ in the double sum are the covariances between X_i and X_j. If the X_is are all mutually independent, then these covariances are zero, and $\mathrm{Var}[L]$ simplifies to the simple sum $\sum c_i^2 \mathrm{Var}[X_i]$.

1.2 FAMILIES OF DISCRETE DISTRIBUTIONS

Much of our effort in describing statistical methods for environmental data will center on outcome variables associated with specific distributional families. These families typically possess one or more unknown **parameters** that describe the mean, variance, shape, and/or other distributional characteristics. (It is often the goal of an environmental study to estimate the unknown parameters of a random variable, and possibly to compare them with parameters from other, associated variables.) Thus it is common to refer to a parametric family of distributions when specifying the p.m.f. or p.d.f. of an outcome variable. We begin with the discrete families.

1.2.1 Discrete distributions: Bernoulli and binomial

Perhaps the simplest form of random outcome is one where the outcome variable is dichotomous, taking on only two possible values. Denote these two values as 'success' and 'failure.' A 'success' refers to some characteristic of interest; for example, observing a tumor in a carcinogenicity experiment. Quantified, we let the random variable X equal 1 if the outcome is a success, and let X equal 0 if the outcome is a failure. Then, let π be the probability of success and let $1 - \pi$ be the probability of failure. Any such random variable is said to comprise a **Bernoulli trial** with success probability π.

If we observe N independent Bernoulli trials, X_i, each with the same success probability, π, then the sum $Y = \sum_{i=1}^{N} X_i$ of these N Bernoulli trials is said to take a **binomial distribution** with parameters N and π. The associated p.m.f. is given by

$$p_Y(y) = \binom{N}{y} \pi^y (1 - \pi)^{N-y}, \quad (1.9)$$

for $y = 0, 1, \ldots, N$, where

$$\binom{N}{y} = \frac{N!}{y!\,(N-y)!}$$

is the **binomial coefficient** and $n!$ is the factorial operator: $n! = n(n-1)(n-2)\cdots(2)(1)$, defined for any positive integer n. For convenience, we also define $0! = 1$. In (1.9), the term $\pi^y(1-\pi)^{N-y}$ is the probability associated with any sequence of trials resulting in y successes and $N - y$ failures, while the binomial coefficient counts the number of arrangements of N trials that result in y successes and $N - y$ failures. The notation to indicate this distribution is $Y \sim \mathcal{Bin}(N,\pi)$. (The tilde symbol, ~, is read as 'is distributed as.') In the special case where only $N = 1$ Bernoulli trial is observed, we often say that the single dichotomous outcome has a **Bernoulli distribution**.

The binomial distribution is common in settings where the response is the number of trials that exhibit some characteristic of interest (technically, where the outcomes are whole numbers bounded above by some known integer, N). In this situation, the number of trials, N, is fixed in advance while the number of successes, Y, is the random variable. The response is often taken as a proportion of successful outcomes, Y/N, and the binomial distribution is a common first choice for describing the random variability of data in the form of proportions. This distribution is the basis of the probit and logistic regression models we will introduce in Chapters 7 and 8.

The mean of a binomial random variable is $E[Y] = N\pi$, the fraction of expected successful outcomes multiplied by the number of Bernoulli trials. The corresponding variance is $Var[Y] = N\pi(1-\pi)$. Notice that for fixed N, this implies that the expected proportion is $E[Y/N] = \pi$, with variance $Var[Y/N] = \pi(1-\pi)/N$. As can be seen in Fig. 1.1, the shape of the binomial distribution is unimodal, exhibiting a positive skew for small π and negative skew for large π. Further, for π near 0.5, the binomial distribution is roughly symmetric. This latter observation relates to the common use of the normal distribution (Section 1.3.4) as an approximation to the binomial when $\pi \approx 0.5$ (Moore and McCabe, 1993, Section 5.1).

Binomial random variables possess a special form of closure within their distributional family: a sum of independent binomials is also binomial when all of the components of the sum have the same success probability. That is, suppose we observe n independent binomial random variables, $X_1, \ldots, X_n$, where $X_i \sim$ (indep.) $\mathcal{Bin}(N_i,\pi)$. Then, the random variable $Y = \sum_{i=1}^{n} X_i$ is also binomially distributed: $Y \sim \mathcal{Bin}(\sum_{i=1}^{n} N_i,\pi)$. This is a useful property; it allows for 'pooling' of binomials when singly each represents a success count with the same success probability, π.

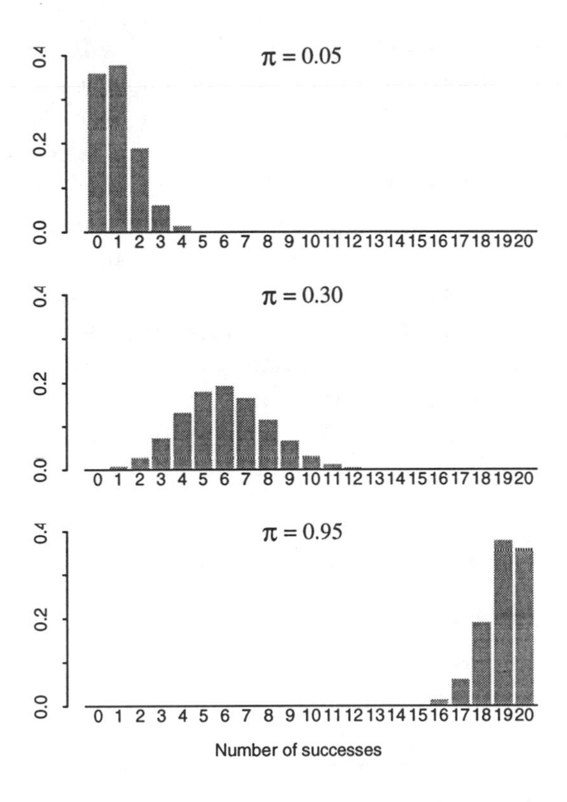

Fig. 1.1 Probability mass functions for the binomial distribution (n = 20).

1.2.2* Discrete distributions: Overdispersion and the beta-binomial distribution

The binomial distribution is most useful when data are recorded in the form of proportions, say Y/N, where Y is the number of successes out of N observed trials. A crucial feature of the binomial assumption for Y is, however, that the variance of Y is a strict function of the success probability, π: $\mathrm{Var}[Y] = N\pi(1 - \pi)$. When this restriction is thought to impose too severe a constraint on the variability of the observations, we can update the distributional assumption and include an additional variance parameter. For instance, consider the form $\mathrm{Var}[Y] = N\pi(1 - \pi)\{1 + \varphi(N - 1)/(1 + \varphi)\}$, where φ is viewed as a **dispersion parameter** that allows for **extra-binomial variability**. (Technically, a dispersion parameter quantifies the percentage increase in variability over and above the simpler variance model, so that here, the entire quantity $\{1 + \varphi(N - 1)/(1 + \varphi)\}$ is the true dispersion parameter. For simplicity, however, we will refer to φ as the dispersion parameter for this distribution.) Notice that at $\varphi = 0$ we recover the binomial variance, while for $\varphi > 0$, variability exceeds that predicted under the binomial model. This effect is referred to as **overdispersion**. A common experimental situation in which extra-binomial variability is observed is a teratogenicity study, where interest is in studying the proportion of offspring in a litter that exhibit some congenital malformation, such as cleft palate. In such an experiment, the assumption of independence of response among pups from the same litter may be unwarranted, and some form of overdispersed binomial model must be used (Haseman and Piegorsch, 1994); see Example 3.6.

A specific statistical distribution that achieves this form of extra-binomial variability is known as the **beta-binomial distribution**, with p.m.f.

$$p_Y(y) = \binom{N}{y} \frac{\Gamma[y + (\pi/\varphi)]\ \Gamma[N - y + (1 - \pi)/\varphi]\ \Gamma(1/\varphi)}{\Gamma(\pi/\varphi)\ \Gamma[(1 - \pi)/\varphi]\ \Gamma[(1/\varphi) + N]} . \quad (1.10)$$

In (1.10), $\Gamma(a)$ is the **gamma function**:

$$\Gamma(a) = \int_0^\infty x^{a-1} e^{-x}\, dx$$

for any $a > 0$ (Spouge, 1994). This function is tabulated in sources such as Abramowitz and Stegun (1972), and in many computer packages.

The beta-binomial p.m.f. (1.10) is defined for any $y = 0, 1, \ldots, N$, with $\varphi > 0$. The limiting form when $\varphi = 0$ recovers the binomial p.m.f. in (1.9).

The beta-binomial has mean $E[Y] = N\pi$, with $0 \le \pi \le 1$. Notice that this is independent of φ and is in fact identical to the mean of the simple binomial distribution. Thus, for fixed N the expected proportion remains $E[Y/N] = \pi$. As noted above, however, the variance of the beta-binomial distribution includes the dispersion parameter φ, and so the variance of the proportion Y/N increases to $\text{Var}[Y/N] = \pi(1 - \pi)\{(1 + \varphi N)/(1 + \varphi)\}/N$, where we assume $\varphi > 0$.

The beta-binomial distribution can be derived as a mixture of binomial distributions, as discussed in Section 1.1.4. That is, if we view the sampling frame in a hierarchical manner and assume that both Y and its success probability p are random variables, then it is possible to show that (1.10) is the marginal p.m.f. from a hierarchical model with $Y|p \sim \mathcal{Bin}(N,p)$ and with p distributed as a beta distribution over the interval $0 \le p \le 1$. (We describe the beta distribution in Section 1.3.1.) Thus the beta-binomial p.m.f. can be viewed as a beta mixture of binomial p.m.f.s. In cases where data are recorded in the form of proportions, but where some random variability may be ascribed to the success probability, p, this hierarchical interpretation for the beta-binomial distribution provides important motivation for its use.

Before closing, we note that one can model **underdispersion** by letting $\varphi < 0$, such that variability is decreased below that assumed under simple binomial sampling. This case is not as common in environmental biology, however, and we will not emphasize it here.

1.2.3 Discrete distributions: Geometric and negative binomial

Consider again a set of independent trials in which two outcomes are possible on any trial, and in which the probability of success, π, is the same from trial to trial. Instead of counting the number of successes in a specified number of trials, consider counting the number of trials *until* a specified number of successes is observed. This is common when, for example, estimating the size of wildlife populations using inverse sampling/capture-recapture techniques (Scheaffer *et al.*, 1996, Section 10.3): a first sample of animals is captured, tagged and released. Then a second sample is obtained by operating until a fixed number of tagged animals are recaptured. Here, the size of the second sample is a random variable. If we take the simplest case where we sample only until one tagged animal is obtained, then the random variable X = {number of animals captured until one tagged animal is observed} has the p.m.f. $p_X(x) = \pi(1-\pi)^x$ for any nonnegative integer $x = 0,1,\ldots$. In such a sampling scenario, the distribution of the number of trials up to (but not including) the first success is called the **geometric distribution**.

The mean of a geometric random variable is $E[X] = (1-\pi)/\pi$, while the variance is $Var[X] = (1-\pi)/\pi^2$. Its c.d.f. is simple to derive, and it possesses some useful properties. From (1.1), the c.d.f. is $F_X(x) = P[X \leq x]$ or for $x = 0,1,\ldots$

$$F_X(x) = \sum_{i=0}^{x} p_X(i) = \sum_{i=0}^{x} \pi(1-\pi)^i = 1-(1-\pi)^{x+1}$$

using the fact that a finite **geometric series** $\sum_{i=0}^{m} r^i$ equals the closed form $\{1-r^{m+1}\}/(1-r)$ for any $|r| < 1$ and $m > 0$.

Now, for a geometric random variable, X, suppose we consider the cumulative complementary event $\{X > u\}$, with $P[X > u] = (1 - \pi)^{u+1}$. This is the probability that X will exceed some count value, u. (In Chapter 11, we view this quantity as a **survival probability**, i.e. the probability of 'surviving' past $X = u$ trials.) Further, the conditional probability that X exceeds some value u, given that it has already exceeded some lesser value v is $P(X > u \mid X > v)$, which from the definition of conditional probability becomes $P(X > u \text{ and } X > v)/P(X > v)$. Notice, however, that for $v < u$, the joint event $\{X > u \text{ and } X > v\}$ simplifies to the event $\{X > u\}$, so that $P(X > u \mid X > v)$ collapses to $P(X > u)/P(X > v)$. For the geometric distribution, this becomes $P(X > u \mid X > v) = (1 - \pi)^{u+1}/(1 - \pi)^{v+1} = (1 - \pi)^{u-v}$, i.e. $P(X > u \mid X > v) = P(X > u - v)$. Thus the probability of exceeding u, given that one has already exceeded v is dependent upon v only through the difference $u - v$, and not on v directly. We refer to this effect as a **memoryless** property: the geometric distribution exhibits a lack of memory as to where is has been, and it will 'go' to a particular value with probability dependent only upon how far away that point is from where the variable has already been.

A natural extension of the simple geometric distribution is to consider the number of trials to the rth success, for $r \geq 1$. It is more convenient, however, to consider instead the associated random variable $Y = \{$number of failures before the rth success$\}$. This random variable has a **negative binomial distribution**, with p.m.f.

$$p_Y(y) = \binom{r + y - 1}{y} \pi^r (1 - \pi)^y$$

for any nonnegative integer $y = 0,1,\ldots$. The associated random variable has mean $E[Y] = r(1 - \pi)/\pi$, and variance $Var[Y] = r(1 - \pi)/\pi^2$. Notice that at $r = 1$ we recover a geometric random variable. It is also the case that a sum of N independent identically distributed geometric random variables (each with the same success probability parameter π) is distributed as negative binomial with parameters N and π.

The negative binomial p.m.f. often is written in a different form, in order to extend its applicability. Let $\mu = r(1 - \pi)/\pi$ and $\delta = 1/r$. The p.m.f. is then expressible as

$$p_Y(y) = \frac{\Gamma[y + (1/\delta)]}{y!\, \Gamma(1/\delta)} \left(\frac{\delta\mu}{1 + \delta\mu}\right)^y \frac{1}{(1 + \delta\mu)^{1/\delta}}, \tag{1.11}$$

where $\mu > 0$, and we extend the definition of the distribution to allow δ to vary over any positive value (not just the reciprocal positive integers). Here, $E[Y] = \mu$ and $Var[Y] = \mu + \delta\mu^2$. We will adopt the negative binomial p.m.f. in (1.11) as the standard form for this probability function. When a random variable possesses it, we write $Y \sim \mathcal{NB}(\mu, \delta)$.

1.2.4 Discrete distributions: Hypergeometric

A somewhat more complex model for the number of successes in a fixed number of trials is the **hypergeometric distribution**, defined as follows: in a finite population with N elements, suppose R of the elements are successes, and $N - R$ are failures. If we select K elements at random and without replacement, the random outcome is $X = \{$number of successes out of $K\}$ and we write $X \sim \mathcal{hg}(R,N,K)$. The p.m.f. is

$$p_X(x \mid R,N,K) = \frac{\binom{R}{x}\binom{N-R}{K-x}}{\binom{N}{K}} \tag{1.12}$$

for any $x = 0, 1, \ldots, R$. We can view the hypergeometric distribution as an analog to the binomial distribution when the probability of success changes from trial to trial, due to sampling without replacement.

In order for (1.12) to define a true p.m.f., we require $x \le R$ and $K - x \le N - R$. This can be expressed more compactly as $R - (N - K) \le x \le R$. If $X \sim \mathcal{hg}(R,N,K)$, it can be shown that $\mathrm{E}[X] = RK/N$, while

$$\mathrm{Var}[X] = K\{R(N - R)(N - K)\}/\{N^2(N - 1)\}.$$

The mean E[X] is the number of trials, K, times the proportion of successes in the population; this is similar to the mean of a binomial random variable. Also, Var[X] is the number of trials, K, times the population proportion of successes, R/N, times the population proportion of failures, $(N - R)/N$, times an additional factor. This additional term, $(N - K)/(N - 1)$, serves as a form of finite-population correction factor (Scheaffer *et al.*, 1996, Section 4.3). When the population size, N, is much larger than the sample size K, the finite-population correction factor is approximately 1, and the hypergeometric can be approximated by a binomial. This is illustrated in Fig. 1.2.

1.2.5 Discrete distributions: Poisson

We complete our introductory survey of discrete distributions with one of the most common distributions seen with biological data, the **Poisson distribution**. This is a simple model for data in the form of unbounded counts, and serves as a basic default distribution for most count data.

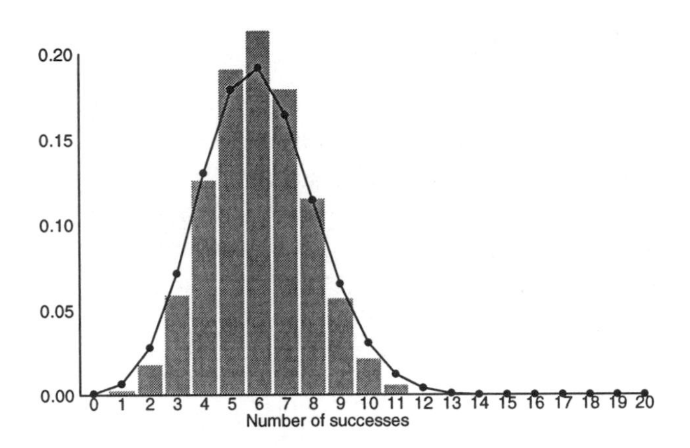

Fig. 1.2 Probability mass function for the hypergeometric distribution (R = 20, N = 100, K = 30) with the binomial distribution (n = 20, π = 0.30) superimposed as a frequency polygon.

The Poisson model is based on a set of elementary conditions that are satisfied in many experimental settings. These are often called the **Poisson postulates** and in environmental contexts they have interpretation from both spatial and temporal perspectives. Summarized using the spatial perspective, one considers some fixed, well-defined spatial region in which random events occur at a rate of μ per unit area of the region. This can have widespread application; for example, radiation-induced mutations ('events') on a chromosome (the 'region'). The postulates are:

1. Start with no event occurrences in the region.
2. Occurrences in disjoint spatial sub-regions are independent.
3. The number of occurrences in different spatial sub-regions depends only upon each sub-region's area.
4. Occurrence probability is proportional to spatial area of occurrence (in a limiting sense, as the area goes to zero).
5. There are no exactly simultaneous occurrences.

A temporal perspective for the Poisson postulates produces a similar set of conditions (Casella and Berger, 1990, Chapter 3; Khuri, 1993, Section 4.5.3). When the five postulates hold, the random count of occurrences is taken as a Poisson random variable with mean μ. One encounters the five Poisson postulates in various forms in a number of environmental settings; they take on different realizations depending on the specific application, but always retain the basic formulation summarized above.

The p.m.f. of a Poisson random variable is defined in terms of its mean parameter, $\text{E}[X] = \mu$: $p_X(x) = \mu^x e^{-\mu}/x!$, for any nonnegative integer $x = 0,1,\ldots$ We write $X \sim \mathcal{Poisson}(\mu)$. For small μ, the Poisson p.d.f. has a positive skew, while for large μ it becomes symmetric and unimodal. As with the binomial distribution, the family of Poisson distributions is closed under addition: if $X_i \sim$ (indep.) $\mathcal{Poisson}(\mu_i)$, then $Y = \sum_{i=1}^{n} X_i$ is also Poisson: $Y \sim \mathcal{Poisson}(\sum_{i=1}^{n}\mu_i)$.

An important characteristic of the Poisson distribution is that its variance equals its mean: $\text{Var}[X] = \sigma_X^2 = \mu_X$. We use this feature to test if unbounded count data appear to exhibit Poisson variability, that is, whether or not their variability is of the same order as their mean; see Section 6.5.3.

When observed variability among unbounded count data is so large that it exceeds its mean value, then the Poisson distribution is contraindicated. Indeed, the mean–variance equality required under Poisson sampling may be too restrictive in some environmental settings, a problem seen also with the binomial distribution for proportion data. In similar form to the binomial, we can consider overdispersed distributions for count data that exhibit **extra-Poisson variability**. We need not look far: we saw previously that the negative binomial p.m.f. in (1.11) gives a variance that is quadratic in its mean for unbounded counts: when $Y \sim \mathcal{NB}(\mu, \delta)$, $\text{Var}[Y] = \mu + \delta\mu^2$. Thus the negative binomial variance is always larger than its mean when the dispersion parameter, δ, is positive. (Technically, δ is not a true dispersion parameter, since it cannot be written as a percentage increase in variability over the simpler Poisson variance. The parameter does quantify the departure from the Poisson model, however, and for simplicity we will continue to refer to it as a dispersion parameter.)

At $\delta = 0$, the negative binomial variance returns to mean–variance equality. In fact, the Poisson distribution is a limiting form: as $\delta \to 0$ the negative binomial c.d.f. approaches the Poisson c.d.f.

The negative binomial p.m.f. may be constructed also as an extension of the Poisson p.m.f., based on a hierarchical model formulation. Specifically, if we take $X|\mu \sim \mathcal{Poisson}(\mu)$ and also take μ as random with some continuous distribution over $0 < \mu < \infty$, that is, a **mixture distribution**, the resulting marginal distribution for X will be overdispersed. A specific choice for $f_\mu(\mu)$ that brings about the negative binomial distribution (1.11)

$\mu \sim \mathcal{G}amma(r,[1-\pi]/\pi)$, for some π between 0 and 1; we will introduce the Gamma distribution in Section 1.3.2.

The Poisson may be used also as a mixing distribution in its own right, as the following example illustrates.

Example 1.1 (continued) Hierarchical count data for insect offspring survival

Recall that in our example of insect offspring survival we constructed a hierarchical model based on the random variables $X = \{$number of surviving insect offspring$\}$ and $Y = \{$total number of eggs laid by the insect$\}$. Since X is a bounded count variable, we might consider the conditional assignment $X|y \sim \mathcal{B}in(Y,\pi)$, and then model Y hierarchically as $Y \sim \mathcal{P}oisson(\mu)$, for $\mu > 0$. This would be a **Poisson–binomial mixture**, where the Poisson mixing distribution is acting on the binomial sample size parameter.

For this particular model, the marginal p.m.f. of X (after mixing over all possible values of Y) takes the form $p_X(x) = (\pi\mu)^x e^{-\pi\mu}/x!$, for any positive integer $x = 0,1,\ldots$ (Casella and Berger, 1990, Section 4.4). This is itself a Poisson p.m.f., with marginal mean $E[X] = \pi\mu$. Thus, mixing a Poisson distribution on a binomial sample size parameter recovers a Poisson distribution. Solomon (1983) describes other mixture distributions for selected biological and ecological applications. ❑

1.3 FAMILIES OF CONTINUOUS DISTRIBUTIONS

1.3.1 Continuous distributions: Uniform and beta

Perhaps the simplest form of a continuous random variable is a **uniform distribution** over some interval, $a \le x \le b$. The corresponding p.d.f. is

$$f_X(x) = \frac{1}{b-a} I_{[a,b]}(x),$$

where the notation $I_{\mathcal{A}}(x)$ is used to express the simple **indicator function**: $I_{\mathcal{A}}(x) = 1$ if x is contained in the set $\mathcal{A}$, and $I_{\mathcal{A}}(x) = 0$ otherwise. For this distribution, $\mathcal{A} = \{x: a \le x \le b\}$. We write $X \sim \mathcal{U}[a,b]$.

The uniform distribution has mean $E[X] = (a+b)/2$ and variance $Var[X] = \frac{1}{12}(b-a)^2$. In the special case of $a = 0$ and $b = 1$ – i.e. sampling over the unit interval – these simplify to $E[X] = 1/2$ and variance $Var[X] = 1/12$. $X \sim \mathcal{U}[0,1]$ is a common probability model in computer simulation studies, where it is used for pseudo-random number generation.

A somewhat more flexible distribution for use over the unit interval is the **beta distribution**, $X \sim \mathcal{Beta}(\alpha,\beta)$, with p.d.f.

$$f_X(x) = \frac{\Gamma(\alpha+\beta)}{\Gamma(\alpha)\,\Gamma(\beta)} x^{\alpha-1}(1-x)^{\beta-1} I_{[0,1]}(x) \,,$$

and parameters $\alpha > 0$ and $\beta > 0$. This has mean $E[X] = \alpha/(\alpha+\beta)$, and variance $\mathrm{Var}[X] = \alpha\beta/\{(\alpha+\beta)^2(\alpha+\beta+1)\}$. Notice that since α and β are both positive, the mean must be a quantity between 0 and 1.

The beta distribution provides a flexible model for describing variability of a random variable over the unit interval. The p.d.f. can be unimodal, ∪-shaped, asymmetric, or symmetric. (The latter case occurs when $\alpha = \beta$.) It can even give a uniform specification: $X \sim \mathcal{Beta}(1,1)$ is identical to setting $X \sim \mathcal{U}[0,1]$. A variety of beta distributions are illustrated in Fig. 1.3; see also Casella and Berger (1990, Figs 3.2.3–4).

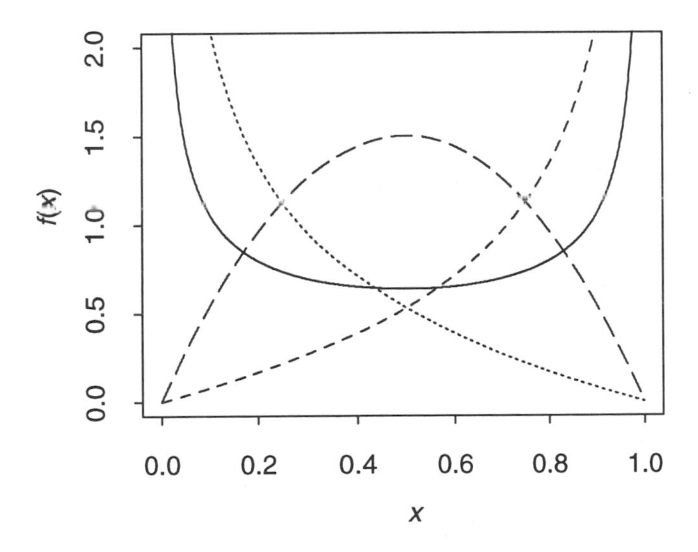

Fig. 1.3 Probability density functions for the beta distribution: $\alpha = \beta = 0.5$ (solid line); $\alpha = 0.5$, $\beta = 2$ (dotted line); $\alpha = 2$, $\beta = 0.5$ (short-dashed line); and $\alpha = \beta = 2$ (long-dashed line).

As suggested above, the beta distribution is a common choice for a mixture distribution over the unit interval. For example, if $X \mid p \sim \mathcal{Bin}(N,p)$ and $p \sim \mathcal{Beta}(\pi/\varphi,[1-\pi]/\varphi)$, then the marginal distribution of X after mixing over the possible values of p in the unit interval is beta-binomial with parameters π and φ in (1.10).

1.3.2 Continuous distributions: Exponential, gamma and χ^2

A number of distributions are available to describe the variability of a positive-valued random variable, $X > 0$. One of the simplest of these is the **exponential distribution**, with p.d.f. $f_X(x) = \beta^{-1}e^{-x/\beta}$ over $x > 0$. (Recall that $\beta^{-1} = 1/\beta$.) The parameter $\beta > 0$ describes the spread of the distribution and is often called the scale parameter: as β increases, the p.d.f. narrows, and the majority of the probability density is concentrated over a narrow scale. The opposite effect occurs as β decreases. (In some models, $\lambda = 1/\beta$ is also referred to as the scale parameter, since it has a comparable, if opposite, 'scaling' effect on the p.d.f.)

The exponential distribution has mean $E[X] = \beta$, and variance $Var[X] = \beta^2$. Its c.d.f. is straightforward to compute, via integral calculus:

$$F_X(x) = P[X \le x] = \int_0^x \beta^{-1}e^{-t/\beta}dt = \left[-e^{-t/\beta}\right]_0^x = 1 - e^{-x/\beta}.$$

From this, the survival probability $P[X > u]$ is just $e^{-u/\beta}$. This is the probability that a exponential random variable, X, will exceed (or 'survive' past) some positive value, u. If we also consider the conditional probability that X exceeds some value u, given that it has already exceeded some lesser value v, we have $P(X > u \mid X > v) = P(X > u)/P(X > v)$. For the exponential case, $P(X > u \mid X > v) = \exp\{-(u - v)/\beta\}$, that is, the probability of exceeding u, given that one has already exceeded v is once again dependent upon v only through the difference $u - v$, and not on v directly. This is the same effect we saw with the geometric distribution in Section 1.2.3: a memoryless property in the cumulative probabilities.

We can extend the exponential distribution to a richer, more flexible family, via the addition of a shape parameter, $\alpha > 0$. This is the **gamma distribution**, $X \sim \mathcal{G}amma(\alpha,\beta)$, with p.d.f.

$$f_X(x) = \frac{1}{\Gamma(\alpha)\beta^\alpha}x^{\alpha-1}e^{-x/\beta}I_{(0,\infty)}(x).$$

The gamma distribution has mean $E[X] = \alpha\beta$ and variance $Var[X] = \alpha\beta^2$. At $\alpha = 1$, it simplifies to the exponential distribution above. As is illustrated in Fig. 1.4, this p.d.f. is skewed to the right, with a long tail decaying exponentially as $x \to \infty$.

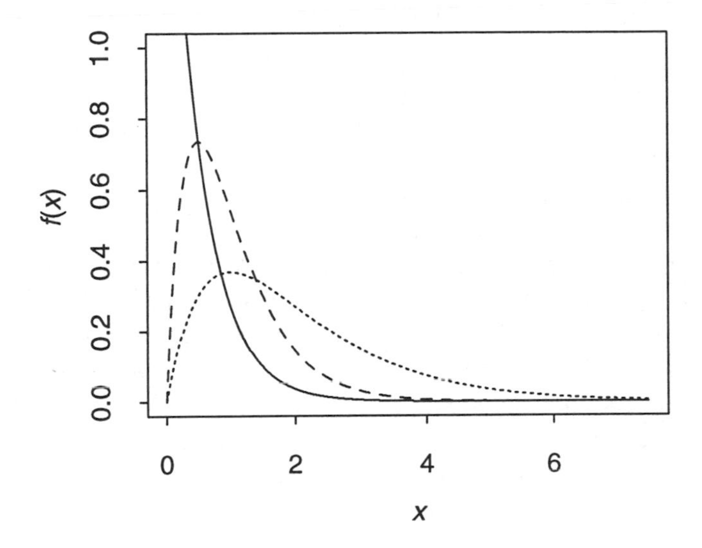

Fig. 1.4 Probability density functions for the gamma distribution: $\alpha = 2$, $\beta = 1$ (solid line); $\alpha = 1$, $\beta = 2$ (dotted line); and $\alpha = \beta = 2$ (dashed line).

Another special case of the gamma distribution that plays an important role in statistical inference is the **chi-square distribution**: $X \sim \chi^2(\nu)$. This is a one-parameter distribution that corresponds to $X \sim \mathcal{G}amma(\nu/2, 2)$, with $E[X] = \nu$ and $Var[X] = 2\nu$. The positive integer parameter ν is known as the distribution's **degrees of freedom** (*df*).

1.3.3 Continuous distributions: Weibull and extreme-value

The exponential distribution may be extended in a number of ways, due in part to its elegant simplicity. As we will see in Chapter 11, an extension of the exponential different from the gamma but nonetheless very useful in modeling time-to-event data is the **Weibull distribution**, with p.d.f. $f_X(x) = (\alpha\gamma)x^{\gamma-1}e^{-\alpha x^{\gamma}}I_{(0,\infty)}(x)$. Notice that the p.d.f. is similar to that of the gamma distribution, with perhaps the most crucial difference being that it carries a power of x in the exponent for e. In the special case of $\gamma = 1$, we recover the exponential distribution with scale parameter $\beta = 1/\alpha$

The Weibull mean and variance are somewhat complex:

$$\mathrm{E}[X] = \frac{\Gamma\left(\frac{\gamma+1}{\gamma}\right)}{\alpha^{1/\gamma}} \quad \text{and} \quad \mathrm{Var}[X] = \frac{\Gamma\left(\frac{\gamma+2}{\gamma}\right) - \Gamma^2\left(\frac{\gamma+1}{\gamma}\right)}{\alpha^{2/\gamma}}.$$

Using integral calculus, one can show that the corresponding c.d.f. is $F_X(x) = 1 - \exp\{-\alpha x^\gamma\}$. The two parameters α and γ provide great flexibility; α acts as a scale parameter, and γ acts as a shape parameter that determines the skew of the distribution. If $Y = X^\gamma$, it can be shown that Y is again exponential with scale parameter $\beta = 1/\alpha$.

The Weibull model is related to another useful model, known as the **extreme-value distribution**. If X is distributed as Weibull with parameters α and γ, then $T = \log_e(X)$ is distributed as a two-parameter extreme-value distribution. Note that we use the natural (base e) logarithm to transform the Weibull to an extreme-value variate. Unless otherwise indicated, we will use $\log(a)$ throughout to represent $\log_e(a)$.

The extreme-value parameters are written commonly as transformations of α and γ: $\delta = 1/\gamma$ and $\lambda = -\gamma^{-1}\log(\alpha)$. The p.d.f. is

$$f_T(t) = \frac{1}{\delta}\exp\left\{\left(\frac{t-\lambda}{\delta}\right) - \exp\left(\frac{t-\lambda}{\delta}\right)\right\} I_{(0,\infty)}(t).$$

As functions of the transformed parameters, the extreme-value mean and variance are $\mathrm{E}[T] = \lambda - \gamma_E\delta$ and $\mathrm{Var}[T] = \delta^2\pi^2/6$, where $\gamma_E = -\Gamma'(1) = 0.5722$ is Euler's constant and, in $\mathrm{Var}[T]$, $\pi = 3.14159\ldots$.

1.3.4 Continuous distributions: Normal and lognormal

An important distribution used in statistics is the **normal distribution**, also known as the **Gaussian distribution**. The normal p.d.f. is parameterized in terms of its mean, μ, and its variance σ^2:

$$f_X(x) = \frac{1}{\sigma\sqrt{2\pi}}\exp\left\{-\frac{1}{2}\left(\frac{x-\mu}{\sigma}\right)^2\right\}.$$

To denote this we write $X \sim N(\mu,\sigma^2)$. Normal random variables have unimodal, symmetric p.d.f.s that possess a 'bell' shape, centered at μ, and with spread governed by σ^2; cf. Moore and McCabe (1993, Figs 1.13, 1.15).

An important feature of the normal is that if $X \sim N(\mu,\sigma^2)$, the transformed variable $Z = (X - \mu)/\sigma$ is again normal, with zero mean and unit variance; i.e. $Z \sim N(0,1)$. We say Z is a **standard normal random variable**, and we will see that it is a reference distribution for many of the test statistics considered in the chapters below. Unfortunately, the standard normal c.d.f. has no closed-form expression, and can be written only as

$$\Phi(z) = \int_{-\infty}^{z} \frac{1}{\sqrt{2\pi}} e^{-t^2/2}\, dt. \tag{1.13}$$

Notice use of the special symbol $\Phi(z)$ for the standard normal c.d.f. This function takes any real value, z, to a probability between 0 and 1. It has been tabulated extensively in numerous introductory and advanced statistics texts; see, for example, Casella and Berger (1990, Table 1), Moore and McCabe (1993, Table A), or Neter *et al.* (1996, Table B.1).

We will also make use of the inverse of the standard normal c.d.f., $\Phi^{-1}(p)$. This inverse function gives the unique real number that achieves a specified cumulative probability, p. This quantity, or this quantity plus a constant, is also known as the **probit function, normal equivalent deviate (NED)**, or **normit function**, associated with the probability p. From this function, we define an **upper-α critical point** as the real value $\mathfrak{z}_\alpha$ whose upper tail area is α. Since this must require that the point's lower tail area is $1 - \alpha$, the upper-α critical points are defined by the relationship

$$\Phi(\mathfrak{z}_\alpha) = 1 - \alpha. \tag{1.14}$$

Standard normal critical points are available in many sources, such as Moore and McCabe (1993, Table D).

A positive-valued random variable associated with the normal distribution is the **lognormal distribution**. If $\log(X) \sim N(\mu,\sigma^2)$, then X has the lognormal distribution, with mean $E[X] = e^{\mu+(\sigma^2/2)}$ and variance $Var[X] = e^{2(\mu+\sigma^2)} - e^{2\mu\sigma^2}$. The p.d.f. is

$$f_X(x) = \frac{1}{x\,\sigma\sqrt{2\pi}} \exp\left\{-\frac{(\log(x) - \mu)^2}{2\sigma^2}\right\} I_{(0,\infty)}(x).$$

This density function is skewed to the right, and is very similar in shape to the gamma p.d.f.; cf. Casella and Berger (1990, Fig. 3.2.6). As with the Weibull distribution, the lognormal distribution is encountered frequently in the analysis of time-to-event data (see Section 11.2.4), and in a variety of other environmental applications (Armstrong, 1992; Stoline, 1993).

1.3.5 Distributions derived from the normal: Chi-square, *t*, and *F*

Part of the important role the normal distribution plays in statistics is that it is the basis of a number of derived distributions. These distributions are, in turn, central to many of the basic statistical methods we will describe in the chapters below. We have already encountered one such distribution: the chi-square distribution, $X \sim \chi^2(\nu)$, from Section 1.3.2. Although we noted that the chi-square is a special form of the gamma distribution, it is also true that the chi-square can be recovered from the normal: if $X \sim N(\mu, \sigma^2)$, then we know $Z = (X - \mu)/\sigma \sim N(0,1)$, and from this it is the case that $Z^2 \sim \chi^2(1)$. That is, the square of a standard normal random variable is distributed as χ^2 with 1 *df*. Further, the χ^2 family exhibits closure under addition: if $Y_i \sim$ (indep.) $\chi^2(\nu_i)$, then $\sum_{i=1}^{n} Y_i \sim \chi^2(\sum_{i=1}^{n} \nu_i)$.

In similar fashion to the standard normal distribution, we define the upper-α critical points of the $\chi^2(\nu)$ distribution as those values $\chi^2_\alpha(\nu)$ that satisfy $P[\chi^2(\nu) > \chi^2_\alpha(\nu)] = \alpha$, for any α between 0 and 1. These critical points are tabulated in various sources, including Casella and Berger (1990, Table 3), Neter *et al.* (1996, Table B.3), and Moore and McCabe (1993, Table G). Alternatively, one can also compute the upper tail areas, $P[\chi^2(\nu) > c]$, for some fixed value of c. We will see that these values have useful interpretations in a number of statistical inferences.

The χ^2 distribution is a crucial component of another derived distribution, the ***t* distribution**. We will see in Chapter 4 and later that the t distribution plays an important role in statistical inference. Its derivation dates back to the seminal work of W.S. Gossett, who wrote under the pseudonym 'Student' (Student, 1908). In its simplest form, the t random variable with ν *df* may be constructed as a ratio of a standard normal random variable to the square root of an independent $\chi^2(\nu)$ random variable which is itself divided by its *df*, i.e. if $Z \sim N(0,1)$ is independent of $Y \sim \chi^2(\nu)$, then $t = Z/(\nu^{-1}Y)^{1/2}$ is distributed as $t(\nu)$. The t distribution is similar to the standard normal distribution. Both are unimodal, symmetric, and centered at zero; however, the t distribution possesses heavier tails than the standard normal.

Notice that if we start with some general normal random variable, $X \sim N(\mu, \sigma^2)$, the standardized variable $Z = (X - \mu)/\sigma$ is then N(0,1). Thus a t random variable can be constructed via further division by $(Y/\nu)^{1/2}$. This operation is known as *Studentizing* the original random variable, X, in honor of Gossett's original work.

We define the upper-α critical points of the $t(\nu)$ distribution as those values $t_\alpha(\nu)$ that satisfy $P[t(\nu) > t_\alpha(\nu)] = \alpha$, for $0 < \alpha < 1$. These critical points are tabulated in various sources, including Casella and Berger (1990, Table 2), Neter *et al.* (1996, Table B.2), and Moore and McCabe (1993, Table E). Alternatively, one can compute the upper tail areas, $P[t(\nu) > u]$, for some fixed value of u.

Before moving on we should note that as the *df* of a t random variable approach infinity, the distribution approaches a standard normal; that is, loosely speaking, as $\nu \rightarrow \infty$, $t(\nu) \rightarrow \mathrm{N}(0,1)$. (Indeed, in most tables of t critical points $t_\alpha(\nu)$, a final row is included with $df = \infty$, corresponding to standard normal critical points, z_α. These include all the sources noted above.)

One additional distribution derived from the χ^2 family is the ***F* distribution**, defined as the ratio of two independent χ^2 random variables over the ratio of their corresponding *df*. That is, if $U_1 \sim \chi^2(\nu_1)$ is independent of $U_2 \sim \chi^2(\nu_2)$, then the ratio $F = (U_1/\nu_1)/(U_2/\nu_2)$ possesses an F distribution with parameters (also called **degrees of freedom**) ν_1 and ν_2. Upper-α F critical points $F_\alpha(\nu_1, \nu_2)$, are available in all the sources noted above.

The F, t, and χ^2 distributions are all related to the standard normal distribution, $Z \sim \mathrm{N}(0,1)$, by varying the *df* value(s). Figure 1.5 illustrates the effect. (See also Leemis (1986) for a more extensive graphic on the relationships among various random variables.)

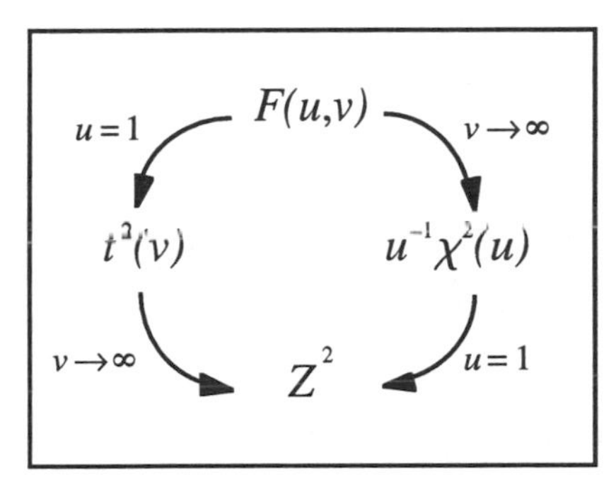

Fig. 1.5 Relationship between F, t, χ^2 and squared normal, Z^2, random variables.

1.3.6* Continuous distributions: Logistic

The logistic distribution is encountered often in environmental biology as a model for population growth if resource constraints are present (Finney, 1952, Section 17.10). It is used also for certain models with binary data, as we will see in Chapters 6–8. The c.d.f. is

$$F(x) = \frac{1}{1 + \exp\left\{-\frac{x - \alpha}{\beta}\right\}}.$$

An important transformation in the context of the logistic distribution is the **logit transformation** or **logit function**. The logit of p is defined as $\text{logit}(p) = \log\{p/(1-p)\}$. For the c.d.f. above, $\text{logit}\{F(x)\} = (x-\alpha)/\beta$. This transformation in effect linearizes the c.d.f. for a logistic distribution.

1.4 THE EXPONENTIAL CLASS OF DISTRIBUTIONS

Many of the univariate distributions described in Sections 1.2 and 1.3 may be collected together within a single family of probability functions, known as the **one-parameter exponential class**. This class of distributions is extensive enough to accept both discrete and continuous probability functions, and it will prove useful in developing the general methodology for statistical inferences that we consider in Chapter 8.

Suppose a random variable X is characterized by a single parameter θ. We say its distribution is a member of the exponential class if its p.m.f. or p.d.f. can be written in the following form:

$$\left.\begin{array}{c} p_X(x) \\ \text{or} \\ f_X(x) \end{array}\right\} = \gamma(x)\mathrm{B}(\theta)e^{T(x)\omega(\theta)},$$

or, by writing $\mathrm{B}(\theta) = e^{-b(\theta)}$ and $\gamma(x) = e^{c(x)}$, and then absorbing the functions into the exponent,

$$\left.\begin{array}{c} p_X(x) \\ \text{or} \\ f_X(x) \end{array}\right\} = \exp\{T(x)\omega(\theta) - b(\theta) + c(x)\}. \tag{1.15}$$

In this exponential class factorization, each individual function is dependent upon x or θ, *but not both*. (Known constants may appear throughout any function, however.) Notice also that for (1.15) to be a valid p.m.f. or p.d.f., we require $T(x)\omega(\theta) \geq 0$, for all x and all θ.

In some settings, the function $\omega(\theta)$ is called the **natural parameter** of the distribution, and written simply as ω. Mathematical operations on the natural parameter are often simpler to complete than those on the base parameter θ (Lehmann, 1983, Section 1.4), although we will not take explicit advantage of this feature.

A surprising number of distributions are members of the exponential class, including both discrete and continuous forms. Here are some examples:

Example 1.2 Poisson distribution as member of the exponential class
Consider first the discrete p.m.f. from the Poisson distribution in Section 1.2.5:

$$p_X(x) = \frac{\mu^x e^{-\mu}}{x!} I_{\{0,1,\ldots\}}(x) \,,$$

where $\mu > 0$ is the Poisson mean parameter. If we recall the relationship $b^x = e^{x\log(b)}$, we see that the Poisson p.m.f. may be re-expressed in the factored form

$$\frac{1}{x!} I_{\{0,1,\ldots\}}(x) \exp\{x \log(\mu) - \mu\}.$$

Then, it is clear that this has the exponential class form (1.15), where $e^{c(x)} = (x!)^{-1} I_{\{0,1,\ldots\}}(x)$, $T(x) = x$, $\omega(\mu) = \log(\mu)$, and $b(\mu) = \mu$. The natural parameter is the natural logarithm of the mean: $\omega = \log(\mu)$. □

Notice the explicit inclusion of the indicator function, $I_{\{0,1,\ldots\}}(x)$, in Example 1.2, and how it was involved in the function $c(x)$. This was a crucial step in the identification of the Poisson p.m.f. as an exponential class form. Indeed, any random variable that is defined over a region which is itself a function of the unknown parameter(s) cannot be a member of the exponential class of distributions. If we use the indicator function to write the p.m.f. or p.d.f. in its fully expressed form, we can check quickly whether the support of the p.m.f. or p.d.f. is dependent upon any unknown parameters. If it is, then the random variable is not part of the exponential class. The uniform distribution with unknown upper or lower limits (or both) is an example of such a nonexponential class p.d.f.

Example 1.2 also illustrates a special kind of exponential class p.m.f., where the function $T(x)$ is the simple identity function, $T(x) = x$. When this occurs for any member of the exponential class, we say the associated random variable is in a **canonical form**.

Example 1.3 Exponential distribution as member of exponential class
Perhaps the simplest distribution that stands as a member of the exponential class is the similarly-named exponential distribution from Section 1.3.2. (We should emphasize, however, that the class is not named for this particular distribution, even though the latter is an element of the former, as follows.) The exponential p.d.f. is

$$f_X(x) = \frac{e^{-x/\beta}}{\beta} I_{(0,\infty)}(x)$$

where $\beta > 0$ is the exponential mean. Re-expressing this as $f_X(x) = I_{(0,\infty)}(x)\exp\{-x\beta^{-1} - \log(\beta)\}$ shows that this p.d.f. is also a canonical member of the exponential class, with $e^{c(x)} = I_{(0,\infty)}(x)$, $b(\beta) = \log(\beta)$, $T(x) = x$, and $\omega(\beta) = 1/\beta$. The natural parameter is the reciprocal of the mean: $\omega = 1/\beta$. □

1.5 FAMILIES OF MULTIVARIATE DISTRIBUTIONS

The distributional families presented in the preceding sections represent only a limited selection from the many possible distributions used in statistics, although we hope those presented give a useful review of the various ways random variability may be modeled for univariate observations, X. Readers interested in further details on statistical distributions may explore the multi-volume set by Johnson *et al.* (1992; 1994; 1995), and the useful summary article by Leemis (1986).

It is possible to extend the basic distributions given above to cases where a random variable takes on more than a simple univariate form, i.e. where the observation appears as a multivariate array or **vector** (denoted via a bold-face Roman letter: $\mathbf{X} = [X_1\ X_2\ \ldots\ X_M]$). Each component of a random vector is itself assumed to be a random variable, but there may be some special relationship among the components that can be modeled within some multivariate family of probability distributions. For example, there may exist some form of structural, nonzero correlation among the X_is so that they are not independent, or there may be some constraint in effect among the X_is, such as that they all must sum to some constant, etc.

Just as in the univariate case, there are many possible forms of multivariate random vectors. Here, we introduce a select handful of multivariate distributions – both discrete and continuous – many of which we employ in the chapters below. Although these distributions are often useful in practice, some readers may not require the full structural details, and they may wish to proceed forward to the summary in Section 1.6.

1.5.1* Multivariate distributions: Multinomial

Recall that a binomial random variable, Y, may be viewed as a count of the number of successes observed out of a total number of trials, N. This setting imposes only two possible outcomes for each trial, success and failure. A natural multivariate extension of this construction allows for more than two possible outcomes, say $M > 2$, where the success probability for each possible outcome is denoted by π_i, $i = 1, ..., M$. (Notice that the sum of the π_is must be 1.) Then, let Y_i = {number of trials that result in outcome label i}, where the sum of the Y_is equals the total number of trials, N. The collection of M such counts is a discrete multivariate random vector, $\mathbf{Y} = [Y_1\ Y_2\ ...\ Y_M]$, with joint p.m.f.

$$p_{\mathbf{Y}}(y_1, y_2, ..., y_M) = \binom{N}{y_1 \cdots y_M} \prod_{i=1}^{M} (\pi_i)^{y_i}, \tag{1.16}$$

where the first term in the p.m.f. is the **multinomial coefficient**,

$$\binom{N}{y_1 \cdots y_M} = \frac{N!}{\prod_{i=1}^{M}(y_i!)}.$$

These coefficients are also seen in the **multinomial theorem**, which is a result from basic algebra: for any real values π_i, $i = 1, ..., M$, the sum of the π_is raised to the Nth power is

$$(\pi_1 + \pi_2 + \cdots + \pi_M)^N = \sum \binom{N}{y_1 \cdots y_M} \prod_{i=1}^{M} (\pi_i)^{y_i},$$

where the sum is taken over all combinations of y_is such that $\sum_{i=1}^{M} y_i = N$. By use of this theorem, one can show that the probabilities in (1.16) do indeed sum to 1 under the requirement that $\sum_{i=1}^{M} \pi_i = 1$. When $M = 2$, the multinomial theorem reduces to the **binomial theorem** (Exercise 1.59). This is useful in establishing fundamental features of the binomial distribution from Section 1.2.1; see Casella and Berger (1990, Section 3.1).

Under the joint p.m.f. (1.16), we say $\mathbf{Y}$ has a **multinomial distribution**, and write $\mathbf{Y} \sim \mathcal{M}ultinom(N, \{\pi_i\})$. This distribution may be viewed as an extension from the two-outcome binomial setting to the multiple-outcome case.

Example 1.4. Genotypes and multinomial distributions

In a simple example from genetics and heredity, suppose a single genetic locus controls the expression of some expressed trait. If two heterozygotes were mated (say, $Aa \times Aa$) the offspring of such a mating could have one of three possible genotypes: homozygous dominant (AA), heterozygous (Aa), or homozygous recessive (aa). To analyze these offspring for their genotypes at this locus, a probability model for the response is needed. A natural model is the multinomial distribution with $M = 3$ levels. In this example, y_1 is the number of offspring that are homozygous dominant with

π_1 equal to the probability of observing a homozygous dominant offspring, y_2 is the number of offspring that are heterozygous with π_2 equal to the probability of observing a heterozygous offspring, and y_3 is the number of offspring that are homozygous recessive with π_3 equal to the probability of observing a homozygous recessive offspring. ❑

Viewed on a per-component level, the multinomial collapses back into the binomial: if $\mathbf{Y} \sim \mathcal{Multinom}(N,\{\pi_i\})$, then the marginal distribution of any single Y_i is binomial (Exercise 1.61); we write $Y_i \sim \mathcal{Bin}(N,\pi_i)$ for all $i = 1, \ldots, M$. As such, the means and variances of the Y_is may be determined from their marginal distributions: $\mathrm{E}[Y_i] = N\pi_i$, and $\mathrm{Var}[Y_i] = N\pi_i(1 - \pi_i)$. Note, however, that the multinomial requires the sum of the Y_is to equal N. In effect, this induces a negative correlation among the individual Y_is, expressed via their covariances as $\mathrm{Cov}[Y_i, Y_j] = -N\pi_i\pi_j$ $(i \neq j)$.

One important feature of a multinomial random variable is that it can be related to conditional Poisson random variables. Specifically, given a set of M independent Poisson variables, $Y_i \sim$ (indep.) $\mathcal{Poisson}(\mu_i)$, $i = 1, \ldots, M$, suppose we condition on the total of the Y_is. That is, suppose that $Y_+ = \sum_{i=1}^{M} Y_i$ is held fixed at some specific value, say y_+. (The notation Y_+ indicates summation over the index in which the '+' sign is active; here $Y_+ = Y_1 + Y_2 + \cdots + Y_M$.) Then, the resulting joint conditional distribution of $\mathbf{Y} = [Y_1\ Y_2 \cdots\ Y_M]$ is multinomial (Exercise 1.62):

$$\mathrm{P}(Y_1 = y_1, Y_2 = y_2, \ldots, Y_M = y_M \mid Y_+ = y_+) = \binom{y_+}{y_1 \ \cdots \ y_M} \prod_{i=1}^{M} \left(\frac{\mu_i}{\mu_+}\right)^{y_i},$$

where $\mu_+ = \sum_{i=1}^{M} \mu_i$. To express this relationship, we write $\mathbf{Y} \mid y_+ \sim \mathcal{Multinom}(y_+, \{\mu_i/\mu_+\})$. (The conditional multinomial probabilities, μ_i/μ_+, still sum to 1, as required.) As we will see in Sections 5.5 and 9.2, this feature allows for simplified manipulation of probability relationships when operating with Poisson-distributed data.

It is also possible to extend the multinomial distribution to account for possible excess variability, as was done with the beta-binomial distribution in Section 1.2.2. The multivariate extension is the **Dirichlet-multinomial distribution**, which can be used to account for **extra-multinomial variability** in multiple-category data (Johnson and Kotz, 1969, Section 11.8).

1.5.2* Multivariate distributions: Multivariate hypergeometric

Similar to the multinomial extension of the binomial distribution is the multivariate extension of the univariate hypergeometric distribution from

Section 1.2.4. Returning to our finite-population model for discrete random variables, suppose that N population elements represent $M > 2$ different categories of response, so that the population contains R_m elements of each category ($m = 1, \ldots, M$). (At $M = 2$ we recover the univariate hypergeometric.) Select K observations from the population at random and without replacement, and let X_m = {number of observations out of K in category m}. These individual random variables comprise a random vector $\mathbf{X} = [X_1\ X_2\ \ldots\ X_M]$ with joint hypergeometric p.m.f.

$$p_{\mathbf{X}}(\mathbf{x}\,|\,\mathbf{R},N,K,M) = \frac{\binom{R_1}{x_1}\binom{R_2}{x_2}\cdots\binom{R_M}{x_M}}{\binom{N}{K}},$$

where we constrain $N = \sum_{m=1}^{M} R_m$ and $K = \sum_{m=1}^{M} x_m$. To denote this, we write $\mathbf{X} \sim \mathcal{hg}_M(\mathbf{R},N,K)$. As with the multinomial distribution in Section 1.5.1, the individual components of a multivariate hypergeometric random vector are themselves hypergeometric. That is, the marginal distribution of any X_m is again univariate hypergeometric: $X_m \sim \mathcal{hg}_1(R_m,N,K)$, for all $m = 1, \ldots, M$.

1.5.3 Multivariate distributions: Multivariate normal

One of the most important multivariate distributions used in environmental statistics is the multivariate extension of the normal distribution from Section 1.3.4. It is common to encounter the multivariate normal at the foundation of many test statistics, at least as a large-sample approximation. (We discuss this in greater detail in Chapter 2.)

Consider first the simplest form of a multivariate normal random vector: the bivariate normal. We say $\mathbf{X} = [X_1\ X_2]$ has a **bivariate normal distribution** when its joint p.d.f. takes the form

$$f_{\mathbf{X}}(x_1,x_2) = \frac{1}{2\pi\sigma_1\sigma_2(1-\rho^2)^{1/2}} \exp\left\{-\frac{1}{2(1-\rho^2)}\left[\left(\frac{x_1-\mu_1}{\sigma_1}\right)^2 - 2\rho\left(\frac{x_1-\mu_1}{\sigma_1}\right)\left(\frac{x_2-\mu_2}{\sigma_2}\right) + \left(\frac{x_2-\mu_2}{\sigma_2}\right)^2\right]\right\}. \quad (1.17)$$

where the parameters μ_1, μ_2, σ_1^2, and σ_2^2 represent the marginal means and variances, respectively, of X_1 and X_2. An additional parameter, ρ, is used to define the **correlation coefficient** between X_1 and X_2: $\rho = \mathrm{Cov}[X_1,X_2]/(\sigma_1\sigma_2)$, where $-1 \le \rho \le 1$.

The unknown parameters μ_1, μ_2, σ_1^2, σ_2^2, and ρ may be collected together into a pair of arrays: first, corresponding to $\mathbf{X}$, the mean vector is $\boldsymbol{\mu} = [\mu_1\ \mu_2]$. Then, for the variation parameters, we construct a **variance-covariance matrix**, which is a square array consisting of the two variance parameters and the covariance $\text{Cov}[X_1,X_2] = \sigma_{12} = \rho\sigma_1\sigma_2$:

$$\mathbf{V} = \begin{bmatrix} \text{Var}[X_1] & \text{Cov}[X_1,X_2] \\ \text{Cov}[X_2,X_1] & \text{Var}[X_2] \end{bmatrix}$$

$$= \begin{bmatrix} \sigma_1^2 & \sigma_{12} \\ \sigma_{21} & \sigma_2^2 \end{bmatrix} = \begin{bmatrix} \sigma_1^2 & \rho\sigma_1\sigma_2 \\ \rho\sigma_1\sigma_2 & \sigma_2^2 \end{bmatrix}.$$

By convention, the diagonal elements of the variance-covariance matrix contain the variances, and the off-diagonal elements contain the covariances. Notice that since $\text{Cov}[X_1,X_2] = \rho\sigma_1\sigma_2 = \text{Cov}[X_2,X_1]$, the matrix $\mathbf{V}$ has an upper off-diagonal element that is identical to the lower off-diagonal element. (We say $\mathbf{V}$ is **symmetric** about its diagonal.) Using this matrix notation, we denote the bivariate normal representation as $\mathbf{X} \sim \text{N}_2(\boldsymbol{\mu},\mathbf{V})$.

An important characteristic of the bivariate normal is that X_1 and X_2 are independent if *and only if* $\rho = 0$. Recall that if two random variables are independent, their correlation is zero; with the bivariate normal the reverse is true as well.

Viewed on a per-component level, the bivariate normal collapses back to the univariate normal: if $\mathbf{X} \sim \text{N}_2(\boldsymbol{\mu},\mathbf{V})$, then the marginal distribution of any single component is $X_i \sim \text{N}(\mu_i,\sigma_i^2)$, for $i = 1,2$.

The **multivariate normal distribution** is a straightforward multivariate analog of the bivariate normal. For a random vector $\mathbf{X} = [X_1\ X_2\ \cdots\ X_M]$, the joint p.d.f. is an extension of the bivariate p.d.f. in (1.17). The mean vector is $\boldsymbol{\mu} = [\mu_1\ \mu_2\ \cdots\ \mu_M]$, and the variance-covariance matrix, $\mathbf{V}$, has ith diagonal element σ_i^2 and (i,j)th off-diagonal element $\text{Cov}[X_i,X_j] = \rho_{ij}\sigma_i\sigma_j$, where ρ_{ij} is the correlation coefficient between X_i and X_j $(i \neq j)$. Once again, since $\text{Cov}[X_i,X_j] = \rho_{ij}\sigma_i\sigma_j = \text{Cov}[X_j,X_i]$, the upper off-diagonal elements are identical to their lower off-diagonal counterparts, so the matrix $\mathbf{V}$ is symmetric about its diagonal. Then, we can write $\mathbf{X} \sim \text{N}_M(\boldsymbol{\mu},\mathbf{V})$ to denote the M-variate normal random vector. As in the bivariate case, if $\mathbf{X} \sim \text{N}_M(\boldsymbol{\mu},\mathbf{V})$, the marginal distributions are all univariate normal: $X_i \sim \text{N}(\mu_i,\sigma_i^2)$, for $i = 1, \ldots, M$.

By standardizing each component of a multivariate normal vector by its mean and standard deviation, we achieve a standard multivariate normal vector. Take $Z_i = (X_i - \mu_i)/\sigma_i$ for each $i = 1, \ldots, M$, and collect the Z_is together into the vector **Z**. Notice that now the covariances among the Z_is are $\mathrm{Cov}[Z_i, Z_j] = \rho_{ij}\,(1)(1)$, i.e. the variance-covariance matrix of **Z** is just the matrix of correlations (with 1 at each diagonal element). In this special case, we write the variance-covariance matrix as a correlation matrix, **R**, and we say $\mathbf{Z} \sim \mathrm{N}_M(\mathbf{0}, \mathbf{R})$, where **0** is simply an M-vector of all zeros.

As with the univariate normal, many multivariate test statistics we will encounter will have exact or at least approximate multivariate normal distributions. When this occurs, we will employ multivariate analogs of the univariate normal critical points, based on the maximum of the Z_is, $Z_{(M)} = \max\{Z_1, \ldots, Z_M\}$, or on the maximum of the absolute values of the Z_is, $|Z|_{(M)} = \max\{\,|Z_1|, \ldots, |Z_M|\,\}$. Since both $Z_{(M)}$ and $|Z|_{(M)}$ depend upon the underlying correlation matrix, **R**, we denote the upper-α critical points of their distributions as $\mathfrak{z}^{(\alpha)}_{M,\mathbf{R}}$ and $|\mathfrak{z}|^{(\alpha)}_{M,\mathbf{R}}$, respectively. In the special case that the ρ_{ij} are all equal to the same value, say ρ, we write $\mathfrak{z}^{(\alpha)}_{M,\rho}$ and $|\mathfrak{z}|^{(\alpha)}_{M,\rho}$, respectively. For this equicorrelated case, tables of the upper-α critical points are available in Hochberg and Tamhane (1987, Tables 2–3). That source also contains greater detail on the multivariate normal (Hochberg and Tamhane, 1987, Appendix 3.1), as does, for example, Johnson and Kotz (1972, Chapter 35).

1.5.4* Multivariate distributions: Multivariate t and other Studentized distributions

The multivariate t distribution is an extension of the univariate t in Section 1.3.5. Indeed, it is constructed in much the same manner: start with a standard M-variate normal random vector, **Z**, with correlation matrix **R**, and take a χ^2 random variable $Y \sim \chi^2(\nu)$ which is statistically independent of **Z**. Then, just as the univariate random variable $t_i = Z_i/(\nu^{-1}Y)^{1/2}$ is distributed as $t(\nu)$, the random M-vector $\boldsymbol{t} = [t_1\ t_2\ \ldots\ t_M]$, is distributed as multivariate t, with parameters M, ν, and **R** (Dunnett and Sobel, 1954): $\boldsymbol{t} \sim t_M(\nu, \mathbf{R})$ Other characterizations are also possible (Johnson and Kotz, 1972, Chapter 37).

Upper-α critical points of this distribution are based on the maximum of the t_is, $t_{(M)} = \max\{t_1, \ldots, t_M\}$, or on the maximum of the absolute values, $|t|_{(M)} = \max\{\,|t_1|, \ldots, |t_M|\,\}$, as in the preceding section. We denote these critical points as $\mathcal{T}^{(\alpha)}_{M,\nu,\mathbf{R}}$ and $|\mathcal{T}|^{(\alpha)}_{M,\nu,\mathbf{R}}$, respectively. In the special case that the ρ_{ij} are all equal to the same value, say ρ, we write $\mathcal{T}^{(\alpha)}_{M,\nu,\rho}$ and

$|\mathcal{T}|^{(\alpha)}_{M,\nu,\rho}$, respectively. For this equicorrelated case, tables of the upper-α critical points are available in Hochberg and Tamhane (1987, Tables 4–5). Note also that as $\nu \to \infty$, the M-variate t distribution approaches an M-variate standard normal distribution with the same correlation matrix, $\mathbf{R}$. In this case, we write $|\mathcal{T}|^{(\alpha)}_{M,\infty,\mathbf{R}} = |\mathfrak{z}|^{(\alpha)}_{M,\mathbf{R}}$.

When the correlation matrix $\mathbf{R}$ contains constant correlations, ρ, and when these correlations are all identically equal to zero, $\mathbf{R}$ is particularly simple in form: it has only ones on its diagonal, and only zeros off the diagonal (both below and above). This special form of square matrix is known as the **identity matrix** (with M rows and M columns). It is denoted by $\mathbf{I}_M$, or simply by $\mathbf{I}$ if the number of rows and columns is clear. In this case, we can write $\boldsymbol{t} \sim t_M(\nu, \mathbf{I})$. The distribution of $t_{(M)} = \max\{t_1, \ldots, t_M\}$ is then known as the **Studentized maximum distribution** with parameters M and ν, while the distribution of

$$|t|_{(M)} = \max\left\{ |t_1|, \ldots, |t_M| \right\}$$

is the **Studentized maximum modulus** distribution, also with parameters M and ν (Hochberg and Tamhane, 1987, Appendix 3.1). For this special case, we write the upper-α critical points of the Studentized maximum distribution as $\mathcal{M}^{(\alpha)}_{M,\nu}$ and the upper-α critical points of the Studentized maximum modulus distribution as $|\mathcal{M}|^{(\alpha)}_{M,\nu}$. Tables of these values are available in Hochberg and Tamhane (1987, Tables 6–7).

Before closing, we note that it is also possible to extend the exponential class to the multivariate setting, and to multiple parameters. We will not discuss these extensions, but instead refer interested readers to the more extensive discussions in, for example, Lehmann (1983, Section 1.4), Efron (1986), or Casella and Berger (1990, Section 3.3).

1.6 SUMMARY

In this chapter, fundamental concepts of probability and statistics are summarized, including the definitions of simple, joint, conditional, and marginal probabilities. It is shown how these definitions relate to basic operations on probabilities, such as addition, subtraction, and multiplication. Random variables are introduced as quantifications of some experimental outcome, and independence among random variables is presented. Summary measures of a random variable are developed, based on an expectation operator. For instance, the expected value of the random variable – commonly called the mean – is a measure of central tendency of the distribution. A measure of variability is the variance, which is seen to be a slightly more complex expected value operation. Other expected value operations produce measures of covariance and correlation between pairs of random variables.

Various families of probability distributions are described that characterize mathematically the probabilistic features of random variables; both univariate and multivariate forms of probability distributions are considered. It is seen that a random variable can be summarized via one or more unknown parameters, and that these parameters typically are associated with the mean and variance of the random variable. In addition, the exponential class of probability functions is presented as a unifying family of distributions that includes many different parametric families.

EXERCISES

(Exercises with asterisks, e.g. 1.11*, typically are more advanced, or require greater knowledge and understanding of probability distributions and/or calculus.)

1.1. Describe the sample space for the following setting: we observe the number of eggs laid by a certain bird species in an ecosystem affected by some toxic damage.

1.2. Describe the sample space for the following setting: we observe the concentration of an environmental toxin in eggshells of a certain bird species.

1.3. Identify if the following random variables are discrete or continuous:

- a. Number of eggs laid by birds of a certain species in an ecosystem affected by some toxic damage.
- b. Concentration of an environmental toxin in eggshells of a certain bird species.
- c. Number of mutations in a bacteria culture after exposure to ultraviolet light.
- d. Blood concentration of a toxic metabolite in hamsters after acute exposure to the toxin.
- e. Weight of mice after 15-day exposure to an environmental toxin.
- f. Time to recover to full mobility in insects after exposure to near-freezing temperatures for 2 hours ('cold shock').
- g. Whether or not a mouse can maneuver successfully through a simple maze after exposure to a neurotoxin.

h. Time for a mouse to maneuver successfully through a simple maze after exposure to a neurotoxin.

i. Proportion of N animals that exhibit liver cancer after 2-year, chronic exposure to an environmental carcinogen.

1.4. For the following pairs of events, decide whether or not the events are independent, and explain why:

a. Count number of eggs laid by a single bird in an ecosystem affected by some toxic damage, and also measure the concentration of the toxin in this bird's eggshells.

b. Observe occurrence of lung cancer in a mouse after exposure to an environmental toxin, and also measure mutation frequency in yeast cultures after exposure to the same toxin.

c. Observe occurrence of lung cancer in two different mice after exposure to the same toxin.

d. Observe occurrence of lung cancer and liver cancer in the same mouse after exposure to an environmental toxin.

1.5. Do the following sets of values represent valid probability mass functions (p.m.f.s)? Why or why not?

a.

x	0	2	4	6	8
$P[X=x]$	0.2	0.2	0.2	0.2	0.2

b.

w	0	2	4	6	8
$P[W=w]$	0.05	0.3	0.5	0.1	0.05

c.

u	−2	−1	1	2
$P[U=u]$	0.22	0.31	0.23	0.24

d.

v	0	1	2	3	4	5	6
$P[V=v]$	0.01	0.01	0.01	0.01	0.01	0.01	0.01

e.

t	0	1/10	3/10	5/10	7/10	9/10
$P[T=t]$	0.1	0.1	0.3	0.3	0.1	0.1

f.

z	0	1	2	3	4	5	6
$P[Z=z]$	0.06	0.21	0.33	0.31	0.20	−0.19	0.08

1.6. For those functions in Exercise 1.5 that are valid p.m.f.s, give the associated cumulative distribution functions (c.d.f.s). Also, sketch the c.d.f.s associated with the valid p.m.f.s.

1.7. Suppose X is a continuous random variable with probability density function (p.d.f.) $f(x)$.

a. Is it possible for $f(x) < 0$? Why or why not?

b. Is it possible for $f(x) > 1$? Why or why not?

1.8. Are the following sets of joint functions valid *bivariate* p.m.f.s? Why or why not?

a.

		x: 0	1	2	3
y	0	0.025	0.04	0.05	0.055
	1	0.07	0.08	0.08	0.09
	2	0.09	0.085	0.08	0.07
	3	0.065	0.05	0.04	0.03

b.

		x: 0	1	2	3
y	0	0.055	0.055	0.055	0.055
	1	0.077	0.077	0.077	0.077
	2	0.058	0.058	0.058	0.058
	3	0.06	0.06	0.06	0.06

c.

		x: 0	1	2	3
y	0	0.035	0.035	0.035	0.045
	1	0.022	0.078	0.08	0.091
	2	0.09	0.08	0.087	0.022
	3	0.065	0.092	0.04	0.024

1.9 For those bivariate functions in Exercise 1.8 that are valid p.m.f.s, find:

(i) The marginal p.m.f. of X.

(ii) The marginal c.d.f. of X.

(iii) The conditional p.m.f. of $X \mid Y = 1$.

(iv) The conditional c.d.f. of $X \mid Y = 1$.

(v) The conditional p.m.f. of $Y \mid X = 0$.

(vi) The conditional c.d.f. of $Y \mid X = 0$.

1.10. For those bivariate functions in Exercise 1.8 that are valid p.m.f.s, identify if X and Y are independent.

1.11*. Let X have p.m.f. as in Exercise 1.5a, and let W have p.m.f. as in Exercise 1.5b. Find the p.m.f. of the mixture distribution of $Z = \omega X + (1 - \omega) W$ for the following mixture parameters:

a. $\omega = 0.5$.

b. $\omega = 0.27$.

c. $\omega = 1$.

1.12. What is the resulting mixture distribution of Z when the two mixing distributions of X and W are the same? Is this dependent on the mixing parameter ω?

1.13. Find the population means and population variances for the p.m.f.s in Exercises 1.5a and 1.5b.

1.14. Evaluate the following expectations, $\mathrm{E}[g(X,Y)]$, for each of the valid bivariate p.m.f.s from Exercise 1.8:

a. $g(X,Y) = X$.
b. $g(X,Y) = (X - \mathrm{E}[X])^2$.
c. $g(X,Y) = Y$.
d. $g(X,Y) = X - Y$.
e. $g(X,Y) = XY$.

1.15. Evaluate $\mathrm{Cov}[X,Y]$ for each of the valid bivariate p.m.f.s from Exercise 1.8. Also, calculate the associated population correlation coefficient, ρ_{XY}.

1.16. Given the random variables X_1, X_2, X_3, X_4, which of the following are valid linear combinations of the X_is? Of these, are any valid contrasts?

a. $X_1 + X_2 + X_3 + X_4$.
b. $\frac{1}{4}X_1 + \frac{1}{4}X_2 + \frac{1}{4}X_3 + \frac{1}{4}X_4$.
c. $X_1 - X_2 - X_3 + X_4$.
d. $3X_1 - 4X_2 + 2.2X_3 - 4.7X_3X_4$.
e. $3X_1 - 4X_2 + 3X_3 - 1.1X_4^2$.
f. $X_1 - 4.2X_2 + 3.1X_3 + 0.1X_4$.

1.17. Suppose each X_i in Exercise 1.16 has an identical population mean $\mathrm{E}[X_i] = \mu$. For each of the valid linear combinations, find the population mean, $\mathrm{E}[L]$.

1.18. Consider a **contrast**, $L = a_1X_1 + \cdots + a_nX_n$, such that $a_1 + \cdots + a_n = 0$, and where each X_i has identical population mean $\mathrm{E}[X_i] = \mu$. Show that $\mathrm{E}[L] = 0$.

1.19. Describe an application (from your own field of study, as appropriate), that generates a binomial random variable, that is, a count of N independent Bernoulli (success vs. failure) trials where the success probability, π, is constant ($0 \le \pi \le 1$).

1.20. Calculate the following values, or state why they are not calculable:

a. $1!$
b. $(-2)!$
c. $2!$
d. $4.7!$
e. $7!$
f. $0!$
g. $10!$
h. $68!$

1.21. Calculate the binomial coefficient

$$\binom{N}{i}$$

for the following values:

a. $N = 3, i = 1.$ b. $N = 9, i = 9.$

c. $N = 7, i = 4.$ d. $N = 9, i = 0.$

e. $N = 12, i = 9.$ f. $N = 5, i = 4.$

1.22. Show that

a. $\binom{N}{0} = \binom{N}{N}$ for any $N = 0,1,\ldots.$

b. $\binom{N}{1} = \binom{N}{N-1}$ for any $N = 1,2,\ldots.$

1.23. Calculate and graph the p.m.f. of the following binomial random variables. Comment on the symmetry of the graphed p.m.f.:

a. $X \sim \mathcal{B}in(4,0.5).$ b. $X \sim \mathcal{B}in(4,0.4).$

c. $X \sim \mathcal{B}in(5,0.1).$ d. $X \sim \mathcal{B}in(5,0.9).$

e. $X \sim \mathcal{B}in(2,0.01).$

1.24. For each binomial distribution in Exercise 1.23, find the population mean E[X] and population variance Var[X].

1.25*. If $X \sim \mathcal{B}in(N,\pi)$, show that $N - X \sim \mathcal{B}in(N,1 - \pi)$.

1.26. Describe an application (from your own field of study, as appropriate) where extra-binomial variability occurs (for instance, where the underlying Bernoulli trials are not independent).

1.27*. Given $X \sim \mathcal{B}in(N,p)$ and hierarchically $p \sim \mathcal{B}eta(\alpha,\beta)$, show that the marginal p.m.f. of X has the beta-binomial form given in (1.10). What is the necessary relation between (α,β) and (π,φ)?

1.28. Identify a situation (from your own field of study, as appropriate) where the memoryless property of the geometric distribution is useful.

1.29*. Use the features of a finite geometric series (i.e. $r^0 + r^1 + \cdots + r^m = \{1 - r^{m+1}\}/(1 - r)$ for any $|r| < 1$ and $m > 0$) to derive the c.d.f. of a geometric random variable.

1.30*. Show that the memoryless property of the geometric distribution, that $P(X > u \mid X > v)$ is only a function of the difference $(u - v)$, does not apply to the negative binomial distribution.

1.31. Recall the tag-recapture example leading to a geometric random variable in Section 1.2.3. Describe how to extend this construction so as to induce a negative binomial random variable.

1.32*. Show that the representation of the negative binomial p.m.f. using the (r,π) parameterization is equivalent to the (μ, δ) form in (1.11), when $\mu = r(1 - \pi)/\pi$ and $\delta = 1/r$.

1.33. Calculate and plot the p.m.f. of X of the following hypergeometric distributions:

a. $X \sim \mathcal{hg}(5,10,5)$. b. $X \sim \mathcal{hg}(3,5,2)$.

c. $X \sim \mathcal{hg}(1,4,1)$. d. $X \sim \mathcal{hg}(4,6,2)$.

e. Compare the plot of the p.m.f. of $X \sim \mathcal{hg}(5,10,5)$ with a plot of the p.m.f. of $Y \sim \mathcal{Bin}(5,0.5)$. Comment on any differences between the plots.

1.34. Describe an application (from your own field of study, as appropriate), that generates a Poisson random variable, that is, a count of outcomes that satisfies the five Poisson postulates, where the mean rate of occurrence, μ, is a positive constant.

1.35. Extra-Poisson variability that leads to overdispersion can occur in many different ways. For instance, disruption of any of the five Poisson postulates can lead to extra variability. Among these five postulates, which is perhaps the most susceptible to violation leading to extra-Poisson variability?

1.36. Consider a set of independent, identically distributed Poisson random variables, each with the same mean μ. What is the distribution of the sample sum $X_1 + X_2 + \cdots + X_n$?

1.37*. Given $Y \mid \lambda \sim \mathcal{Poisson}(\lambda)$ and, hierarchically, $\lambda \sim \mathcal{Gamma}(\alpha, \beta)$, show that the marginal p.m.f. of Y has a negative binomial p.m.f.

1.38*. As in Example 1.1, supposing $X \mid N \sim \mathcal{Bin}(N,\pi)$ and, hierarchically, $N \sim \mathcal{Poisson}(\mu)$, show that the marginal p.m.f. of X takes the form $p_X(x) = (\pi\mu)^x e^{-\pi\mu}/x!$, for any positive integer $x = 0,1,\ldots.$

1.39*. Let $X \sim \mathcal{Beta}(\alpha, \beta)$. Using integral calculus, show that the expected value, E[X], is $\alpha/(\alpha + \beta)$. (Recall from (1.7b) that the expected value of a continuous random variable is the integral of the product $x f_X(x)$ over the entire range of X.)

1.40. Let $X \sim \mathcal{Beta}(\alpha, \alpha)$. Show that the p.d.f. is symmetric about $x = 1/2$. Find E[X] and Var[X].

1.41*. Suppose $X \sim \mathcal{Exp}(\beta)$.

a. Use integral calculus to show that E[X] = β.

b. Use integral calculus to find E[X^2]. Use this result to show that Var[X] = β^2.

1.42. Consider the following statement about use of the exponential distribution to model survival times: 'when it comes to the exponential model, survival time of a mouse is identical to operating time of a transistor.' Explain.

1.43. Suppose $X \sim \mathcal{Gamma}(\alpha, \beta)$. For what specific value(s) of α and/or β does this reduce to $X \sim \mathcal{Exp}(\beta)$?

1.44. Use the fact that if $\alpha = \nu/2$ and $\beta = 2$, the gamma p.d.f. reduces to the $\chi^2(\nu)$ distribution, to find the $\chi^2(\nu)$ p.d.f.

1.45. Suppose $X \sim N(\mu, \sigma^2)$. Find the distribution of:

a. $W = X - \mu$.
b. $U = X/\sigma$.
c. $Z = (X - \mu)/\sigma$.
d. $V = (X + \mu)/\sigma$.

1.46. Suppose $X \sim N(0,1)$. Use a table of standard normal probabilities to find the upper-α critical points $z_{0.10}$, $z_{0.05}$, $z_{0.025}$, and $z_{0.01}$.

1.47*. If $\Phi(z)$ is the standard normal c.d.f, use the symmetry of the normal distribution about its population mean to show that $\Phi(-z) = 1 - \Phi(z)$, for $z > 0$.

1.48. Suppose $\log(X) \sim N(0,1)$. What are the population mean and variance of X?

1.49. Describe an experimental outcome, and the associated response variable, that might be modeled using each of the following distributions:

a. Geometric.
b. Gamma.
c. Negative binomial.
d. Extreme-value.
e. Hypergeometric.
f. Normal.
g. Uniform.
h. Lognormal.
i. Exponential.
j. Logistic.

1.50. Find a journal article in which each of the distributions in Exercise 1.49 was used to model the random event of interest. Provide the complete article reference along with a summary of how each distribution is used.

1.51. Consider a set of independent, identically distributed χ^2 random variables, X_i, each with the same mean ν. What is the distribution of the sample sum $X_1 + X_2 + \cdots + X_n$?

1.52*. Return to the logistic distribution from Section 1.3.6.

a. Using differential calculus, find the logistic p.d.f.

b. Using integral calculus, find the population mean.

1.53. Another use of the logit function is to model the natural logarithm of the odds of an event's occurrence, as a function of a **predictor variable**, C. For instance, consider a mortality study in which p = probability of survival, and the log-odds of survival is modeled as a logistic function of the concentration, C, of some toxin. We describe the resulting prediction equation as a **logistic regression model**:

$$\log\left(\frac{p}{1-p}\right) = \beta_0 + \beta_1 C.$$

a. Describe how the LC_{01}, the concentration leading to 1% probability of survival, would be estimated under this logistic regression model.

b. Describe how any LC_q, i.e. the concentration leading to q% probability of survival, would be estimated under this logistic regression model.

1.54. Which of the distributions in Exercise 1.49 are members of the one-parameter exponential class of distributions? Are any of these in canonical form?

1.55*. Show that the exponential class characterization $\gamma(x)B(\theta)e^{T(x)\omega(\theta)}$ is equivalent to that in (1.15), by writing $B(\theta) = e^{-b(\theta)}$ and $\gamma(x) = e^{c(x)}$.

1.56*. In terms of the original parameter(s), find the natural parameters of any distributions in Exercise 1.49 that are members of the exponential class of distributions.

1.57. Describe an application (from your own field of study, as appropriate), that generates a multinomial random variable, that is, M category counts from N independent trials where the success probability, π_i, is constant within each category ($i = 1, \ldots, M$).

1.58. Show that if $M = 2$ in $\mathbf{Y} \sim \mathcal{M}ultinom(N,[\pi_1, \ldots, \pi_M])$, the p.m.f. collapses to $Y_1 \sim \mathcal{B}in(N,\pi_1)$ and $N - Y_1 \sim \mathcal{B}in(N,\pi_2)$.

1.59*. Write the simplified form of the multinomial theorem in the binomial case, i.e. with $M = 2$. Show how this implies that the binomial p.m.f. in (1.9) sums to 1 over its sample space.

1.60. As in Example 1.4, suppose we undertake a simple genetic experiment to produce offspring from a cross between two heterozygous parents, $Aa \times Aa$. The resulting N offspring possess one of $M = 3$ genotypes: AA, Aa, or aa. From simple

Mendelian theory, the expected genotypic proportions among these three categories are 0.25 : 0.50 : 0.25, respectively. Let Y_1 be the number of *AA* offspring, Y_2 the number of *Aa* offspring, and Y_3 the number of *aa* offspring from a mating producing N total offspring. Calculate the individual multinomial probability of observing the following genotypic outcomes among $N = 6$ offspring:

a. $Y_1 = 2,\ Y_2 = 2,\ Y_3 = 2$. b. $Y_1 = 1,\ Y_2 = 4,\ Y_3 = 1$.

c. $Y_1 = 2,\ Y_2 = 4,\ Y_3 = 0$. d. $Y_1 = 3,\ Y_2 = 0,\ Y_3 = 3$.

1.61*. Let $\mathbf{Y} \sim \mathcal{M}ultinom(N,[\pi_1, \ldots, \pi_M])$. Use the multinomial theorem to show that the marginal distribution of any Y_i is $\mathcal{B}in(N,\pi_i)$.

1.62*. Suppose we observe M independent Poisson variables, each with identical mean μ: $Y_i \sim$ (indep.) $\mathcal{P}oisson(\mu)$, $i = 1, \ldots, M$. We noted that if we condition the statistical analysis on the observed total, Y_+, the distribution of the vector of counts $\mathbf{Y}$ given $Y_+ = y_+$ is multinomial: $\mathbf{Y} \mid y_+ \sim \mathcal{M}ultinom(y_+,\{\mu/\mu_+\})$. (What is μ_+?) Find the associated multinomial p.m.f. (conditional on y_+), $P(Y_1 = y_1, Y_2 = y_2, \ldots, Y_M = y_M \mid Y_+ = y_+)$.

1.63. A **diagonal matrix** is a square matrix whose off-diagonal elements are all zero. What condition on the bivariate normal distribution from Section 1.5.3 is necessary to make its variance-covariance matrix diagonal?

1.64. Extending Exercise 1.49f, describe an application (from your own field of study, as appropriate) where $M = 2$ (or more) continuous measurements are taken that have a bivariate (or multivariate) normal distribution. For the bivariate case, is the associated correlation coefficient zero or nonzero?

1.65*. What happens to the bivariate normal p.d.f. in (1.17) as $\rho \to 1$?

1.66*. If $\mathbf{X} \sim N_M(\boldsymbol{\mu},\mathbf{V})$ as in Section 1.5.3, how many different possible correlation coefficients are there among the M marginal variables?

1.67. Suppose all the correlation coefficients, ρ_{ij}, in a multivariate normal random vector are constant, say $\rho_{ij} = \rho$. Does this ensure that the covariances, $\mathrm{Cov}[X_i,X_j]$, are all constant? Why or why not?

2

Fundamentals of statistical inference

Probability distributions used to describe variation in environmental data are associated generally with one or more unknown parameters. The parameters represent fundamental features of these distributions, and we often wish to estimate these unknown values, or to test certain features about them, based upon observed data. For instance, we might

- estimate the proportion of rodents exhibiting a toxic response after exposure to an environmental chemical;
- estimate the rate of mutation in mammalian cell cultures after the cells are exposed *in vitro* to ionizing radiation; or
- test to see if the mean survival rates of two species of fish differ after some environmental change in the ecosystem.

In addition to estimating or testing features of a particular environmental response or process, it is important to quantify the level of uncertainty associated with these efforts. In the examples above, and in many others we will illustrate in the following chapters, estimation of unknown parameters – such as mean rates, proportions, and other measures – requires application of statistical methods. Inferences about estimated parameters are based on the form of underlying statistical distribution that is assumed for the data, and also on the type of estimation method used. Our goal in this chapter is to introduce and review some general methods of parametric statistical inference, and where possible recommend those with good precision and sensitivity. Applied to problems in modern environmental biology, the statistical complexity of these methods can vary from fairly straightforward to mathematically quite difficult. We will require some of the more complex forms for use in later chapters, and these will be noted by asterisks (*) in the following sub-section headings. Readers wishing only a brief introduction or review may wish to move past these sections upon first encountering them, returning as necessary when called upon in later chapters.

2.1 INTRODUCTORY CONCEPTS IN STATISTICAL ESTIMATION

2.1.1 Vocabulary of parameter estimation

As we saw in the previous chapter, the statistical distribution used to characterize a set of data typically is dependent on one or more unknown parameters. The unknown parameters are quantifications of some attribute of the underlying **population** being studied, such as the proportion of responding subjects in a toxicity study, or the rate of survival in a mortality study. In general, we denote unknown parameters via lower-case Greek letters or other type of special symbol; for example, in Chapter 1 we denoted the mean of a population by the Greek letter μ.

To estimate unknown parameters, we take a **sample** from the underlying population of interest, producing a data set of values, say $X_1, \ldots, X_n$. If the sample is taken wholly at random from this population, we say it is a **random sample**. (We discuss selected issues of random sampling and experiment design in Chapter 3.) Here, n is the **sample size**. A **statistic** is a value *calculated from the data* that is used to estimate the unknown parameter, or perhaps some function of it. In general, statistics are denoted using upper-case Roman letters, such as the sample mean $\overline{X} = \sum_{i=1}^{n} X_i/n$, or in some cases with a circumflex ('hat') above the parameter symbol, e.g. $\hat{\mu}$. The former case is more typical for the common estimators such as $\overline{X}$, especially when the estimator has a closed-form expression.

Notice that since all statistics are based on a random sample of data, they are also random quantities, and thus their realized values vary over different samples selected from the same population. As such, a statistic may be viewed as a random variable with its own distribution. For example, given a set of $n > 1$ data points, X_i, suppose the original data were distributed independently and identically as normal with mean μ and variance σ^2, i.e. $X_i \sim$ i.i.d. $N(\mu,\sigma^2)$. (The notation **i.i.d.** stands for 'independent and identically distributed.') Since the X_is will vary from sample to sample, so will their sample mean, $\overline{X}$. In this sense, $\overline{X}$ also has some statistical distribution by which it may be characterized. The statistical distribution associated with any statistic from a random sample is called the **sampling distribution** of the statistic. Since any statistic has a distribution over repeated samples, we can consider properties of this distribution, such as the theoretical mean or variance of the statistic. The standard deviation of a statistic, another property associated with the sampling distribution of a statistic, is called the **standard error (s.e.** or ***se***) of the statistic. For example, when the underlying parent distribution is normal, i.e. when $X_i \sim$

i.i.d. $N(\mu,\sigma^2)$, it is known that $\overline{X}$ is also normally distributed, with mean μ and variance σ^2/n – i.e. $\overline{X} \sim N(\mu,\sigma^2/n)$ – and with $se[\overline{X}] = (\sigma^2/n)^{1/2}$. As we note in upcoming sections, the concept of the sampling distribution of a statistic is the foundation upon which statistical estimation and formal hypothesis testing are built.

2.1.2 The central limit theorem

An important feature of the sample mean, $\overline{X}$, is that even if the underlying parent distribution is not normal, the distribution of $\overline{X}$ is *approximately* normally distributed. This result is known as the **central limit theorem** (Moore and McCabe, 1993, Section 5.2): take any random sample, $X_1, \ldots, X_n$, from some parent distribution with finite mean μ and finite variance σ^2. Recall that $E[\overline{X}] = \mu$ and $Var[\overline{X}] = \sigma^2/n$. The central limit theorem gives the distribution of the sample mean as approximately normal, with mean μ and with variance σ^2/n. As n increases, this approximation improves. Thus we write $\overline{X} \dot{\sim} N(\mu,\sigma^2/n)$ for any underlying parent distribution for the X_is. (The tilde symbol with a dot above it, $\dot{\sim}$, is read as 'is approximately distributed as'.) For continuous parent distributions whose p.d.f.s are roughly symmetric and unimodal, the central limit theorem yields a good approximation for sample sizes as low as $n = 10$. For parent distributions that are more skewed and/or discrete, however, the sample size must grow larger, upwards of 30 or 40, for the theorem to provide a good approximation of the sampling distribution of the sample mean. (In some extreme cases, where the underlying parent distribution is extremely skewed or with discrete random variables that assume only a small number of possible values, it can take $n = 100$ or more for the central limit theorem's effect to become evident.)

The central limit theorem is not restricted to the sample mean. Numerous versions of the theorem exist for other forms of statistics, and we will discuss a few of these below. In general, the development of theorems that describe the approximate nature of a statistic's sampling distribution as the sample size grows large is known as **large-sample theory** or **asymptotic theory** (Lehmann, 1983, Chapter 5; Casella and Berger, 1990, Section 8.4). As we will see in the various chapters below, sampling distributions for most of the statistical estimators we encounter are mathematically intractable or are not known exactly in small samples. In these cases, we must invoke a limit theorem or other limiting result that approximates the large-sample distribution of the estimator, for use in performing statistical inferences on the unknown parameter(s).

2.1.3 Test statistics

A **test statistic** is a special kind of statistic used to test certain feature(s) of a population parameter. It is best thought of as a quantity, calculated from the data, that has a special form when the unknown parameter takes on a particular value. We discuss the role of test statistics within the general framework of statistical hypothesis tests in Section 2.4.

2.2 NATURE AND PROPERTIES OF ESTIMATORS

Suppose we are sampling observations from a parent population where the attribute in the population is described by a distribution with unknown parameter θ. We make the following technical distinction: an **estimator** of θ, say $\hat{\theta}$, is a statistic based on the data whose value is used to estimate θ, while an **estimate** of θ is the realized value of the estimator from a particular sample. In a sense, an estimator is the random variable whose form specifies the way θ will be estimated, while an estimate is a specific quantity arrived at by applying the estimator to the sample of data.

There are two general forms of estimators (and, once calculated, of estimates): a **point estimator** is a single quantity used to estimate an unknown parameter, while an **interval estimator** is a interval of values that provides some plausible range in which the unknown parameter may lie. Point estimates are useful for providing a single, 'spot' value for an unknown parameter θ, while interval estimates provide a greater level of information about θ.

If an estimator, $\hat{\theta}$, is such that the mean of its sampling distribution is exactly the parameter being estimated – that is, if the expected value $E[\hat{\theta}]$ equals θ exactly – we say that the estimator is **unbiased** for θ. This is a sort of 'scientific objectivity' of the estimator: it is expected to equal, on average, the quantity that it is estimating. A common estimator that is unbiased is the sample mean, $\bar{X}$. When constructed for a sample from any underlying parent population, the sample mean is known to be unbiased for the underlying population mean (if it exists), μ: $E[\bar{X}] = \mu$.

In some complex cases, however, it may be impossible to find an estimator for θ that is unbiased in small samples. If so, we may nonetheless be able to achieve a form of large-sample regularity. For instance, if as $n \to \infty$, $\hat{\theta}$ approaches the true value of θ, we would still consider using $\hat{\theta}$ to estimate θ, even though it is not necessarily unbiased in small samples. This is known as **consistency** of an estimator. Consistent estimators are not uncommon. Intuitively, one expects that an estimator will converge to the quantity it estimates as n grows large, and failure to do so generally suggests an inherent weakness. The sample mean is an example of a consistent estimator. (Technically, $\bar{X}$ satisfies $P[|\bar{X} - \mu| > \varepsilon] \to 0$ as $n \to \infty$ for any $\varepsilon > 0$, which is one way to express consistency of an estimator.)

Another important characteristic often desired in an estimator is that it vary as little as possible, relative to other competitors. That is, an estimator of some parameter θ whose sampling variance, $\mathrm{Var}[\hat{\theta}]$, is as small as possible is more desirable than one whose variance is large. Here again, this is an intuitively sensible feature: the less variable an estimator is, the more precise will be any inferences about the value it estimates. Unfortunately, it is often impossible to identify a unique estimator whose sampling variance is smaller than any other at every possible sample size, n. In these cases, we appeal instead to asymptotic arguments, and ask whether an estimator exists whose large-sample variance is as low as possible. If so, we say the estimator is **asymptotically efficient**. In combination, consistency and asymptotic efficiency are important large-sample qualities for an estimator: when they hold, at least in very large samples, the estimator approaches the parameter of interest, and it does so with as little variability as possible.

2.3 TECHNIQUES FOR CONSTRUCTING STATISTICAL ESTIMATORS

2.3.1 Method of moments

Perhaps the simplest method for constructing a statistical estimator of an unknown parameter is known as the **method of moments** (MoM). Simply put, it equates the sample moments with the corresponding population moments for as many parameters as are being estimated (up to n) from the data. If the sample moments are denoted by $M_k = n^{-1}\sum_{i=1}^{n} X_i^k$, and the population moments are $E[X^k]$, then to estimate a set of K unknown parameters $\theta_1, \ldots, \theta_K$, set $M_k = E[X^k]$ for $k = 1, \ldots, K$, and solve for each θ_j. We call the equation or equations that form an estimating relationship **estimating equations**. The K-dimensional solution to the MoM estimating equations $M_k = E[X^k]$ provides a set of K estimators for the θ_js.

Example 2.1 Estimating π from a binomial distribution (random sample from a Bernoulli distribution)

Suppose an experiment studies the response of laboratory rats to an environmental toxin. A random sample of $N > 1$ independent rats generates a count of dead animals, where each individual animal has constant probability π of dying from exposure to the toxin. Each animal's outcome can be viewed as a Bernoulli trial with $X_i = 1$ when the ith animal dies and $X_i = 0$ when the ith animal survives. The X_is are i.i.d. Bernoulli random variables, each with p.m.f. $p_X(1) = \pi$ and $p_X(0) = 1 - \pi$. To estimate π via the MoM, we equate E[X] = π (at $K = 1$) with $M_1 = \sum_{i=1}^{N} X_i/N$. This gives $\hat{\pi}_{\text{MOM}} = \sum_{i=1}^{N} X_i/N$, or for $Y = \sum_{i=1}^{N} X_i$, $\hat{\pi}_{\text{MOM}} = Y/N$. This estimator is simply the sample proportion of animals dying.

Alternatively, from Section 1.2.1 we know the sum, say Y, of the N independent and identically distributed Bernoulli variables has a binomial distribution with parameters N and π: $Y \sim \mathcal{Bin}(N,\pi)$. To estimate the single response parameter π via the method of moments, we equate the first moment, E[Y] = $N\pi$, with Y itself, and then solve for π. Again, the result is the MoM estimator: $\hat{\pi}_{\text{MOM}} = Y/N$. □

A concern with the MoM approach is that it need not produce unique estimators of the unknown parameters, and that it is often possible to construct other estimators whose sampling variances are smaller. This depends upon the relationships among the moments, and the effect can vary across parent distributions. Nonetheless, for very complex parent distributions, the MoM can provide useful point estimators.

2.3.2 Least squares

Another traditional method for identifying a statistical estimator for an unknown parameter is the **method of least squares**. This approach minimizes the squared deviation between the estimator and the quantity it estimates: for some parameter θ, find an estimator $\hat{\theta}$ that minimizes the sum of squared deviations

$$\mathcal{Q} = \sum_{i=1}^{N} (X_i - \theta)^2.$$

Example 2.1 (continued) Estimating π from a binomial distribution (random sample from a Bernoulli distribution)

Consider again our rodent toxicity example, in which each animal was viewed as a Bernoulli trial with a random variable X_i indicating whether the ith animal dies. To estimate the mortality probability π using least squares, the criterion minimizes

$$\mathcal{Q} = \sum_{i=1}^{N} (X_i - \pi)^2.$$

To find the minimum, apply differential calculus: differentiate Q with respect to π, set this derivative equal to 0, then solve for π. In this example, the derivative is $\partial Q/\partial\pi = -2\sum_{i=1}^{N}(X_i - \pi)$; set equal to zero, we solve for π:

$$\hat{\pi}_{LS} = \sum_{i=1}^{N} \frac{X_i}{N}.$$

Recall that when the first derivative of a function is zero, the function is reaching a **stationary point** of zero slope. This can be either a minimum, a maximum, or some sort of inflection point. To verify that the solution to $\partial Q/\partial\pi = 0$ is a minimum, the second derivative must be positive at the stationary point. Here, the second derivative of Q is $\partial^2 Q/\partial\pi^2 = 2N > 0$, so the point is indeed a minimum. For this example, the LS estimator and the MoM estimator coincide. ❑

Least squares (LS) estimation can be extended to minimize the squared difference between the X_is and some function of K parameters, $g(\theta_1, \theta_2, \ldots, \theta_K)$, resulting in a K-dimensional vector of estimators for the θ_js.

Finding LS estimators can require a fair amount of computational effort, and in many settings LS estimators are no easier to calculate than those from other statistical methods. In some of these instances, however, the LS estimator collapses to the same value as that from a more complex method, suggesting a sort of dual optimality for the resulting quantity. Even so, we will not place too great an emphasis on identifying LS estimators.

2.3.3 Maximum likelihood

Perhaps the most useful general method for identifying a statistical estimator is based on the concept of maximizing a quantity known as the likelihood function. This approach states essentially that the most likely value of an unknown parameter, θ, is the value that led to the particular configuration of data that was actually observed. The concept is by nature quite complex. It was developed in the early twentieth century by the British statistician R.A. Fisher (1912; 1922), and has been shown since then to beget many interesting statistical estimation concepts. An intermediate-level discussion is available in Casella and Berger (1990, Section 6.2). A more advanced exposition is given by Barndorff-Nielsen (1988).

The **likelihood function** is simply the joint p.m.f. or p.d.f. of the random sample of observations, $p_{\mathbf{X}}(x_1, \ldots, x_n \mid \boldsymbol{\Theta})$ or $f_{\mathbf{X}}(x_1, \ldots, x_n \mid \boldsymbol{\Theta})$, viewed however as a function of the unknown parameter vector $\boldsymbol{\Theta} = [\theta_1\ \theta_2\ \ldots\ \theta_K]$. That is, consider the joint p.m.f. or p.d.f. as a function of

$\mathbf{\Theta}$, rather than as a function of $\mathbf{X}$. Then, maximize this function with respect to the individual parameters θ_j ($j = 1, \ldots, K$).

Symbolically, we write the likelihood function as $\mathcal{L}(\theta_1, \ldots, \theta_K \,|\, \mathbf{X})$. If the data come from a simple random sample, we can assume they are statistically independent. Then, as in (1.3), the joint p.m.f. or p.d.f. will factor into the individual probability components for each X_i, which simplifies the likelihood function. For the discrete case this produces

$$\mathcal{L}(\theta_1, \ldots, \theta_K \,|\, \mathbf{X}) = \prod_{i=1}^{n} p_{X_i}(x_i \,|\, \mathbf{\Theta}). \tag{2.1}$$

For the continuous case, replace the individual p.m.f.s $p_{X_i}(x_i \,|\, \mathbf{\Theta})$ with the p.d.f.s $f_{X_i}(x_i \,|\, \mathbf{\Theta})$ in (2.1). In either case, to find the **maximum likelihood (ML) estimators** of $\mathbf{\Theta}$, we maximize $\mathcal{L}(\theta_1, \ldots, \theta_K \,|\, \mathbf{X})$ with respect to each component θ_j.

As might be expected, maximizing the likelihood function is generally a difficult analytic task. Given certain regularity conditions on the parent distribution (Greenwood and Nikulin, 1996, p.115) the ML estimator and MoM estimator coincide, making the estimation process somewhat simpler. (We illustrate this effect in Example 2.1, continued below.) More generally, however, we find ML estimators directly by recognizing that the likelihood function in (2.1) attains a maximum wherever the **log-likelihood function**

$$\ell(\theta_1, \ldots, \theta_K) = \sum_{i=1}^{n} \log\{p_{X_i}(x_i \,|\, \mathbf{\Theta})\} \tag{2.2}$$

also is maximized. (For continuous distributions, the p.m.f.s in (2.2) are replaced by the corresponding p.d.f.s.) For most statistical distributions, the log-likelihood in (2.2) is easier to maximize than the likelihood function in (2.1).

To find the ML estimators of the θ_js via (2.2), it is usually sufficient to appeal to methods of differential calculus, and set the first derivative of the log-likelihood with respect to each θ_j equal to zero for all $j = 1, \ldots, K$. (Again, this point can be either a minimum, a maximum, or an inflection point. When dealing with statistical log-likelihood functions, these stationary points are generally maximum points, although technically one should verify this in each case.) That is, the ML estimator is found by solving the set of K estimating equations

$$\frac{\partial \ell(\theta_1, \ldots, \theta_K)}{\partial \theta_j} = 0 \tag{2.3}$$

for each θ_j. In some cases the log-likelihood derivatives in (2.3) are straightforward to calculate. In other cases, however, it may be necessary to use a computer and solve (2.3) numerically. We will see that this latter case occurs with many data-analytic settings in environmental biology.

Even though computation of ML estimators can require extensive calculation, the benefits accrued from this effort are numerous. The estimators possess many important qualities. One of these is a form of **functional invariance**: if $\hat{\theta}$ is the ML estimator for θ, then $h(\hat{\theta})$ will be the ML estimator of $h(\theta)$ for any function $h(\cdot)$. This allows for easy determination of ML estimators for a wide variety of parameterizations.

Another important result is that as the sample size grows large, the distribution of the ML estimator $\hat{\theta}$ takes a special form. Under fairly mild regularity conditions, the sampling distribution of $\hat{\theta}$ approaches a normal distribution as $n \to \infty$, with large-sample mean θ and with large-sample variance dependent upon the log-likelihood function. In particular, the large-sample variance is the reciprocal of the **Fisher information number**

$$\mathfrak{F}_\theta = \mathrm{E}\left[\frac{-\partial^2 \ell(\theta)}{\partial \theta^2}\right]. \tag{2.4}$$

In (2.4), the expectation is taken with respect to the original random variables, $\mathbf{X}$. Thus we write the large-sample variance as $\mathrm{Var}[\hat{\theta}] = 1/\mathfrak{F}_\theta$. For simplicity, we often write the reciprocal of the Fisher information number as σ^2_θ, i.e. $\sigma^2_\theta = 1/\mathfrak{F}_\theta = \mathrm{Var}[\hat{\theta}]$, and so $\hat{\theta} \mathrel{\dot\sim} \mathrm{N}(\theta, \sigma^2_\theta)$.

Example 2.1 (continued) Estimating π from a binomial distribution (random sample from a Bernoulli distribution)

Consider again our rodent toxicity example with $X_i \sim$ i.i.d. $\mathcal{Bin}(1,\pi)$ for $i = 1, \ldots, N$. Recall that this implies $Y = \sum_{i=1}^N X_i \sim \mathcal{Bin}(N,\pi)$. We illustrate estimation of the ML estimator for π by first building the likelihood function associated with the random sample of Bernoulli observations. From (2.1) the likelihood function is

$$\mathcal{L}(\pi \mid \mathbf{X}) = \prod_{i=1}^{N} p_{X_i}(x_i \mid \pi) = \prod_{i=1}^{N} \pi^{x_i}(1-\pi)^{1-x_i} = \pi^y(1-\pi)^{N-y},$$

and from (2.2) we see that the corresponding log-likelihood function is

$$\ell(\pi) = \sum_{i=1}^{N} \log\{p_{X_i}(x_i \mid \pi)\} = \left(\sum_{i=1}^{N} x_i\right)\log\left\{\frac{\pi}{1-\pi}\right\} + N\log(1-\pi).$$

Alternatively, viewing $\sum_{i=1}^N X_i$ as a single binomial observation, Y, the likelihood function is just the p.m.f. viewed as a function of π: $\mathcal{L}(\pi \mid y) =$

$N!\,\pi^y(1-\pi)^{N-y}/\{y!(N-y)!\}$. The corresponding log-likelihood is proportional to

$$\begin{aligned}\ell(\pi) &= Y\log(\pi) + (N-Y)\log(1-\pi)\\ &= Y\log\left\{\frac{\pi}{1-\pi}\right\} + N\log(1-\pi).\end{aligned}$$

(For simplicity we have removed all separate terms that do not involve π; these terms are constants whose derivatives with respect to π will be zero. Hence, they will not influence calculation of the ML estimator.) Thus the log-likelihood for π is the same regardless of whether the data are viewed as a set of independent Bernoulli trials or as a single binomial experiment.

For the ML estimator, we find the derivative of $\ell(\pi)$ with respect to π:

$$\frac{\partial\ell(\pi)}{\partial\pi} = \frac{Y}{\pi} - \frac{N-Y}{1-\pi},$$

which simplifies to $(Y-N\pi)/(\pi-\pi^2)$. Setting this equal to zero produces a single estimating equation for π: $Y - N\pi = 0$. Solving gives the ML estimator $\hat{\pi} = Y/N$, which is once again the sample proportion. For the binomial distribution, then, the MoM, LS, and ML estimators agree (but this is not always the case for other parent distributions).

To verify that the solution to $\partial\ell(\pi)/\partial\pi = 0$ is in fact a maximum, again apply differential calculus: recall that if the second derivative of a function is negative at a stationary point, then that point is a maximum. Here, $\partial^2\ell(\pi)/\partial\pi^2 = -\{\pi^{-2}Y + (1-\pi)^{-2}(N-Y)\}$, which is negative for all $Y = 0,\ldots,N$ and all π between 0 and 1. So, $\hat{\pi}$ does indeed achieve the maximum of the (log-)likelihood.

To derive the large-sample distribution of $\hat{\pi}$, we must find the Fisher information number for π. From (2.4), this is

$$\mathfrak{F}_\pi = \mathrm{E}\left[\frac{-\partial^2\ell(\pi)}{\partial\pi^2}\right] = -\mathrm{E}\left[-\frac{Y}{\pi^2} - \frac{N-Y}{(1-\pi)^2}\right] = \frac{\mathrm{E}[Y]}{\pi^2} + \frac{N-\mathrm{E}[Y]}{(1-\pi)^2}$$

which for $\mathrm{E}[Y] = N\pi$ simplifies to $\mathfrak{F}_\pi = N/\{\pi(1-\pi)\}$. Thus we say that, in large samples, $\hat{\pi} \mathrel{\dot\sim} N(\pi, \pi(1-\pi)/N)$. □

Example 2.2 Estimation of π from a binomial distribution

Suppose that $N = 20$ animals are evaluated in the rodent toxicity study described in Example 2.1. Further, assume that $y = 10$ animals died. Our maximum likelihood estimate of π is $\hat{\pi} = y/N = 10/20 = 0.5$ with an associated standard error taken as the square root of the estimated variance: $(\hat{\pi}\{1-\hat{\pi}\}/N)^{1/2} = (0.5\{1-0.5\}/20)^{1/2} = 0.1118$. □

2.3.4* Maximum likelihood with multiple parameters

For more than one unknown parameter, large-sample normality of the ML estimator still applies, except that the variance term is somewhat more complex. For any vector of values $\hat{\mathbf{\Theta}} = [\hat{\theta}_1 \ldots \hat{\theta}_K]$ satisfying the conditions in (2.3), it is the case that as $n \rightarrow \infty$ the limiting distribution of $\hat{\mathbf{\Theta}}$ is multivariate normal, as in Section 1.5.3. (Again, a set of fairly mild regularity conditions must hold as well.) The large-sample mean is simply $\mathbf{\Theta}$, while the large-sample variance-covariance matrix, $\mathbf{V}_{\mathbf{\Theta}}$, is found as the inverse of the matrix of Fisher information numbers. (An **inverse matrix** is an array of values that combines with another matrix to produce an identity matrix. The effect is similar to multiplying a number by its reciprocal to get 1. Thus for some matrix $\mathbf{A}$, we write $\mathbf{A}^{-1}$ as it inverse. Then, $\mathbf{A}\mathbf{A}^{-1} = \mathbf{A}^{-1}\mathbf{A} = \mathbf{I}$, where $\mathbf{I}$ is the identity matrix from Section 1.5.4. Matrix inverses may be calculated by hand (Searle, 1982, Chapter 5), but the computations usually are left to a computer.)

To find $\mathbf{V}_{\mathbf{\Theta}}$, we must identify the individual Fisher information numbers. As in (2.4), information is calculable for any single parameter via the expected negative second derivatives of ℓ: $\mathfrak{F}_{jj} = \mathrm{E}[-\partial^2\ell(\theta_1, \ldots, \theta_K)/\partial\theta_j^2]$, $j = 1, \ldots, K$. In this multi-parameter setting, however, information is also available across parameters via the expected negative mixed partial derivatives

$$\mathfrak{F}_{jk} = \mathrm{E}\left[\frac{-\partial^2\ell(\theta_1, \ldots, \theta_K)}{\partial\theta_j\partial\theta_k}\right] \tag{2.5}$$

(for $j \neq k$). These values are all collected together into a square matrix, with the jth diagonal element given as $\mathfrak{F}_{jj}$, and with the (j,k)th off-diagonal elements given by (2.5). For simplicity of notation, we write $\mathbf{F}_{\mathbf{\Theta}} = \{\mathfrak{F}_{jk}\}$, and call this the **Fisher information matrix**. The large-sample variance-covariance matrix for $\hat{\mathbf{\Theta}}$ is then the inverse of this $\mathbf{F}_{\mathbf{\Theta}}$ matrix: $\mathbf{V}_{\mathbf{\Theta}} = \mathbf{F}_{\mathbf{\Theta}}^{-1}$. In terms of the notation of Section 1.5.3, we say $\hat{\mathbf{\Theta}} \mathrel{\dot{\sim}} N_K(\mathbf{\Theta}, \mathbf{V}_{\mathbf{\Theta}})$.

Many of the settings we will discuss below involve multiple parameters, and when estimating these parameters we will center much of our discussion around ML estimation and its large-sample properties. When appropriate, we will display some of the pertinent formulas and calculations in each of these settings, but generally we will leave the more intense computations aside. Whenever possible in these more complex cases, we will appeal to the computer, and note whether any standard computer programs or packages can perform these computations as part of their programming.

2.4 STATISTICAL INFERENCE – TESTING HYPOTHESES

The basic features of hypothesis testing are well known, as provided, for example, in the excellent introduction by Moore and McCabe (1993, Section 6.2). Here we simply review elementary details and establish notation for the general problem of assessing the significance of hypotheses about unknown parameters.

Given a parameter or vector of parameters θ associated with the parent distribution under study, suppose interest centers upon a particular value of θ, say θ_0. We write the **null hypothesis** as $H_0: \theta = \theta_0$. This is also called the **no-effect hypothesis**, since θ_0 often indicates some lack of effect, such as no difference between two groups, or no increase in dose response to some environmental toxin, etc. The **alternative hypothesis** or **research hypothesis**, H_a, is a specification for θ that the investigator feels is a plausible alternative to H_0. (Some authors denote the alternative hypothesis as H_1.) The research hypothesis often motivates the original experimental inquiry. By defining the research question operationally in H_a, the null hypothesis is then set up as a statement of no effect. For instance, $H_a: \theta \neq \theta_0$ suggests that the true value of θ differs in either direction (higher or lower) from the null hypothesized value.

Within this context, there are five basic steps involved when testing hypothesis:

1. Specify the null and alternative hypotheses, H_0 and H_a.
2. Set the **significance level**, α, at which the test will operate. This is also known as the **Type I error rate**, and can be expressed as the largest desired probability of rejecting H_0 when it is true: $P[\text{reject } H_0 \mid H_0]$. We will also refer to this error rate as the **false positive error rate**, since it represents the frequency with which an investigator will reach a false indication that H_a is true.
3. Calculate a **test statistic** that measures the departure from H_0 evidenced in the data. Refer this statistic to a **null reference distribution**, which is the sampling distribution of the test statistic when H_0 is true.
4. Construct a **rejection region**, which is the set of values for the test statistic over which H_0 is to be rejected.
5. If the test statistic falls within the rejection region, reject H_0 and conclude that significant evidence exists in the data to rule in favor of H_a.

In steps 4 and 5, above, one can also calculate a measure that helps indicate the strength of the departure from H_0, called the ***P*-value**. The P-value is defined as the probability of recovering a response as extreme as or more extreme than that actually observed, when H_0 is true. (Note that 'more extreme' is defined in the context of H_a. For example, when testing $H_0: \theta = \theta_0$ vs. $H_a: \theta > \theta_0$, 'more extreme' corresponds to values of the test statistic larger than that actually observed.) The P-value is calculated by appealing to the null reference distribution (step 3, above), and calculating the probability under this distribution of obtaining a result as extreme as or more extreme than the actual value of the test statistic. Small P-values indicate departure from H_0, since they imply that the probability of observing a result at least as extreme as that seen with the given data is lower than might be expected by chance. For a fixed significance level α, reject H_0 when $P < \alpha$.

This latter approach to testing hypotheses is also known as **significance testing**, since it uses the P-value to measure the significance of H_0; Goodman (1993) gives an interesting exposition on significance testing and hypothesis tests. In what follows, we will often emphasize the calculation of P-values, since these have become common in environmental toxicology.

One additional concept we need to review is that of the **power** (or **sensitivity**) of a hypothesis test, which is the probability of rejecting H_0 when it is indeed false: $P[\text{reject } H_0 \mid H_a]$. This form of rejection probability should be larger than the nominal significance level α, and preferably close to 1. We refer to it as a **power function**, since it is a function of the unknown parameter(s), θ. Tests will be viewed as optimal when they possess large power functions for values of θ consistent with H_a. For example, we say a test with a power function that is larger than any other test's power function across all possible values of θ represented by H_a is **uniformly most powerful** (UMP). Unfortunately, UMP tests do not always exist; when they do, they are recommended for use.

A **locally most powerful test** is a test whose power function is larger than any other test's power function across a relevant ('local') subset of values for θ. Locally powerful tests generally depend upon the specific features of the data, and may yield lower power than a UMP test, if one exists.

2.4.1 Likelihood ratio tests

As might be expected, the construction of a test statistic and its associated null reference distribution will depend on a number of factors, including the hypotheses being tested, the form of underlying parent distribution assumed for the data, the sample size, etc. It is possible, however, to formulate some general constructions based on the likelihood function. These may be applied to almost any data-analytic scenario in which a specific parametric likelihood is used. We begin with a ratio-based form: the **generalized likelihood ratio test**.

Suppose we have two hypotheses about an unknown parameter, θ, say $H_0: \theta = \theta_0$ vs. $H_a: \theta \neq \theta_0$. Given a likelihood function, $\mathcal{L}(\theta \mid \mathbf{X})$, departure from H_0 in favor of H_a is indicated when $\mathcal{L}(\theta \mid \mathbf{X})$ under H_0 is smaller than most other values of $\mathcal{L}(\theta \mid \mathbf{X})$. To summarize this departure we find $\mathcal{L}(\theta_0 \mid \mathbf{X})$, the 'maximum' value of $\mathcal{L}(\theta \mid \mathbf{X})$ under H_0, and compare it to the maximum value of $\mathcal{L}(\theta \mid \mathbf{X})$ over all possible values of θ. For the latter case, recall that the ML estimate, $\hat{\theta}$, maximizes $\mathcal{L}(\theta \mid \mathbf{X})$, so the maximum value must be $\mathcal{L}(\hat{\theta} \mid \mathbf{X})$. A comparative measure of the two values is then the **likelihood ratio** (LR),

$$\Lambda = \frac{\mathcal{L}(\theta_0 \mid \mathbf{X})}{\mathcal{L}(\hat{\theta} \mid \mathbf{X})} .$$

When this ratio is small, departure from H_0 is evidenced.

An important result in statistical theory is that as $n \to \infty$, the null reference distribution of $G^2 = -2\log\{\Lambda\}$ is known. (Recall that the logarithm is taken with respect to the natural base, e.) Specifically, $G^2 \mathrel{\dot\sim} \chi^2(\nu_0)$, where the *df* ν_0 is the number of parameters that are specified in H_0. We have described the case for $\nu_0 = 1$ here, but this method extends to the multi-parameter case in a straightforward fashion: simply replace the single parameter θ with the vector of parameters $\boldsymbol{\Theta}$. Some minor corrections are necessary for the numerator of Λ if only a subset of all possible parameters is being tested, but the methodology remains essentially the same; see Casella and Berger (1990, Section 8.2) or Cox (1988) for greater detail.

For the LR test, an approximate P-value for testing H_0 is $P \approx \mathrm{P}[\chi^2(\nu_0) > G^2_{\text{calc}}]$, where G^2_{calc} is the value of G^2 calculated from the data. (We use the subscript 'calc' in a generic sense, indicating a statistic that can be calculated fully from the data. In some instances below we may use other subscripts to emphasize a certain feature of the test statistic, but in general 'calc' will represent a calculable statistic used for testing hypotheses.) The χ^2 approximation improves as the sample size, n, grows large. Alternatively, one can construct a rejection region: reject H_0 when the LR statistic G^2_{calc} exceeds an upper-α critical point of the $\chi^2(\nu_0)$ distribution, $\chi^2_\alpha(\nu_0)$.

Notice, by the way, that the likelihood ratio G^2 is actually a difference in maximized *log*-likelihoods, so that we can write $G^2 = -2\{\ell(\theta_0) - \ell(\hat{\theta})\} = 2\ell(\hat{\theta}) - 2\ell(\theta_0)$. Differences in twice log-likelihoods are also known as **deviance differences**, and we will employ them in Chapters 7 and 8.

Example 2.3 Testing π from a binomial distribution
Consider again the rodent toxicity experiment from Example 2.1, where we observe $Y \sim \mathcal{Bin}(N,\pi)$. Suppose we wish to test whether π takes on some specific value, say $H_0:\pi = 0.6$ vs. $H_a:\pi \neq 0.6$. From Example 2.1, we know $\mathcal{L}(\pi \mid Y) = N!\,\pi^Y(1-\pi)^{N-Y}/\{Y!(N-Y)!\}$. We also saw that the ML estimator of π was $\hat{\pi} = Y/N$, the sample proportion. Thus, the LR statistic for this testing problem is

$$G^2 = -2\log\left\{\frac{(0.6)^Y(0.4)^{N-Y}}{\hat{\pi}^Y(1-\hat{\pi})^{N-Y}}\right\}.$$

(Notice that the incidental terms not involving π cancel out of the ratio.) This simplifies to

$$G^2 = -2\left\{Y\log\left(\frac{0.6}{\hat{\pi}}\right) + (N-Y)\log\left(\frac{0.4}{1-\hat{\pi}}\right)\right\}.$$

Suppose our random sample yields $y = 10$ deaths among $N = 20$ exposed animals, so that $\hat{\pi} = 0.5$. The calculated LR test statistic becomes $G^2_{\text{calc}} = -20\{\log(1.2) + \log(0.8)\} = 0.8164$. The corresponding P-value for testing H_0 is approximately $P[\chi^2(1) > 0.8164]$. We can use the statistical computing package SAS® (SAS Institute Inc., 1989) to compute P-values from selected reference distributions such as χ^2. SAS code and (edited) output for the P-value $P[\chi^2(1) > 0.8164]$ appear in Fig. 2.1, where the actual value is seen to be $P = 0.366$. (In some versions of SAS, the statement `run;` may be required at the end of the SAS code. We will not use it here, however.) Hence with $\hat{\pi} = 10/20$, the evidence is weak that the true π deviates from 0.6.

The P-value also may be calculated using the `pchisq` function in the statistical computer package S-PLUS® (StatSci Division of MathSoft Inc., 1995). In particular, the command

```
1 - pchisq(.8164,1)
```

yields the P-value $P[\chi^2(1) > 0.8164] = 0.36623$. Other calculations, such as normal-, t-, or F-based P-values or critical points, can be conducted in SAS or S-PLUS using analogous functions.

Failure to attain significance in this example may be due to a number of reasons, the most prominent of which is that π is not truly different from 0.6. Another possibility, however, is that the limited sample size of only $N = 20$ exposed animals does not provide sufficient power to detect a true difference. For instance, if the true π were only slightly different from 0.6, say 0.55, it may require many more observations to detect the slight departure from H_0. Sample size selection for proper statistical inference is an important issue in environmental analyses, and we discuss in further detail the selection of sample sizes in Sections 3.4 and 4.1.5. More general expositions are given in Moore and McCabe (1993, Sections 6.1, 8.1) and Neter *et al.* (1996, Chapter 26), and selected illustrations are found, for example, in Lachenbruch (1992), Oris and Bailer (1993), and the references therein. ❑

```
* SAS code to find chi-square P-value;
data pval;
  cdf = probchi(.8164 , 1);
  pval = 1 - cdf;
cards;
proc print;
```

```
                The SAS System
      OBS           CDF            PVAL
        1       0.63377         0.36623
```

Fig. 2.1 SAS program and output (edited) to find χ^2 *P*-value.

2.4.2 Standard errors and Wald tests

An alternative to the test in Section 2.4.1 based on a ratio of likelihoods is to work directly with the large-sample distribution of the ML estimator, following a construction proposed originally by Wald (1943). Recall from Section 2.3.3 that the ML estimator, $\hat{\theta}$, for an unknown parameter θ is approximately normal. Its mean is θ and its variance $\sigma^2_\theta = \mathfrak{F}^{-1}_\theta$ is the reciprocal of the Fisher information number (2.4). As such, the distribution of the standardized quantity $z = (\hat{\theta} - \theta)/\sigma_\theta$ approaches standard normal as n grows large. To construct a calculable test statistic from this, we replace θ in the numerator of z with some hypothesized value θ_0, corresponding to a test of the hypotheses $H_0: \theta = \theta_0$ vs. $H_a: \theta \neq \theta_0$.

Unfortunately, the standard deviation of $\hat{\theta}$ in the denominator of the z-value may still depend upon θ; if so, it must be estimated to make the test statistic calculable. Since an estimator is a statistic with some associated sampling distribution, we refer to the standard deviation associated with an estimator as its **standard error**. We also define an estimate of this standard error as the **standard error of the estimator**. Given an ML estimator of θ, we write

$$se[\hat{\theta}] = \sigma_{\hat{\theta}} = \mathfrak{F}_{\hat{\theta}}^{-1/2},$$

where the subscript indicates evaluation at the ML estimate $\hat{\theta}$.

Wald's test uses the standard error of $\hat{\theta}$ as the denominator in the statistic

$$W_{\text{calc}} = \frac{\hat{\theta} - \theta_0}{se[\hat{\theta}]}. \tag{2.6}$$

This ratio is known as a **Wald statistic**. Under H_0, the sampling distribution of W_{calc} will still be approximately standard normal, $W_{\text{calc}} \mathrel{\dot\sim} N(0,1)$, so that the square of this quantity, W^2_{calc}, is approximately $\chi^2(1)$. From this, approximate P-values or rejection regions may be constructed for H_0 in a similar fashion as for the LR test. For example, the approximate Wald test P-value is $P \approx P[\chi^2(1) > W^2_{\text{calc}}]$. Equivalently, one can appeal directly to the standard normal reference distribution for Wald test P-values or rejection regions. For instance, against the two-sided alternative $H_a{:}\,\theta \neq \theta_0$, the approximate P-value, $P[\chi^2(1) > W^2_{\text{calc}}]$, is identical to $2P[Z > |W_{\text{calc}}|]$, where $Z \sim N(0,1)$. Against a **one-sided alternative** hypothesis, such as $H_a{:}\,\theta > \theta_0$, the P-value is found using only the standard normal reference distribution: $P \approx P[Z > W_{\text{calc}}]$. For $H_a{:}\,\theta < \theta_0$, take $P \approx P[Z < W_{\text{calc}}]$. Extensions to multiple parameters are also possible; see, for example, Cox (1988).

Example 2.3 (continued) Testing π from a binomial distribution

For the rodent toxicity experiment in this example, we had $Y \sim \mathcal{Bin}(N,\pi)$, and we considered the hypotheses $H_0{:}\pi = 0.6$ vs. $H_a{:}\pi \neq 0.6$. The squared Wald statistic (2.6) for this problem requires the ML estimator of π and also its large-sample standard error $se[\hat{\pi}]$. We saw in Example 2.1 that the ML estimator was $\hat{\pi} = Y/N$, and that its large-sample distribution was approximately normal: $\hat{\pi} \mathrel{\dot\sim} N(\pi, \pi(1 - \pi)/N)$. Thus the standard error of the ML estimator is $se[\hat{\pi}] = \{\hat{\pi}(1 - \hat{\pi})/N\}^{1/2}$. From these, the squared Wald statistic under $H_0{:}\pi = 0.6$ is

$$W^2_{\text{calc}} = \frac{\left(\hat{\pi} - 0.6\right)^2}{\frac{1}{N}\hat{\pi}(1 - \hat{\pi})}.$$

For a random sample with $y = 10$ deaths among $N = 20$ exposed animals, the components of the Wald statistic are $\hat{\pi} = 0.5$ and $se[\hat{\pi}] = \sqrt{0.0125}$, producing $W^2_{\text{calc}} = (-0.1)^2/(0.0125) = 0.80$. The corresponding P-value for testing H_0 is, approximately, $P[\chi^2(1) > 0.80]$, which can be found to be $P \approx 0.371$. Notice the similarity to the LR results, above. Here again, we see that $\hat{\pi} = 10/20$ is weak evidence that the true π deviates from 0.6. □

One caveat we raise with use of the Wald test is that it is known to be sensitive to the model parameterization (Væth, 1985). For example, suppose the mean parameter from an exponential distribution, β, were under study. If for some reason interest were directed at the reciprocal of β, $\lambda = 1/\beta$, then we could reparameterize the likelihood in terms of λ instead of β. Under this new parameterization, one would expect that any inference on λ would translate to equivalent inferences on β through the transformation $\beta = 1/\lambda$. Indeed, the LR test from Section 2.4.1 does possess such an **invariance** to reparameterization. Unfortunately, however, the Wald test (and Wald confidence intervals; see Section 2.5.2) do not exhibit this invariance, so that inferences made on λ are not guaranteed to correspond to equivalent inferences on $\beta = 1/\lambda$.

In some experimental settings this lack of invariance is not critical, and Wald tests/Wald intervals are perfectly reasonable and acceptable approaches to statistical inference. The investigator must be careful to consider potential parameter invariance, however, when choosing among various forms for use in statistical inference.

2.4.3* Likelihood-based score tests

A third possible approach to test statistic construction based on the likelihood function is known as the score statistic, named after the **likelihood scores** or **efficient scores** (Rao, 1947) of the log-likelihood:

$$U_j(\theta_1, \ldots, \theta_K) = \frac{\partial \ell(\theta_1, \ldots, \theta_K)}{\partial \theta_j} \tag{2.7}$$

($j = 1, \ldots, K$). The notation in (2.7) emphasizes that the scores are functions of the unknown parameters $\theta_1, \ldots, \theta_K$, although for simplicity we will write U_j for $U_j(\theta_1, \ldots, \theta_K)$. Notice that the U_j are also functions of the data, $\mathbf{X}$. Indeed, they are the quantities that are set equal to zero to produce the ML estimating equations; cf. (2.3).

The likelihood scores from (2.7) are useful for the following reason: as $n \to \infty$, the distribution of the vector of scores, $\mathbf{U} = [U_1 \ \ldots \ U_K]$, is approximately normal. Its mean vector is $\mathbf{0}$ and its large-sample variance-covariance matrix is *equal* to the Fisher information matrix $\mathbf{F}_\Theta$, given by

(2.5). Thus, using the notation from Section 1.5.3, we write $\mathbf{U} \stackrel{.}{\sim} N_K(\mathbf{0}, \mathbf{F}_\Theta)$.

In the simple one-parameter case, $K = 1$, there is only one score: $U = \partial\ell(\theta)/\partial\theta$. As n grows large, U is approximately normal, with mean $E[U] = 0$ and with variance $\mathrm{Var}[U] = \mathfrak{F}_\theta$, where the latter quantity is given by (2.4). Thus the standardized score is $U/\mathfrak{F}_\theta^{1/2} \stackrel{.}{\sim} N(0,1)$. We can derive a test statistic from this relationship by evaluating the standardized score under the null hypothesis H_0: $\theta = \theta_0$. This is $T_{calc} = U(\theta_0)/\{\mathfrak{F}_\theta^{(0)}\}^{1/2}$, where $\mathfrak{F}_\theta^{(0)}$ is the Fisher information evaluated at $\theta = \theta_0$. As n grows large, the distribution of T_{calc} is approximately standard normal under H_0, and as such, its square is approximately $\chi^2(1)$. The result is a calculable test statistic which we refer to as the **score statistic**:

$$T^2_{calc} = \frac{\{U(\theta_0)\}^2}{\mathfrak{F}_\theta^{(0)}}. \tag{2.8}$$

In the econometrics literature, score tests using the statistic in (2.8) are also known as **Lagrange multiplier tests**, as given initially by Aitchison and Silvey (1958). Under either rubric, extensions of (2.8) to the multi-parameter case are also possible, using the vector of scores **U** and the inverse of the Fisher information matrix; see, for example, Cox (1988).

Tests of H_0 based on (2.8) are conducted in a similar fashion as for the LR test or the Wald test in the preceding sections. For instance, the approximate score test P-value is $P \approx P[\chi^2(1) > T^2_{calc}]$.

Example 2.3 (continued) Testing π from a binomial distribution
Continuing with the rodent toxicity experiment, we have $Y \sim \mathcal{Bin}(N,\pi)$, and H_0:$\pi = 0.6$ vs. H_a:$\pi \neq 0.6$. The score statistic (2.8) for this problem is straightforward to compute. Begin with the log-likelihood, which we saw in Example 2.1 to be essentially $\ell(\pi) = Y\log(\pi) + (N - Y)\log(1 - \pi)$. The corresponding score was $U(\pi) = \partial\ell(\pi)/\partial\pi = (Y - N\pi)/(\pi - \pi^2)$. Also, the Fisher information number for π was given by

$$\mathfrak{F}_\pi = E\left[\frac{-\partial^2\ell(\pi)}{\partial\pi^2}\right] = -E\left[-\frac{Y}{\pi^2} - \frac{N - Y}{(1 - \pi)^2}\right] = \frac{E[Y]}{\pi^2} + \frac{N - E[Y]}{(1 - \pi)^2}$$

which for $E[Y] = N\pi$ we saw simplified to $\mathfrak{F}_\pi = N/(\pi - \pi^2)$.

With these quantities, the score statistic is the square of $U(\pi_0)$, divided by $\mathfrak{F}_\pi^{(0)}$. For the binomial likelihood this simplifies to

$$T^2_{calc} = \frac{\left\{\dfrac{(Y - N\pi_0)}{\pi_0(1 - \pi_0)}\right\}^2}{\left\{\dfrac{N}{\pi_0(1 - \pi_0)}\right\}},$$

or simply $T^2_{calc} = N\{\hat{\pi} - \pi_0\}^2/\{\pi_0(1 - \pi_0)\}$, where $\hat{\pi} = Y/N$ is the sample proportion. For the random sample with $y = 10$ deaths among $N = 20$ exposed animals, the score statistic is $T^2_{calc} = 0.833$. The corresponding P-value for testing H_0 is approximately $P[\chi^2(1) > 0.833]$, which can be found to equal 0.361. Here again, we see that $\hat{\pi} = 10/20$ is weak evidence that the true π deviates from 0.6.

Note that for this binomial example the form of the score statistic T^2_{calc} is almost identical to the form of the Wald statistic; T^2_{calc} differs from W^2_{calc} only by its use of $\pi_0(1 - \pi_0)/N$ instead of $\hat{\pi}(1 - \hat{\pi})/N$ in its simplified denominator. The former quantity is the large-sample variance $\sigma^2_{\hat{\pi}}$, evaluated at the hypothesized value π_0. Since $(\hat{\pi} - \pi)/\sigma_{\hat{\pi}}$ is asymptotically standard normal, both $\pi_0(1 - \pi_0)/N$ and $\hat{\pi}(1 - \hat{\pi})/N$ estimate $\sigma_{\hat{\pi}}$ in the denominator when $H_0{:}\pi = \pi_0$ is true. In this sense, both T^2_{calc} and W^2_{calc} are valid approaches to constructing a test statistic. □

As can be seen throughout Example 2.3, the three different statistics G^2, W^2, and T^2 all operate in a fairly similar manner. For the one-parameter testing problem, they all refer to a large-sample $\chi^2(1)$ distribution – in the K-parameter testing problem they would all refer to a $\chi^2(K)$ distribution; cf. Cox (1988) – and, at least in the example we considered, their calculated values are all similar. Indeed, for most experimental settings, the limiting value and large-sample reference distribution for each of these three test statistics will be the same; this is a form of **asymptotic equivalence** among the three test methods. When n is very large, the three methods will provide essentially the same inference.

In small samples, however, differences can occur among the three likelihood-based methods. Each has its own strengths and weaknesses – for example, the score statistic does not usually require calculation of the ML estimate of θ, but it does require calculation of possibly complicated derivatives – and these differences change with the underlying parent distribution of the data and with the parameterization under study. As we discuss different settings where parametric hypotheses are to be tested, we will consider use of these likelihood-based methods, and note any cases where one of the three forms G^2, W^2, or T^2 may be more desirable than the others.

2.5 STATISTICAL INFERENCE – CONFIDENCE INTERVALS

A second form of statistical inference that is related to hypothesis testing involves construction of a range of plausible values for the unknown parameter. In a sense, information about the variability in the data is used to expand the point estimator, and suggest an **interval estimator** within which the unknown parameter, θ, may lie. The particular form of interval estimator we consider is known as a **confidence interval** for θ. It is based on the reference distribution of some statistic, and is constructed by rewriting the probability statements so that they describe an interval of values for θ.

A general method for constructing a confidence interval is known as **inverting a rejection region**. Take any realized value of a test statistic, say $Z = z_{calc}$, to assess the validity of the null hypothesis H_0: $\theta = \theta_0$, the rejection region is constructed by referring Z to some null distribution. Commonly, if z_{calc} is larger than the upper-α critical point of this distribution we reject H_0, realizing that the probability of falsely doing so (the Type I error rate) is α. It is important to recognize that this construction is dependent upon the hypothesized value of θ_0. Different values of θ_0 will yield different rejection regions, and from this one can construct a set of θ_0-values over which a given set of data will lead to rejection of H_0.

We can invert this strategy, and construct the set of θ-values for which H_0 is not rejected, the **acceptance region**. For most forms of test statistics, this results in a interval of values that constitute the interval estimator for θ. The probability of correctly 'accepting' H_0 is $1 - \alpha$, and hence we say that the interval has a **confidence level** or a **confidence coefficient** of $1 - \alpha$ for containing θ. This coefficient is interpreted as a relative frequency: how often the process of constructing confidence intervals captures the true value of the parameter over repeated experiments.

We should emphasize here that confidence is *not* probability. The probability that a calculated interval actually contains the true value of θ is not $1 - \alpha$; it is either 0 or 1. That is, once we have specified that, say, $1.27 < \theta < 2.99$, the probability that θ is truly in this interval is 0 (if it is not) or 1 (if it is). Obviously, we cannot known which of these cases is true, but since we ascribe no random variability to θ, neither can we place any other form of probability on the statement $1.27 < \theta < 2.99$.

Instead of a probabilistic interpretation, the confidence level $1 - \alpha$ has what is called a **frequentist interpretation**. If, over repeated samplings we were able actually to count the number of times one covered the true value of θ with the calculated confidence interval, we would find that about $100(1 - \alpha)\%$ of the intervals covered correctly. Thus confidence is a measure of the frequency of correct coverage of the unknown parameter by the interval; cf. Moore and McCabe (1993, Fig. 6.2).

There are many forms and methods available for constructing confidence intervals on an unknown parameter. When appropriate, we will describe specific approaches suitable for each of the environmental settings discussed in the chapters below. Before doing so, however, we will describe some general methods that correspond to the likelihood-based test procedures discussed in the preceding sections.

2.5.1 Likelihood-based inference: Likelihood ratio intervals

As we noted above, any test can have its rejection region inverted to produce confidence limits on an unknown parameter. Applied to the likelihood ratio (LR) test from Section 2.4.1, this process is straightforward. Given an LR statistic G^2_{calc}, the large-sample rejection region is $G^2_{\text{calc}} > \chi^2_\alpha(1)$. The corresponding acceptance region contains all values of θ that satisfy

$$G^2_{\text{calc}} \leq \chi^2_\alpha(1),$$

where $\chi^2_\alpha(1)$ is the upper-α critical point of the $\chi^2(1)$ distribution. This produces a $1-\alpha$ likelihood ratio confidence interval for θ.

Notice that this construction is based on large-sample arguments. In small samples there is no guarantee that the interval will have exact confidence level $1-\alpha$. In some instances, the actual coverage coefficient attained by these intervals may be larger than the nominal level of $1-\alpha$. This is not a major concern; the investigator is still assured of achieving at least $1-\alpha$ coverage, although the interval may be slightly longer than necessary (corresponding to the increase in the actual confidence level). The result is then viewed as a **conservative confidence interval**. Contrastingly, for very small samples the true confidence level of the large-sample LR intervals may be below the nominal level of $1-\alpha$: an **anti-conservative** or **radical confidence interval**. This is an especially undesirable characteristic: an investigator employing a radical interval is given no assurance that the desired confidence level is even being met.

Within this context, we note that in small samples LR-based confidence intervals offer no assurance of conservative confidence levels, and can be quite unstable. In select instances, however, their true confidence levels can be very close to $1-\alpha$, depending on the underlying parent distributions. As $n \to \infty$ the true LR confidence level will approach $1-\alpha$.

Unfortunately, while the LR confidence interval is simple to describe, it is not always simple to implement. Computer calculation is common for this method, due to its inherent complexity. In some cases, however, it can simplify, as the next example illustrates.

Example 2.4 LR-based confidence interval for a binomial π
Return to the binomial testing problem with $Y \sim \mathcal{Bin}(N,\pi)$. In Example 2.3, the likelihood ratio statistic for testing $H_0:\pi = \pi_0$ was given by

$$G^2 = -2\left\{Y\log\left(\frac{\pi_0}{\hat{\pi}}\right) + (N - Y)\log\left(\frac{1 - \pi_0}{1 - \hat{\pi}}\right)\right\},$$

where $\hat{\pi} = Y/N$ is the sample proportion. Rejection occurs when $G^2 > \chi^2_\alpha(1)$. The corresponding acceptance region that defines the LR confidence interval is the set of all values for π such that $G^2 \le \chi^2_\alpha(1)$. This inequality can be written as

$$Y\log(\pi) + (N - Y)\log(1 - \pi) \ge -\tfrac{1}{2}\chi^2_\alpha(1) + Y\log(\hat{\pi}) + (N - Y)\log(1 - \hat{\pi})$$

or

$$\pi^Y(1 - \pi)^{N-Y} - \exp\left\{-\tfrac{1}{2}\chi^2_\alpha(1) + Y\log(\hat{\pi}) + (N - Y)\log(1 - \hat{\pi})\right\} \ge 0. \quad (2.9)$$

The left side of (2.9) defines a polynomial in π, which when solved for its real roots produces $1 - \alpha$ limits on π. In most cases, the polynomial will be of sufficiently large order to make computer solution necessary for the real roots. In some cases, however, some simplification into a lower-order polynomial is possible, leading to an algebraic solution.

For instance, suppose we observe the data from Example 2.3: $y = 10$ and $N = 20$. Then, (2.9) simplifies to

$$\pi - \pi^2 - \exp\left\{-\tfrac{1}{20}\chi^2_\alpha(1) + 2\log(0.5)\right\} \ge 0.$$

At $\alpha = 0.05$, $\chi^2_{0.05}(1) = 3.84$, yielding the quadratic inequality $\pi^2 - \pi + 0.2063 \le 0$. Solving this for π produces the two roots 0.291 and 0.709. Thus an approximate 95% LR confidence interval on π based on these data is $0.291 < \pi < 0.709$. This is illustrated graphically in Fig. 2.2, where $G^2 = -2\{\log(\pi_0) - \log(\hat{\pi})\}$ is plotted versus $\hat{\pi}$. A horizontal reference line is drawn corresponding to $\chi^2_{0.05}(1) = 3.84$, that is, the distance 3.84 units from the value of G^2 evaluated at the ML estimator. (Changing the confidence level $1 - \alpha$ changes $\chi^2_\alpha(1)$ and hence the location of the reference line.)

In Fig. 2.2, the G^2 curve intersects the $\alpha = 0.05$ horizontal reference line at the endpoints of the confidence interval for π, here $0.291 < \pi < 0.709$. □

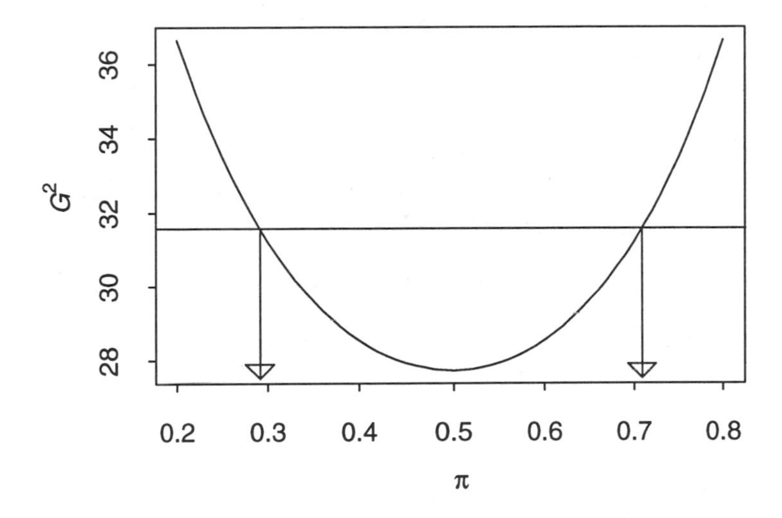

Fig. 2.2 Plot of G^2 versus π for a binomial experiment with $N = 20$ trials yielding $y =$ 10 successes. A horizontal reference line is drawn corresponding to $\chi^2_{0.05}(1) = 3.84$. Vertical reference lines identify the endpoints of an LR-based confidence interval for π.

2.5.2 Likelihood-based inference: Wald intervals

Inversion of a rejection region into a confidence interval may be accomplished from any test statistic. For example, one could construct a score-based confidence interval from the score statistic in Section 2.4.3 (cf. Section 2.6.1), or a Wald confidence interval from the Wald statistic in Section 2.4.2. In particular, the Wald intervals usually are quite simple to define. Recall that the Wald statistic for testing $H_0: \theta = \theta_0$ is based on a standardized ratio $\{\hat{\theta} - \theta\}/se[\hat{\theta}]$, where $\hat{\theta}$ is the ML estimator, and $se[\hat{\theta}]$ is the corresponding standard error. In large samples this Wald ratio has approximately a standard normal reference distribution. The ratio's simple form admits a fairly straightforward closed-form expression for $1 - \alpha$ confidence limits on θ, as follows: Start with the probability expression

$$\mathrm{P}\left[-\mathfrak{z}_{\alpha/2} < \frac{\hat{\theta} - \theta}{se[\hat{\theta}]} < \mathfrak{z}_{\alpha/2}\right] \approx 1 - \alpha,$$

where z_a is the upper-a standard normal critical point from (1.14), for $a \leq 0.5$. (For $a > 0.5$ one finds critical points via the symmetry of the standard normal distribution: $z_{1-a} = -z_a$ for any a between 0 and 1. Thus in the expression above if $\alpha = 0.05$, $z_{0.025} = 1.96$ corresponds to $-z_{0.975} = 1.96$.) Since this is based on a large-sample result, we know the approximation improves as $n \rightarrow \infty$. That is, we expect the Wald ratio to rest between $\pm z_{\alpha/2}$ approximately $100(1 - \alpha)\%$ of the time as the sample size grows large. (This is literally the acceptance region of the Wald test for θ.)

Now, operate on the relationship inside the probability brackets by (a) multiplying all of the inequalities by $se[\hat{\theta}]$,

$$-z_{\alpha/2}\, se[\hat{\theta}] < \hat{\theta} - \theta < z_{\alpha/2}\, se[\hat{\theta}]$$

(note that since $se[\hat{\theta}] > 0$, the operation preserves the direction of the inequalities); and then (b) solving for θ on both sides of the inequalities,

$$\hat{\theta} - z_{\alpha/2}\, se[\hat{\theta}] < \theta < \hat{\theta} + z_{\alpha/2}\, se[\hat{\theta}]. \tag{2.10}$$

Equation (2.10) is the $1 - \alpha$ Wald confidence interval for θ.

The exceedingly simple closed-form of the confidence interval in (2.10) leads to its ubiquitous use, as we will see in later chapters. Here is an example (continuing) with the binomial distribution:

Example 2.4 (continued) Wald interval for a binomial π

Continuing with the binomial setting, $Y \sim \mathcal{Bin}(N,\pi)$, we saw in Example 2.3 that the ML estimator for π is $\hat{\pi} = Y/N$, with large-sample standard error $se[\hat{\pi}] = \{\hat{\pi}(1 - \hat{\pi})/N\}^{1/2}$. Substituting these quantities into (2.10) yields the Wald interval for π:

$$\hat{\pi} - z_{\alpha/2}\sqrt{\hat{\pi}(1 - \hat{\pi})/N} < \pi < \hat{\pi} + z_{\alpha/2}\sqrt{\hat{\pi}(1 - \hat{\pi})/N}. \tag{2.11}$$

As expected, increasing N forces the confidence limits to contract.

To illustrate, suppose we observe the rodent toxicity data from Example 2.3: $y = 10$ and $N = 20$. The components of the Wald interval are $\hat{\pi} = 0.5$ and $se[\hat{\pi}] = \sqrt{0.0125} = 0.1118$. At $\alpha = 0.05$, $z_{0.025} = 1.96$, so (2.11) becomes simply $0.5 \pm (1.96)(0.1118)$, or $0.281 < \pi < 0.719$.

Notice that the calculated limits for the Wald interval are very similar to the LR-based interval for π. Unfortunately, this similarity is not always assured. For example, suppose instead that $y = 2$ successes are observed among the $N = 20$ trials. The Wald interval is then based on $\hat{\pi} = 0.1$ and $se[\hat{\pi}] = 0.067$, yielding the 95% confidence interval $0.1 \pm (1.96)(0.067) = 0.1 \pm 0.135$, or $-0.031 < \pi < 0.231$. In contrast, the LR-based confidence interval for these data can be found as $0.018 < \pi < 0.278$.

The latter calculations illustrate a potential shortcoming with the Wald procedure in this setting. Clearly, the negative lower endpoint for π, –0.031, is senseless. (In practice we would simply truncate the lower endpoint at π = 0.) The negative endpoint is result of appealing to a weak normal approximation for $\hat{\pi}$, and really should not be applied here. For these data, the LR-based method yields more sensible intervals. □

Since all the likelihood-based methods (LR, Wald and score) are asymptotically equivalent, we expect that confidence limits constructed from them will converge to similar values as $n \rightarrow \infty$, and this is indeed the case. In small samples the three approaches can give different values, however, so it is of some interest to consider the other methods as well. Indeed, as we note in Section 2.6.1, a modification of the score test produces one of the best performing forms of confidence interval for a binomial parameter π.

2.5.3* Bootstrap-based confidence intervals

As seen in the preceding section, likelihood-based confidence intervals are quite useful. When the parent distribution of the data is known, it is common to employ them as the basis for statistical interval estimators. This is not to say, however, that likelihood approaches are the only method(s) from which an interval estimator may be constructed. Many other approaches are possible, and some are even available when the underlying parent distributions are not fully known. (In this latter case, likelihood-based methods are unavailable, since it is impossible to construct a complete likelihood for the unknown parameter.) These employ computer-intensive methods, such as **bootstrap resampling** (Babu and Rao, 1993; Efron and Tibshirani, 1993) or other **Monte Carlo methods** (Garthwaite and Buckland, 1992) to calculate approximate confidence limits on some measure of the data. This measure can be parametric, such as a mean or variance, or semi- or nonparametric, such as a median. Such generality provides these computer-intensive methods with great applicability, although their natural computational intensity restricts their use to settings where appropriate computer resources are available.

As an illustration, we introduce a common form of bootstrap-based confidence interval construction. The bootstrap method is based upon an elegantly simple idea: since sampling distributions for statistics are based upon repeated samples with replacement (or **resamples**) from the same population, one can use the computer to simulate repeated sampling, calculating the statistic for each simulated sample. The resulting, simulated sampling distribution for the statistic is used to approximate the true sampling distribution of the statistic of interest, leading to approximate interval estimates. The approximation improves as the number of simulated samples increases.

The empirical distribution of the observed data is used as the basis for the simulated resamples, to provide insight into the form of the underlying sampling distribution. Simulated resamples are drawn by computer from a theoretical distribution that matches the empirical distribution of the data, and the statistic of interest is calculated for each simulated resample. The resampled values of the statistic provide an approximate distribution from which to construct confidence intervals and/or hypothesis tests. This empirical simulation of the sampling distribution for the statistic of interest captures the essence of bootstrapping.

Let us consider some of the details for the procedure. Suppose a random sample, $X_1, \ldots, X_n$ is obtained from some population with unknown c.d.f. $F_X(x)$. Further suppose we are interested in obtaining an interval estimate of some unknown parameter θ based upon an estimator $\hat{\theta}$ calculated from the data. We can estimate $F_X(x)$ from our data via the empirical c.d.f.

$$\hat{F}_X(x) = \frac{\{\#(X_i) \le x\}}{n},$$

where $\#(X_i)$ denotes the number of X_is. From this, assuming the data are discrete, we can construct the corresponding empirical p.m.f., $p_X(x)$, as

$$\hat{p}_X(x) = \frac{\{\#(X_i) = x\}}{n}, \quad \text{where } x = x_1, \ldots, x_n.$$

If $\hat{F}_X(x)$ is a good estimate of $F_X(x)$ and $p_X(x)$ is a good estimate of the true population p.m.f., then we have all that we need to generate the bootstrap samples. Essentially, the following recipe is followed:

1. Generate a bootstrap sample, say $X_1^*, \ldots, X_n^*$, at random from the empirical p.m.f. $\hat{p}_X(x)$.
2. Calculate the statistic of interest, say $\hat{\theta}$, from the bootstrap sample.
3. Repeat steps 1 and 2 a large number of times, say $B \ge 1000$. (Babu and Singh (1983) give theoretical arguments for $B = n(\log\{n\})^2$, although if $n < 60$ a common recommendation is to set at least $B = 1000$ for confidence interval construction.) Collect together all the statistics that were calculated from each of the bootstrap samples. Let $\hat{\theta}_1^*, \ldots, \hat{\theta}_B^*$ be the collection of all of these statistics.

We tabulate the collection of all the bootstrap statistics in order to form an empirical estimate of the sampling distribution of $\hat{\theta}$. One approach for obtaining a confidence interval for θ is known as the **percentile method.** This method is based upon selecting specified quantiles from the empirical distribution of $\hat{\theta}$ formed from the collection of all bootstrap statistics. For example, suppose a 95% confidence interval is desired for θ. For the lower bound of this confidence interval, the percentile method takes the 2.5th percentile of the empirical distribution of $\hat{\theta}$, i.e. the value that lies above 2.5% and below 97.5% of the ordered values of $\hat{\theta}_1^*, \ldots, \hat{\theta}_B^*$. For the upper bound, the percentile method takes the 97.5th percentile of the empirical distribution of $\hat{\theta}$.

Other formulations are possible, as well. Interested readers may explore further bootstrap methods of confidence interval construction in, for example, Efron and Tibshirani (1993).

Example 2.4 (continued) Bootstrap-based confidence interval for a binomial π

Continuing with the binomial estimation problem, recall that in $N = 20$ trials, $y = 10$ successes were observed. Our goal is to obtain an interval estimate for the population probability of success. We can imagine the data as a vector of ten zeros and ten ones, and so write the data in vector form as $\mathbf{X}$ = [0 0 0 0 0 0 0 0 0 0 1 1 1 1 1 1 1 1 1 1]. A nonparametric bootstrap involves resampling from this data vector and generating bootstrap samples of size 20, $\mathbf{X}^* = [X_1^* \ldots X_{20}^*]$, where the X_i^*s are obtained by sampling with replacement from the data vector $\mathbf{X}$. A sample proportion, y/N, is generated from each of these bootstrap samples.

With $N = 20$, the Babu–Singh recommendation is to use $B = 20(\log\{20\})^2 = (20)(2.996)^2 \approx 180$ bootstrap samples. For added computational accuracy, however, we chose to simulate a larger set of B = 1000 bootstrap samples. We generate a sample proportion from each of the 1000 resamples, and then construct a confidence interval by selecting particular quantiles from the bootstrap-based sampling distribution of the sample proportion. Our computer results yielded a 95% percentile method confidence interval for π of $0.30 < \pi < 0.75$. A second set of computer-simulated 1000 bootstrap samples yielded a similar 95% percentile method confidence interval for π of $0.30 < \pi < 0.70$.

These intervals were obtained from applying a small S-PLUS function that is presented in Fig. 2.3. (The S-PLUS analysis is interactive, so program code and results occur on the same output. In our displays, we separate programming from output results with grayed lines.) The input to the function is the observed data, $\mathbf{X}$ (`n.one` is the number of successes, and `nsize` is the number of trials); the number of bootstrap samples desired, B

```
npar.boot.bin<-function(n.one,nsize,
                        nboots=1000,conf.coef=.95) {
   # observed data
     data.vec <- c(rep(0,nsize-n.one),rep(1,n.one))

       phat <- n.one/nsize
     pboots <- matrix(0,nrow=nboots,ncol=1)
   for (ii in 1:nboots) {
      data.boot.vec <- sample(data.vec,replace=T)
       pboots[ii,1] <- sum(data.boot.vec)/nsize
  }

   alpha <- (1-conf.coef)
   list(phat=phat,confid.coef=conf.coef,
       CI=quantile(pboots[,1],c(alpha/2,1-alpha/2)))
}
 npar.boot.bin(n.one=10,nsize=20,nboots=1000,conf.coef=.95)
```

```
$phat:
[1] 0.5

$confid.coef:
[1] 0.95

$CI:
  2.5% 97.5%
   0.3   0.75
```

```
 npar.boot.bin(n.one=10,nsize=20,nboots=1000,conf.coef=.95)
```

```
$phat:
[1] 0.5

$confid.coef:
[1] 0.95

$CI:
  2.5% 97.5%
   0.3   0.7
```

Fig. 2.3 S-PLUS function for bootstrap intervals on a binomial π.

(`nboots`, which defaults to 1000); and the confidence coefficient for the interval estimate, $1-\alpha$ (`conf.coef`, which defaults to 0.95). The key element of this function is the sampling with replacement from the observed data vector (as implemented by the `sample` function with the `replace=T` switch activated). The output of this function echoes the observed sample proportion, and displays the confidence coefficient along with the calculated bootstrap confidence interval. □

2.6 CONFIDENCE INTERVALS FOR SOME SPECIAL DISTRIBUTIONS

Even when the likelihood is fully specified, there are cases where other forms of confidence intervals compete favorably with the LR or Wald intervals. The following sections identify some examples, with emphasis on one-parameter distributions seen commonly in environmental analyses. Wherever possible, we will highlight closed-form intervals that are both simple to evaluate and that have established, stable coverage properties.

2.6.1 Confidence intervals for a binomial success probability

We have used the binomial distribution, where $Y \sim \mathcal{B}in(N,\pi)$, as our primary example throughout this chapter, including likelihood-based confidence intervals in Example 2.4, above. Indeed, the Wald interval from (2.11) is one of the most basic ways to construct a confidence interval on π; cf. Moore and McCabe (1993, Section 8.1) As is illustrated in Example 2.4 when $y = 2$ and $N = 20$, however, the approach is known to suffer in the case of small samples: a general rule of thumb for its use is to require $N\hat{\pi} \geq 5$ and $N(1 - \hat{\pi}) \geq 5$ as an indicator of a 'large' sample. (Notice that for $y = 2$ and $N = 20$, $N\hat{\pi} = 2 < 5$.) When these criteria are not met, the small-sample coverage properties of (2.11) are questionable, and its use is inadvisable.

A **continuity correction** is available that adjusts the Wald interval for the discrete nature of the binomial when referring the construction to a (continuous) standard normal distribution: in $se[\hat{\pi}]$, simply add the correction $1/(2N)$ to $\hat{\pi}(1 - \hat{\pi})$ before dividing by N. This can stabilize the Wald interval somewhat in small samples, but it can still yield suspect inferences when $N\pi$ is very small.

As an alternative, one can take advantage of an equivalence relationship that exists between binomial and F distribution c.d.f.s. By using integral calculus to manipulate selected equalities between sums of binomial probabilities, it can be shown that if $Y \sim \mathcal{B}in(N,\pi)$ and if V is distributed as an F distribution with df equal to $2(N - Y + 1)$ and $2Y$, then $P[Y \geq k] = P[V > k(1 - \pi)/\{\pi(N - k + 1)\}]$; see Leemis and Trivedi (1996, Appendix

A). Note that this is an exact relationship, not a large-sample approximation. We can re-express this equality as a set of probability statements that bound π by a set of random critical points of the F distribution. This results in the $1 - \alpha$ confidence interval

$$\frac{1}{1 + \dfrac{N - y + 1}{y} F_{\alpha/2}(2\{N - y + 1\}, 2y)} < \pi < \frac{\dfrac{y + 1}{N - y} F_{\alpha/2}(2\{y + 1\}, 2\{N - y\})}{1 + \dfrac{y + 1}{N - y} F_{\alpha/2}(2\{y + 1\}, 2\{N - y\})} \qquad (2.12)$$

where y is the observed value of Y, and $F_{\alpha/2}(\nu_1, \nu_2)$ is the upper-$(\alpha/2)$ critical point from an F distribution with ν_1 and ν_2 *df*. (If $y = 0$, set the lower limit to 0; if $y = N$, set the upper limit to 1.) This F-equivalence interval is generally conservative for π, although in most cases it competes favorably with the approximate Wald interval (2.11). If F distribution critical points are readily available, this interval can be recommended for general use (but, see also the interval given in (2.14) below).

Vollset (1993) has surveyed a number of other alternatives for confidence limits on π, including the Wald intervals (with and without a continuity correction) and LR intervals. He concludes that neither the Wald nor the LR interval is as stable as other possible methods, some of which are almost as simple to compute as (2.11). One of the best performers is based on the score statistic from Section 2.4.3. From (2.8), the score statistic for the binomial testing problem may be written as

$$T^2 = \frac{(\hat{\pi} - \pi)^2}{\left\{\dfrac{\pi(1 - \pi)}{N}\right\}},$$

referenced in large samples to a $\chi^2(1)$ distribution; cf. Example 2.3. To construct a confidence interval for π, invert this test and find the acceptance region, that is, the set of all π such that $T^2 \le \chi^2_\alpha(1)$. This is

$$N(\hat{\pi} - \pi)^2 \le \chi^2_\alpha(1)\pi(1 - \pi),$$

which at equality is seen to be a quadratic equation in π (Wilson, 1927). The interrelationship between the $\chi^2(1)$ and standard normal distributions allows us to write $\{\chi^2_\alpha(1)\}^{1/2} = z_{\alpha/2}$, so collecting terms and solving for π produces the two solutions

$$\frac{Y + \frac{\chi^2_\alpha(1)}{2} \pm z_{\alpha/2}\left(Y - \frac{Y^2}{N} + \frac{\chi^2_\alpha(1)}{4}\right)^{1/2}}{N + \chi^2_\alpha(1)} . \tag{2.13}$$

As with the Wald interval (2.11), the score interval in (2.13) can exhibit confidence levels that drop below the nominal level of $1 - \alpha$, due to the discrete nature of the parent binomial distribution. This instability is most pronounced in small samples. Strong improvement in stability is available, however, by incorporating a continuity correction:

$$\frac{\left(Y \pm \frac{1}{2}\right) + \frac{\chi^2_\alpha(1)}{2} \pm z_{\alpha/2}\left\{\left(Y \pm \frac{1}{2}\right) - \frac{\left(Y \pm \frac{1}{2}\right)^2}{N} + \frac{\chi^2_\alpha(1)}{4}\right\}^{1/2}}{N + \chi^2_\alpha(1)} . \tag{2.14}$$

This continuity-corrected score interval is slightly conservative in small samples, but possesses one of the best combinations of stability and ease of calculation available for bounding π. Because of this, we recommend it along with the exact interval in (2.12) for use with the binomial distribution.

Additional refinement in a confidence interval for π is available via an algorithmic approach due to Casella (1986). His method can be applied to any confidence interval, including the recommended procedures (2.12) or (2.14). It takes advantage of the discrete nature of the binomial distribution, and has the effect of uniformly shortening any interval to which it is applied, while still retaining at least $1 - \alpha$ confidence. The method cannot be written to produce a closed-form expression for the confidence limits, however, so in general its use is limited to settings where sufficient computational resources are available. To facilitate use of the method, Casella gives a table of refined interval endpoints on π for the cases $\alpha = 0.05$ and $\alpha = 0.01$, and for N in the range $N = 6, \ldots, 30$.

Example 2.4 (continued) Confidence interval for a binomial π: $\hat{\pi} = 0.5$
Continuing with the rodent toxicity example, the data were $y = 10$ and $N = 20$. For the F-equivalence intervals from (2.12), we find $2\{N - y + 1\} = 2\{y + 1\} = 22$, and $2\{N - y\} = 2y = 20$. Thus the appropriate F distribution for both upper and lower endpoints is $F(22,20)$. At $\alpha = 0.05$, the associated upper-($\alpha/2$) critical point is $F_{0.025}(22,20) = 2.4337$. Applied to (2.12), we have

$$\frac{1}{1 + \frac{11}{10}(2.4337)} = 0.272 < \pi < 0.728 = \frac{\frac{11}{10}(2.4337)}{1 + \frac{11}{10}(2.4337)} .$$

Turning to the continuity-corrected score interval from (2.14), we have $\chi^2_{0.05}(1) = 3.84$ and $z_{0.025} = 1.96$. Then the 95% limits on π are

$$\frac{\left(10 \pm \frac{1}{2}\right) + 1.92 \pm 1.96\left\{\left(10 \pm \frac{1}{2}\right) - \frac{1}{20}\left(10 \pm \frac{1}{2}\right)^2 + 0.960\right\}^{1/2}}{20 + 3.84},$$

or simply $0.279 < \pi < 0.722$.

Additional refinement of these limits is possible via the Casella algorithm. From his Table 1 (and using his notation), we find the lower 95% limit on π as $l_{10} = 0.293$. The corresponding upper limit is $u_{20} = 1 - l_{20-10} = 0.707$, i.e. $0.293 < \pi < 0.707$. Note that the refined interval produces the narrowest limits among any of the fixed methods calculated above, and they do so while still assuring minimal 95% confidence. □

Example 2.4 (continued) Confidence interval for a binomial π: $\hat{\pi} = 0.1$
Suppose in our rodent toxicity example that instead of 10 deaths, we observe $y = 2$ out of $N = 20$ deaths. As found previously, the Wald-based confidence interval for π is $-0.031 < \pi < 0.231$ (which in practice would be reported as $0 < \pi < 0.231$), while the LR-based confidence interval is $0.018 < \pi < 0.278$. For the F-equivalence intervals from (2.12), we find $2\{N - y + 1\} = 2\{18 + 1\} = 38$, while $2\{N - y\} = 2(18) = 36$. At $\alpha = 0.05$, the upper-($\alpha/2$) critical point is $F_{0.025}(2\{y + 1\}, 2\{N - y\}) = F_{0.025}(6,36) = 2.7845$ and the lower-($\alpha/2$) critical point is $F_{0.025}(2\{N - y + 1\}, 2y) = F_{0.025}(38,4) = 8.4191$. Applied to (2.12), we find

$$\frac{1}{1 + \frac{19}{2}(8.4191)} = 0.012 < \pi < 0.317 = \frac{\frac{3}{18}(2.7845)}{1 + \frac{3}{18}(2.7845)}.$$

Turning to the continuity-corrected score interval from (2.14), we have $\chi^2_{0.05}(1) = 3.84$ and $z_{0.025} = 1.96$. Then the 95% limits on π are

$$\frac{\left(2 \pm \frac{1}{2}\right) + 1.92 \pm 1.96\left\{\left(2 \pm \frac{1}{2}\right) - \frac{1}{20}\left(2 \pm \frac{1}{2}\right)^2 + 0.960\right\}^{1/2}}{20 + 3.84},$$

or simply $0.018 < \pi < 0.331$. □

2.6.2 Confidence intervals for a Poisson mean

In the case of sampling from a Poisson distribution, many different methods can be applied for building confidence intervals on μ (Sahai and Khurshid,

1993a). Of these, only a few combine ease of use with good stability. We recommend two: the first is a method that takes advantage of an equivalence relationship between Poisson and χ^2 c.d.f.s, similar in nature to the binomial–F distribution equivalence that led to (2.12). Noted originally by Przyborowski and Wilenski (1935) and also by Garwood (1936), the intervals are constructed from the following equivalence relationship: if $Y \sim \mathcal{Poisson}(\mu)$ then $P[Y > y] = P[V < 2\mu]$, where $V \sim \chi^2(2y)$. (As with the binomial–F equivalence, the derivation of this Poisson–χ^2 equivalence takes advantage of certain relationships between sums and integrals; see, for example, McCulloch and Casella (1983).) From this, we can derive the following, simple confidence limits on μ:

$$\frac{1}{2}\chi^2_{1-(\alpha/2)}(2y) < \mu < \frac{1}{2}\chi^2_{\alpha/2}(2\{y+1\})\,. \tag{2.15}$$

In (2.15), y is the observed value of Y, and, for example, $\chi^2_{\alpha/2}(2y)$ is the upper-$(\alpha/2)$ critical point from a χ^2 distribution with $2y$ *df*. Note that if $y = 0$, we set the lower endpoint to zero.

The χ^2-equivalence interval in (2.15) is generally conservative for μ, although it can achieve very close to nominal $1 - \alpha$ coverage in some cases; cf. Fig. 1 in Casella and Robert (1989). It is possible to derive a slightly more accurate procedure, however, by inverting the score statistic, T^2, for testing μ. Recall that the score statistic requires calculation of the derivative of the log-likelihood; for the Poisson case $\ell(\mu) = \log\{e^{-\mu}\mu^Y/Y!\} = Y\log(\mu) - \mu - \log(Y!)$. The likelihood score is then $U_\mu = \partial\ell(\mu)/\partial\mu = (Y/\mu) - 1$. We also require calculation of the Fisher information value for μ:

$$\mathfrak{F}_\mu = \mathrm{E}\left[-\frac{\partial^2\ell(\mu)}{\partial\mu^2}\right] = \mathrm{E}[-(-\mu^{-2}Y)] = \mu^{-2}\mathrm{E}[Y] = \mu^{-2}\mu = \frac{1}{\mu}.$$

Then, from (2.8), the score statistic may be written as the ratio of $\{(Y/\mu) - 1\}^2$ to μ^{-1}. After some algebra, this simplifies to $T^2 = (Y - \mu)^2/\mu$, which is referenced in large samples to a $\chi^2(1)$ distribution. To construct a confidence interval for μ, invert this test and find the acceptance region, that is, the set of all μ such that $T^2 \le \chi^2_\alpha(1)$. This is $(Y - \mu)^2 \le \mu\chi^2_\alpha(1)$, which at equality is again a quadratic score equation. Collecting terms and solving for μ produces two solutions which are the upper and lower endpoints of the score interval:

$$Y + \frac{\chi^2_\alpha(1)}{2} \pm \mathfrak{z}_{\alpha/2}\sqrt{Y + \frac{\chi^2_\alpha(1)}{4}}\,.$$

Unfortunately, for small values of Y this score-based confidence interval can yield confidence levels that are below the nominal level of $1 - \alpha$, due to the discrete nature of the underlying Poisson distribution. To correct this instability, we apply a continuity correction, similar to that used with the binomial confidence limits in (2.14):

$$\left(Y \pm \tfrac{1}{2}\right) + \frac{\chi^2_\alpha(1)}{2} \pm \mathfrak{z}_{\alpha/2}\sqrt{\left(Y \pm \tfrac{1}{2}\right) + \frac{\chi^2_\alpha(1)}{4}} \tag{2.16}$$

(Casella and Robert, 1989). As in (2.13), the interrelationship between the $\chi^2(1)$ and standard normal distributions allows us to write $\{\chi^2_\alpha(1)\}^{1/2} = \mathfrak{z}_{\alpha/2}$. The limits in (2.16) are slightly conservative, but also exhibit good stability. We recommend them for use in practice.

As discussed in the binomial setting in Section 2.6.1, it is possible to apply a refining algorithm to the confidence limits in (2.16), based on computer-intensive calculations. Casella and Robert present such an algorithm, and also give a table (their Table 1) of 95% confidence limits for μ over the range $Y = 0, \ldots, 49$. For larger values of Y, the refined limits will be essentially similar to (2.16). For small values of Y, however, the refinement can improve the conservative nature of the continuity-corrected score limits in (2.16).

We should note that it is straightforward to modify these results when observing a random sample of observations $Y_1, \ldots, Y_n$ from a Poisson distribution. In Section 1.2.5 we noted that the Poisson distribution is closed under addition; that is, if we observe $Y_i \sim$ i.i.d. $\mathcal{Poisson}(\mu)$, the corresponding sample sum, $Y_+ = \sum_{i=1}^{n} Y_i$, is distributed as Poisson with mean $n\mu$: $Y_+ \sim \mathcal{Poisson}(n\mu)$. From this fact, we simply apply (2.16) to Y_+, and calculate $1 - \alpha$ confidence limits on $n\mu$. When divided by n, these produce $1 - \alpha$ confidence limits on μ.

Example 2.5 Aphid infestation data

Consider the following example on pest control in a particular commercial ecosystem, modified from Casella and Berger (1990, Chapter 9). Suppose an investigator is interested if a certain pesticide will reduce aphid infestation on commercial potatoes. As part of a preliminary sample to study the aphid concentrations on potatoes, the investigator takes $n = 6$ randomly selected field potatoes and in the laboratory counts the number of aphids per potato. The data are:

$$Y_1 = 2,\ \ Y_2 = 0,\ \ Y_3 = 0,\ \ Y_4 = 1,\ \ Y_5 = 0,\ \ Y_6 = 1.$$

The sum of these observations is $Y_+ = 4$. A point estimate of the mean rate of per-potato aphid infestation, μ, is $\overline{Y} = 0.667$. Assuming a Poisson distribution for the number of aphids per potato, let us calculate a 95% confidence interval for μ.

We begin by recognizing that if $Y_i \sim \mathcal{P}oisson(\mu)$, then $Y_+ \sim \mathcal{P}oisson(6\mu)$. To calculate a χ^2-equivalence interval on 6μ from (2.15), we require the critical points $\chi^2_{0.975}(8) = 2.18$ and $\chi^2_{0.025}(10) = 20.48$. Then, (2.15) gives $1.09 < 6\mu < 10.24$. Dividing by six implies $0.182 < \mu < 1.707$.

For the more stable continuity-corrected score interval from (2.16), we require the critical points $\chi^2_{0.05}(1) = 3.84$ and $z_{0.025} = 1.96$. Then, 95% corrected score limits on 6μ are

$$\left(4 \pm \tfrac{1}{2}\right) + \frac{3.84}{2} \ \pm\ 1.96\sqrt{\left(4 \pm \tfrac{1}{2}\right) + \frac{3.84}{4}}$$

or simply $1.28 < 6\mu < 11.00$. From these, a 95% confidence interval on μ is $0.213 < \mu < 1.833$.

Refined 95% Poisson limits from Casella and Robert (1989, Table 1) may be read as $1.37 < 6\mu < 10.04$, so that our best interval estimate of the mean per-potato infestation rate is $0.228 < \mu < 1.673$. □

2.6.3 Confidence intervals for an exponential mean

Among the continuous distributions we introduced in Section 1.3, perhaps the simplest was the exponential distribution from Section 1.3.2. Recall that if X has an exponential distribution with parameter $\beta > 0$, then the c.d.f. of X is $\mathrm{P}[X \le x] = 1 - e^{-x/\beta}$. Also, the mean of X is $\mathrm{E}[X] = \beta$. For this distribution, β has dual interpretation as both a mean rate of occurrence and as a scale parameter. To construct a confidence interval on β, we appeal to the c.d.f. Notice that for a continuous random variable, X, any probability statement of the form $\mathrm{P}[a \le X \le b]$ is equivalent to $\mathrm{P}[X \le b] - \mathrm{P}[X \le a]$. Applying this fact to the exponential distribution, it is a simple calculation to show that the probability

$$\mathrm{P}\left[-\beta\log\left\{1 - \frac{\alpha}{2}\right\} \ \le\ X \ \le\ -\beta\log\left\{\frac{\alpha}{2}\right\}\right]$$

is equal to $1 - \alpha$. (A minor technical constraint here is that $\alpha < 0.5$, which is common in practice.) From this, a $1 - \alpha$ confidence interval on β is given by

$$\frac{X}{-\log\left(\frac{\alpha}{2}\right)} \le \beta \le \frac{X}{-\log\left(1 - \frac{\alpha}{2}\right)} \qquad (2.17)$$

(George and Elston, 1993). We should emphasize here that this interval is exact, in the sense that its confidence coefficient is always $1 - \alpha$. (This contrasts with the case for discrete random variables such as the binomial or Poisson, above. The continuity of the exponential distribution's c.d.f. assures us that the interval in (2.17) has exact $1 - \alpha$ coverage at all values of X and β.)

If we were to observe a random sample of exponential observations, $X_i \sim$ i.i.d. $\mathcal{Exp}(\beta)$, then it is slightly more difficult to construct an exact confidence interval for β. As we will see, however, the form of the resulting interval is essentially the same as that in (2.17). To begin, we employ two facts from distribution theory. First, recall from Section 1.3.2 that the exponential is a special case of the gamma distribution with first parameter set to 1: if $X \sim \mathcal{Exp}(\beta)$, then we may also write $X \sim \mathcal{Gamma}(1,\beta)$. It is the case then that this form of gamma distribution exhibits closure under addition: if $X_i \sim$ (indep.) $\mathcal{Gamma}(1,\beta)$, then $X_+ = \sum_{i=1}^{n} X_i \sim \mathcal{Gamma}(n,\beta)$. That is, a sum of independent exponential random variables is distributed as gamma.

The second fact of which we take advantage is another special relationship, this time between gamma and χ^2 random variables: if $X_+ \sim \mathcal{Gamma}(n,\beta)$, where n is a positive integer, then $V = 2X_+/\beta \sim \chi^2(2n)$; see Casella and Berger (1990, Section 9.2). That is, a gamma random variable with a whole-number first parameter can be scaled into a χ^2 random variable. This may seem at first to have little consequence, but note the following: suppose $V \sim \chi^2(2n)$. Then by definition the critical points $\chi^2_{\alpha/2}(2n)$ and $\chi^2_{1-(\alpha/2)}(2n)$ satisfy

$$P[V > \chi^2_{\alpha/2}(2n)] = \frac{\alpha}{2} \quad \text{and} \quad P[V > \chi^2_{1-(\alpha/2)}(2n)] = 1 - \frac{\alpha}{2},$$

respectively. This can be written as

$$P[\chi^2_{1-(\alpha/2)}(2n) \le V \le \chi^2_{\alpha/2}(2n)] = 1 - \frac{1}{2}\alpha - \frac{1}{2}\alpha = 1 - \alpha.$$

Replacing V with $2X_+/\beta$, as suggested by the second fact above, produces the probability statement

$$P\left[\chi^2_{1-(\alpha/2)}(2n) \le \frac{2X_+}{\beta} \le \chi^2_{\alpha/2}(2n)\right] = 1 - \alpha.$$

From this, an exact $1 - \alpha$ confidence interval on β is given by

$$\frac{2X_+}{\chi^2_{\alpha/2}(2n)} \le \beta \le \frac{2X_+}{\chi^2_{1-(\alpha/2)}(2n)}. \tag{2.18}$$

2.6.4* Confidence intervals for two-parameter distributions

Before concluding, we should note that a large literature exists for constructing confidence intervals on parameters from distributions with two (or more) unknown parameters. Unfortunately, with two parameters, say θ_1 and θ_2, greater difficulties are encountered in constructing the confidence limits. Suppose, for example, we were interested in an interval estimator for θ_1. It is often the case that θ_2 is required in the interval for θ_1, and must either be known or estimated concurrently to calculate θ_1 limits. When θ_2 is unknown and estimated, however, the limits for θ_1 may require adjustment, in order to account for the additional variability in the estimated value for θ_2. In such a case, we say θ_2 is a form of **nuisance parameter** in terms of performing statistical inferences on θ_1.

The best-known case of this occurs with the normal distribution, $Y \sim N(\mu, \sigma^2)$, where a confidence interval for the mean μ requires knowledge about the variance σ^2. The confidence interval for μ is based on the t distribution from Section 1.3.5: given a simple random sample of observations $Y_1, \ldots, Y_n$ from $N(\mu, \sigma^2)$, the **sample variance,**

$$s^2 = \frac{1}{n-1} \sum_{i=1}^{n} (Y_i - \overline{Y})^2, \tag{2.19}$$

is used to estimate the nuisance parameter σ^2. The $1 - \alpha$ confidence interval for μ is then

$$\overline{Y} \pm t_{\alpha/2}(n-1)\frac{s}{\sqrt{n}},$$

where $t_\alpha(n-1)$ is the upper-α critical point of the $t(n-1)$ distribution. Pearson and Turton (1993, Chapter 6) and Moore and McCabe (1993, Chapter 7) discuss this interval in greater detail.

Similar constructions are necessary with other forms of two-parameter distributions. We will not discuss interval estimators for these cases in any detail here, and interested readers may consult the following sources as necessary: Wypij and Santner (1990) for the two-parameter beta-binomial distribution; Lau and Lau (1991) for the two-parameter beta distribution; Lawless (1982, Section 5.1.2) or Shiue and Bain (1990) for the two-parameter gamma distribution; Mann and Fertig (1977) or Lawless (1982, Section 4.1.2) for the two-parameter Weibull distribution and/or the two-parameter extreme-value distribution; and Dahiya and Guttman (1982) or Lawless (1982, Section 5.2.2) for the two-parameter lognormal distribution. Details on parameter estimation for many of these distributions are available also in the multi-volume set by Johnson *et al*. (Johnson and Kotz, 1972; Johnson *et al.*, 1992; 1994; 1995).

2.7 SEMI-PARAMETRIC INFERENCE

The bulk of material in this chapter has dealt with inferences for parametric families of statistical distributions. If an investigator is able to identify or assume a specific parametric form for the distribution of the experimental data, it is possible to construct a log-likelihood function and then proceed with any of the estimators, tests, and confidence intervals described above. Inferences for situations where the parent distribution of the data is not known fully are, however, more difficult to achieve. In some of these cases, it may be known that the mean response and even the variance of the data have a specific form, but that the actual distribution from which they are derived is unknown. When this is true, we turn to a form of **semi-parametric inference**.

In this section, we describe selected examples of semi-parametric inferences similar in form to the likelihood-based methods introduced above. The material is by necessity somewhat complex, and readers who do not require or plan to employ these more advanced estimation methods may wish to move forward to the summary in Section 2.8. In any case, we note that it is not our goal in this section to survey this field of study. Instead, we only review the concepts behind basic semi-parametric methodologies, in preparation for their occasional use in future chapters of this text. For more technical discussions of modern semi-parametric modeling, see, for example, Godambe (1991), or Severini and Staniswallis (1994).

2.7.1* Maximum quasi-likelihood estimation

When the precise form of the likelihood for a set of data cannot be specified, it is still possible to estimate features of the data, using a **quasi-likelihood function**. Coined by Wedderburn (1974), this term refers to a function constructed to be similar in nature to the likelihood for estimating a mean parameter when the actual likelihood function is unknown. The methodology is particularly useful when some basic feature of the data is known, but it is unclear how to form a fully parametric model. For example, a set of count data may be known to exhibit variability in excess of the usual Poisson model for counts, but it may not be clear if a negative binomial model or some other overdispersed model for the count data is valid.

In the general case, suppose we have a set of independent observations Y_i, and we are interested in estimating the means, μ_i, of the Y_i, $i = 1, \ldots, n$. For maximum flexibility, we allow each Y_i to have a possibly different mean, μ_i, but assume the existence of some vector of $P < n$ unifying parameters $\boldsymbol{\beta} = [\beta_1 \ldots \beta_P]$, relating all the μ_is. For simplicity of notation, we write the unifying function as $\mu_i = \mu_i\{\boldsymbol{\beta}\}$. For instance, the simplest case ($P = 1$) models all the μ_is as equal, and sets μ_i equal to the single parameter β. More complex model formulations are also possible, but these always have the feature that since $P < n$, we are in effect decreasing the number of parameters to be estimated via the unifying functions $\mu_i\{\boldsymbol{\beta}\}$.

Suppose also that we are willing to specify the exact parametric nature of the variance of Y_i via some function, say

$$\text{Var}[Y_i] = \mathcal{V}(\mu_i, \varphi),$$

where φ is an additional, possibly unknown dispersion parameter. Notice that this variance specification allows $\text{Var}[Y_i]$ to be a function of μ_i, and hence of $\boldsymbol{\beta}$. Traditionally, the dispersion parameter is taken to factor out of the variance function,

$$\text{Var}[Y_i] = \varphi \mathcal{V}(\mu_i),$$

as given by Wedderburn (1974). We do not require this, however, and our formulation is more in keeping with an extended form of quasi-likelihood, discussed by Nelder and Pregibon (1987).

Together, the functions $\mu_i\{\boldsymbol{\beta}\}$ and $\mathcal{V}(\mu_i, \varphi)$ specify a mean–variance parameterization for the observations, but they do not go so far as to specify the full parametric form of the likelihood. Thus the model is semi-parametric, as opposed to 'fully' parametric.

In order to estimate $\boldsymbol{\beta}$ under this mean–variance specification, the quasi-likelihood approach constructs a set of estimating equations for $\boldsymbol{\beta}$ similar in nature to those in (2.3). Recall that the derivatives of the log-likelihood in (2.3) are simply the scores U_j from (2.7). Of interest in quasi-likelihood estimation is the development of quasi-score functions that mimic the likelihood scores. One possibility is the **quasi-scores**

$$V_q(\boldsymbol{\beta}) = \sum_{i=1}^{n} \frac{Y_i - \mu_i\{\boldsymbol{\beta}\}}{\mathcal{V}(\mu_i\{\boldsymbol{\beta}\}, \varphi)} \frac{\partial \mu_i}{\partial \beta_q}, \tag{2.20}$$

$q = 1, \ldots, P$. This construction is motivated in part by the fact that when the observations are from a parent distribution in the exponential class, the true likelihood scores also have the form in (2.20); see Exercise 2.37. McCullagh and Nelder (1989, Section 9.2) show that the quasi-scores in (2.20) exhibit other similarities to true likelihood scores; in particular, when set equal to zero the quasi-scores produce a set of P **quasi-likelihood estimating equations**

$$\sum_{i=1}^{n} \frac{Y_i - \mu_i\{\boldsymbol{\beta}\}}{\mathcal{V}(\mu_i\{\boldsymbol{\beta}\},\varphi)} \frac{\partial \mu_i}{\partial \beta_q} = 0, \qquad (2.21)$$

$q = 1, \ldots, P$.

The solution to this system of P equations in P unknowns produces $\hat{\boldsymbol{\beta}}$, a **maximum quasi-likelihood** (MQL) estimator of $\boldsymbol{\beta}$. The corresponding MQL estimators of μ_i are found by applying $\hat{\boldsymbol{\beta}}$ to the μ_is, via $\mu_i\{\hat{\boldsymbol{\beta}}\}$.

The effort required to solve the quasi-score equations can be substantial; in many cases the system of equations (2.21) is so complex that the solution can be found only via iterative computer calculation. This is, in a sense, a necessary compromise when the likelihood cannot be specified. Hastie and Pregibon (1992) describe a computerized function, `quasi`, for calculating quasi-likelihood estimators with the statistical computer package S-PLUS.

Note that the system of MQL estimating equations in (2.21) is constructed under the assumption that the dispersion parameter, φ, is known. If φ is unknown, it too must be estimated. A common approach is to iterate between solving the quasi-likelihood estimating equations (2.21) to estimate $\boldsymbol{\beta}$ and appealing to the method of moments (MoM) to estimate φ. That is, suppose for any fixed initial value of φ one calculates the MQL estimates $\mu_i\{\hat{\boldsymbol{\beta}}_\varphi\}$. The difference between an observation, Y_i, and its predicted value under some model, here $\mu_i\{\hat{\boldsymbol{\beta}}_\varphi\}$, is called a **raw residual**, $Y_i - \mu_i\{\hat{\boldsymbol{\beta}}_\varphi\}$. Residuals are used to quantify variability in the model fit, and to diagnose model adequacy. To estimate the dispersion parameter φ under this MQL model, we square the raw residuals and divide by their estimated variance functions, creating a set of squared **generalized Pearson residuals**, $(Y_i - \mu_i\{\hat{\boldsymbol{\beta}}_\varphi\})^2/\mathcal{V}(\mu_i\{\hat{\boldsymbol{\beta}}_\varphi\},\varphi)$, $i = 1, \ldots, n$. (The term honors the statistician Karl Pearson, who first suggested a similar form of residual quantity; see Section 9.3.1.) A measure of variation providing information on φ is then the sum of these generalized residuals,

$$\sum_{i=1}^{n} \frac{\left(Y_i - \mu_i\{\hat{\boldsymbol{\beta}}_\varphi\}\right)^2}{\mathcal{V}(\mu_i\{\hat{\boldsymbol{\beta}}_\varphi\},\varphi)}.$$

To solve for φ, set this sum equal to its expected value, which is approximately equal to $n - P$. This is an MoM estimating equation for φ,

$$\sum_{i=1}^{n} \frac{\left(Y_i - \mu_i\{\hat{\boldsymbol{\beta}}_\varphi\}\right)^2}{\mathcal{V}(\mu_i\{\hat{\boldsymbol{\beta}}_\varphi\},\varphi)} = n - P. \tag{2.22}$$

The final MQL/MoM estimators for the $P + 1$ unknown parameters $\boldsymbol{\beta}$ and φ are found by iterating between (2.21) and (2.22) until convergence is achieved.

Example 2.6 MQL estimators for overdispersed count data
To illustrate the MQL approach, consider the following example. Suppose data are collected as counts, Y_i, of number of bacterial mutants on each of n Petri dishes after the plated microbes were exposed to some environmental mutagen. It may be known that the data exhibit variability in excess of the Poisson model for counts, i.e. if $\mathrm{E}[Y_i] = \mu_i$, it may be thought that $\mathrm{Var}[Y_i] > \mu_i$. This is a form of overdispersion, as discussed in Section 1.2. If the investigator is unwilling to specify fully a distribution for the counts, such as negative binomial, the MQL approach for modeling and parameter estimation may be appropriate. Suppose the data are all from a homogeneously treated population, so that it is reasonable to model a constant mean, $\mu_i = \beta > 0$ for all $i = 1, \ldots, n$ (i.e. $P = 1$). To model the overdispersion, let each Y_i have a variance function of the form $\mathcal{V}(\beta, \varphi) = \beta + \varphi\beta^2$, i.e. quadratic in β. Under this model specification, there are $P + 1 = 2$ unknown parameters to estimate: β and φ.

To set up the MQL estimating equation for β, we form the quasi-score from (2.20)

$$V(\beta) = \sum_{i=1}^{n} \frac{Y_i - \beta}{\mathcal{V}(\beta,\varphi)} \frac{\partial \mu_i}{\partial \beta} = \sum_{i=1}^{n} \frac{Y_i - \beta}{\beta + \varphi\beta^2},$$

and set it equal to zero. For $\varphi > 0$, the estimating equation simplifies to $\sum_{i=1}^{n}\{Y_i - \beta\} = 0$. (Notice that this is independent of φ.) The resulting MQL solution for β is the sample mean $\hat{\beta}_{\mathrm{MQL}} = \overline{Y}$.

To estimate the dispersion parameter, we apply the MoM estimating equation, (2.22):

$$\sum_{i=1}^{n} \frac{\left(Y_i - \hat{\beta}_{\mathrm{MQL}}\right)^2}{\hat{\beta}_{\mathrm{MQL}} + \varphi\hat{\beta}_{\mathrm{MQL}}} = n - 1.$$

Since $\hat{\beta}_{\mathrm{MQL}} = \overline{Y}$, this simplifies to $s^2 = \overline{Y} + \varphi\overline{Y}^2$, where s^2 is the sample variance from (2.19); i.e. $\hat{\varphi}_{\mathrm{MOM}} = (s^2 - \overline{Y})/\overline{Y}$.

An unfortunate feature of $\hat{\varphi}_{\mathrm{MOM}}$ is that $\overline{Y}$ may be greater than s^2. This would force $\hat{\varphi}_{\mathrm{MOM}} < 0$, violating our assumption that φ be positive. To correct for this, it is possible to extend the MQL approach and form a quasi-score equation for φ. (The MQL estimator for β remains $\overline{Y}$.) The corresponding MQL estimating equation for φ is

$$\sum_{i=1}^{n}\left\{\varphi^{-2}\log\left(\frac{1+\varphi\overline{Y}}{1+\varphi Y_i}\right)-\frac{Y_i}{1+\varphi Y_i}+\frac{1+6Y_i}{2(\varphi+6+6\varphi Y_i)}\right\}=\frac{n}{2(\varphi+6)} \quad (2.23)$$

(Clark and Perry, 1989). Unfortunately, this equation has no closed-form solution and must be solved via computer iteration. Together with $\hat{\beta}_{\mathrm{MQL}} = \overline{Y}$, the result provides the MQL estimators for β and φ. □

In large samples, the MQL estimators posses distributional characteristics that mimic those of the ML estimators discussed in Section 2.3.3. Most useful is the fact that as $n\rightarrow\infty$, the MQL estimator is distributed as approximately multivariate normal: $\hat{\boldsymbol{\beta}} \mathrel{\dot\sim} \mathrm{N}_P(\boldsymbol{\beta}, \mathbf{C}_{\boldsymbol{\beta}}^{-1})$, where $\mathbf{C}_{\boldsymbol{\beta}}$ is the $P\times P$ matrix whose (q,r)th element is

$$c_{qr} = \sum_{i=1}^{n}\frac{1}{\mathcal{V}(\mu_i\{\boldsymbol{\beta}\},\varphi)}\frac{\partial\mu_i}{\partial\beta_q}\frac{\partial\mu_i}{\partial\beta_r}$$

(McCullagh and Nelder, 1989). In many ways $\mathbf{C}_{\boldsymbol{\beta}}$ is analogous to the Fisher information matrix $\mathbf{F}_{\boldsymbol{\Theta}}$ from a formal likelihood analysis: it is inverted to find the large-sample variance-covariance matrix for $\hat{\boldsymbol{\beta}}$, and in practice, it is often a function of the β_qs. When this occurs, we replace the unknown parameters with their MQL estimates to calculate an estimated variance-covariance matrix, $\hat{\mathbf{C}}_{\boldsymbol{\beta}}^{-1}$.

The large-sample normality of $\hat{\boldsymbol{\beta}}$ facilitates inference on the individual β_q-parameters. For instance, the Wald test of $\mathrm{H}_0{:}\,\beta_q = 0$ uses the statistic $W_{\mathrm{calc}} = \hat{\beta}_q/se[\hat{\beta}_q]$, where $se[\hat{\beta}_q]$ is the square root of the qth diagonal element of $\hat{\mathbf{C}}_{\boldsymbol{\beta}}^{-1}$. Under H_0, the large-sample distribution of W_{calc} is standard normal: $W_{\mathrm{calc}} \mathrel{\dot\sim} \mathrm{N}(0,1)$. Inferences on β_q follow from this; for example, the two-sided P-value for testing H_0 is $P \approx 2\{1 - \Phi(|W_{\mathrm{calc}}|)\}$. Reject H_0 when this value drops below some α, say $\alpha = 0.05$ or $\alpha = 0.01$. Alternatively, a large-sample $1-\alpha$ Wald interval on β_q is simply $\hat{\beta}_q \pm \mathfrak{z}_{\alpha/2}se[\hat{\beta}_q]$. Other constructions, such as tests based on the quasi-likelihood scores, are also possible; see, for example, Breslow (1990).

If the dispersion parameter φ is unknown, concurrent estimation of it does not adversely affect our approximations for the distribution of $\hat{\boldsymbol{\beta}}$. Large-sample inferences may proceed as noted above. (Technically, the estimator $\hat{\varphi}$ must approach the true value of φ as $\sqrt{n}\rightarrow\infty$; cf. Moore (1986). In many instances, the MoM estimator satisfies this requirement.)

We should note that although the MQL approach does allow the investigator to reduce the number of assumptions necessary for parameter estimation and testing, it nonetheless requires that the mean-variance assumptions be accurate. That is, the functional relationships for the mean and variance represented via $\mu_i\{\boldsymbol{\beta}\}$ and $\mathcal{V}(\mu_i\{\boldsymbol{\beta}\},\varphi)$ must be essentially correct for the actual underlying effects they purport to model. This is especially true for $\mathcal{V}(\mu_i\{\boldsymbol{\beta}\},\varphi)$: if the variance function is misspecified, the MQL estimators for $\boldsymbol{\beta}$ and $\mu_i\{\boldsymbol{\beta}\}$ will not suffer greatly, but the associated variance estimate will be incorrect. (Technically, if $\mathcal{V}(\mu_i\{\boldsymbol{\beta}\},\varphi)$ is incorrectly specified, $\hat{\boldsymbol{\beta}}$ will still approach the true value of $\boldsymbol{\beta}$ as $n\to\infty$, but estimates of $\mathbf{C}_{\boldsymbol{\beta}}$ based on $\hat{\boldsymbol{\beta}}$ may not necessarily approach $\mathbf{C}_{\boldsymbol{\beta}}$.) Inefficient and possibly incorrect inferences on $\boldsymbol{\beta}$ may result. For settings where the form of the variance function may be in question, more robust estimators of variance for MQL models have been proposed; see, for example, Moore and Tsiatis (1991) for the special case of data in the form of counts or proportions, or Davidian (1990) for the general problem of variance estimation with nonnormal data.

2.7.2* Generalized estimating equations for correlated data

The MQL methodology from Section 2.7.1 can be extended to construct generalized sets of estimating equations, that provide a more robust approach to semi-parametric estimation and testing. As with the MQL model above, one begins with a set of observations whose means are modeled via some function of unifying parameters, $\boldsymbol{\beta}$. This mean specification forms the basis of a set of **generalized estimating equations** (GEEs) that estimate the $P < n$ components of the vector $\boldsymbol{\beta}$.

As proposed by Liang and Zeger (1986), the GEE method generalizes the MQL approach by viewing each independent observation as a vector of correlated measurements, as might be encountered when the data are **repeated measures** (Carr and Chi, 1992) from the same subject, or when the data are taken in a **longitudinal** fashion over time (Zeger and Liang, 1992). Write the observed vector for each independently sampled subject as $\mathbf{Y}_i$, a set of J_i correlated observations $[Y_{i1}\ Y_{i2}\ \ldots\ Y_{iJ_i}]$, $i = 1, \ldots, n$. This formulation allows each subject (i) the opportunity to have a different number of measurements (J_i). Then, for each $\mathbf{Y}_i$, model the correlations, $\mathrm{Corr}[Y_{ij}, Y_{ik}]$, among the observations as some parametric function $\rho_{ijk}\{\boldsymbol{a}\}$, where $\boldsymbol{a} = [a_1\ a_2\ \ldots\ a_M]$ is another vector of unifying parameters common across all observations $\mathbf{Y}_i$. The correlations are collected together for the ith subject into a **working correlation matrix**, $\mathbf{R}_i(\boldsymbol{a})$, of dimension $J_i \times J_i$. The diagonal elements of $\mathbf{R}_i(\boldsymbol{a})$ are 1 (since $\mathrm{Corr}[Y_{ij}, Y_{ij}]$ is 1), and the off-diagonal elements are the modeled functions $\rho_{ijk}\{\boldsymbol{a}\}$. If we also define the variance function for Y_{ij} as $\mathcal{V}(\mu_{ij}\{\boldsymbol{\beta}\},\varphi)$, the corresponding variance-covariance matrix for each $\mathbf{Y}_i$ is then of the form $\mathbf{V}_i = \{v_{ijk}\}$, where

$$v_{ijk} = \begin{cases} \mathcal{V}(\mu_{ij}\{\boldsymbol{\beta}\},\varphi) & \text{if } j = k \\ \rho_{ijk}\{\boldsymbol{a}\}\sqrt{\mathcal{V}(\mu_{ij}\{\boldsymbol{\beta}\},\varphi)\,\mathcal{V}(\mu_{ik}\{\boldsymbol{\beta}\},\varphi)} & \text{if } j \neq k \end{cases}$$

and where j and k range over $1, \ldots, J_i$.

Given these functional models for the mean, variance, and correlations, one constructs the GEEs

$$\sum_{i=1}^{n}\sum_{j=1}^{J_i}\sum_{k=1}^{J_i} \frac{\partial \mu_{ij}}{\partial \beta_q} v_i^{(jk)}(Y_{ik} - \mu_{ik}) = 0 \tag{2.24}$$

for every $q = 1, \ldots, P$. In (2.24), $v_i^{(jk)}$ denotes the (j,k)th element of $\mathbf{V}_i^{-1}$, the inverse variance-covariance matrix for the ith subject.

In most cases, solving the P equations (2.24) for the P unknowns in $\boldsymbol{\beta}$ requires computer iteration, although Carr and Chi (1992) give some special cases where the GEE estimator has a closed form. As with the MQL estimator in Section 2.7.1, the GEE estimator possesses a large-sample multivariate normal distribution: $\hat{\boldsymbol{\beta}}_{\text{GEE}} \stackrel{\cdot}{\sim} N_P(\boldsymbol{\beta}, \mathbf{D}_{\boldsymbol{\beta}}^{-1})$, where the matrix $\mathbf{D}_{\boldsymbol{\beta}}$ has (q,r)th element

$$d_{qr} = \sum_{i=1}^{n}\sum_{j=1}^{J_i}\sum_{k=1}^{J_i} v_i^{(jk)} \frac{\partial \mu_{ij}}{\partial \beta_q}\frac{\partial \mu_{ik}}{\partial \beta_r} \tag{2.25}$$

$(q,r = 1, \ldots, P)$. As with $\mathbf{C}_{\boldsymbol{\beta}}$ in Section 2.7.1, $\mathbf{D}_{\boldsymbol{\beta}}$ is analogous to the Fisher information matrix $\mathbf{F}_{\boldsymbol{\Theta}}$. Since the d_{qr} quantities in (2.25) generally will be functions of $\boldsymbol{\beta}$, we estimate them by replacing $\boldsymbol{\beta}$ with $\hat{\boldsymbol{\beta}}_{\text{GEE}}$, and write the estimated variance-covariance matrix as $\hat{\mathbf{D}}_{\boldsymbol{\beta}}^{-1}$.

The large-sample normality of $\hat{\boldsymbol{\beta}}_{\text{GEE}}$ facilitates construction of confidence intervals and/or hypothesis tests for $\boldsymbol{\beta}$. For example, a $1 - \alpha$ Wald interval on β_q is $\hat{\beta}_q \pm z_{\alpha/2} se[\hat{\beta}_q]$, where $se[\hat{\beta}_q]$ is found as the square root of the qth diagonal element of $\hat{\mathbf{D}}_{\boldsymbol{\beta}}^{-1}$.

An important feature of the GEE approach is that the large-sample properties of the estimator $\hat{\boldsymbol{\beta}}_{\text{GEE}}$ are not adversely affected if the form for the working correlation functions $\rho_{ijk}\{\boldsymbol{a}\}$ are misspecified. If $\rho_{ijk}\{\boldsymbol{a}\}$ is only approximately correct, $\hat{\boldsymbol{\beta}}_{\text{GEE}}$ will still approach $\boldsymbol{\beta}$ as $n \to \infty$. This gives the GEE approach great applicability. Unfortunately, the variance-covariance matrix for $\hat{\boldsymbol{\beta}}_{\text{GEE}}$ does not share this large-sample robustness; under a misspecified correlation function, the matrix $\mathbf{D}_{\boldsymbol{\beta}}$ is no longer a valid representation of $\mathrm{Var}[\hat{\boldsymbol\beta}_{\mathrm{GEE}}]$ (Liang and Zeger, 1986). To adjust $\mathbf{D}_{\boldsymbol\beta}$, we form a new matrix, $\mathbf{D}^*_{\boldsymbol\beta}$, as follows: begin with the quantities

$$A_q = \sum_{i=1}^{n}\sum_{j=1}^{J_i}\sum_{k=1}^{J_i} \frac{\partial \mu_{ij}}{\partial \beta_q} v_i^{(jk)}(Y_{ik} - \mu_{ik})$$

($q = 1, \ldots, P$), which are simply the generalized scores used for the GEEs in (2.24). Take the products $u_{qr} = A_q A_r$, and collect these u_{qr} terms together into a $P \times P$ matrix, say $\mathbf{U}$. Invert the matrix $\mathbf{U}$ and denote the elements of $\mathbf{U}^{-1}$ as $u^{(qr)}$. Then, the (s,t)th element of the adjusted matrix $\mathbf{D}^*_{\boldsymbol\beta}$ is $d^*_{st} = \sum_{q=1}^{P}\sum_{r=1}^{P} d_{qt} u^{(qr)} d_{sr}$ ($s,t = 1, \ldots, P$), where the d_{qr} values are from the original $\mathbf{D}_{\boldsymbol\beta}$ matrix in (2.25). The adjusted matrix $\mathbf{D}^*_{\boldsymbol\beta}$ formed from the elements d^*_{st} is called an **information sandwich** (Lin and Wei, 1989; Boos, 1992), since the $u^{(qr)}$-values are 'sandwiched' between information-like quantities from $\mathbf{D}_{\boldsymbol\beta}$. From these quantities, the large-sample variance-covariance matrix of $\hat{\boldsymbol\beta}_{\mathrm{GEE}}$ is found as $\mathrm{Var}[\hat{\boldsymbol\beta}_{\mathrm{GEE}}] = (\mathbf{D}^*_{\boldsymbol\beta})^{-1}$. With $\boldsymbol\beta$ replaced by $\hat{\boldsymbol\beta}_{\mathrm{GEE}}$, this is a valid estimator of $(\mathbf{D}^*_{\boldsymbol\beta})^{-1}$ even when the $\rho_{ijk}\{\boldsymbol a\}$ function is misspecified (although there may be a slight loss in efficiency).

Since the GEE estimator for $\boldsymbol\beta$ is robust to the correlation specification, it is common to employ very simple forms for $\rho_{ijk}\{\boldsymbol a\}$, such as the equicorrelated form defined for any $i = 1, \ldots, n$ as

$$\rho_{ijk}\{a\} = \begin{cases} a & \text{if } j \neq k \\ 1 & \text{if } j = k \end{cases}$$

for $-1 < a < 1$. This can simplify the calculations required to perform inferences on $\boldsymbol\beta$. If the actual values of $\boldsymbol a$ and/or φ are unknown, estimating them via MoM (or other) equations does not adversely affect the large-sample normality of $\hat{\boldsymbol\beta}_{\mathrm{GEE}}$, provided that the estimators for $\boldsymbol a$ and φ approach the true values of $\boldsymbol a$ and φ as $\sqrt{n} \to \infty$. For example, suppose the variance function is modeled simply as $\mathrm{Var}[Y_{ij}] = \varphi\mathcal{V}(\mu_{ij}\{\boldsymbol\beta\})$, and for $j \neq k$ take $\rho_{ijk}\{\boldsymbol a\} = a$ as constant. Then, moment equations for a and φ are

$$\hat{a} = \hat{\varphi}^{-1} \frac{\sum_{i=1}^{n}\sum_{j=1}^{n}\hat{R}_{ij}}{J_+ - P} \tag{2.26}$$

and

$$\hat{\varphi} = \frac{\sum_{i=1}^{n}\sum_{j=1}^{J_i-1}\sum_{k=j+1}^{J_i}\sqrt{\hat{R}_{ij}\hat{R}_{ik}}}{\sum_{i=1}^{n}\frac{J_i(J_i-1)}{2} - P}, \tag{2.27}$$

where $J_+ = \sum_{i=1}^{n} J_i$ is the total number of observations and $\hat{R}_{ij}$ is a generalized Pearson residual: $\hat{R}_{ij} = (Y_{ij} - \mu_{ij}\{\hat{\boldsymbol{\beta}}_{\text{GEE}}\})^2 / \hat{\varphi}\mathcal{V}(\mu_{ij}\{\hat{\boldsymbol{\beta}}_{\text{GEE}}\})$. To find the GEE/MoM estimates, iterate among the $P + 3$ estimating equations (2.24), (2.26) and (2.27) until convergence is achieved.

2.8 SUMMARY

In this chapter, methods are described for constructing statistics that estimate and test unknown parameters in basic sampling scenarios. When the parent distribution of the data is known completely, estimation proceeds by viewing the joint probability function of the data as a function of the unknown parameters. Called the likelihood function, this function is maximized to find the most 'likely' estimates of the unknown parameters, given the data. A variety of estimators and statistical tests can be considered using the fully parametric form of the likelihood function, and these are reviewed.

For cases where the fully parametric form of the likelihood cannot be specified, semi-parametric forms are proposed as alternative modeling paradigms. These include bootstrap resampling methods that operate from the empirical distribution of the data, and quasi-likelihood models that mimic the complete likelihood, but require specification of only the mean and variance of the observations. It is seen that the maximum quasi-likelihood estimators behave similarly to the full maximum likelihood estimators.

Semi-parametric generalizations are given also for the case where nonzero correlations are observed among the observations. As with the quasi-likelihood models, the generalized estimating equations (GEEs) developed from these models require specification of only the mean and variance, but now also require the correlation function(s) among the correlated measurements. Although computer-intensive, the GEE calculations produce parameter estimates for the mean response that can exhibit strong robustness to misspecification in the variance and correlation models, making them useful in a number of experimental settings.

EXERCISES

Unless otherwise indicated, all indices range over $i = 1, \dots, n$. (Exercises with asterisks, e.g. 2.14*, typically are more advanced, or require greater knowledge and understanding of probability distributions and/or calculus.)

2.1. Besides the sample mean, $\overline{X}$, name some other statistics discussed in this chapter. If possible, give their sampling distributions.

2.2. We saw in Section 2.1.1 that i.i.d. stands for 'independent and identically distributed.' Can you envision an application (from your own field of study, as appropriate) where observations are not

a. identically distributed?

b. independent?

Explain.

2.3. We saw in Section 2.1.1 that if $X_i \sim$ i.i.d. $N(\mu,\sigma^2)$, then $\overline{X} \sim N(\mu,\sigma^2/n)$. What is the standard deviation of X_i? How does this differ from the standard error of $\overline{X}$, $se[\overline{X}]$?

2.4. Assuming that the sample size is sufficiently large, the central limit theorem implies that for any random sample of size n from some parent distribution with mean μ and (finite) variance σ^2, $\overline{X} \stackrel{.}{\sim} N(\mu,\sigma^2/n)$. How is this different from the statement about the sampling distribution of $\overline{X}$ in Exercise 2.3? Which statement is stronger?

2.5. In the following settings, which population parameter is being estimated?

a. The estimator is an observed sample mean $\overline{X}$.

b. The estimator is an observed sample variance s^2.

c. The estimator is an observed sample standard deviation s.

2.6. Let $X_i \sim$ i.i.d. $N(\mu,1)$. Derive the method of moments estimator for μ.

2.7. Let $X_i \sim$ i.i.d. $N(\mu,\sigma^2)$. Derive the method of moments estimators for μ and σ^2. Apply your results to the following data on weight increases (in grams) for $n = 5$ female mice after 16-day exposure to the environmental toxin 2-mercaptobenzothiazole (note that a weight loss would be viewed as a negative increase): $X_1 = 1.09$, $X_2 = 1.33$, $X_3 = 0.33$, $X_4 = -0.19$, $X_5 = 0.49$.

2.8. Let $Y_i \sim$ i.i.d. $\mathcal{Poisson}(\mu)$. Derive the method of moments estimator for μ. Apply your results to both of the following data sets on the numbers of marine species, Y_i, identified in a cluster of $n = 6$ sediment samples from a polluted harbor (Green *et al.*, 1993):

a. number of arthropods: $Y_1 = 2$, $Y_2 = 0$, $Y_3 = 0$, $Y_4 = 1$, $Y_5 = 2$, $Y_6 = 0$.

b. number of mollusks: $Y_1 = 66$, $Y_2 = 84$, $Y_3 = 58$, $Y_4 = 3$, $Y_5 = 7$, $Y_6 = 18$.

2.9. In reproductive toxicity assays using the aquatic organism *Ceriodaphnia dubia*, a common measure for study is the total number of *C. dubia* offspring produced in three broods (Bailer and Oris, 1993). (In fact, this endpoint serves often as the basis of regulations on effluent release into aquatic ecosystems.) Suppose $n = 10$ animals are placed on test to establish a baseline for the particular laboratory running the assay, yielding the following offspring counts: 20, 15, 21, 17, 31, 21, 23, 18, 18, 19. Assume that these counts represent a random sample from a Poisson distribution with mean μ, and derive the method of moments estimator for μ. Further, derive the least squares estimator for μ.

2.10. Let $Y_i \sim$ i.i.d. $\mathcal{NB}(\mu,\delta)$, as given in (1.11). Derive the method of moments estimators for μ and δ. Apply your results to the following data on numbers of mutated bacterial colonies counted on $n = 3$ plates after gaseous exposure to the environmental toxin 1,3-butadiene: $Y_1 = 100$, $Y_2 = 92$, $Y_3 = 89$.

2.11. Let $U_i \sim$ i.i.d. $\mathcal{Exp}(\beta)$. Derive the method of moments estimator for β.

2.12. Let $U_i \sim$ i.i.d. $\mathcal{Gamma}(\alpha,\beta)$. Derive the method of moments estimators for α and β. Apply your results to the following data on mutation frequencies ($\times 10^5$) of a transfected bacterial gene in $n = 6$ samples from specially bred transgenic mice after exposure to the potential mutagen hydroxyurea (Piegorsch *et al.*, 1994): $U_1 = 2.257$, $U_2 = 8.183$, $U_3 = 12.880$, $U_4 = 6.133$, $U_5 = 8.112$, $U_6 = 6.020$.

2.13*. Let $X_i \sim$ i.i.d. $N(\mu,1)$. Construct the log-likelihood function for μ and from this:

a. Show that the maximum likelihood (ML) estimator for μ is the sample mean $\overline{X}$.

b. Find the Fisher information and the associated large-sample variance for the ML estimator, $\overline{X}$. How does this result compare to what the central limit theorem says about the distribution of $\overline{X}$?

2.14*. Let $Y_i \sim$ i.i.d. $\mathcal{Poisson}(\mu)$. Construct the log-likelihood function for μ, and from this:

a. Obtain the likelihood $\mathcal{L}(\mu \mid \mathbf{X})$ and the log-likelihood $\ell(\mu)$. Further, plot $\ell(\mu)$ versus μ and speculate on the value of the maximum likelihood (ML) estimate of μ based on this plot.

b. Obtain the score equation, $\partial \ell / \partial \mu = 0$, and show that the ML estimator for μ is the sample mean $\overline{X}$.

c. Find the Fisher information and the large-sample variance for the ML estimator, $\overline{X}$. How does this result compare with what the central limit theorem says about $\overline{X}$?

d. Apply your results to estimate the mean response for the marine species data sets in Exercise 2.8.

e. Apply your results to estimate the mean response for the *C. dubia* data in Exercise 2.9.

2.15*. Let $X_i \sim$ i.i.d. $N(\mu, \sigma^2)$. Derive the maximum likelihood estimators for μ and σ^2. Apply your results to estimate μ and σ^2 for the murine weight increase data in Exercise 2.7.

2.16*. Consider the following applications of the binomial sampling scenario illustrated in Example 2.1. For each, assume $Y \sim \mathcal{Bin}(N, \pi)$, and find the maximum likelihood estimator of the underlying proportion response, π. Also calculate the Fisher information and the associated large-sample standard error, $se[\hat{\pi}]$.

a. An experiment is conducted to study the carcinogenicity of laboratory mice to the environmental toxin 1,3-butadiene. At an inhalation exposure of 625 ppm, a random sample of $N = 49$ independent mice generates a count of animals with lung tumors as $y = 12$.

b. In part (a), a second random sample of $N = 50$ independent mice at an inhalation exposure of 1250 ppm generates a count of animals with lung tumors as $y = 23$.

c. A laboratory study is conducted to study induction of micronucleation in blood cells after exposure to the mutagen methyl methane sulfonate. One finds $y = 204$ micronucleated cells out of $N = 8000$ cells studied.

d. A laboratory experiment is conducted to study the genotoxic potential of the toxin ethyl carbamate in fruit flies. Of $N = 2197$ flies exposed to the toxin, $y = 117$ exhibit a lethal mutation.

2.17. Let $X \sim \chi^2(\nu)$. Use a computer program to find the following tail probabilities:

a. $P[X > 3.9]$ when $\nu = 1$.
b. $P[X > 12.4]$ when $\nu = 10$.
c. $P[X > 6.0]$ when $\nu = 2$.
d. $P[X > 36.2]$ when $\nu = 20$.
e. $P[X > 10.8]$ when $\nu = 4$.
f. $P[X \leq 36.2]$ when $\nu = 20$.

2.18. Let $T \sim t(\nu)$. Use a computer program to find the following probabilities:

a. $P[T > 3.8]$ when $\nu = 1$. b. $P[\,|T|\, > 3.8]$ when $\nu = 1$.

c. $P[T > 2.56]$ when $\nu = 10$. d. $P[T \leq 1.72]$ when $\nu = 20$.

e. $P[T > 2.56]$ when $\nu = 14$. f. $P[T > 1.72]$ when $\nu = 20$.

g. $P[\,|T|\, > 2.6]$ when $\nu = 10$.

2.19*. Let $Y_i \sim$ i.i.d. $\mathcal{NB}(\mu, \delta)$, as given in (1.11). Construct the log-likelihood function for μ and δ, and from this:

a. Give equations for the maximum likelihood (ML) estimators for μ and δ (Piegorsch, 1990).

b. Find the Fisher information and the associated large-sample variances and covariances for the ML estimators.

c. Apply your results to estimate μ for the mutagenicity data in Exercise 2.10.

2.20*. Let $U_i \sim$ i.i.d. $\mathcal{Gamma}(\alpha, \beta)$. Construct the log-likelihood function for α and β, and from this:

a. Give equations for the maximum likelihood (ML) estimators for α and β.

b. Find the Fisher information and the associated large-sample variances and covariances for the ML estimators.

c. Apply your results to estimate α and β for the mutant frequency data in Exercise 2.12.

d. Use the functional invariance property of the ML estimators to determine the ML estimator of the mean response, $E[U]$. Apply your result to the mutant frequency data in Exercise 2.12.

2.21*. Let $U_i \sim$ i.i.d. $\mathcal{Exp}(\beta)$. Construct the log-likelihood function for β, and from this:

a. Find the maximum likelihood (ML) estimator for β.

b. Find the Fisher information and the associated large-sample variance for the ML estimator.

2.22*. Let X_i ($i = 1, \ldots, n$) be a random sample of counts of surviving insect offspring in a experimental field plot after exposure to an environmental pesticide. Consider the Poisson–binomial mixture distribution to model X (i.e. the marginal distribution of X, number of surviving insect offspring, after mixing over all values of Y, total number of eggs laid by the insect), with p.m.f. given in Example 1.1. Derive maximum likelihood estimators for μ and π.

2.23*. A continuous distribution useful in certain spatial applications with three-dimensional directional data is the **Maxwell distribution**. We say $X \sim \mathcal{Maxwell}(\theta)$, with parameter $\theta > 0$, and with p.d.f.

$$f_X(x \mid \theta) = \left(\frac{2}{\pi}\right)^{1/2} \theta^{-3} x^2 \exp\left\{-\frac{x^2}{2\theta^2}\right\} I_{(0,\infty)}(x) .$$

Construct the log-likelihood function for θ, and from this:

a. Determine if this p.d.f. is a member of the exponential class from Section 1.4.

b. Give an equation for the maximum likelihood (ML) estimator for θ.

c. Find the Fisher information and the associated large-sample variance for the ML estimator.

d. The population mean for the Maxwell distribution is $\mathrm{E}[X] = \theta\sqrt{8/\pi}$. Use the functional invariance property of the ML estimators to determine the ML estimator of $\mathrm{E}[X]$.

2.24. Suppose $Y \sim \mathcal{Bin}(N,\pi)$. Viewing π as a success probability, it is common in some applications to calculate the **odds** of success $\theta = \pi/(1 - \pi)$. For example, one may wish to test if the odds are 'even', that is, if $\theta = 1$. (Notice that this is equivalent to $\pi = 0.5$.) The corresponding null hypothesis is $\mathrm{H}_0: \theta = 1$. How would you convert this into a null hypothesis with some quantity set equal to zero? (That is, what operations on θ produce a zero argument in the null specification?) What would you call the resulting parameter?

2.25*. Let $Y_i \sim$ i.i.d. $\mathcal{Poisson}(\mu)$. Based on your results from Exercise 2.14, construct a test of the hypotheses $\mathrm{H}_0: \mu = \mu_0$ vs. $\mathrm{H}_a: \mu \neq \mu_0$ using:

a. A Wald test.

b. A score test.

c. A likelihood ratio test.

2.26. Apply your results in Exercise 2.25 to the *C. dubia* data in Exercise 2.9, and test $\mathrm{H}_0: \mu = 18$ vs. $\mathrm{H}_a: \mu \neq 18$. Can you calculate the *P*-value associated with this test? Compare this result to that obtained from testing these hypotheses based upon the sampling distribution of the sample mean. How, if at all, do these results differ?

2.27*. Let $Y_i \sim$ i.i.d. $\mathcal{NB}(\mu,\delta)$. Assume δ is a known constant.

a. Derive a score test to test the hypotheses $\mathrm{H}_0: \mu = \mu_0$ vs. $\mathrm{H}_a: \mu \neq \mu_0$.

b. Repeat part (a) using a likelihood ratio test.

c. Perform a likelihood ratio test of $H_0: \mu = 100$ vs. $H_a: \mu \neq 100$ for the mutagenicity data in Exercise 2.10, assuming $\delta = 0.0075$. Set your Type I error rate to $\alpha = 0.01$.

2.28*. Let $U_i \sim$ i.i.d. $\mathcal{E}xp(\beta)$. Based on your results from Exercise 2.21, construct a test of the hypotheses $H_0: \beta = \beta_0$ vs. $H_a: \beta \neq \beta_0$ using:

a. A score test.

b. A likelihood ratio test.

2.29*. Let $U_i \sim$ i.i.d. $\mathcal{G}amma(\alpha, \beta)$.

a. Based on your results from Exercise 2.20, test the hypotheses $H_0: \alpha = \beta = 1$ vs. $H_a: \{\alpha \neq 1 \text{ or } \beta \neq 1\}$ using a score test.

b. Repeat part (a) using a likelihood ratio test.

c. Perform a likelihood ratio test of $H_0: \alpha = \beta = 1$ vs. $H_a: \{\alpha \neq 1 \text{ or } \beta \neq 1\}$ for the mutant frequency data in Exercise 2.12. At least approximately, what is the associated P-value?

2.30. Use the results from Example 2.4 to calculate a 95% likelihood ratio confidence interval for π in each of the data settings from Exercise 2.16.

2.31. Use the results from Example 2.4 to calculate a 95% Wald-based confidence interval for π in each of the data settings from Exercise 2.16. Compare the results with those from Exercise 2.30.

2.32. Calculate 95% confidence intervals for π in each of the data settings from Exercise 2.16, using both the F-equivalence method and the continuity-corrected score interval. Compare the results with those from Exercises 2.30 and 2.31.

2.33*. Let $Y_i \sim$ i.i.d. $\mathcal{P}oisson(\mu)$.

a. Using your results from Exercises 2.14 and 2.25 (as necessary), develop a $1 - \alpha$ confidence interval for μ based on a Wald test.

b. Repeat part (a) based on a score test.

c. Repeat part (a) based on a likelihood ratio test.

d. Calculate a 90% confidence interval for μ with the *C. dubia* data in Exercise 2.9, based on

(i) Wald-based methods.

(ii) The likelihood ratio.

(iii) The continuity-corrected score interval.

(iv) Bootstrap resampling.

e. What is the probability that μ is contained in each of the intervals you calculated in part (d)?

2.34. Let $U_i \sim$ i.i.d. $\mathcal{Exp}(\beta)$. Using your results from Exercises 2.21 and 2.28 (as necessary), develop a $1 - \alpha$ confidence interval for λ based on the following methods. Where possible, compare the result with the method given in Section 2.6.3.

a. A Wald interval.

b. A score interval.

c. A likelihood ratio interval.

2.35*. Return to Example 2.6 where maximum quasi-likelihood estimating equations were derived for overdispersed count data. The data are counts, Y_i, whose means are all assumed constant: $E[Y_i] = \beta > 0$, $i = 1, \ldots, n$. Instead of using the quadratic variance function in that example, consider the linear (or 'proportional') variance function $\mathcal{V}(\beta, \varphi) = \varphi\beta$ for $\varphi > 1$.

a. Find the quasi-likelihood estimating equation for β. Compare your result with that in Example 2.6.

b. Assume φ is unknown. Suggest a method of moments estimating equation for it as well.

c. Apply your results in parts (a) and (b) to the mutagenicity data from Exercise 2.10.

2.36*. Suppose one observes independent data in the form of proportions, Y_i/N_i $(i = 1,\ldots,n)$, where some form of overdispersion invalidates use of the binomial distribution model for Y. Assume that each Y_i has mean $E[Y_i] = N_i\pi$, so that each proportion is an unbiased estimator of the unknown parameter π. Consider the variance function $\text{Var}[Y_i] = \mathcal{V}(\pi, \varphi) = N_i\pi(1 - \pi)\{1 + \varphi(N_i - 1)/(1 + \varphi)\}$, as in Section 1.2.2 ($\varphi > 0$). For simplicity, write this as $\mathcal{V}(\pi,\rho) = N_i\pi(1 - \pi)\{1 + \rho(N_i - 1)\}$, for $\rho = \varphi/(1 + \varphi)$.

a. Find the quasi-score equation for estimating π.

b. Assume ρ is unknown. Suggest a method of moments estimating equation for it as well.

c. Apply your results in parts (a) and (b) to the following $n = 22$ proportions of rodent fetuses that die after maternal exposure to the toxin 1,3-butadiene (Piegorsch, 1994, Table 2):

$$\frac{0}{13}, \frac{0}{14}, \frac{0}{15}, \frac{0}{3}, \frac{2}{14}, \frac{0}{12}, \frac{1}{9}, \frac{4}{14}, \frac{0}{10}, \frac{0}{12}, \frac{0}{17},$$

$$\frac{0}{14}, \frac{1}{10}, \frac{0}{16}, \frac{5}{16}, \frac{1}{13}, \frac{0}{9}, \frac{1}{10}, \frac{0}{15}, \frac{0}{16}, \frac{2}{15}, \frac{0}{15}.$$

2.37*. Take a random sample Y_i, $i = 1, \ldots, n$, from $N(\mu, \sigma^2)$, with σ as a known positive constant.

a. Show that the log-likelihood may be written as

$$\ell(\mu) = -\frac{n}{2}\log(2\pi) - \frac{n}{2}\log(\sigma^2) - \frac{1}{2}\sum_{i=1}^{n}\frac{(y_i - \mu)^2}{\sigma^2}.$$

b. Suppose we write μ as some function of a unifying parameter, β, i.e. $\mu = \mu(\beta)$. Show that the derivative of $\ell(\mu)$ with respect to β takes the form

$$\frac{\partial \ell(\mu)}{\partial \beta} = \sum_{i=1}^{n}\frac{(y_i - \mu)}{\sigma^2}\frac{\partial \mu}{\partial \beta}.$$

c. Show that under this model, the result in part (b) is equivalent to the quasi-likelihood estimating equation in (2.21).

2.38*. Commonly, in linear regression analyses where some covariate x is associated with a response variable Y, Y is assumed normally distributed: $Y_i \sim N(\beta_0 + \beta_1 x_i, \sigma^2)$ for $i = 1, \ldots, n$. In many situations (especially biological contexts), however, nonnormal response models for Y are more appropriate. As an example, suppose $Y_i \sim \mathcal{Poisson}(\beta_0 + \beta_1 x_i)$ for $i = 1, \ldots, n$.

a. Use maximum likelihood techniques to obtain estimates of the intercept and slope of this Poisson regression model. It is sufficient to stop once the first derivative score equations are obtained.

b. Explain how a test of the significance of the relationship between Y and x might be obtained for the Poisson regression context.

c. Suppose $(b_0 + b_1 x_i) < 0$ for some x_i, where b_0 and b_1 are estimates of the slope and the intercept. What would this mean for the predictions from this model? Suggest another parameterization of the relationship between x and the mean parameter of the Poisson distribution that would not create such a problem.

3

Fundamental issues in experiment design

Whether conducting a laboratory experiment, an observational study, or some other kind of environmental study, an environmental scientist must recognize the importance of design issues and accord proper consideration to their study. Experiment design is a critical step in the formulation of any environmental investigation, and many modern texts are available that describe in detail important forms, structures, and facets of proper experiment design. Our goal is not to replicate these readily available details; rather, we will introduce basic design terminology and discuss briefly some of the fundamental aspects of experiment design that arise in environmental studies. For good introductions to experiment design, we refer the reader to Moore and McCabe (1993, Chapter 3) or Green (1979). For more technical discussions on design issues, see Hinkelmann and Kempthorne (1994), Kuehl (1994), or the classic text by Cochran and Cox (1957).

3.1 BASIC TERMINOLOGY IN EXPERIMENT DESIGN

We begin with a review of basic design terminology. An **experiment** is a controlled application, introduction, or intervention of some external stimulus on a set of individual sampling units. The sampling units are the basic material of the design: individual animals or human subjects; containers of cellular, microbial, or ecological material; field plots; cordoned or delineated sections of a marsh, pond or other ecosystem; etc. We often distinguish a controlled experiment from an **observational study**; the latter is a directed examination – possibly retrospective – of the effects of some stimulus on a group of subjects, but does not share the element of control that we assume is intrinsic to a true experiment.

We will assume that the sampling units are derived from a larger, possibly infinite, **population** of subjects. It is to this population that all statistical inferences are directed. This is an important characterization: proper identification of the study population is the first essential step in any scientific investigation. The investigator must ask: to or for whom do I wish to describe environmental effects and draw inferences?

Example 3.1 Chromosomal damage
Laboratory studies of chromosome damage after exposure to some toxic pollutant often employ *in vitro* cellular cultures. The damage is identified after exposing the cells to the toxin and examining their chromosomes for aberrations or other detrimental effects. A common material for such a study is Chinese hamster ovary cells (Galloway *et al.*, 1985; Margolin *et al.*, 1986). Technically, the population in such a study is that of all Chinese hamster ovary cells which can be cultured, and all inferences are directed solely to this population. If inferences are to be made to cells from a different species, or to cells *in vivo* or in another study condition, then the experiment must employ material from those different species, different conditions, etc. For instance, if inferences to humans are intended, then human cells should be studied directly. Further, if the study goal is to make *in vivo* inferences about chromosomal aberration in humans, then, technically, the investigator must study human subjects exposed directly to the toxin, and not, for example, their cells exposed in culture. For direct *in vivo* exposure, of course, ethical considerations prohibit direct experimental human exposure. Indeed, for outcomes involving human exposures, chromosomal studies often take on an observational nature, rather than a controlled experimental nature; see, for example, Galloway *et al.* (1986).

Attempting to make inferences about toxic *in vivo* human chromosomal response based on data from *in vitro* cell cultures is a form of **extrapolation** from one study population to another. In general, simple extrapolation is a weak basis for scientific inquiry, and the current example illustrates this: few would expect chromosomal responses of hamster cells in culture to predict accurately the complex *in vivo* response of cells and their chromosomes inside the hamster body, or, for that matter, inside the human body. (There are, however, approaches in **quantitative risk assessment** available for using data from one population to make scientific inferences to a more complex population; see Portier (1989).) ❑

3.1.1 Local control

Virtually all experiments or observational studies direct attention to some introduction or intervention of external stimuli on a population's members. Typically, we refer to the particular stimulus of interest as the **treatment**, particularly in toxicity experiments or clinical studies.

The goal of a proper experiment design is to eliminate any external variables unrelated to the treatment that may systematically or deterministically affect the experimental outcome(s). Common sense provides guidance for many of the strategies used in eliminating external variables. Kuehl (1994) refers to this elimination as an attempt to achieve **local control** in order to increase the sensitivity of an experiment and better detect systematic, treatment-related effects.

The basic principle underlying local control is that the experimental units should be as similar as possible before treatment intervention. For example, placing fish on test that are sampled from the same stream, during the same season, and from the same species size class helps provide assurance that the animals are from the same population (i.e. the same stream), with similar metabolisms (temperatures frequently vary with season and metabolism frequently varies with temperature), and are physiologically similar (since the same size class may have similar physiological features: similar fat content, muscle mass, etc.). Of course, one must also monitor closely the experimental conditions and specimen preparation. Proper local control requires that all experimental conditions must be as similar as possible, in order to avoid **confounding** the inferences by external factors such as temperature, humidity, vibration, etc.

Example 3.2 Local control in an aquatic toxicity study
Suppose a study is conducted to evaluate the toxicity of copper on aquatic organisms (Weber *et al.*, 1989), such as fathead minnows, *Pimephales promelas*. Fish are placed into one of two types of aquaria: one with a relatively high copper level and a second with a standard, background level of copper. The study's goal is to compare the mortality experience of these fish between exposure levels, as an indicator of responses to environmental pollution in the wild.

For this experiment, local control is achieved at a number of levels. First, the water chemistry of the aquaria must be monitored closely so that the only systematic differences between the aquaria are related to copper levels. If, for example, the water hardness or temperature differed substantively between the two types of aquaria, then any differences observed in the mortality experience between fish in the two aquaria may be attributable to those external factors, and not to differences in copper toxicity.

Second, this experiment should be conducted with essentially equal numbers of fish in each type of aquarium, or equal fish densities if the aquaria are of different sizes. Higher density or loading of fish in a static exposure system such as an aquarium may increase fish mortality, due, for example, to higher waste production by the fish. (This may not be of concern if the exposures involved flow-through water conditions.) Here again, improperly controlled experimental conditions may underlie observed differences in mortality, confounding any attempts to assess differences in copper toxicity. ❑

Another important aspect of local control is consistency in the data recording process. That is, the process of measuring responses in the experimental units should be performed using the same chemical preparations, measurement devices, background conditions, etc. In addition, the investigators themselves must be careful not to bias measurements as they record them. This latter point may seem obvious and simple to attain; it is, in fact, overlooked in many environmental studies. For instance, suppose pathologists are recording the results from a study of tumor onset after laboratory animals are exposed to an environmental carcinogen. The pathologists read tissue sections on slides from animals in each of the dose groups. Suppose a pathologist believes that increased exposures to the carcinogen are likely to increase tumorigenicity. Although perhaps reasonable, this belief should be established from the experimental data and not vice versa. Nonetheless, the pathologist may examine slides from animals exposed to higher doses of the carcinogen with greater attention than those from animals exposed to lower doses, and possibly report greater frequencies of cancer in the high-dose groups wholly as a result of increased attention to those groups. This is not necessarily a conscious attempt to bias the observations, but it can still lead to serious biases in the recorded data, and hence spurious statistical conclusions.

A simple remedy is available for the problem of investigator bias: **blinding** or **masking** the information on exposure or treatment levels during data recording. There are a number of possible levels of this blinding. In a **single-blind study**, the experimental unit does not know which treatment it receives. (This level of blinding is more relevant in human clinical trials than in, say, rodent or fish toxicity experiments.) In a **double-blind study**, the single-blind conditions are extended so that the investigator does not know which subjects (slides, in the pathology example above) came from which treatment group. The data are coded so that treatment information is not immediately obvious. Lastly, **triple-blind studies** extend double-blinding to the data analyst. That is, double-blind data are analyzed using the codes from the double-blind readings (or even using new codes). The results are then decoded by an ostensibly unbiased supervisor to yield appropriate inferences regarding the treatment. (Tripleblinding is not practical, however, if the treatment represents some quantitative variable such as dose of an environmental toxin. If the quantitative features of the variable are to be employed in the statistical analysis, the analyst must be aware of this.)

In essence, blinding attempts to shield the observation process from biases that may be introduced by human predispositions. Although blinding is more common in clinical trials research, its implementation is just as relevant in environmental studies, as we see in the next example.

Example 3.3 Blinding in a sediment toxicity study
Sediment toxicity studies are important for evaluating the health of both freshwater and marine water systems. In freshwater sediment testing, a common experimental paradigm introduces tiny animals, *Hyallela azteca*, into containers whose sediment is contaminated with varying amounts of a potential environmental toxin. After a few days, the number of survivors in each sediment container are counted. Unfortunately, *H. azteca* ingests and processes the sediment in which it lives, resulting in coloration changes that allow the animal to present the same color as the surrounding sediment. This makes the counting of the survivors a difficult task. Expecting, a priori, more survivors to be found in low-toxin sediments, an investigator counting *H. azteca* survivors may focus more on the low-toxin containers, resulting in an observer bias toward greater survivorship at lower toxic exposures. One mechanism to avoid this bias is to 'blind' the data counter as to which container contains which sediment. (For example, have a co-investigator label the containers with noninformative codes and keep the codes hidden from the data counter until all measurements are completed.) □

3.1.2 Control groups for comparison

When external, systematic components known to influence the experimental subjects/material cannot be eliminated via local control efforts, they must be controlled via incorporation into the study design. For instance, when the experiment has a comparative nature, it is often important to record or observe responses from a group of subjects that have received no treatment or intervention. We call this a concurrent **control group**. Any experiment intended to make inferences about a treatment's or intervention's effect on a population of interest should always include a concurrent control. This group serves as a basis or foundation from which to make descriptive and inferential comparisons.

Although not always possible to achieve (for example, a no-intervention group in an environmental toxicity study with human subjects may not be ethically feasible), the goal in control group specification is for the corresponding treatment group(s) to differ in only one critical aspect from the control group: that being the exposure, treatment, intervention, etc. Comparisons between the treatment group and the control group then represent as purely as possible the treatment's effect on the experimental material.

There are a number of different ways to construct a control group. The simplest is to leave a group of subjects completely untreated or unaffected: an **untreated control**. This is perhaps the most common form of control group in environmental studies. For example, in sediment toxicity screening studies sediment samples from a suspect waste site comprise the 'treatment' group, while the comparison or control group corresponds to unrelated sediment that contains as little as possible of the toxic waste.

Three other forms of control group are possible in environmental studies. In the first form, the control group may reflect a well-established standard. Comparisons made to a **standard control group** are common with biological subjects, where use of an untreated group is not ethical, given the availability of an efficacious alternative. (This is quite common in biomedical studies of pharmaceutical preparations.)

A second form of control group is a **positive control**, which provides a condition that is expected to elicit as strong a response as possible. Use of a positive control may seem counter-intuitive, but it has an important consequence: it can identify if a complex experimental system is operating properly. For instance, is a mass spectrometer measuring correct chemical responses of the experimental material, or does it require recalibration? The positive control group provides a performance check on the experimental mechanisms and components expected not to vary from study to study.

A third form of control group is used when comparisons are made among subjects handled in a consistent, systematic fashion (except for exposure to the stimulus or intervention). In this case, a control group may receive the same vehicle used to administer the treatment, or receive the same intervention without an active ingredient, etc. We call this group the **sham control/placebo**, **placebo control**, or **vehicle control**, depending on the experimental context. Vehicle or sham controls are especially important when the action of administering an environmental stimulus to the subject (without toxic exposure) may induce a response. The vehicle control group helps in this case to identify or adjust for such nonexposure-related effects.

Example 3.4 Control groups in laboratory animal experiments
In laboratory carcinogenesis experiments, chronic exposure regimens are employed to study the long-term exposure effects of environmental carcinogens on animal subjects (Haseman, 1984; Robens *et al.*, 1989). Different routes of exposure are applied for different chemicals, depending on the expected context of human exposure to the compound. For most exposures via feed, water, or inhalation, it is appropriate to employ an untreated control group. In some experiments, however, it is important to control precisely the amount of chemical the animal receives. In these cases, the chemical is often delivered in a corn oil suspension via esophageal gavage.

Consider now specification of the control group in a carcinogenicity study using gavage exposures. In this case, an untreated control group differs from the treatment group not only in the presence or absence of the treatment, but also in the presence or absence of the corn oil vehicle. Such a comparison confounds the question of whether the corn oil, the carcinogen, or even the process of handling the animals for gavage is the source of any observed increase in carcinogenicity in the treatment group. Indeed, evidence has appeared that suggests corn oil may cause pseudo-carcinogenic responses in certain animals (Haseman *et al.*, 1985). This does not invalidate the material for use, but it does require the investigator to design carefully the type of control group(s) used for comparison with the treatment group(s).

We say the corn oil gavage is the 'vehicle' here, and use as a vehicle control group animals administered the corn oil gavage without any added chemical carcinogen. The base comparison group upon which to gauge the carcinogenic effects of the chemical is this concurrent vehicle control, since it represents animals exposed to the same systematic experimental conditions as those in the treatment group, *sans* the carcinogen exposure. Any significant differences observed between the two groups are then attributable ostensibly to the carcinogen.

It is also possible to employ multiple control groups, such as an untreated and a vehicle control (Haseman *et al.*, 1986; Margolin, 1987), in experimental or observational studies. Rosenbaum (1987) gives greater details on the use of multiple control groups. □

3.1.3 Treatments, factors, and levels

When there is no true treatment applied to the population under study, the term 'treatment' is used in a generic sense to indicate the stimulus or exposure regimen. A well-defined, designed source of explainable variability may more generally be referred to as an experimental **factor**. We say a factor is composed of individual **levels**. A factor whose levels differ only in a nominal fashion is called a **qualitative factor**, while a factor whose levels differ in quantitatively meaningful ways is called a **quantitative factor** or, in some settings, a **covariate** or **explanatory variable**.

For our purposes, it is assumed almost exclusively that factor levels correspond exactly to the levels of scientific interest. The factors are then viewed as **fixed effects**. An alternative to fixed-effect factors that we will discuss only in passing views a factor's levels as a sample of all possible levels for that particular factor. This is called a **random effect** (Neter *et al.*, 1996, Section 24.1); see Section 10.2. In a simple, fixed-effects, exposure-vs.-control experiment, the factor is presence or absence of the exposure, and we say it has two levels: one level is the presence and the other level is the absence of the exposure. In a multiple-level experiment of some quantifiable environmental stimulus, the factor is the particular stimulus, and the levels are the various quantified doses of that stimulus, including a zero-dose control.

Multi-factor experiments can involve complicated designs where the individual factors are **crossed**, so that each level of one factor is observed at all levels of the other factor(s). This produces a **factorial design**, and it is often of interest to study complex, multiple **interactions** between the various factors in such studies; see, for example, the discussion in Neter *et al.* (1996, Chapter 23). In factorial experiments, a treatment received by an individual experimental unit corresponds to a combination of levels of various factors.

Example 3.5 A factorial study of snake velocity
Suppose an ecologist is interested in the forward velocity of snakes under different light and temperature conditions. Exposure conditions of full light and shaded light are crossed with temperatures of 10, 20, and 30°C. There are two factors here: exposure condition and temperature.

In this example, light exposure is a qualitative factor while temperature is a quantitative factor. Light exposure possesses two levels and temperature possesses three levels. Table 3.1 illustrates the factorial design.

The factorial combination of levels from these two factors defines six unique treatment combinations: (shaded light, 10°C), (full light, 10°C), (shaded light, 20°C), (full light, 20°C), (shaded light, 30°C), and (full light, 30°C). The velocity of snakes in this study would then be observed at each of these six treatment combinations. ❑

Table 3.1 2×3 factorial design for snake velocity study

		Temperature °C		
		10	20	30
Light	Shaded	Shaded, 10	Shaded, 20	Shaded, 30
	Full	Full, 10	Full, 20	Full, 30

3.1.4 Blocking

In many experimental and observational settings, it is known that a specific factor will affect the population in a specific, deterministic manner. There is no interest in studying the effects of this factor, but because it is a known source of variability in the experiment, it must be controlled. In this case, we call the known source of variability a **blocking factor**, or simply a **block**. When left unrecognized, a blocking factor contributes in a systematic fashion to experimental error in the observations. By recognizing and incorporating blocking factors into the design, we improve the efficiency and sensitivity of the experiment to detect true differences among the treatment effects (Samuels *et al.*, 1994; Neter *et al.*, 1996, Chapter 27). In its simplest form, the design and analysis of an experiment that includes one or more blocks can proceed in roughly similar fashion to the design and analysis of any multi-factor experiment – in fact, the design for the simple paired t-test (Moore and McCabe, 1993, Section 7.1) is a form of blocked experiment – but very complex block designs are also possible (Federer, 1976). For a greater exposition on blocking factors, factorial designs, and other associated issues, we refer the reader to Cochran and Cox (1957) or Hinkelmann and Kempthorne (1994).

3.2 THE EXPERIMENTAL UNIT

We now introduce an important definition. For any study or experiment, the **experimental units** are the largest self-contained sampling units to which the external treatment, stimulus, or observational criterion is applied. When experimental units are humans, these study units are frequently called **subjects**. (Notice that we have used these terms throughout Section 3.1.)

Statistical analyses are based generally on the information provided by the experimental units, and it is the population from which these units are drawn to which inferences are directed. The identification of experimental units is related to the concept of deriving information from independently responding units. This distinction may seem obvious, but it often slips by unwary investigators. Here is a classic example arising in environmental teratology.

Example 3.6 Litter effects

In laboratory studies of the toxic effects of environmental agents, a common endpoint of interest is the damaging effect of the agent on developing embryos and/or fetuses of an exposed parent. The induction of damage to the developing embryo or fetus is called **teratogenesis**. When induced chemically, this is a form of **developmental toxicology**, and study of it poses clear environmental health concerns (Schwetz and Harris, 1993). In a typical teratogenesis study, pregnant female rodents (or **dams**) are administered the agent of interest during gestation. The animals are later sacrificed and examined to determine whether or not they exhibit exposure-related effects on a fetal endpoint of interest, such as losses in uterine implantation, or increases in mortality and/or malformations. Although the toxin is administered to the adult animal, the primary variables of interest involve fetal response. Thus, the question arises: what is the appropriate experimental unit upon which to base the statistical analysis?

As we noted above, the experimental unit is the largest sampling unit to which the treatment is applied. For the developmental toxicity paradigm described here, the pregnant dam is exposed to the toxin. Hence, the dam is seen as the experimental unit. In this case, fetuses sampled from an individual dam represent multiple observations on a single experimental unit, and it is likely that the per-fetus responses will be correlated (Haseman and Hogan, 1975). If, for example, the observations are simple dichotomous outcomes, say, whether each fetus died or not *in utero*, the per-fetus responses represent **correlated binary data**. The associated **intralitter correlation** – a form of **intralevel** or **intraclass correlation** – adds excess variability to the per-fetus observations; this is known as a **litter effect**. The resulting increase in variability is a form of **overdispersion**. Failure to recognize overdispersion can have adverse ramifications on statistical inferences; we will investigate a few of these ramifications in Example 6.9. ❑

In many experiments, the experimental unit is straightforward to characterize and identify. As the litter effect example shows, however, environmental investigators must avoid cavalier specification of the experimental unit. Here is another example:

Example 3.7 Salmonella *mutagenesis*
Consider again the laboratory determination of the toxic effects of some environmental pollutant. Now, however, suppose interest concerns genetic damage in the form of DNA mutation. Studies of toxicity-related **mutagenesis** possess numerous similarities to those of toxicity-related teratogenesis, as in Example 3.6, but from both biological and statistical perspectives, they also possess some inherent differences (Lovell, 1996).

Perhaps the most common mutagenesis experiment involves the *Salmonella* mutagenicity assay (Ames *et al.*, 1975), employing the bacteria *Salmonella typhimurium* to identify damage to DNA after exposure to environmental stimuli. Bacterial or microbial systems such as the *Salmonella* assay combine ease of use and lower costs (Zeiger *et al.*, 1985) with shorter time-scales for study (since, for example, bacteria reproduce rapidly, providing information on mutagenic response over a period of days or weeks, rather than months or years as in the case of multi-cellular animals).

The *Salmonella* assay proceeds by seeding a million or more bacteria on or into a microenvironment, such as a Petri dish or test tube. The bacteria are exposed to the toxin in order to study their mutagenic response. Here again, specification of the experimental unit requires careful regard: the largest sampling unit to which the treatment is applied is the Petri dish or test tube, *not* the individual bacterial cells. Hence the plate or tube is the experimental unit, and the observed mutant counts or mutant frequencies are taken at a per-plate or per-tube level. □

3.3 RANDOM SAMPLING AND RANDOMIZATION

Two essential ingredients of a properly conducted experiment are the use of a representative sample of experimental units from a particular population of interest, and the careful allocation of these units to the various levels of the treatments or factors. The first ingredient requires consideration of simple random sampling, and the second ingredient requires consideration of the randomization of experimental units to treatment conditions. We explain both these features in this section.

In most cases, it is impossible or prohibitive to observe responses from every member of a population under study, so instead we **sample** experimental units from the population. As suggested at the start of this chapter, the obtained sample should represent the study population of interest, and it is important not to overlook this fundamental feature of study design. For example, if a study of web spinning is conducted with spiders found in a garage in Ohio, would the sample of spiders necessarily be representative of spiders that live throughout the United States? Perhaps not; indeed, studies of spiders in southwest Ohio typically do not include conditions of limited moisture; the corresponding data provide inferences on web spinning that are of little value to arachnologists in, say, central Nevada.

Left unrecognized, haphazard sampling can introduce severe biases, limiting the scope of statistical inferences an experiment can provide. In order to avoid these sorts of systematic bias in the sampling process, one must sample the units randomly, by collecting a **random sample** of experimental units from an appropriate target population.

In its simplest form, we say a random sample is a sampling of experimental units from a larger population where each unit has an equal chance of being selected. This is a **simple random sample**. Simple random sampling carries two important repercussions. First, knowledge of the probability of inclusion in a sample provides the foundation for inferences back to the population. Second, if members of a population are being selected totally at random with equal probability, then we expect each sampled unit not to impact or influence any other sampled unit's observed response to the treatment, intervention, etc. This is a form of statistical **independence** among the observations, as introduced in Section 1.1. As noted there and in Section 2.3, independence is critical to the construction of statistical likelihoods that describe the mathematical properties of the unknown parameters. It is a crucial consequence of a properly implemented random sample. Details on simple random sampling, as well as more complicated sampling strategies, can be found in texts devoted to sample survey methodology, such as Scheaffer *et al*. (1996).

In previous sections of this chapter, we discussed the imposition of local control techniques to avoid the impact of extraneous environmental factors, blinding to avoid potential investigator bias, and blocking to control known sources of variability. Unfortunately, even with all of this care and effort, control of all possible extraneous factors often cannot be achieved. Simply put: experimental units will differ! The mechanism we employ to balance out uncontrolled systematic effects in the experiment is called **randomization**, the random assignment of experimental units to treatment conditions. The concept is simple: before imposition of some treatment or intervention, the experimental units are allocated to different treatment levels in such a way that each experimental unit has the same chance of being assigned to each treatment level. This process is appropriate for any designed experiment, such as laboratory studies of toxic effects described in the examples above. (If a study is observational, however, it is often difficult to apply complete randomization.)

Example 3.8 Randomization of fish in an aquatic toxicity study
Water quality is often assessed by evaluating the growth and survival of some harbinger species – such as fathead minnows, *P. promelas* – after exposure to environmental agents (Weber *et al.*, 1989). *P. promelas* represents an important trophic level in the freshwater ecosystem, and is a useful species to study for environmental effects. In addition, it can be readily maintained and studied under controlled conditions, making it particularly effective for laboratory investigations.

In these experiments, the fish are assigned to large aquaria in which they are exposed to a certain concentration of the agent. (Each aquarium contains a different dose of the agent.) To begin, let us imagine that fish were not randomized in their assignment to treatment concentrations. Suppose that the first 50 fish caught from a large holding tank are assigned to the zero-concentration control aquarium, the next 50 fish caught assigned to the lowest concentration, the next 50 fish caught to the next lowest concentration, etc., until the last 50 fish caught are assigned to the aquarium that receives the highest concentration. Now, suppose that slow fish are caught first and that, unknown to the investigator, the slowest fish are generally weaker and not as healthy as faster fish. As a result of the naive allocation of fish to aquaria, the sickest fish will be exposed to the lowest concentration of the agent, somewhat less sick fish will be exposed to intermediate concentrations of the agent, and this will continue in a systematic fashion until the healthiest fish are exposed to the highest concentration of the agent. The probable result of this assignment is that healthier fish will respond less severely to the toxic insult, while sicker fish may die more quickly, even after receiving limited exposures. This will grossly underestimate the agent's ability to induce toxicity in the fish.

Clearly, this nonrandom allocation scheme is foolhardy, and may produce misleading scientific results. A better scheme would employ complete randomization, where fish are assigned randomly to the various aquaria. (Recognize, by the way, that if hundreds of fish are to be studied, complete randomization may produce a logistical nightmare. Some compromise might be reached in this case, for instance, by randomizing nets of five fish to each concentration. In any case, however, some form of randomization is necessary to avoid sampling biases of the sort described in this example.) ❑

In general, to randomly assign experimental units to the different treatment levels (and, if a blocking factor is also considered, to the block levels, etc.) one must use a random device such as a random number table. From these, one assigns labels or identifiers to each experimental unit, and uses these labels to assign each unit randomly to a treatment group. Random number tables, and more details on randomization, are available in a variety of sources (Snedecor and Cochran, 1980, Table A1; Fleiss, 1981, Table A4; Moore and McCabe, 1993, Table B).

3.4 SAMPLE SIZES AND OPTIMAL ANIMAL ALLOCATION

When designing environmental studies, it is a basic experimental precept that one identify whenever possible the number of experimental units to be allocated to the treatment groups, so as to minimize resource requirements. In the aquatic toxicity illustration from Example 3.8, the investigator chose to employ 50 fish in each aquarium, perhaps because of limitations in material costs. The use of multiple independent experimental units in each treatment group is a form of **replication**, and the independent experimental units are called **replicates**. This is yet another critical principle in experiment design: by employing a large enough set of replicates and allocating them randomly to each treatment or study group, any random differences that exist among the similarly treated experimental units in each group are averaged out, increasing the statistical sensitivity to any systematic treatment effects.

In Example 3.8, the assignment of the same number of fish to each aquarium is called **equi-replication**. Equi-replication **balances** the experimental resources among all the treatment levels, and this balance can lead to simplifications in the statistical analysis (Neter *et al.*, 1996, Appendix D). We often design studies in environmental biology to exhibit balance, as with, for example, the need for density balance in Example 3.2.

We call the total number of experimental units employed in the study the **sample size**. It is often desired to determine sample size requirements for the study before generating the random sample(s). Too small a sample size can lower the experiment's ability to identify significant differences among the treatment groups, while too large a sample size wastes resources that may be applied in another study or to some other important environmental endpoint. Indeed, in cases where the study involves exposure to an environmental toxin, a secondary benefit of a priori sample size determination is the exposure of as few subjects as possible to the potentially damaging effects of the toxin (Muller and Benignus, 1992).

An additional concern in some environmental studies is the loss of experimental units due to external factors: so called **drop-outs**. These are not uncommon in field experiments or in observational studies with human subjects, but they can be controlled to some extent in laboratory experiments. Nonetheless, organisms can die prior to the end of a study, Petri dishes can be dropped inadvertently, etc., and some consideration to account for drop-outs may be necessary when selecting sample sizes. In addition, if the drop-out is related mechanistically to the response, for example, an animal dies from overt toxicity to an environmental agent before it can develop a cancer, then some adjustment is necessary in the statistical analysis. We illustrate such a case in Examples 8.7 and 11.4.

The probability of detecting treatment differences when such differences exist may be used to quantify the sensitivity of a statistical test procedure. This probability is called the **power** of the test procedure. Power is a function of the sample size: a properly constructed statistical test will increase in power as sample size increases. (This is an intuitively sensible feature!) By reversing this relationship, one can ask what sample size specification is required to generate a pre-selected level of power.

Besides power, sample size is a function of the size of the treatment difference (relative to the standard deviation σ) that we wish to detect. Large differences between treatments are easier to detect than smaller differences, so at a fixed level of power, the sample size decreases as the size of the desired detectable treatment effect increases. Sample size is also a function of the false positive (Type I) error rate, α. (Recall from Section 2.4 that a false positive error occurs when we conclude that a difference between treatment groups exists, when in fact it does not.) As α is lowered, the sample size required to detect a specific treatment effect increases.

Combining the three components (a) desired power, (b) detectable treatment difference(s), and (c) Type I error rate, we can calculate an optimum sample size for the experiment. To simplify the calculation, we often employ statistical tables, graphs, and formulas that provide guidelines for sample size selection. For each different form of experimental response, underlying treatment effect, desired level of power, etc., a different sample size table or formula is required, however. In the chapters below, we will describe sample size formulas for some selected experimental outcomes common in environmental biology. More generally, and for further sample size tables, a number of useful sources may be recommended, including Bock and Toutenburg (1991), Casagrande *et al.* (1978), Fleiss (1981, Chapter 3), Haseman (1978), Kastenbaum *et al.* (1970a; 1970b), and Lachenbruch (1992). For an interesting case study of the issues encountered in the calculation of sample size in the context of reproductive inhibition in aquatic experiments, see Oris and Bailer (1993).

3.5 DOSE SELECTION

A critical, yet often overlooked aspect of an experiment's design is the selection and specification of the actual treatment levels. When the experiment concerns the dose-related effects of some quantifiable stimulus such as an environmental toxin, it is important to select the dose levels with control and forethought. Here is an important, motivating example from regulatory toxicology.

Example 3.9 Observed-effect levels
Identification of the lowest or least potent concentration of a chemical at which toxicity is observed is an important issue when setting regulatory exposure standards for environmental toxins. The lowest concentration where a toxic effect is detected is called the **lowest-observed-effect concentration** (LOEC), or **lowest-observed-effect level** (LOEL); the LOEL is sometimes called the **least effective dose** (LED). Lying below the LOEL or LOEC is the highest concentration where no toxicity is observed: the **no-observed-effect concentration** (NOEC), or **no-observed-effect level** (NOEL). Related to these is the no-observed-adverse-effect level (NOAEL). These quantities are determined statistically by comparing each concentration with the zero-concentration control group (Yanagawa *et al.*, 1994). The issues of sample size and power discussed in Section 3.4 are important for the determination of LOEC or NOEC; for example, a design that allocates too few experimental units to the exposure levels may overestimate or even fail to identify the LOEC.

Also, an important concern raised when calculating an NOEC or LOEC is that they must, by definition, correspond to actual levels of the quantitative factor under study. Thus, the spacing of test doses is tied intimately to these summary values, and ultimately to the resulting exposure standards based upon them. Dose selection must be performed carefully, and include both a dense enough grid of dose levels and adequate resources at each level, to estimate LOEC and/or NOEC with sufficient sensitivity. These difficulties have led to concern that observed-effect levels are poor summary statistics for dose-response data (Chapman *et al.*, 1996; van der Hoeven, 1997).

Detailed methods for comparing different treatment levels – including multiple treatment-vs.-control comparisons – follow in Chapters 4 and 5. □

3.5.1 Nonlinear dose response

When selecting doses in a toxicity study, it is important to recognize that the experimental units may exhibit a dose response that is nonlinear, i.e. that deviates from a simple, easily interpreted straight line. For instance, a **threshold model** occurs when doses below a threshold point fail to elicit any substantive toxic response (a form of NOEC). It is critical to include dose levels above the threshold; if not, no data will be produced that illustrate the toxic effect and the statistical analysis will be pointless.

Of course, if threshold levels exist in a particular experimental setting, they are usually unknown. Hence the dose selection must proceed with some preliminary information on the nature of the dose response. Previous experiments, or data on chemical disposition and pharmacokinetic effects, may prove useful in this case (Reitz *et al.*, 1988).

In other settings, the dose response may be extremely nonlinear, and even inflect and change direction. Here are a few examples.

Example 3.10 Temperature optimality in spider web weight
Performance of animals in meeting their dietary requirements is related to a wide range of environmental factors. Suppose an ecologist is interested in the quality of webs, as measured by web weight, produced by spiders for catching prey. An important research question is whether an optimal ambient temperature exists for web production.

For estimation of the optimal temperature, the spacing of candidate temperatures ('doses' in the language of this section) is critical. For example, if all temperatures are positioned below or above the true optimal temperature, it will be difficult to estimate the optimum. (But not impossible: a design that misses the true optimal temperature still provides data for extrapolation beyond the range of observed temperatures. As in Example 3.1, however, extrapolation can lead to imprecise or inaccurate estimates of the parameter of interest, and it is not recommended.) Or, if the temperature range contains the optimal point but is also too widely spaced, the estimated optimal temperature may be far from the true optimum, and any interval estimates based on this estimate may be too wide for scientific value. □

Example 3.7 (continued) Salmonella *mutagenesis*
Consider again the problem of identifying DNA damage in the bacterium *S. typhimurium*, after exposure to some environmental toxin. The *Salmonella* assay is based on development of auxotrophic strains of the bacterium that are unable to synthesize histidine, an amino acid required for growth. This

production deficiency can be reversed into a production capability via point mutations at selected sites on the bacterial genome. DNA damage is then indicated by mutation of the bacteria from the auxotrophic state to the prototrophic, self-sustaining state. In effect, mutated cells will *grow* in a Petri dish containing only limited amounts of histidine; greater mutant yield at higher exposures to the environmental chemical suggests that mutagenesis increases with increasing dose. Observational accuracy of the assay is enhanced by use of a selective medium for the growth environment, so that only prototrophic mutant colonies may grow after exposure to the environmental toxin.

As an illustration, Table 3.2 displays mutagenic response data from a *Salmonella* assay of the chemical 1,3-butadiene. This environmental agent is an intermediate in industrial polymer production, and is also present in a variety of airborne compounds, including cigarette smoke, gasoline vapor, and automobile exhaust (National Toxicology Program, 1984). Of interest is its mutagenic potential after gaseous exposure. In a study conducted by the US National Toxicology Program (NTP), *S. typhimurium* bacteria were exposed to 1,3-butadiene via gaseous contact. Six dose levels were used: a zero-dose control and five test groups exposed to between 0.002 and 0.030 ppm of the chemical. Notice that at each dose level an equal number of (three) replicate plates is used. As we saw above, these plates represent the experimental units. The arithmetic averages of each per-dose triplet are given in the last column of the table, to give an indication of the nonlinear nature of the dose response.

The response in Table 3.2 exhibits an increase over the zero-dose control as the dose increases. Closer examination of the dose response indicates, however, a curious downturn at higher doses. This form of **nonmonotone dose response** is common in the *Salmonella* assay, and is observed often in other environmental mutagenesis assays (Margolin, 1985; Piegorsch, 1992). (We discuss methods for analysis of nonmonotone dose response in Sections 6.7 and 7.1.3.)

Table 3.2 *Salmonella* assay results for 1,3-butadiene in strain TA1535

Dose (ppm)	*Replicate mutants per plate*			*Average plate count*
0.000	20	31	27	26.0
0.002	100	92	89	93.7
0.007	147	123	178	149.3
0.014	216	170	181	189.0
0.020	176	154	183	171.0
0.030	154	153	149	152.0

Data from NTP (1984).

In *S. typhimurium*, the downturn phenomenon is driven by a number of possible mechanisms, the most common of which is a consequence of the experimental scheme employed to generate mutations. Since mutagenesis is identified by growth of the prototrophs on a selective medium, any other mechanism that hinders growth will compete with the endpoint of interest. Thus, for example, high exposures of the environmental stimulus can lead to cell death or perhaps chemical/threshold-induced increases in DNA repair. These **competing risks** may act to reduce the yield of mutated cells by killing or neutralizing the microbes before mutations can be observed. The result is a downturn at the higher doses, that is, an **umbrella response** (Mack and Wolfe, 1981). This nonmonotonicity motivates careful dose selection: if the point at which the downturn begins (called an **umbrella point** or an **umbrella index**) is not reached, then the full mutagenic potential of the toxin may not be observed or appreciated. Indeed, if only the initial, limited portion of the upturn in dose response is observed, and if this upturn is not strong enough to be significant in the statistical analysis, the toxin's mutagenicity may not be recognized. This is not an error in the statistical analysis, rather a failure to select the dose levels properly.

Contrastingly, extending the dose levels too far past the umbrella point is also undesirable, especially when the downturn drops below the control response. This is possible if cellular toxicity is the driving force behind the downturn: the bacteria literally die before they mutate, even if the mutation is spontaneous. Such a response pattern provides no information about the mutagenic effect, and wastes resources. In fact, an optimally designed experiment with the *Salmonella* assay aims to set the maximum dose level just past the umbrella point, so as to capture the full extent of the mutagenic increase while expending only limited resources on the downturn. □

Example 3.11 Nutrient effects preceding toxicity

Nonmonotone downturns in response are observed in many different toxicology studies. As mentioned earlier, a variety of tests are used to evaluate the impact of chemicals on freshwater ecosystems (Weber *et al.*, 1989). These impacts are not always detrimental: some chemicals are *nutritive* at low concentrations, enhancing growth, survival and/or reproduction of organisms exposed at these concentrations. As concentrations increase beyond the nutritive levels, however, toxic response may take over. An upward trend in the concentration response, followed by a toxic downturn pattern, results naturally from the organisms' physiological processing of the chemical. The effect is known as **hormesis** (Stebbing, 1982), and we discuss a model for it in Section 7.3.1. Here again, careful dose selection is required if precise estimation of the umbrella point is desired. ❑

3.5.2 Maximum tolerated doses in chronic exposure studies

In practice, doses at which downturns, thresholds, or other nonlinear phenomena occur are not known in advance. To select experimental dose levels to include these important points, investigators must employ some prior information on the dose response. This can include prior results on a similar chemical or exposure regimen, concurrent laboratory results with similar strains or species, or, as is common with chronic, long-term exposure studies, use of preliminary, sub-chronic studies to determine the appropriate dose range for the chronic study. From these various sources, one attempts to estimate the maximum dose at which the experimental units tolerate exposure to a toxic stimulus, express their maximal response before a downturn, pass a threshold event, or achieve some other critical milestone.

For example, in long-term, chronic exposure studies of laboratory animal carcinogenicity, dose selection centers around estimation of the **maximum tolerated dose** (MTD). This is the highest dose of the carcinogen the animals can accept that, when applied throughout the chronic exposure period, is not expected to shorten their longevity due to any noncarcinogenic toxic effects (Gart *et al.*, 1979; Robens *et al.*, 1989). Unfortunately, the MTD may be difficult to estimate from sub-chronic or preliminary exposure data, due to differences in the animals' responses between short-term and long-term exposures, *in vivo* metabolic changes in the environmental carcinogen over longer periods, cellular damage thresholds that are reached after extended exposures, etc. Modern quantitative methods aid in this estimation process, and some works have appeared that employ various parameters from the sub-chronic data to estimate the MTD and explore its interrelationship with carcinogenesis (Gombar *et al.*, 1991; Rosenkranz and Klopman, 1993; Haseman and Lockhart, 1994).

The US NTP, when estimating MTD for its two-year chronic exposure carcinogenesis studies in rodents, has employed two pre-chronic studies: a 14-day repeated dose study and a 90-day sub-chronic exposure study (Huff *et al.*, 1986). Large decreases (over 10%) at high doses in body weight gain or survivorship of animals usually suggest extreme toxicity, and the MTD is often taken as the highest dose below which these extreme effects are observed. We should warn, however, that this is an oversimplification of the effort and attention required to estimate an MTD. Indeed, multi-disciplinary coordination is required in MTD selection based on pre-chronic data; the required input includes pathology on the nature of any pre-neoplastic lesions observed at high doses, pharmacology and biochemistry on any metabolic or molecular changes the chemical undergoes after exposure or ingestion, toxicology on any chemical-related effects seen in similar species or with other endpoints such as environmental mutagenesis, and, of course, statistics to guide analysis and interpretation of the pre-chronic data. Indeed, modern toxicological testing requires interaction among all these disciplines in order to design a useful long-term study. Dose selection is an important consequence of this interaction.

Once estimated, the MTD is employed as the highest dose in the chronic exposure study. Lower doses are usually taken at fractional values of the MTD, for example, at one-half and one-fourth the MTD. Coupled with a zero-dose control, this would achieve a four-level dosage regimen. This sort of four-dose design is common in long-term, environmental carcinogenesis studies (Portier and Hoel, 1983), including many conducted by the US NTP. Sample sizes for such a study often involve balanced designs, with typically 50 animals randomly allocated to each of the four dose levels.

3.6 SUMMARY

In this chapter, basic concepts for the conduct and design of experiments are summarized. Control of extraneous variables, inclusion of comparison or control groups, identification of experimental units, sampling of observations, randomization of experimental units to treatment conditions, determination of appropriate sample sizes, and spacing of quantitative factor levels are some of the issues addressed. The motivation for considering these issues prior to conducting a study can be stated simply: no amount of sophisticated statistical analysis can salvage a poorly designed and/or poorly conducted experiment.

It is worth reiterating that experimental designs can take on more complicated forms than those illustrated within this chapter. Experimental factors can be much more complex than, say, simply crossing fixed factors with potential blocking variables. Common variations encountered in environmental studies include factors nested within other factors, where unique levels of one factor occur within levels of another factor, or repeated measures experiments where each experimental unit is measured repeatedly or over time. Readers interested in these advanced topics are encouraged to explore texts devoted specifically to the design and analysis of experiments, such as Hinkelmann and Kempthorne (1994) or Kuehl (1994).

EXERCISES

3.1. What is the population to which inferences are directed in the following studies? (Note that more than one population may be reasonable for these studies.)

a. Laboratory experiment to examine the ability of an environmental toxin to induce cancer in female mice, where the mice are exposed chronically (five times per week) to the same dose of the toxin over a 104-week period.

b. Field study in a coastal marsh of growth in marsh grass over different levels of water temperature and salinity.

c. Retrospective study of cases of lung cancer in navy ship builders exposed to asbestos while performing construction duties 50 years ago.

d. Laboratory experiment to examine the ability of an environmental toxin to induce chromosome loss (a form of *aneuploidy*) in cultured yeast cells, where the cultures are exposed once to the toxin.

e. Laboratory experiment to examine the ability of river sewage waste to induce death in small freshwater snails, where the snails are exposed three times to the same dose of the toxin over a 1-week period.

f. Field study to quantify the relationship between growth rates of fecal coloform bacteria in beach water and rainfall (measured at storm drain overflows) in a metropolitan area.

3.2. To what population are inferences directed in Example 3.2?

3.3. In each of the settings from Exercise 3.1, is there a treatment factor? If so, what is it?

3.4. In each of the settings from Exercise 3.1, can the study or experiment be conducted in a single- or double-blind fashion? If so, describe how.

3.5. In each of the settings from Exercise 3.1, describe an appropriate control group, if one exists.

3.6. The chemical ethylnitrosourea (ENU) is a potent inducer of mutations in a variety of animal and microbial systems. Suppose, in an experiment to study the mutagenicity of a related chemical, say ethyl methane sulfonate, a group of experimental units is exposed to ENU at a highly potent dose. Is this group a vehicle, untreated, or positive control? Why?

3.7. Describe which of the following control groups is appropriate for each of the following environmental carcinogenesis experiments in laboratory mice: an untreated, vehicle, or placebo control. Support your choice.

a. Exposure to 2-mercaptobenzothiazole via esophageal gavage.
b. Exposure to 1,2-epoxybutane via inhalation.
c. Exposure to L-ascorbic acid via feed.
d. Exposure to 1,4-dicholopropane via esophageal gavage.
e. Exposure to sodium fluoride via drinking water.
f. Exposure to acrylamide via epidermal contact application.
g. Exposure to benzene via intraperitoneal injection.

3.8. In a study of pesticide run-off toxicity, small aquatic crustacea are exposed to the following materials

(i) fresh water.
(ii) 3 mg/l pesticide in fresh water.
(iii) 6 mg/l pesticide in fresh water.

What is the treatment here? What are the corresponding levels? Is the treatment quantitative or qualitative?

3.9. Suppose, in the pesticide run-off experiment from Exercise 3.8, the exposure regimen was extended to include

(iv) 1 mg/l treated sewage in fresh water.
(v) 5 mg/l treated sewage in fresh water.
(vi) 1 mg/l treated sewage and 3 mg/l pesticide in fresh water.
(vii) 5 mg/l treated sewage and 3 mg/l pesticide in fresh water.
(viii) 1 mg/l treated sewage and 6 mg/l pesticide in fresh water.
(ix) 5 mg/l treated sewage and 6 mg/l pesticide in fresh water.

Have the treatment(s) changed for this experiment? If so, describe any new treatments and their corresponding levels.

3.10. An extension of the simple teratogenesis experiment discussed in Example 3.6 is a heritable toxicity experiment, where *male* mice are exposed to a toxin, then mated with untreated female mice. The resulting offspring are studied for inherited damage, ostensibly due to the toxin's effects on the male parent's germ cells. What is the experimental unit in this experiment?

3.11. Suppose, in the heritable toxicity experiment from Exercise 3.10, that the female mice were also exposed to the toxin. Does the experimental unit change? If so, what is it?

3.12. Allocate $n = 15$ experimental units evenly to three levels of one treatment (i.e. a balanced randomization scheme) using the following set of random numbers to perform the randomization:

78 54 28 62 72 16 58 75 88 03 58 84 11 58 45 68 26
41 19 42 92 06 34 09 77 62 37 03 38 59 68 94 16 54

3.13. From the set of random numbers given in Exercise 3.12, allocate $n = 20$ experimental units to a control and two levels of one treatment, with twice as many units in the control group as in the other two groups.

3.14. Sample size, n, statistical power (or sensitivity) of the test, β', and Type I error rate, α, are interrelated. Thus, we can write sample size as a function of the other two quantities, say $n = f(\alpha, \beta')$. In fact, since sample size increases as false positive errors decrease, we can write $f(\alpha, \beta') < f(\alpha', \beta')$, such that $\alpha' < \alpha$. We have suggested that as sample size increases, so does statistical power. Express this as an inequality similar to the one given above.

3.15. Referring to the sample size 'function' $f(\alpha, \beta')$ from Exercise 3.14, relate via inequalities the following quantities, where $\alpha' > \alpha$ or where $\beta'' < \beta'$:

a. $f(\alpha, \beta')$ vs. $f(\alpha', \beta')$.

b. $f(\alpha, \beta')$ vs. $f(\alpha, \beta'')$.

3.16. Associated with observed effect levels such as NOEC or LOEC (Example 3.9) is a more central measure, known as **median effective dose**, or ED_{50}. This is the dose level that produces a 50% response in the experimental material. For example, in a simple toxicity experiment it is the dose that leads 50% of the experimental units to die, an LD_{50} or **median lethal dose**. Would you expect ED_{50} to be larger or smaller than LOEC from the same dose-response experiment? Why?

3.17. Can you envision a *sensible* setting from your own field of study in which a dose response includes a threshold? Explain.

3.18. Can you envision a *sensible* setting from your own field of study in which a dose response is nonmonotone? Explain.

4

Data analysis of treatment-versus-control differences

Perhaps the most fundamental statistical inference made in environmental sciences is the comparison in response between a treatment group and a concurrent, independent, control group. As in earlier chapters, 'treatment' is used here in a generic sense, indicating some exposure, intervention, observational condition, etc., being applied to the experimental units. The response in the concurrent control represents the state of the experimental units after no treatment. As noted in Chapter 3, however, it may involve a vehicle control or sham exposure to allow for proper comparability between the two groups. Of interest in such a comparison is the identification of some significant difference between the treated experimental units, relative to the controls. If exhibited, this difference implies that the treatment has some substantive effect on the population(s) under study.

In general, we compare a treatment group with a control group via specific population parameters, for example, differences in means, $\mu_0 - \mu_1$, or in rates of response, $\pi_0 - \pi_1$, or in ratios of responses, π_0/π_1, etc. The goal is then to construct confidence limits on this new comparison parameter, or, equivalently, to test whether the observed difference is significantly different from 0 or whether the observed ratio is different from 1. These constructions will change, however, depending on the nature of the observations and their underlying statistical distributions. We describe in this chapter some standard approaches for settings common to environmental biology, focusing on the simplest case of a two-sample comparison.

4.1 TWO-SAMPLE COMPARISONS – TESTING HYPOTHESES

The comparison of a single treatment's effects with a concurrent control group is a special case of the general **two-sample** or **two-group** setting, where the quantitative features of two independent samples or groups are compared statistically. We assume the control group consists of observations Y_{0j}, where the index j represents independent control replicates: $j = 1, \ldots, n_0$. Similarly, the independent observations from the treatment group are Y_{1j} ($j = 1, \ldots, n_1$). Of interest is a test for any significant difference between the two groups, as represented by some difference in their underlying population parameters.

How we approach comparison of the two populations' parameters depends on the underlying distributions for the observations, Y_{ij}. For example, if the observations are **continuous measurements** such as distance, time, weight, or change (before and after treatment) in any of these quantities, we often assume that the normal distribution holds for Y_{ij}. Other distributional forms for continuous data are also possible; these include gamma distributions for responses exhibiting constant coefficient of variation (Chapter 8) or exponential or Weibull distributions for time-to-event data (Chapter 11).

If, however, the observations are **counts** of some phenomenon, then a normal distribution may not apply. A common alternative to consider for count data is the Poisson distribution. If these counts are bounded above by some known, fixed integer, say N_i ($i = 0,1$), or if they represent the sum of dichotomous indicators over N_i independent trials of a binary outcome, then the observed **proportions**, Y_i/N_i, may be of interest. In this case, the binomial distribution is often considered for the Y_is.

In all these cases, tests for significant differences between the two groups are based upon differences in the sample means or proportions. By comparing the observed differences with their sampling variability, we arrive at test statistics that assess the degree of departure from a null hypothesis of no difference between the two groups. Perhaps the best-known example of this is the t-statistic for the case of normal distribution sampling. We begin our presentation of significance tests for two-group differences by reviewing the details for this special, well-established setting.

4.1.1 Two-sample testing for the normal distribution – Equal variances

When the independent observations are recorded as continuous measurements, the normal distribution is a common model used to explain the random variability. Under normal sampling, we write $Y_{ij} \sim$ (indep.) $N(\mu_i, \sigma^2)$, $i = 0,1$; $j = 1, \ldots, n_i$. This is a simple model: the means μ_i are the parameters of interest, and the population variances for each group are assumed equal, although possibly unknown. This equal-variance assumption is critical for the construction of many test statistics used in the analysis of two (or more) groups' response. We refer to it as **variance homogeneity** or **homoscedasticity**.

For independent, homoscedastic, normally distributed observations, the t-statistic is used to test the null hypothesis $H_0{:}\mu_0 = \mu_1$ vs. the **two-sided alternative hypothesis** $H_a{:}\mu_0 \neq \mu_1$. The statistic can be obtained as a simplification of the likelihood ratio test of H_0 vs. H_a (Casella and Berger, 1990, Chapter 8), or from the sampling distribution of the difference in sample means (Moore and McCabe, 1993 Section 7.2). The statistic is

$$t_p = \frac{\overline{Y}_1 - \overline{Y}_0}{s_p \left(\dfrac{1}{n_0} + \dfrac{1}{n_1}\right)^{1/2}}, \tag{4.1}$$

where $\overline{Y}_i$ is the arithmetic mean of the n_i observations from the ith sample and $s_p = \sqrt{(s_p^2)}$ is a **pooled estimator** of σ based on

$$s_p^2 = \frac{1}{n_0 + n_1 - 2}\left\{\sum_{j=1}^{n_0}\left(Y_{0j} - \overline{Y}_0\right)^2 + \sum_{j=1}^{n_1}\left(Y_{1j} - \overline{Y}_1\right)^2\right\}.$$

(Use of the subscript 'p' emphasizes the pooled nature of the variance estimator.) Under H_0, $t_p \sim t(n_0 + n_1 - 2)$. Reject H_0 in favor of H_a when $|t_p| > t_{\alpha/2}(n_0 + n_1 - 2)$.

Recall from Section 2.4 that we can also use ***P*-values** to make decisions about H_0. For the pooled t-test the P-value is $P = 2\mathrm{P}\{t(n_0 + n_1 - 2) > |t_p|\}$. A statistically significant difference between μ_0 and μ_1 is implied when P drops below a pre-specified significance level, α. (Note that the order of the difference in the numerator of t_p is arbitrary for testing against $H_a{:}\mu_0 \neq \mu_1$ under either the rejection region or the P-value approach; however, recognition of the order, $\overline{Y}_1 - \overline{Y}_0$ or $\overline{Y}_0 - \overline{Y}_1$, is critical when calculating P-values for one-sided alternatives, as noted below.)

To test H_0 against the **one-sided alternative** that the mean in the treated group exceeds the mean in the control group, $H_a{:}\mu_0 < \mu_1$, one rejects H_0 in favor of H_a when $t_p > t_\alpha(n_0 + n_1 - 2)$. This is often of interest when prior expectations suggest that the treatment can only act to increase the mean response, for example, if exposure to an environmental toxin is thought to increase toxic response in the experimental units under study.

Notice the use of α instead of $\alpha/2$ in the t distribution critical point for the one-sided construction, reflecting the allocation of error probability to the entire upper tail of the t distribution. (Also, notice the lack of the absolute value in calculation of the test statistic; order is now important, since we specify the direction of the effect in $H_a: \mu_0 < \mu_1$.) The P-value in this case is $P = P\{t(n_0 + n_1 - 2) > t_p\}$. To test H_0 against the other one-sided alternative $H_a: \mu_0 > \mu_1$, reject H_0 when $t_p < -t_\alpha(n_0 + n_1 - 2)$. The P-value is then $P = P\{t(n_0 + n_1 - 2) < t_p\}$.

4.1.2 Two-sample testing for the normal distribution – Unequal variances

The various assumptions made as part of the construction of the t-statistic are not always met in practice. In particular, it is not uncommon to find that the underlying control group variance, σ_0^2, is not equal to the treatment group variance, σ_1^2. In this case, we say there is **variance heterogeneity** or **heteroscedasticity**, and we write $Y_{ij} \sim$ (indep.) $N(\mu_i, \sigma_i^2)$ $(i = 0,1; j = 1, \ldots, n_i)$. If the sample sizes are approximately equal ($n_0 \approx n_1$), then the pooled variance t-test from (4.1) is fairly robust to violations of the variance homogeneity assumption up to ratios of variances of about 4.0 (Miller, 1986, Section 2.3). Conversely, the t-test using the pooled variance estimator, s_p^2, is most adversely affected by variance heterogeneity if n_0 is much smaller than n_1, but σ_0 is much larger than σ_1 (or vice versa).

Note that some investigators may be tempted to precede a test of $H_0: \mu_0 = \mu_1$ using (4.1) with a test of $H_0: \sigma_0 = \sigma_1$, to assess variance homogeneity prior to conducting a test that requires equal variances; Moore and McCabe (1993, Section 7.3) give computational details. If it is truly unknown whether the variances are equal, and that the necessary assumptions of parent normality are met, this is a reasonable strategy. Unfortunately, common tests of variances are highly sensitive to the normal distribution assumptions (Bailer, 1989), and we do not recommend applying such an approach if it is unclear whether the normal parent assumptions are valid. If in this case there is concern about the validity of the homogeneous variance assumption, then t_p may not be appropriate and the alternative procedure described below should be considered.

When substantial variance heterogeneity is present, or if it is uncertain whether the variances are equal, a useful alternative to the pooled variance t-statistic (4.1) employs the per-group sample variances, s_i^2, in

$$t_{\text{calc}} = \frac{\bar{Y}_1 - \bar{Y}_0}{\sqrt{\dfrac{s_0^2}{n_0} + \dfrac{s_1^2}{n_1}}}. \tag{4.2}$$

(Recall that the 'calc' subscript is generic, indicating any test statistic fully calculable from the data.) The sampling distribution of (4.2) is no longer a t distribution, however. In fact, the distribution of t_{calc} is quite complex, and requires special tables for use (Fisher and Yates, 1957, Table VI). This distribution was first discovered by Behrens (1929), and later developed by Fisher (1939). As such, the problem of comparing two normal means when the unknown population variances are unequal is often referred to as the **Behrens–Fisher problem**.

A common alternative to the Behrens–Fisher sampling distribution is to find an approximation for the distribution of t_{calc} using the t distribution. Smith (1936) and later Welch (1938) found that $t_{calc} \dot{\sim} t(\nu_S)$, where for $u_i = s_i^2/(n_i - 1)$, $i = 0,1$,

$$\nu_S = \frac{(u_0 + u_1)^2}{\dfrac{u_0^2}{n_0 - 1} + \dfrac{u_1^2}{n_1 - 1}}. \tag{4.3}$$

(When ν_S is not a whole number, round *down* to the nearest integer.) The approximation for the *df* in (4.3) is often called a **Welch–Satterthwaite correction** for the heterogeneous variance setting, since it is a special case of a general method for moment-based *df* estimation described by Satterthwaite (1946). It is usually quite adequate (Wang, 1971), and it improves as the sample sizes get large. In practice, (4.3) may be recommended for use when $n_i > 5$ for both $i = 0$ and $i = 1$. With it, rejection regions and P-values of the same form as those described above may be employed for testing H_0, although it must be kept in mind that these provide only close approximations to the true statistical inferences. For example, an approximate test of H_0 against the one-sided alternative $H_a: \mu_0 < \mu_1$ rejects H_0 when $t_{calc} > t_\alpha(\nu_S)$; the P-value is, approximately, $P[t(\nu_S) > t_{calc}]$.

4.1.3 Two-sample testing for the normal distribution – Transformations

Another alternative method for handling variance heterogeneity is to apply a transformation of the Y_{ij}s that makes their variances at least approximately equal. This approach is especially useful when the heterogeneity is encountered in tandem with another assumption violation: nonnormality in the parent distributions of the original variables. In cases where Y_{ij} is only approximately normal, transformation can help drive the distribution closer to normal as well. For example, a common statistical distribution seen in environmental settings for positive-valued data, $Y_{ij} > 0$, is the **lognormal distribution**, where the transformation $V_{ij} = \log(Y_{ij})$ produces exactly a normal random variable. Then, exact calculations may proceed on V_{ij} using normal-based methods.

It is important to emphasize, however, that transformation of data in order to support inferences on some population effect must be motivated from previous experimental or scientific evidence. Unless determined a priori, transforms can be misused to inflate or mitigate observed significance in a spurious fashion. In such cases, it may be difficult to persuade scientists and regulators that the statistical results have any merit, and the entire experimental effort may go to waste.

In general, when the departure from normality is such that the parent distributions are skewed strongly to the left or right, a transformation can reduce the skew. Since the t-statistic is robust to departures from normality that exhibit symmetric variation about their means μ_i, a skew-reducing transformation may be a desirable precursor to calculation of t_{calc} or t_p. In fact, even if there is substantial skew in the parent distributions of the original variables, or if there is substantial variance heterogeneity, the t-test again is fairly robust when the sample sizes are at least approximately equal (Kendall and Stewart, 1967; Miller, 1986; Posten, 1992).

A useful parametric family of transformations that can often reduce skew, stabilize variance, achieve approximate normality, and is appropriate in certain environmental applications (Stoline, 1991; Razzaghi and Kodell, 1992), is the **power transformation**

$$V = \frac{Y^\lambda - 1}{\lambda}$$

(Box and Cox, 1964). This includes the logarithmic transform as a limiting case when $\lambda \to 0$. Other common transforms contained in this family are the square root $\lambda = 1/2$ (i.e. $V = 2Y^{1/2} - 2$), the quadratic $\lambda = 2$ (i.e. $V = \{Y^2 - 1\}/2$), and the reciprocal $\lambda = -1$ (i.e. $V = 1 - Y^{-1}$). Notice that these are actually linear combinations of the more common square root ($\sqrt{Y}$), quadratic (Y^2), and reciprocal ($1/Y$) transformations.

When λ is unknown, it can be estimated via iterative inspection: select a range of values for λ, and then select the value that stabilizes the variances, produces the most nearly normal transformation (indicated via a roughly linear normal probability plot), or maximizes the likelihood function for λ. Computer calculation is typically necessary for these complex calculations, however, and this may deter use when computer resources are limited.

Unfortunately, estimation of λ can induce substantial correlation between the estimate, $\hat{\lambda}$, and the transformed variates, complicating further statistical manipulations. A common solution in this case is to standardize the transformed variable using the per-group geometric means,

$$\tilde{Y}_i = \left(\prod_{j=1}^{n_i} Y_{ij} \right)^{1/n_i} .$$

This produces

$$V_{ij} = \frac{Y_{ij}^{\hat{\lambda}} - 1}{\hat{\lambda}\, \tilde{Y}_i^{\hat{\lambda}-1}} \tag{4.4}$$

At $\hat{\lambda} = 0$, use $V_{ij} = \tilde{Y}_i \log(Y_{ij})$. Extensions of (4.4) to improve the transformation's effect on the parent distributions are also available; see, for example, Kingman and Zion (1994) or the text by Carroll and Ruppert (1988). Once constructed, the transformed variables V_{ij} are employed as the basis for all t-test calculations.

A concern with these sorts of transformations, however, is that information on the transformed scale may not be interpretable, and that it may be more appropriate to perform statistical calculations in the original scale. While we do not wholly support this concern – for example, many transformations may be inverted to recover information on the original scale – we do agree that operations in the original scale can retain appeal in many instances. Analyses on transformed scales can cloud interpretation of the magnitude of an effect, as reflected, say, in a confidence interval containing 0, or containing 1, etc. Analyses on the scale of observation are more readily interpreted. Indeed, motivation for normalizing transformations was developed before computer resources and, in some cases, appropriate statistical methods were available for calculations with nonnormal data. At that time, there was a need to refer the statistical calculations to standard normal or t distribution tables. With the advent of modern statistical procedures and wide distribution of high-speed computing resources, however, this need is greatly reduced. In general, we will present our analyses of nonnormal data in the original scale and work with the original parent distributions. When advantageous to the statistical inferences, however, we will describe selected transformations for a priori use as well.

The test procedures presented here for testing equality of means from normal populations may serve well in other distributional contexts. If the parameter estimators under the nonnormal population are some form of sums of the Y_{ij}s, then various extensions of the central limit theorem often yield sampling distributions that are well approximated by the normal. Thus, a limiting form for the t-statistic can be constructed using these estimators that will be approximately normal when the sample sizes are large. (Since the large-sample reference distribution is normal, we denote the new statistic as Z or z_{calc}.) This result is the basis for the constructions we present in the next sub-section for testing differences between two binomial populations.

4.1.4 Two-sample testing for the binomial distribution – Large-sample tests

When the independent observations are recorded as discrete counts of some environmental phenomenon, the normal distribution is no longer a valid distributional model. Besides the obvious recognition that the data are no longer continuous (a basic characteristic of the normal distribution), it is also the case that variability of count data often changes as a function of the underlying population mean. This is a form of variance heterogeneity, and as noted above, variance heterogeneity can lead to invalid statistical inferences under standard t-tests.

For example, consider the case where the discrete observations are counts of how many independent experimental units in a group or sample respond to a specific environmental stimulus. We often view each unit's response as a success or failure after exposure to the stimulus. Hence the response is **dichotomous**, that is to say, taking on one of only two values: success/failure, yes/no, dead/alive, on/off, etc. For instance, the data might be numbers of cultured hamster cells that exhibit chromosomal damage after the cells were exposed *in vitro* to an environmental toxin. Each cell's response is categorized as either damaged or not damaged, and the outcome of interest is the probability that any one cell exhibits chromosomal damage. For homogeneously treated cells, it is expected that these probabilities are constant, that is, each cell in a homogeneously treated group or sample should possess the same underlying probability of exhibiting chromosomal damage.

In general, we refer to these individual probabilities as the **response probabilities** or, if no confusion exists over definition of a success, as the **success probabilities**. The response probability is estimated by the corresponding **observed proportion** of experimental units exhibiting the response. In a two-group (control vs. treated) experiment, it is of interest to compare the response probabilities between the control group and the treated group, in order to identify whether any significant differences are evidenced between them.

The simplest model for the statistical distributions of the successes Y_i in each group is the binomial: $Y_i \sim \mathcal{Bin}(N_i,\pi_i)$, $i = 0,1$. The π_is are the response probabilities; these are estimated by the observed proportions $p_i = Y_i/N_i$. The expected value of p_i is $\mathrm{E}[p_i] = \pi_i$ (i.e. p_i is unbiased for π_i), and the variance is $\mathrm{Var}[p_i] = \pi_i(1-\pi_i)/N_i$. Notice that the variances are functions of each π_i, and of the group sizes N_i. Thus, variance heterogeneity between the p_is will be present if either the π_is or the N_is differ between the two groups.

It is possible to formulate transformations that bring the data closer to normality and/or stabilize variance. A common stabilizing transformation for binomial proportions is the **arc sine transform** $V_i = \arcsin\{(Y_i/N_i)^{1/2}\}$ (Anscombe, 1948). The units of the transformed scale are angular (from 0° to 90° or from 0 to $\pi/2$ radians) associated with the sample proportion. A slightly more stable extension of the arc sine transform is

$$V_i' = \arcsin\left\{\left(\frac{Y_i + \frac{3}{8}}{N_i + \frac{3}{4}}\right)^{1/2}\right\}.$$

Under a binomial parent distribution, the distribution of V_i' is approximately normal, with variance $1/(4N_i + 2)$ no longer depending upon π_i. The approximation improves as the quantities N_i and $N_i - Y_i$ grow large. In this case, approximate inferences between the treatment groups may proceed using normal distribution methods on V_i'.

More generally, however, comparison of two groups in which the data are binomial counts $Y_i \sim \mathcal{Bin}(N_i,\pi_i)$, $i = 0,1$, is performed directly on the observed proportions. To do so, the null hypothesis, $H_0{:}\pi_0 = \pi_1$, is compared against the two-sided alternative $H_a{:}\pi_0 \neq \pi_1$. (We discuss one-sided alternatives below.) A simple statistic for testing these hypotheses that is analogous to the two-sample t-statistic from (4.2) is based on the difference in observed proportions $p_1 - p_0$, where $p_i = Y_i/N_i$, $i = 0,1$. To construct it, we begin with the sampling distribution of the estimated difference $p_1 - p_0$. As noted previously, p_0 and p_1 are unbiased estimators of π_0 and π_1, respectively. Thus, the expected value of their difference is $\mathrm{E}[p_1 - p_0] = \pi_1 - \pi_0$. Also, since the two samples are assumed independent, the variance of the difference is the sum of the individual variances, $\mathrm{Var}[p_1 - p_0] = \{\pi_1(1-\pi_1)/N_1\} + \{\pi_0(1-\pi_0)/N_0\}$. Recall that any binomial random variable can be viewed as the sum of independent Bernoulli random variables; here, we view each Y_i as the sum of N_i

independent Bernoulli variables, so that p_i is an arithmetic mean of N_i Bernoulli variables. From this, central limit theory suggests that as the N_is grows large, the sampling distribution of p_i will be approximately normal. Since linear combinations of normal random variables are themselves normal, we have $p_1 - p_0 \,\dot{\sim}\, N(\pi_1 - \pi_0, Var[p_1 - p_0])$.

At this point, we are tempted to standardize this quantity and use standard normal critical values for hypothesis testing. $Var[p_1 - p_0]$ is a function of the unknown parameters π_1 and π_0, however, so that dividing $(p_1 - p_0) - (\pi_1 - \pi_0)$ by its standard deviation, $\{Var[p_1 - p_0]\}^{1/2}$, produces a statistic that is also a function of the unknown parameters. To form a calculable statistic, we replace these unknown quantities with their unbiased estimators, that is, replace π_i with p_i. Then, the standard error of $p_1 - p_0$ is

$$se[p_1 - p_0] = \left[\frac{p_0(1 - p_0)}{N_0} + \frac{p_1(1 - p_1)}{N_1}\right]^{1/2} . \tag{4.5}$$

Now, for testing $H_0: \pi_0 = \pi_1$ vs. $H_a: \pi_0 \neq \pi_1$, notice that under H_0 the response probabilities share a common value, say π. Then, the variance of the difference $p_1 - p_0$, simplifies to $Var_0[p_1 - p_0] = \pi(1 - \pi)\{(1/N_0) + (1/N_1)\}$, where the subscript '0' on the variance indicates estimation under H_0. The corresponding standard error is based on a **pooled estimator** of the response probability

$$\bar{p} = \frac{N_0 p_0 + N_1 p_1}{N_0 + N_1} = \frac{Y_0 + Y_1}{N_0 + N_1} , \tag{4.6}$$

and substituting $\bar{p}$ for p_0 and p_1 in (4.5) yields

$$se_0[p_1 - p_0] = \left[\bar{p}(1 - \bar{p})\left(\frac{1}{N_0} + \frac{1}{N_1}\right)\right]^{1/2} . \tag{4.7}$$

The test statistic is then a form of score statistic; it is computed as the difference in observed proportions divided by the standard error of this difference under H_0, producing

$$z_{calc} = \frac{p_1 - p_0}{\left[\bar{p}(1 - \bar{p})\left(\frac{1}{N_0} + \frac{1}{N_1}\right)\right]^{1/2}} . \tag{4.8}$$

Under H_0, as the $N_i \to \infty$ the sampling distribution of z_{calc} in (4.8) is approximately standard normal, so we write $z_{calc} \,\dot{\sim}\, N(0,1)$. Hence, the associated score test rejects H_0 in favor of H_a when $|z_{calc}| > z_{\alpha/2}$, where $z_{\alpha/2}$ is the upper-$(\alpha/2)$ critical point from the standard normal distribution. The approximate P-value is $2P[|z_{calc}| > z_{\alpha/2}] = 2\{1 - \Phi(|z_{calc}|)\}$, where $\Phi(\cdot)$ is the standard normal c.d.f.

In many instances, it is possible to specify a priori that the only effect of interest in the alternative hypothesis is an increase in the treatment response, relative to the control response. In terms of H_a, this is $H_a{:}\pi_0 < \pi_1$. To test H_0 against this one-sided hypothesis, one mimics the approach taken with the t-test above and, using (4.8), rejects H_0 in favor of H_a when $z_{calc} > z_\alpha$. The corresponding P-value is $P \approx 1 - \Phi(z_{calc})$. Similarly, to test H_0 against the decreasing one-sided hypothesis $H_a{:}\pi_0 > \pi_1$, reject H_0 when $z_{calc} < -z_\alpha$. The corresponding P-value is $P \approx \Phi(z_{calc})$.

4.1.5 Two-sample testing for the binomial distribution – Sample sizes

The results in Section 4.1.4 can be used to select the sample sizes N_0 and N_1, prior to sampling. As discussed in Chapter 3, one asks what sample size specification is required for a pre-selected level of **power** to detect the effect given by H_a.

To test $H_0{:}\pi_0 = \pi_1$ vs. $H_a{:}\pi_0 \neq \pi_1$, suppose we wish to detect a difference between the control and treatment population of size $\delta > 0$, that is, $|\pi_1 - \pi_0| \geq \delta$ if H_a is true. Suppose also that we wish to detect this difference with probability β', when the significance level is set at α. (β' is the power of the test, cf. Section 2.4.) Then, appealing to a large-sample normal approximation for the estimated difference $p_1 - p_0$, and given some pre-supposed guess at the values of π_0 and π_1, the approximate required sample size is

$$N_0 = \frac{\left\{ z_{\alpha/2}\left[2\bar{\pi}(1-\bar{\pi})\right]^{1/2} + z_{1-\beta'}\left[\pi_0(1-\pi_0) + \pi_1(1-\pi_1)\right]^{1/2} \right\}^2}{\delta^2} \tag{4.9}$$

where $\bar{\pi} = (\pi_0 + \pi_1)/2$ (Fleiss, 1981, Chapter 3). For testing against one-sided alternatives, use z_α instead of $z_{\alpha/2}$ in (4.9).

Once calculated, N_0 from (4.9) specifies $N_1 = N_0$ if a balanced design is desired. For unbalanced designs where the ratio $\kappa = N_1/N_0$ is specified in advance, one can extend (4.9) to

$$N_0 = \frac{\left\{ z_{\alpha/2}\left[(1+\kappa)\bar{\pi}(1-\bar{\pi})\right]^{1/2} + z_{1-\beta'}\left[\pi_0(1-\pi_0)\kappa + \pi_1(1-\pi_1)\right]^{1/2} \right\}^2}{\delta^2\kappa}$$

and then take $N_1 = \kappa N_0$ (Bock and Toutenburg, 1991).

4.1.6 Two-sample testing for the binomial distribution – Continuity correction

The score statistic in (4.8) is simple to compute and, since standard normal tables are widely available, it is easy to evaluate. Unfortunately, a major weakness exhibited by (4.8) is that in small samples the standard normal approximation can be quite poor. This is due to the fact that the discrete binomial parent distribution for Y_i induces only a finite number of possible outcomes for z_{calc}. In effect, we are approximating the discrete small-sample distribution of z_{calc} with a continuous large-sample reference distribution (standard normal). When the N_i are fairly large – typical recommendations require $N_i p_i > 5$ and $N_i(1 - p_i) > 5$ for both $i = 0$ and $i = 1$ (Moore and McCabe, 1993, Section 8.2), or $0 \leq p_i \pm 2\{p_i(1 - p_i)/N_i\}^{1/2} \leq 1$ for both $i = 0$ and $i = 1$ (Wackerly *et al.*, 1996, Section 7.5) – the score statistic is usually sufficient for use. Alternatively, modern computer-intensive methods known as **exact tests** – similar to those we describe in Section 4.1.7 – are appropriate at any sample size (Mehta and Patel, 1991; Greenwood and Nikulin, 1996). When these large-sample conditions are not met or when resources for exact tests are unavailable, however, some correction for continuity is in order.

A traditional approach to continuity correction modifies (4.8) to make it exhibit less of a discrete nature: for testing $H_0:\pi_0 = \pi_1$ vs. $H_a:\pi_0 \neq \pi_1$, subtract the correction term $\{(1/N_0) + (1/N_1)\}/2$ from the numerator of z_{calc} and once again refer to the standard normal distribution (Snedecor and Cochran, 1980, Section 7.11). This is often called Yates's **continuity correction**, after Yates's initial work on the topic (Yates, 1934); more recently, see Yates (1984). The test statistic is usually written in terms of z^2_{calc}, based on squaring (4.8); it simplifies in this case to

$$z^2_{cc} = \frac{\left\{N_+ \left|\{Y_0(N_1 - Y_1) - Y_1(N_0 - Y_0)\}\right| - \frac{1}{2}N_+\right\}^2}{N_0 N_1 Y_+(N_+ - Y_+)} . \qquad (4.10)$$

The '+' subscript indicates summation over the index in which the '+' sign is active; for example, $N_+ = N_0 + N_1$. Using this short-hand notation, we can simplify many of our notational conventions. For instance, from (4.6) we can write the pooled estimator of π under H_0 as $\bar{p} = Y_+/N_+$.

Recall from Section 1.3.5 that the square of a standard normal random variable is distributed as $\chi^2(1)$. Thus, with Yates's correction we reject H_0 when z^2_{cc} exceeds an upper-α critical point from $\chi^2(1)$. The P-value is $P \approx P[\chi^2(1) > z^2_{cc}]$.

Much has appeared in the literature on validity of the continuity correction and (4.10) vs. use of the uncorrected statistic in (4.8). An extended review is beyond the scope of this text, however. We note only that the correction is best viewed as a way to improve the approximation of the exact, discrete P-value (Mantel and Greenhouse, 1968). It does not improve the approximation to the $\chi^2_\alpha(1)$ reference distribution, as might be thought at first.

If adopted for use, the continuity correction requires an adjustment to the approximate sample size formula from (4.9). For use with z^2_{cc}, and assuming a balanced design, begin with N_0 from (4.9), and correct it via

$$N_0' = \frac{1}{4} N_0 \left\{ 1 + \sqrt{1 + \frac{4}{N_0 \delta}} \right\}^2$$

(Casagrande *et al.*, 1978). Then, set $N_1' = N_0'$. For an illustration of the use of (4.9) and this continuity-corrected form when designing sediment toxicology experiments, see Bailer and Oris (1996).

4.1.7 Fisher's exact test for 2×2 tables

Yates's continuity correction from Section 4.1.6 is, in fact, a fairly accurate two-sided approximation to a more complex test of differences between two independent binomial populations. The more complex construction is best viewed by presenting the count data from the experiment in a tabular form, called a **contingency table**. For the simple case of comparing two binomial populations, the table separates the two populations into two separate columns, and then separates the successes from the failures into two separate rows. The result is a four-celled, **2×2 contingency table**. Using our notation from above, and including margins for the row totals and for the column totals, we have

	Control	Treatment	TOTAL
Success	Y_0	Y_1	Y_+
Failure	$N_0 - Y_0$	$N_1 - Y_1$	$N_+ - Y_+$
TOTAL	N_0	N_1	N_+

Viewed from the perspective of a 2×2 contingency table, a test of $H_0{:}\pi_0 = \pi_1$ vs. the one-sided alternative $H_a{:}\pi_0 < \pi_1$ can be based upon exact calculations; that is, without call to some large-sample approximation using a standard normal or χ^2 distribution. This approach is called an

exact test. The concept dates back to the theories of R.A. Fisher (1935b). Fisher advocated computing all possible tabular configurations of the observed data that could be generated under the condition that the row and column marginal totals are held fixed. We say then that the test is **conditional** on the observed pattern in the row and column margins. The corresponding conditional P-value is the probability of recovering a tabular configuration as extreme as or more extreme than that actually observed. If $P < \alpha$, reject H_0 in favor of H_a.

To calculate the conditional P, fix the row and column margins in the 2×2 table. Under this assumption, if one knew *and fixed* a value for any cell in the table, say $Y_0 = y$, then all the other values for the cells would follow from the fixed marginal totals. That is, given N_0, N_1, Y_+, $N_+ - Y_+$, and $Y_0 = y$, then $N_0 - Y_0$ must equal $N_0 - y$. But then Y_1 must equal $Y_+ - y$, and so $N_1 - Y_1$ must equal $N_1 - (Y_+ - y)$. Thus each possible 2×2 configuration can be indexed by one and only one of the cells of the table (here, by Y_0). This illustration gives some insight into use of the term **degrees of freedom**, since we say that this table and any statistics generated from it possess 1 *df*.

Formally, the P-value associated with the 1 *df* in a 2×2 table is the conditional probability of each possible tabular configuration being equal to or more extreme than that observed. As we saw above, each of these conditional probabilities may be indexed by Y_0. Under the binomial sampling assumption, conditioning on the marginal totals forces Y_0 to take a **hypergeometric distribution**:

$$\mathrm{P}[Y_0 = y \mid Y_+, N_0, N_1] = \frac{\binom{N_0}{y}\binom{N_1}{Y_+ - y}}{\binom{N_+}{Y_+}} \qquad (4.11)$$

for $y = 0, \ldots, \min\{N_0, Y_+\}$. The one-sided P-value is then $\sum \mathrm{P}[Y_0 = i \mid Y_+, N_0, N_1]$, where the summation is taken over the range $i = y_0, y_0 + 1, \ldots, \min\{N_0, Y_+\}$ and y_0 is the observed value of Y_0.

For computing two-sided P-values, the calculations are more complex. In large samples, however, Yates's continuity correction based on (4.10) provides an accurate approximation for the two-sided P, via $\mathrm{P}[\chi^2(1) > X^2_{cc}]$. Alternatively, one can modify the basic exact test, by identifying some two-sided measure of departure such as z^2_{calc}, and computing this measure for all possible 2×2 tables with the same row and column totals as those observed. The modified P-value is then the sum of the hypergeometric probabilities (4.11) that correspond to all those tables whose departure measures are larger than the observed table's.

Example 4.1 Adrenal cortex adenomas in B6C3F$_1$ mice exposed to 1,3-butadiene

To illustrate the calculations for Fisher's exact test, consider a long-term carcinogenicity study of 1,3-butadiene in B6C3F$_1$ mice performed by the US National Toxicology Program (NTP, 1993a). In each experimental group, 50 female mice were exposed via inhalation to the chemical. We focus on two experimental groups: a 0 ppm control group, and a 200 ppm exposure group. Interest exists in testing the alternative hypothesis that adrenal cortex adenomas are more likely to arise in the exposed group than in the control. The observed data are given in Table 4.1a.

The P-value for Fisher's exact test is the probability that we recover a table as extreme or more extreme in the direction of the alternative hypothesis compared to that in Table 4.1a. The probability for any such table is based upon the hypergeometric distribution in (4.11). In this example, 'more extreme' implies that higher tumor burdens are expected at higher exposures; however, the only configuration 'more extreme' than that in Table 4.1a is one in which no tumors are observed in the control group. Since the marginal totals in Table 4.1a must remain fixed, all other counts are adjusted appropriately. The resulting 2×2 table is given in Table 4.1b.

Table 4.1a Observed counts of tumor-bearing mice from 1,3-butadiene experiment

	Control	*200 ppm*	*TOTAL*
Adenoma present	1	3	4
Adenoma absent	49	47	96
TOTAL	50	50	100

Data from NTP (1993a).

Table 4.1b 'More extreme' configuration of counts for tumorigenicity data from Table 4.1a

	Control	*200 ppm*	*TOTAL*
Adenoma present	0	4	4
Adenoma absent	50	46	96
TOTAL	50	50	100

The P-value for the test of no difference in tumor onset versus an increased tumor burden in the exposed group is the sum of the hypergeometric probabilities from (4.11) corresponding to Table 4.1a and Table 4.1b:

$$P = \frac{\binom{50}{1}\binom{50}{3}}{\binom{100}{4}} + \frac{\binom{50}{0}\binom{50}{4}}{\binom{100}{4}}$$

$$= \frac{(50)(19\,600)}{3\,921\,225} + \frac{(1)(230\,300)}{3\,921\,225}$$

or $P = 0.309$. Since this value is greater than most typical values of α, we conclude there is insignificant evidence that the proportion of tumor-bearing animals in the 200 ppm group differs from the proportion of tumor-bearing animals in the control group. ❑

Tables for selecting sample sizes with Fisher's exact test may be found in several sources, such as Haseman (1978) or Fu and Arnold (1992).

When sample sizes grow very large, the calculations required for the exact P-value can become unwieldy. Computational algorithms are available in this case (Agresti *et al.*, 1979; Mehta and Patel, 1983), and computer implementation is recommended when appropriate resources are available. For example, the one-sided exact test may be performed in SAS, using PROC FREQ with the `/exact` option in the `tables` sub-command, or with the computer package STATXACT (Mehta and Patel, 1991). An alternative methodology for calculating one-sided P-values is described by Berger and Boos (1994), using manipulation of the score statistic in (4.8). Although based on unconditional probability statements, and therefore (technically) different from the conditional aspects of Fisher's exact test, the method has potential for increasing power to detect true differences between π_0 and π_1. Berger (1996) discusses these features, and notes the availability of associated computer code.

Before continuing, it is important to clarify that the discrete nature of the hypergeometric exact test does not ensure a false positive (Type I) error rate of exactly α, even when the test is performed nominally at α level significance (D'Agostino *et al.*, 1988). One can only construct the test to ensure that the true false positive rate, say α_e, is no larger than α; we call this a **conservative test** of H_0, drawing an analogy from the similar effect with confidence intervals. The test is exact for this underlying level of α_e, but is nonetheless conservative since typically $\alpha_e < \alpha$.

Example 4.2 Aneuploidy in hamster cell cultures

As an illustration of these methodologies for comparing two binomial proportions, consider the following data from an experiment in genetic toxicology. Of interest is a chemical's ability to cause genotoxic damage in the form of gain or loss of whole chromosomes. This condition is known as **aneuploidy**, and the processes and mechanisms underlying its induction are of interest to geneticists studying how environmental chemicals damage chromosomes (Whittaker *et al.*, 1989).

Laboratory data on aneuploidy induction often are collected from small mammals such as Chinese hamsters; recorded is the proportion, p_1, of cultured hamster cells that exhibit aneuploidy after the cells are exposed to some toxin. For example, in a study of the genotoxic effects of the synthetic estrogen diethylstilbestrol (DES), hamster cells were exposed *in vitro* for 16 hours to 5 μg/ml of DES, and compared to the response of cells receiving no exposure in a 16-hour control group (Dulout and Natarajan, 1987). The resulting data are presented in Table 4.2, in a 2×2 format.

Table 4.2 Aneuploidy induction in Chinese hamster cells

	Control	*5 μg/ml DES*	*TOTAL*
Aneuploid cells, Y_i	9	36	45
Nonaneuploid cells, $Y_i - N_i$	141	164	205
TOTAL	150	200	350

Data from Dulout and Natarajan (1987).

From the table, the observed proportions are $p_0 = 9/150 = 0.06$ and $p_1 = 36/200 = 0.18$. Of interest in the experiment is whether *in vitro* exposure to DES increases the probability that any individual cell will exhibit chromosomal aneuploidy. This is a one-sided testing question, and we set the hypotheses as $H_0: \pi_0 = \pi_1$ vs. $H_a: \pi_0 < \pi_1$.

The sample sizes in Table 4.2 are large enough that direct calculation of the hypergeometric exact test probabilities becomes tedious. For simplicity, we turn to the computer: Fig. 4.1 gives a short SAS program using PROC FREQ for Fisher's exact test with these data. (In Fig. 4.1, the symbols `@@` in the `input` statement allow for multiple data records per line in the input stream.) Figure 4.2 provides the output (edited for presentation).

In Fig. 4.2, the *P*-value for Fisher's exact test for an increase (the `Left` *P*-value) in aneuploidy induction due to DES exposure is $P = 5.52 \times 10^{-4}$. A strongly significant increase is suggested.

Alternatively, the large-sample conditions $N_i p_i > 5$ and $N_i(1 - p_i) > 5$ for both $i = 0$ and $i = 1$ are met for these data:

$$N_0 p_0 = 9 > 5 \quad \checkmark \qquad N_1 p_1 = 36 > 5 \quad \checkmark$$
$$N_0(1 - p_0) = 141 > 5 \quad \checkmark \qquad N_1(1 - p_1) = 164 > 5 \quad \checkmark$$

(Notice that these are just the four values from the 2×2 table, properly arranged.)

```
* SAS code for Aneuploidy data from Table 4.2;
data table42;
input  apoid trt y @@;
cards;
0 0  9    0 1 36    1 0  141    1 1  164
proc freq;
weight y;
tables apoid*trt/exact;
```

Fig. 4.1 SAS PROC FREQ program for Fisher's Exact test of data in Table 4.2.

```
                    The SAS System
                  The FREQ Procedure

                 TABLE OF APOID BY TRT
         APOID          TRT
         Frequency|
         Percent  |
         Row Pct  |
         Col Pct  |       0|       1|  Total
         ---------+--------+--------+
                0 |      9 |     36 |     45
                  |   2.57 |  10.29 |  12.86
                  |  20.00 |  80.00 |
                  |   6.00 |  18.00 |
         ---------+--------+--------+
                1 |    141 |    164 |    305
                  |  40.29 |  46.86 |  87.14
                  |  46.23 |  53.77 |
                  |  94.00 |  82.00 |
         ---------+--------+--------+
          Total        150      200       350
                     42.86    57.14    100.00

          STATISTICS FOR TABLE OF APOID BY TRT
   Statistic                        DF     Value         Prob
   ----------------------------------------------------------
   Chi-Square                        1    11.016        0.001
   Likelihood Ratio Chi-Square       1    11.916        0.001
   Continuity Adj. Chi-Square        1     9.971        0.002
   Fisher's Exact Test (Left)                        5.52E-04
                       (Right)                          1.000
                       (2-Tail)                      1.06E-03

   Total Sample Size = 350
```

Fig. 4.2 Output (edited) from SAS PROC FREQ: Fisher's exact test of data in Table 4.2.

Sincc these large-sample conditions are verified, we can consider application of the large-sample statistic z_{calc} from (4.8). To do so, we calculate the pooled estimator, $\bar{p}$, from (4.6) as $\bar{p} = (9 + 36)/(150 + 200) = 45/350 = 0.129$. Then, from (4.7), the standard error under H_0 is

$$se_0[p_1 - p_0] = \sqrt{(0.129)(0.871)\left(\frac{1}{150} + \frac{1}{200}\right)} ,$$

or simply 0.036. The resulting statistic is $z_{calc} = (0.18 - 0.06)/0.036 = 3.319$, with one-sided P-value approximated by $1 - \Phi(3.319) = 4.52\times10^{-4}$; this is fairly close to the exact value. Thus, both tests provide a powerful indication that exposure to DES induces an increase in aneuploidy induction in these cultured mammalian cells. □

4.1.8 Two-sample testing for the Poisson distribution

When the data take the form of unbounded counts, the binomial is not generally a useful form for the parent distribution of the observations. Instead, we turn to the well-established Poisson distribution from Section 1.2.5.

For the two-group setting, assume measurements Y_{ij} in both the control ($i = 0$) and treatment ($i = 1$) groups arise from Poisson parent distributions. For the two-sample Poisson model, we write $Y_{ij} \sim$ (indep.) $\mathcal{Poisson}(\mu_i)$, $i = 0,1$; $j = 1, \ldots, n_i$, and consider testing for significant differences between the mean parameters μ_0 and μ_1.

Recall that the variances of the observations under Poisson sampling are identically equal to the means: $\text{Var}[Y_{ij}] = \mu_i$, $i = 0,1$. As such, variance heterogeneity is often a problem, leading to invalid statistical inferences if standard t-tests are used. To correct this, one can transform the Poisson data to bring their parent distributions closer to normality and/or stabilize variance (Anscombe, 1948). The simplest of these is based on use of the square root; a number of variations exist (Sahai and Misra, 1992), including the simple **square root transform** $V_{ij} = 2\sqrt{(Y_{ij})}$ and the more advanced **Freeman–Tukey transform** $V_{ij} = \sqrt{(Y_{ij} + 1)} + \sqrt{(Y_{ij})}$ (Freeman and Tukey, 1950). Although these transformations can prove useful in practice, we will emphasize in this sub-section statistical calculations on the original scale of measurement, and formulate our calculations under the assumption that the Poisson model is valid for the data under study.

The null hypothesis of no difference between populations corresponds to $H_0: \mu_0 = \mu_1$; the two-sided alternative is $H_a: \mu_0 \neq \mu_1$. If we take $Y_{ij} \sim$ (indep.) $\mathcal{Poisson}(\mu_i)$, the sums $Y_{i+} = \sum_{j=1}^{n_i} Y_{ij}$ are again distributed as Poisson: $Y_{i+} \sim$ (indep.) $\mathcal{Poisson}(n_i\mu_i)$, $i = 0,1$. Then, we test H_0 via a conditional argument, in similar fashion to the exact test for binomial proportions described in Section 4.1.7. For the two-sample Poisson setting, the test conditions on the grand sum $Y_{++} = Y_{0+} + Y_{1+}$. This yields a conditional distribution for the treatment sum Y_{1+} as

$$\mathrm{P}[Y_{1+} = y \mid Y_{++}] = \binom{Y_{++}}{y} \pi^y (1-\pi)^{Y_{++}-y}, \tag{4.12}$$

where $\pi = n_1\mu_1/(n_0\mu_0 + n_1\mu_1)$ and $y = 0, \ldots, Y_{++}$ (Lehmann, 1959, Section 4.5). This is recognized as a binomial p.m.f., so $Y_{1+} \mid Y_{++} \sim \mathcal{B}in(Y_{++},\pi)$. Under $\mathrm{H_0}{:}\mu_0 = \mu_1$, we find $\pi = n_1/(n_0 + n_1)$, so testing $\mathrm{H_0}{:}\mu_0 = \mu_1$ vs. $\mathrm{H_a}{:}\mu_0 \neq \mu_1$ is equivalent to testing

$$\mathrm{H_0}\text{: } \pi = \frac{n_1}{n_+} \text{ vs. } \mathrm{H_a}\text{: } \pi \neq \frac{n_1}{n_+}, \tag{4.13}$$

where $n_+ = n_0 + n_1$, and both n_0 and n_1 are known, fixed values. Thus we have reduced the problem from one of testing between two Poisson means to one of testing the value of a single binomial probability parameter.

The two-sided conditional P-value for testing the hypotheses in (4.13) is

$$P = \mathrm{P_0}\left[Y_{1+} = y_1 \mid Y_{++}\right] + 2\left[\min\left\{\mathrm{P_0}\left(Y_{1+} < y_1 \mid Y_{++}\right), \mathrm{P_0}\left(Y_{1+} > y_1 \mid Y_{++}\right)\right\}\right], \tag{4.14}$$

where $\mathrm{P_0}[\cdot \mid Y_{++}]$ indicates calculation of (4.12) evaluated at the value of π under the null hypothesis, $\pi_0 = n_1/n_+$, and y_1 is the observed value of Y_{1+} (Vollset, 1993). This conditional P-value may be calculated exactly from the discrete probabilities based on (4.12). $\mathrm{H_0}$ is rejected in favor of $\mathrm{H_a}$ in (4.13) when $P < \alpha$. Table 4.3 gives selected values of (4.14) for the special case of balanced designs, $n_0 = n_1$, leading to $\pi = 1/2$. (Enter the table with $N = Y_{++}$ and $y = y_1$.)

For one-sided alternatives, the calculations simplify slightly. For example, testing $\mathrm{H_0}$ against an alternative with increasing rate of response, $\mathrm{H_a}{:}\mu_0 < \mu_1$, is equivalent to

$$\mathrm{H_0}\text{: } \pi = \frac{n_1}{n_+} \text{ vs. } \mathrm{H_a}\text{: } \pi > \frac{n_1}{n_+}.$$

The corresponding conditional P-value becomes

$$P = \mathrm{P_0}[Y_{1+} \geq y_1 \mid Y_{++}] = \sum_{k=y_1}^{Y_{++}} \binom{Y_{++}}{k} \left(\frac{n_1}{n_+}\right)^k \left(\frac{n_0}{n_+}\right)^{Y_{++}-k}.$$

Similar modifications occur for the alternative hypothesis of decreasing mean, $\mathrm{H_a}{:}\mu_0 > \mu_1$ (Sahai and Misra, 1992).

Table 4.3 Two-sided P-values for testing $\pi = 1/2$ for $Y \sim \mathcal{B}in(N,\pi)$, with total sample size N and observed value $Y = y$

	N										
y	*2*	*3*	*4*	*5*	*6*	*7*	*8*	*9*	*10*	*11*	*12*
0	0.250	0.125	0.062	0.031	0.016	0.008	0.004	0.002	0.001	*	*
1	1.000	0.625	0.375	0.219	0.125	0.070	0.039	0.021	0.012	0.006	0.003
2	0.250	0.625	1.000	0.687	0.453	0.289	0.180	0.109	0.065	0.039	0.022
3		0.125	0.375	0.687	1.000	0.727	0.508	0.344	0.227	0.146	0.092
4			0.062	0.219	0.453	0.727	1.000	0.754	0.549	0.388	0.267
5				0.031	0.125	0.289	0.508	0.754	1.000	0.774	0.581
6					0.016	0.070	0.180	0.344	0.549	0.774	1.000
7						0.008	0.039	0.109	0.227	0.388	0.581
8							0.004	0.021	0.065	0.146	0.267
9								0.002	0.012	0.039	0.092
10									0.001	0.006	0.022
11										*	0.003
12											*

	N										
y	*13*	*14*	*15*	*16*	*17*	*18*	*19*	*20*	*22*	*24*	*26*
0	*	*	*	*	*	*	*	*	*	*	*
1	0.002	0.001	0.001	*	*	*	*	*	*	*	*
2	0.013	0.007	0.004	0.002	0.001	0.001	*	*	*	*	*
3	0.057	0.035	0.021	0.013	0.008	0.004	0.003	0.001	*	*	*
4	0.180	0.118	0.077	0.049	0.031	0.019	0.012	0.007	0.003	0.001	*
5	0.424	0.302	0.210	0.143	0.096	0.064	0.041	0.027	0.011	0.004	0.002
6	0.791	0.607	0.454	0.332	0.238	0.167	0.115	0.078	0.035	0.015	0.006
7	0.791	1.000	0.804	0.629	0.481	0.359	0.263	0.189	0.039	0.043	0.019
8	0.424	0.607	0.804	1.000	0.815	0.648	0.503	0.383	0.210	0.108	0.052
9	0.180	0.302	0.454	0.629	0.815	1.000	0.824	0.664	0.405	0.230	0.122
10	0.057	0.118	0.210	0.332	0.481	0.648	0.824	1.000	0.678	0.424	0.248
11	0.013	0.035	0.077	0.143	0.238	0.359	0.503	0.664	1.000	0.690	0.442
12	0.002	0.007	0.021	0.049	0.096	0.167	0.263	0.383	0.678	1.000	0.701
13	*	0.001	0.004	0.013	0.031	0.064	0.115	0.189	0.405	0.690	1.000
14		*	0.001	0.002	0.008	0.019	0.041	0.078	0.210	0.424	0.701
15			*	*	0.001	0.004	0.012	0.027	0.093	0.230	0.442
16				*	*	0.001	0.003	0.007	0.035	0.108	0.248
17					*	*	*	0.001	0.011	0.043	0.122
18						*	*	*	0.003	0.015	0.052
19							*	*	*	0.004	0.019
20								*	*	0.001	0.006
21									*	*	0.002
22									*	*	*
23										*	*
24										*	*
25											*

* Asterisks indicate $P < 0.001$.

The calculations for exact conditional P-values become prohibitive, however, as Y_{++} grows larger than about 10. As a simple approximation, one can consider the (conditional) distribution of the maximum likelihood (ML) estimator of π: $p = y_1/Y_{++}$. In large samples, this ML estimator is approximately normally distributed with mean π and variance $\pi\{1 - \pi\}/Y_{++}$. The corresponding large-sample test statistic for (4.13) is

$$z_{\text{calc}} = \frac{\dfrac{y_1}{Y_{++}} - \dfrac{n_1}{n_+}}{\sqrt{\dfrac{1}{Y_{++}}\dfrac{n_1}{n_+}\left(1 - \dfrac{n_1}{n_+}\right)}} \ . \tag{4.15}$$

As $Y_{++} \rightarrow \infty$, $z_{\text{calc}} \mathrel{\dot\sim} \text{N}(0,1)$, so a large-sample approximation to the exact conditional P-value can be calculated. For the two-sided case, $P \approx 2\{1 - \Phi(|z_{\text{calc}}|)\}$ approximates (4.14); for the one-sided case, say against H_a:$\pi > n_1/n_+$, $P \approx 1 - \Phi(z_{\text{calc}})$ approximates the conditional P-value, etc. Sahai and Misra (1992) derive the same statistic for testing H_0 based on an unconditional large-sample argument. For the balanced case where $n_0 = n_1$, the exact small-sample distribution of (4.15) was tabulated by Sichel (1973).

Note that a similar relationship established the large-sample approximation to the statistic in (4.8). As with (4.8), the normal approximation to (4.15) represents a continuous reference distribution for a discrete-valued statistic; however, the approximation suffers from instability when Y_{++} is small. In order for the normal distribution to serve as a good reference approximation for (4.15), the data must satisfy large-sample conditions. We recommend $Y_{++}p = y_1 > 5$ and $Y_{++}(1 - p) > 5$ before turning to use of (4.15).

A continuity correction of similar form to that seen in Section 4.1.6 is possible for (4.15), and this can improve the approximation of the exact P-value. There is some tendency toward over correction, however; see Snedecor and Cochran (1980, Section 7.6) for greater details. We recommend the continuity correction only for very small Y_{++}, at which point the exact binomial probabilities from (4.12) or, in the balanced case, tabulated values (Sichel, 1973) may be used instead.

Example 4.3 Genetic toxicity to HI-6 dichloride

As an illustration of these methods for comparing Poisson populations, consider the following two-group study of genetic toxicity in cultured hamster ovary cells after exposure to the oxime HI-6 dichloride (Putman *et al.*, 1996). Data were recorded from the cell cultures as total number of mutant colonies per Petri dish, Y_{ij}, where $i = 0,1$ indexes the control ($i = 0$) and a 160 μg/ml HI-6 dichloride exposure ($i = 1$) group, and $j = 1, \ldots,$

10 indexes the plates examined in each group. Note that although the experimental material here is based on cultured hamster cells, as in Example 4.2, this assay generates data with structure more similar to the *Salmonella* mutagenesis assay discussed in Example 3.7.

We assume the Y_{ij}s are distributed as independent Poisson random variables with means μ_i, $i = 0,1$. Notice that the design is balanced, with $n_0 = n_1 = 10$. The data appear in Table 4.4.

Table 4.4 *In vitro* genetic toxicity of HI-6 dichloride in Chinese hamster ovary cells

Experimental group	*Mutant colonies,* Y_{ij}					*Total,* Y_{i+}
Control	2	2	0	1	2	
	1	0	2	1	0	11
160 μg/ml HI-6 dichloride	2	1	3	1	4	
	4	3	3	1	2	24

Data from Putman *et al*. (1996).

The concern here is whether HI-6 dichloride induces a significant increase in mutant colonies. This is a one-sided question, and we set the formal hypotheses under study as $H_0{:}\mu_0 = \mu_1$ vs. $H_a{:}\mu_0 < \mu_1$. Under a Poisson assumption on the observed counts, we condition the analysis on the observed grand total $Y_{++} = 35$, producing corresponding hypotheses based on (4.13) as $H_0{:}\pi = \frac{1}{2}$ vs. $H_a{:}\pi > \frac{1}{2}$.

The observed value of the conditional test statistic is $y_1 = 24$. The conditional null distribution of the test statistic is $Y_{1+} \mid Y_{++} \sim \mathcal{Bin}(35, 0.5)$. A significant decrease in rate of response for the exposed group is indicated if the P-value, $P = P_0[Y_{1+} \geq 24 \mid Y_{++} = 35]$, drops too low, say less than $\alpha = 0.05$.

Applying the probability function for a $\mathcal{Bin}(35, 0.5)$ distribution here to find the one-sided P-value is complex, due to the rather large values of Y_{++} and y_1. The calculation is possible: determine the binomial sum

$$\sum_{i=24}^{35} \binom{35}{i} \left(\frac{1}{2}\right)^i \left(\frac{1}{2}\right)^{35-i} = \left(\frac{1}{2}\right)^{35} \sum_{i=24}^{35} \binom{35}{i},$$

which is found to be $P = 0.0205$. A simpler calculation for these data involves, however, the large-sample normal approximation based on (4.15). Our conditions for application of the large-sample statistic are $Y_{++}p > 5$ and $Y_{++}(1-p) > 5$, where $p = y_1/Y_{++}$. These are met for the data in Table 4.4:

$$Y_{++}p = 24 > 5 \;\checkmark \qquad\qquad Y_{++}(1-p) = 11 > 5 \;\checkmark$$

Hence, from (4.15) the test statistic is

$$z_{\text{calc}} = \frac{\frac{24}{35} - \frac{1}{2}}{\sqrt{\frac{1}{35}(0.5)^2}},$$

or $z_{\text{calc}} = 0.1857/0.0845 = 2.197$. The one-sided approximate P-value is $1 - \Phi(2.197) = 0.0140$, corresponding (very) roughly to the exact value found above.

In either case, the evidence provided by this conditional test is significant for an increase in the rate of mutation, since the P-values do not exceed 5%.

Many analysts argue that count data can be analyzed simply by applying a proper transformation, such as the Freeman–Tukey transform noted above: $V_{ij} = \sqrt{(Y_{ij} + 1)} + \sqrt{(Y_{ij})}$. Although generally not as powerful as the conditional test based on (4.14) or, in large samples, (4.15) – the conditional test is UMP among all tests whose minimum power is never lower than α when H_0 is false – the transformation does possess the important characteristic of simplicity of use. For comparison's sake, and to illustrate the methodology, we consider application of the Freeman–Tukey transform to the data in Table 4.4.

To begin, consider the SAS code in Fig. 4.3: PROC TTEST is used for t-tests of differences in mean response between the control and HI-6 exposure groups, using either t_p from (4.1) or t_{calc} from (4.2). (The output includes both tests; the user should decide a priori which is appropriate.) In PROC TTEST, the `var` statement can be used to output t-test results for a number of different variables; we study the transformed values V_{ij} here. The resulting output (edited for presentation) appears in Fig. 4.4.

```
* SAS code for mutation data from Table 4.4;

data table44;
input group y @@;
V = sqrt(y+1) + sqrt(y);
cards;
0 2   0 2   0 0   0 1   0 2   0 1   0 0   0 2   0 1   0 0
160 2   160 1   160 3   160 1   160 4   160 4
160 3   160 3   160 1   160 2

proc ttest;
 class group;
 var V;
```

Fig. 4.3 SAS PROC TTEST code for Freeman–Tukey transformed data from Table 4.4.

In Fig. 4.4, the SAS output provides the two-sided (approximate) P-value for testing the equality of the Freeman–Tukey transformed colony counts (`V`) in the two experimental groups, using t_{calc} (4.2) with a Welch–Satterthwaite correction (4.3). (Here, SAS has calculated $-t_{\text{calc}}$, since it has placed the '0' group before the '160' group.) This is $P \approx 0.0133$, based on $16.9 \approx 17$ *df;* see under `Prob>|T|` associated with the row for `Variances` are `Unequal`. For our one-sided problem, we find $P \approx \text{P}[t(17) \leq -2.7642] = 0.0066$. If it could be assumed a priori that `Variances` are `Equal`, SAS uses $-t_{\text{p}}$ based on (4.1). (A test for equal variances is provided as part of PROC TTEST, although we recommend using this result only when the assumption of normality is valid; see Section 4.1.2.) In Fig. 4.4, the corresponding two-sided P-value is $P \approx 0.0128$. For our one-sided problem, we find $P \approx \text{P}[t(18) \leq -2.7642] = 0.0063$.

```
                     The SAS System
                     TTEST PROCEDURE
Variable: V

GROUP      N            Mean           Std Dev        Std Error
-----------------------------------------------------------------
  0       10      2.28276982        0.94108796       0.29759814
160       10      3.32034578        0.72337426       0.22875103

Variances          T          DF     Prob>|T|
---------------------------------------------
Unequal      -2.7642        16.9       0.0133
Equal        -2.7642        18.0       0.0128

For H0: Variances are equal, F' = 1.69    DF = (9,9)
                            Prob>F' = 0.4452
```

Fig. 4.4 SAS output (edited) for t-tests of Freeman–Tukey transformed data (`V`) from Table 4.4.

Indeed, due in part to the similar variation in response between the two experimental groups on the original scale, even a t-test on the original measurement scale would lead to the same conclusion as that on the transformed scale: a significant difference in mutagenicity exists between the control and 160 μg/ml HI-6 dichloride conditions.

This analysis can be made more accurate by noting that the theoretical variance of a Freeman–Tukey transformed count is $\text{Var}[V] \approx 1$, for all i and j (Freeman and Tukey, 1950). Thus estimation of sample variances is unnecessary here; the variance of V is, in effect, known. With known variances, there is no need to appeal to the t distribution for the test statistic; instead, we use a standard normal reference distribution, via

$$z_{calc} = \frac{\overline{V}_1 - \overline{V}_0}{\sqrt{\dfrac{1}{n_0} + \dfrac{1}{n_1}}} .$$

With this, we reject the null hypothesis that the means of the transformed variates are equal against an increasing one-sided alternative if $z_{calc} \geq z_{\alpha}$. The approximate P-value is $1 - \Phi(z_{calc})$. (Two-sided tests are similar.)

From the output in Fig. 4.4, we find $\overline{V}_1 - \overline{V}_0 = 3.3203 - 2.2827 = 1.0376$. This produces $z_{calc} = 2.3201$, with $P \approx 0.0102$. Once again, we infer that a significant increase in mutagenicity is brought on by HI-6 dichloride exposure. □

4.1.9 Rank-based tests for two sample comparisons

All of the procedures reviewed above for testing differences between two independent samples are based on specification of the underlying parent distribution for the data. In many instances, however, the exact parent distribution may be unknown, so that the procedures above may not be appropriate for use. If sufficient numbers of observations are taken – that is, if the sample sizes, n_0 and n_1, are very large – the central limit theorem suggests that the sample means $\overline{Y}_0$ and $\overline{Y}_1$ are at least approximately normally distributed. In large samples, then, we can apply statistical inferences based on normal distributions with at least approximate validity.

Inferences for situations where the sample sizes are not large, and where the parent distribution of the data is not known fully, are more difficult to achieve. If only the mean and variance are known, then quasi-likelihood methods as described in Section 2.7 can be applied. Since these rely on large-sample approximations for their test statistics to be useful, however, they may not be best for the small-sample setting. In fact, so little may be known about the underlying distributions that only a completely nonparametric, or **distribution-free**, statistical method is possible. In this case, inferences are based often on the ranks of the observed data.

In this section, we will review briefly some basic rank-based methods for testing hypotheses about two populations. It will not be our goal to survey this field, rather to illustrate the concepts behind the basic methodologies, in preparation for their occasional use in future chapters. Indeed, the study of distribution-free statistical methods based on ranks has a rich history, and it is beyond the scope of this text to discuss it in detail. For a good introduction to rank-based methods, see, for example, Pearson and Turton (1993, Chapter 9). For more comprehensive introductions and reviews, see Hollander and Wolfe (1973), Fisher (1983), or van der Laan and Verdooren (1987).

Consider the simple problem of testing whether two populations, a control and a treated group, differ. In particular, is the population of treated experimental units responding at a higher (or lower) level than the control units? Since we cannot specify underlying parent distributions for the two populations, we will assume only that each distribution has some **population median,** $\boldsymbol{m}_0$ or $\boldsymbol{m}_1$. (Recall that the population median, $\boldsymbol{m}$, is the value above which at least 50% of the population's members will respond, and below which at least 50% of the population's members will respond.) Our concern is with the null hypothesis of no difference in location, $\mathrm{H_0}{:}\boldsymbol{m}_0 = \boldsymbol{m}_1$, vs. some form of departure from $\mathrm{H_0}$. For example, $\mathrm{H_a}{:}\boldsymbol{m}_0 < \boldsymbol{m}_1$ represents a positive shift in the median of the treatment relative to the control, $\mathrm{H_a}{:}\boldsymbol{m}_0 > \boldsymbol{m}_1$ a negative shift, or $\mathrm{H_a}{:}\boldsymbol{m}_0 \neq \boldsymbol{m}_1$ any shift.

As above, suppose we observe control data Y_{0j}, and treatment data Y_{1j}, where $j = 1, \ldots, n_i$. To test the null hypothesis that treatment does not shift the location of the data values, consider the ordering indicator function

$$I(x < y) = \begin{cases} 1 & \text{if } x < y \\ \frac{1}{2} & \text{if } x = y \\ 0 & \text{if } x > y \end{cases}$$

and calculate the test statistic

$$U_{\text{calc}} = \sum_{k=1}^{n_1} \sum_{j=1}^{n_0} I(Y_{0j} < Y_{1k}). \qquad (4.16)$$

Essentially, U_{calc} is the number of times responses in the treatment group exceed responses in the control group (one must make a total of $n_0 n_1$ pairwise comparisons!); the use of 1/2 at equality in $I(x < y)$ accounts for any ties among the observations. If the treatment values are consistently higher (lower) than the control values, we expect the value of U_{calc} to be very large (small). By using basic rules of counting and permutations, we can assess how extreme these large (small) values of U_{calc} are, relative to a null hypothesis of no difference between the two underlying populations.

Alternatively, the U-statistic in (4.16) may be constructed as the sum of the treatment group's ranks, say $\mathcal{W}_{calc}$, after pooling both groups together, although it then requires a correction term:

$$\mathcal{W}_{calc} = U_{calc} + \frac{n_1(n_1 + 1)}{2}.$$

$\mathcal{W}$ is often referred to as the **rank-sum statistic**, or also as the two-sample **Mann–Whitney–Wilcoxon statistic**, after its originators (Wilcoxon, 1945; Mann and Whitney, 1947).

To test H_0 against an alternative hypothesis of increased shift in location due to treatment, reject H_0 when $\mathcal{W}_{calc}$ is greater than or equal to an upper-α critical point from the null distribution of $\mathcal{W}$, denoted as $w_\alpha(n_0,n_1)$. Tables of these points (or of associated quantities) are available in sources such as Hollander and Wolfe (1973, Table A.5) or Sen and Krishnaiah (1991, Table 2.2).

To test H_0 against a decreased shift in location, reject when $\mathcal{W}_{calc} \leq n_1(n_+ + 1) - w_\alpha(n_0,n_1)$. To test against a two-sided shift, reject when $\mathcal{W}_{calc} \geq w_{\alpha/2}(n_0,n_1)$ or $\mathcal{W}_{calc} \leq n_1(n_+ + 1) - w_{\alpha/2}(n_0,n_1)$.

In large samples – say, when $n_0 > 10$ and $n_1 > 10$ – the null distribution of $\mathcal{W}_{calc}$ is approximately normal, with mean $n_1(n_+ + 1)/2$ and variance $n_0n_1(n_+ + 1)/12$. The standardized statistic

$$z_{calc} = \frac{\mathcal{W}_{calc} - \frac{1}{2}n_1(n_+ + 1)}{\left\{\frac{1}{12}n_0n_1(n_+ + 1)\right\}^{1/2}}$$

is then approximately standard normal, allowing for straightforward inferences. For example, against an increasing shift, reject H_0 when $z_{calc} > z_\alpha$. The corresponding one-sided P-value is $P \approx 1 - \Phi(z_{calc})$.

Corrections to this large-sample approximation are necessary when ties are observed among the observations. Begin by pooling the data, and then count the number of distinct groups of tied values. Denote this number by G. Let τ_g be number of tied observations in the gth group, $g = 1, \ldots, G$. Then, in the large-sample variance term replace $n_0n_1(n_+ + 1)/12$ with

$$\mathrm{Var}^*[\mathcal{W}] = \frac{n_0n_1}{12}\left(n_+ + 1 - \frac{\sum_{g=1}^{G} \tau_g(\tau_g^2 - 1)}{n_+(n_+ - 1)}\right).$$

If there are no ties among the observations, each distinct value represents a 'group,' so $G = n_+$ and $\tau_g = 1$ for all g. Then, $\mathrm{Var}^*[\mathcal{W}]$ collapses to $n_0n_1(n_+ + 1)/12$. In any case, the corrected test statistic is

$$z_{\text{calc}} = \frac{\mathcal{W}_{\text{calc}} - \frac{1}{2} n_1 (n_+ + 1)}{\sqrt{\text{Var}^*[\mathcal{W}]}}$$

(Hollander and Wolfe, 1973, Section 4.1). Refer z_{calc} to a standard normal reference distribution, as above.

Example 4.4 Genetic toxicity to nitrofurantoin
Consider determination of toxic genetic damage in the bacterium *Salmonella typhimurium* after exposure to the chemical nitrofurantoin. As discussed in Example 3.7, the *Salmonella* mutagenicity assay generates counts of mutated bacterial colonies on a Petri dish, where each colony indicates a bacterial mutation. Since interest concerns increases in mutated colonies, we formulate the hypotheses as $H_0{:}\boldsymbol{m}_0 = \boldsymbol{m}_1$ vs. a positive shift, $H_a{:}\boldsymbol{m}_0 < \boldsymbol{m}_1$.

Data for *Salmonella* strain TA98 in both a control (Y_{0j}) and a nitrofurantoin exposure (Y_{1j}) group consist of three observations per group, so $n_0 = n_1 = 3$. The three control values are $Y_{11} = 46$, $Y_{12} = 43$, and $Y_{13} = 44$. The three exposure group values are $Y_{11} = 67$, $Y_{12} = 77$, and $Y_{13} = 78$. Table 4.5 illustrates the computations for U_{calc}. The final statistic is calculated as $U_{\text{calc}} = 9.0$. With $n_0 = n_1 = 3$, this gives $\mathcal{W}_{\text{calc}} = 9 + (3)(4)/2 = 15$. From a table of rank-sum critical values, such as in Hollander and Wolfe (1973, Table A.5), we find $w_{0.1}(3,3) = 15$. Since our interest is in an increased shift, we reject H_0 at the 5% significance level when $\mathcal{W}_{\text{calc}} \geq w_{0.05}(n_0,n_1) = 15$. Hence a significant departure is evidenced at $\alpha = 0.05$.

Table 4.5 *Salmonella* mutagenesis data and rank-sum computations

j	k	Y_{0j}	Y_{1k}	$I(Y_{0j} < Y_{1k})$	*Cumulative* U_{calc}
1	1	46	67	1	1
1	2	46	77	1	2
1	3	46	78	1	3
2	1	43	67	1	4
2	2	43	77	1	5
2	3	43	78	1	6
3	1	44	67	1	7
3	2	44	77	1	8
3	3	44	78	1	$U_{\text{calc}} = 9$

Data from NTP (1989b).

Although the sample sizes with these data are not large enough to validate the large-sample approximation for the rank-sum test, it is instructive nonetheless to calculate the z-statistic associated with $\mathcal{W}_{\text{calc}}$. The numerator of the statistic is $15 - 10.5 = 4.5$. There are no ties among the observations, so there is no need to correct for ties in the large-sample variance. The variance term equals $(9)(6 + 1)/12 = 5.25$, leading to $z_{\text{calc}} = (15 - 10.5)/\sqrt{5.25} = 1.964$. At $\alpha = 0.05$, the one-sided rejection region is $z_{\text{calc}} > 1.645$, again suggesting a significant positive shift in the treated group. The one-sided P-value is $P \approx 0.025$.

While the calculation of the rank-sum statistic is relatively straightforward with small sample sizes, computer analysis is necessary for larger data sets. For example, SAS computes the rank-sum statistic in PROC NPAR1WAY. To illustrate with the current data, the SAS code given in Fig. 4.5 does the following: (a) sets up the data set using a short do loop to assign the group identifier in the `data` step: `group=0` is the control and `group=1` is the exposure group, (b) prints out the data set as a check of the do-loop calculations, and (c) invokes PROC NPAR1WAY.

In the (edited) output in Fig. 4.6, SAS calculates the rank-sum statistic by incorporating a **continuity correction** into the normal approximation. Specifically, 0.5 is subtracted from (added to) the numerator of z_{calc} prior to dividing by the square root of $\text{Var}^*[\mathcal{W}]$ when z_{calc} is greater (less) than zero. Here, the continuity-corrected statistic is $(15 - 10.5 - 0.5)/\sqrt{5.25} = 1.746$. (In Fig. 4.6, SAS displays -1.746, since by default it uses the rank-sum from the smaller sample size, or from the first group if the sample sizes are equal.) SAS reports the two-sided P-value for testing $H_0{:}\boldsymbol{m}_0 = \boldsymbol{m}_1$ vs. $H_a{:}\boldsymbol{m}_0 \neq \boldsymbol{m}_1$ as `Prob > |Z| = 0.0809`, that is, $2\{1 - \Phi(|1.746|)\} = 2(1 - 0.9595) = 0.081$. For testing H_0 against the one-sided alternative $H_a{:}\boldsymbol{m}_0 < \boldsymbol{m}_1$, the corresponding P-value is $1 - \Phi(1.746) = 0.0405$.

```
* SAS code for Salmonella data from Table 4.5;
data ex44;
     * do-loop to set up control group;
  do i=1 to 3; input count @@;  group=0; output; end;
     * do-loop to set up exposure group;
  do i=1 to 3; input count @@;  group=1; output; end;
cards;
46 43 44   67 77 78

proc print;  *check output/do-loop operation;
proc npar1way;   class group;   var count;
```

Fig. 4.5 SAS PROC NPAR1WAY code for calculating the rank-sum statistic with data from Table 4.5.

```
                    The SAS System

        OBS      I      COUNT     GROUP
         1       1        46        0
         2       2        43        0
         3       3        44        0
         4       1        67        1
         5       2        77        1
         6       3        78        1

      N P A R 1 W A Y   P R O C E D U R E

   Wilcoxon Scores (Rank Sums) for Variable COUNT
           Classified by Variable GROUP

             Sum of      Expected        Std Dev        Mean
GROUP   N    Scores      Under H0        Under H0       Score
  0     3      6.0     10.5000000     2.29128785        2.0
  1     3     15.0     10.5000000     2.29128785        5.0

  Wilcoxon 2-Sample Test (Normal Approximation)
  (with Continuity Correction of .5)
  S =   6.00000   Z = -1.74574   Prob > |Z| = 0.0809
```

Fig. 4.6 SAS output (edited) for calculating the rank-sum statistic with data from Table 4.5.

We note in closing that the complete output from SAS PROC NPAR1WAY is far more extensive than that displayed here; the program can supply an impressive variety of distribution-free statistics. Further details are available in the SAS manual (SAS Institute Inc., 1989). ❑

4.1.10* Permutation tests for two-sample comparisons

Another approach for comparing populations in a distribution-free context is provided by **permutation tests**, also called **randomization tests** (Good, 1994). These methods date back to the 1930s, when an idealized example by Fisher (1935a) illustrated the concepts. Fisher's technique evaluated a claim by a woman who said she was able to detect whether milk was added to her cup either before or after tea was added; a summary is given by Holschuh (1980). In this **lady tasting tea experiment**, the woman was presented with eight randomly ordered cups of tea with milk, where four cups were prepared with first tea then milk, and the other four cups were prepared with first milk then tea. (She was informed of the even four-cup split between milk-then-tea and tea-then-milk, but did not know the order with which these were presented to her.) The woman was asked to declare whether milk was added first or second in each cup. Her ability to classify teas was judged by evaluating how unusual her classification was, assuming she actually had no ability to detect the order of milk added to tea.

To perform the analysis, Fisher calculated all possible permutations of eight cups partitioned into two groups of four each. He recognized that if the woman had no ability to classify milk-then-tea vs. tea-then-milk, then all of these permutations would be equally likely as possible outcomes. Statistically, a P-value is constructed by comparing the proportion of permutations in which the number of correct classifications equals or exceeds the number of observed correct classifications.

This wonderful example has strong similarities to the experimental structure described previously for Fisher's exact test; however, it has potential beyond this application. In particular, we examine how this technique can be used in the general two-group context.

As in previous sections, denote the responses measured on independent samples from two populations by Y_{ij} ($i = 0,1$; $j = 1, \ldots, n_i$). The total sample size is $n_+ = n_0 + n_1$. Note that no specific forms are assumed for the parent populations of the Y_{ij}s; independence is the only requisite assumption for this construction. Let $T(\mathbf{Y}_0, \mathbf{Y}_1)$ denote some statistic designed to measure difference between the observations in the two samples, where $\mathbf{Y}_i$ is vector notation for either group of observations, for example, $\mathbf{Y}_0 = [Y_{01}\ Y_{02} \ldots Y_{0n_0}]$. Any of the previously introduced test statistics could be used for $T(\mathbf{Y}_0, \mathbf{Y}_1)$, for example, t_{calc} in (4.2).

If there is no difference between observations in the two samples, that is, if the null hypothesis of no difference between groups is true, then each permutation of the n_+ observations partitioned into two groups of sizes n_0 and n_1 is equally likely. Notice that there are here

$$\binom{n_+}{n_1}$$

such permutations. Let $\{\mathbf{Y}_0^*, \mathbf{Y}_1^*\}$ represent one of these permutations. To calculate the permutation P-value, we evaluate the test statistic at each of the permutations of the data, $T(\mathbf{Y}_0^*, \mathbf{Y}_1^*)$, and we record the number of permuted samples whose test statistics are as extreme as or more extreme than the observed test statistic. That is, we count the number of times $T(\mathbf{Y}_0^*, \mathbf{Y}_1^*) \geq T(\mathbf{Y}_0, \mathbf{Y}_1)$. This count, divided by the total number of possible permutations, is the permutation P-value. Note that this provides in effect an **empirical *P*-value**, avoiding the need to reference some statistical distribution such as t or χ^2. Manly (1991) gives many excellent examples of permutation testing applied in environmental biology.

4.2 TWO-SAMPLE COMPARISONS – CONFIDENCE INTERVALS

A major criticism leveled against hypothesis testing is that the process reduces the statistical inference to a simple decision for or against rejection of the null hypothesis. Even tests of significance, in which the magnitude of the calculated P-value adds a level of quantification to the decision process, are chastised, since they cannot provide information concerning the magnitude of the difference between the two populations. As an alternative, many statisticians argue for construction and use of confidence intervals on the parameter(s) of interest. Indeed, it is possible in most settings to construct a confidence interval by *inverting* a two-sided hypothesis test, selecting as the interval those values for the parameter of interest that do *not* lead to rejection of H_0. The resulting **acceptance region** is essentially the inverse of the rejection region. From this perspective, there exists a direct relationship between confidence intervals and hypothesis tests, where one can be constructed from the other, and vice versa; see the discussion in Section 2.5.

We present in this section confidence intervals that can be used when comparing a control group with a single treatment group. We limit ourselves to two-sided confidence limits, where the interval has both a lower and an upper bound. These forms relate directly to the two-sided hypothesis tests discussed in the preceding sections. For all of these constructions, one-sided confidence bounds are also possible. These are useful, for example, in cancer risk assessment when considering unit potency estimates of a carcinogen or other environmental toxin; see Kodell and West (1993). We will not emphasize their construction here, however.

4.2.1 Confidence intervals for comparing two normal distributions

Returning to the two-sample setting under normal distribution sampling, we assume continuous observations, Y_{ij}, are recorded with homogeneous variances, such that $Y_{ij} \sim$ (indep.) $N(\mu_i, \sigma^2)$ $(i = 0,1;\ j = 1, \ldots, n_i)$. Of interest is interval estimation of the difference in means $\Delta = \mu_1 - \mu_0$, and we construct the confidence limits by pivoting calculations on the t-statistic

$$t_\Delta = \frac{\overline{Y}_1 - \overline{Y}_0 - \Delta}{s_p \left(\frac{1}{n_0} + \frac{1}{n_1}\right)^{1/2}},$$

where s_p is the pooled estimator of σ. Under homoscedastic normal sampling, the distribution of t_Δ is $t_\Delta \sim t(n_0 + n_1 - 2)$. The corresponding confidence interval for Δ is found by inverting the t-test of $H_0: \Delta = 0$ from Section 4.1.1. This produces limits of the form

$$\overline{Y}_1 - \overline{Y}_0 \pm t_{\alpha/2}(n_0 + n_1 - 2)\, s_p \left(\frac{1}{n_0} + \frac{1}{n_1}\right)^{1/2}, \tag{4.17}$$

where $t_{\alpha/2}(n_0 + n_1 - 2)$ is the upper-$(\alpha/2)$ critical point from the $t(n_0 + n_1 - 2)$ distribution. This interval covers the true value of Δ with exact $1 - \alpha$ confidence. It is related to the homogeneous variance t-test in that for any (null) hypothesized value of Δ, say Δ_0, when the $1 - \alpha$ confidence limits do not contain Δ_0, the t-test will reject $H_0: \Delta = \Delta_0$ in favor of $H_a: \Delta \neq \Delta_0$ at significance level α, and vice versa.

The confidence interval in (4.17) may be constructed in most statistical computer packages. For example, given a variable denoting group membership, say `group`, and a response variable, `Y`, then possible SAS statements for constructing (4.17) are

```
proc glm;
  class group;
  model Y = group;
  means group / t  cldiff;
```

If the homogeneous variance assumptions are not met, that is, if $\sigma_1^2 \neq \sigma_2^2$, then the confidence interval for the difference $\Delta = \mu_1 - \mu_0$ will not be exact. An approximate interval relies instead on a Welch–Satterthwaite correction from Section 4.1.2. We write $Y_{ij} \sim N(\mu_i, \sigma_i^2))$, $i = 0,1$; $j = 1, \ldots, n_i$, and pivot the confidence limits about a statistic based on (4.2), resulting in the interval

$$\overline{Y}_1 - \overline{Y}_0 - t_{\alpha/2}(\nu_S)\sqrt{\frac{s_0^2}{n_0} + \frac{s_1^2}{n_1}} < \Delta < \overline{Y}_1 - \overline{Y}_0 + t_{\alpha/2}(\nu_S)\sqrt{\frac{s_0^2}{n_0} + \frac{s_1^2}{n_1}} \tag{4.18}$$

where ν_S is the Welch–Satterthwaite corrected *df*, given in (4.3). As the $n_i \rightarrow \infty$, the interval covers the true value of Δ with confidence approaching $1 - \alpha$. The approximate coverage is usually quite good even at sample sizes as low as $n_i = 10$.

4.2.2 Confidence intervals for comparing two binomial distributions

In the two-sample binomial setting, data are recorded as bounded counts $Y_i \sim \mathcal{B}in(N_i, \pi_i)$, $i = 0,1$. Interest can center on a variety of quantities: the difference in response probabilities, $\delta = \pi_1 - \pi_0$; or the **risk ratio** of response probabilities, $RR = \pi_1/\pi_0$, also called the **relative risk**; or the **odds ratio** of response probabilities

$$OR = \frac{\left(\frac{\pi_1}{1 - \pi_1}\right)}{\left(\frac{\pi_0}{1 - \pi_0}\right)}$$

(Agresti, 1990, Section 2.2). While δ may be the more familiar effect measure for binomial response data, the *RR* and *OR* measures can be more indicative of differences between the two groups when the response probabilities are close to 0 or 1. For example, suppose $\pi_0 = 0.005$ and $\pi_1 = 0.095$. Then, $\delta = 0.09$ is near zero, but $RR = 19.0$ and $OR = 20.89$. Alternatively, when $\pi_0 = 0.605$ and $\pi_1 = 0.695$, the difference is unchanged at $\delta = 0.09$, but $RR = 1.15$ and $OR = 1.49$ are now closer to 1.0.

Further interest in *RR* and *OR* as measures of effect will arise in later chapters. Here, we consider confidence intervals for each of these three effect measures separately.

Beginning with the difference in proportions, δ, an obvious point estimator is $p_1 - p_0$, where $p_i = Y_i/N_i$. Based on a large-sample normal approximation for the p_i, and mimicking the construction used in (4.18), a simple interval for δ can be constructed by pivoting on the test statistic

$$z_\delta = \frac{(p_1 - p_0) - \delta}{se[p_1 - p_0]} \tag{4.19}$$

where $se[p_1 - p_0]$ is the standard error of $p_1 - p_0$, given in (4.5). This is essentially a Wald-type construction; hence, a large-sample normal approximation for (4.19) provides that $z_\delta \dot{\sim} N(0,1)$ as the $N_i \rightarrow \infty$. From this, an approximate $1 - \alpha$ confidence interval on δ is

$$p_1 - p_0 - z_{\alpha/2}\sqrt{\frac{p_1(1-p_1)}{N_1} + \frac{p_0(1-p_0)}{N_0}}$$

$$< \delta < \qquad (4.20)$$

$$p_1 - p_0 + z_{\alpha/2}\sqrt{\frac{p_1(1-p_1)}{N_1} + \frac{p_0(1-p_0)}{N_0}}$$

where $z_{\alpha/2}$ is the upper-$(\alpha/2)$ standard normal critical point.

Unfortunately, the coverage characteristics of the Wald interval in (4.20) can prove poor in small samples, especially when the true values of the π_i are near to 0 or 1, or when the N_i are very small. In some instances, the actual coverage coefficient attained by these intervals may be larger than the nominal level of $1 - \alpha$. This is not a major concern; the investigator is still assured of achieving at least $1 - \alpha$ coverage, although the interval may be slightly longer than necessary (corresponding to the increase in the actual coverage level). The result is then viewed as a conservative confidence interval, as noted in Section 2.5.

Contrastingly, for very small samples the actual coverage coefficient of the Wald limits given in (4.20) may be well below the nominal level of $1 - \alpha$. We call this an **anti-conservative** or **radical** confidence interval. As discussed in Section 2.5, this is an especially undesirable characteristic, since the investigator employing the radical interval is given no assurance that the desired confidence level is even being met. Because of this potential, (4.20) is recommended only for generously large samples, where $N_i p_i > 20$ and $N_i(1 - p_i) > 20$ for both $i = 0$ and $i = 1$.

For use in moderate to large samples – say, $N_i p_i > 5$ and $N_i(1 - p_i) > 5$ for both $i = 0$ and $i = 1$ – we recommend an alternative method that yields closed-form equations for the confidence limits. The method is often conservative and is far more stable than (4.20) in small samples. Its construction involves a reparameterization of p_0 and p_1 into two other values: the quantity of interest, δ, and what is essentially a nuisance parameter, $\tau = p_1 + p_0$. By recognizing that the theoretical variance of $p_1 - p_0$ can be written as a function of δ and τ, the reparameterization allows for a variety of possible confidence intervals to be constructed. Specifically, one writes $\mathrm{Var}[p_1 - p_0]$ as a function, $V(\delta,\tau)$, where it can be shown that

$$V(\delta,\tau) = \nu_1\left\{(2 - \tau)\tau - \delta^2\right\} + 2\nu_2(1 - \tau)\delta\,,$$

with $\nu_1 = \{(1/N_0) + (1/N_1)\}/4$ and $\nu_2 = \{(1/N_0) - (1/N_1)\}/4$. Although ν_2 can be negative, $V(\delta, \tau)$ is always greater than zero (Exercise 4.30).

Now, square the test statistic z_δ from (4.19) and replace the squared standard error in the denominator with $V(\delta,\tau)$ (since a standard error is an estimate of the square root of a variance). Setting the corresponding ratio equal to a standard normal critical point yields a quadratic relationship for δ:

$$\left\{\delta - (p_1 - p_0)\right\}^2 = z_{\alpha/2}^2 V(\delta,\tau) . \tag{4.21}$$

Solving for δ yields approximate $1 - \alpha$ lower and upper confidence limits which are dependent only on $p_1 - p_0$, $z_{\alpha/2}$, and τ. The general concept of solving a quadratic relationship such as (4.21) to achieve confidence limits is derived from inversion of the score statistic for a binomial parameter, and is due to Wilson (1927).

In most settings, the value of τ is unknown, and must be estimated from the data. Thus, the actual reparameterized limits for δ are found using $V(\delta,\tilde{\tau})$ as a large-sample variance estimate, where $\tilde{\tau}$ is some consistent estimator of τ. Many forms for $\tilde{\tau}$ are available; see Beal (1987). We recommend

$$\tilde{\tau} = \frac{N_0 p_0}{N_0+1} + \frac{N_1 p_1}{N_1+1} + \frac{1}{2}\left\{\frac{1}{N_0+1} + \frac{1}{N_1+1}\right\} .$$

This particular form for $\tilde{\tau}$ is based on the theoretical invariance principles of the statisticians Jeffreys and Perks, and we will refer to the resulting confidence limits for δ under this Wilson-type construction as **JPW limits**. Following Beal (1987), the limits become

$$\frac{1}{1 + z_{\alpha/2}^2 \nu_1}\Big\{p_1 - p_0 + z_{\alpha/2}^2 \nu_2 (1 - \tilde{\tau}) \tag{4.22}$$
$$\pm z_{\alpha/2}\Big\{ V(p_1 - p_0, \tilde{\tau}) + z_{\alpha/2}^2\left[\nu_1^2(2 - \tilde{\tau})\tilde{\tau} + \nu_2^2(1 - \tilde{\tau})^2\right]\Big\}^{1/2}\Big\}.$$

In the balanced case of $N_i = N_0 = N$, (4.22) simplifies greatly, to

$$\frac{p_1 - p_0 \pm c_\alpha \left\{(2-\tilde{\tau})\tilde{\tau}\left[1+c_\alpha^2\right] - (p_1 - p_0)^2\right\}^{1/2}}{1 + c_\alpha^2}, \tag{4.23}$$

where $c_\alpha^2 = (2N)^{-1} z_{\alpha/2}^2$ and $\tilde{\tau} = \{N(p_1 + p_0) + 1\}/(N + 1)$.

In some cases, confidence intervals for δ can be found via exact calculation, in order to achieve coverage as close to the nominal level as possible. Commonly, these are based on inverted hypothesis tests (noted above), and constructed from acceptance regions for δ. As might be expected, however, constructing acceptance regions that lead to $1 - \alpha$ confidence intervals is a complex process, and often requires a computer.

This is the case in the two-sample binomial problem, where small-sample intervals for δ are given only via algorithms for their computation (Santner and Yamagami, 1993). Explicit description of these complex algorithms is beyond the scope of this chapter. Nonetheless, when sample sizes are small – say, $N_i p_i \leq 5$ or $N_i(1 - p_i) \leq 5$ for $i = 0$ or $i = 1$ – we advise caution when applying the large-sample methods discussed here.

Example 4.2 (continued) Aneuploidy in hamster cell cultures
Returning to our example with *in vitro* chromosome loss (aneuploidy) in hamster cells, let us now quantify the extent of the difference between the control group and the DES-exposed group, by constructing confidence limits on the difference in response probability, δ.

Recall that the data from Table 4.2 gave observed proportions of $p_0 = 0.06$, and $p_1 = 0.18$. The estimated difference is $p_1 - p_0 = 0.12$, a potential increase in probability of exhibiting chromosomal aneuploidy after DES exposure of 12%. The sample sizes here are not large enough to satisfy our requirements for use of (4.20), since, for example, $N_0 p_0 = 9 < 20$. Hence we consider use of the more complex JPW limits (4.22): begin by setting $\alpha = 0.05$, so we know $z_{0.025} = 1.96$. Then, $\tilde{\tau}$ is

$$\tilde{\tau} = \frac{36}{201} + \frac{9}{151} + \frac{1}{2}\left(\frac{1}{151} + \frac{1}{201}\right) = 0.179 + 0.060 + 0.006$$

or 0.245. Next, find ν_1 and ν_2:

$$\nu_1 = (0.25)(0.005 + 0.0067) = 0.0022$$
$$\nu_2 = (0.25)(0.005 - 0.0067) = -0.0004.$$

Using these, the estimated variance is

$$V(p_1 - p_0, \tilde{\tau}) = 0.003\{(1.755)(0.245) - 0.014\} + (0.0008)(0.755)(0.12)$$

or simply 0.0011.

There are four main components to the confidence limits in (4.22). For these data, the first is $p_1 - p_0 = 0.12$. The other three are

$$z^2_{\alpha/2}\nu_2(1 - \tilde{\tau}) = (3.84)(-0.0004)(0.755) = -0.0012\,,$$

$$z_{\alpha/2}\Big(V(p_1 - p_0, \tilde{\tau}) + z^2_{\alpha/2}\big\{\nu_1^2(2 - \tilde{\tau})\tilde{\tau} + \nu_2^2(1 - \tilde{\tau})^2\big\}\Big)^{1/2} =$$

$$(1.96)\sqrt{0.0011 + (3.84)(3.651\times10^{-6} + 9.909\times10^{-8})} =$$

$$1.96\sqrt{0.0011 + 1.441\times10^{-5}} = 0.066\,,$$

for the numerator, and $1 + \nu_1 z^2_{\alpha/2} = 1.008$ for the denominator. From (4.22), the JPW confidence limits are then

$$\delta_{\text{lower}} = \frac{0.12 - 0.0012 - 0.066}{1.008} = 0.0519$$

$$\delta_{\text{upper}} = \frac{0.12 - 0.0012 + 0.066}{1.008} = 0.1836$$

and so we write $0.052 < \delta < 0.184$. That is, with 95% confidence, we conclude that DES exposure in this experiment increased the *in vitro* aneuploid response rate by between 5.2% and 18.4%. □

Moving now to odds ratios of response probabilities, this measure is exactly what its name implies, a ratio of the odds of success in group $i = 1$ to the odds of success in group $i = 0$. If π_1 and π_0 are equal, then $OR = 1$, so that tests of $H_0{:}\, OR = 1$ provide an equivalent means of testing $H_0{:}\pi_0 = \pi_1$. An estimate of OR arises quite naturally from substituting the ML estimates of π_1 and π_0 into the formula for OR:

$$\hat{OR} = \frac{Y_1(N_0 - Y_0)}{Y_0(N_1 - Y_1)}.$$

This quantity is essentially a **cross-product ratio** from a 2×2 table of counts. Unfortunately, the sampling distribution of $\hat{OR}$ is highly skewed in small samples (Woolf, 1955), and asymptotic normality arguments to form confidence intervals or test hypotheses are quite unstable. Instead, the sampling distribution of $\log(\hat{OR})$ tends to normality at much smaller sample sizes (Fleiss, 1981, Section 10.2). This log-odds ratio can be written as

$$\log(\hat{OR}) = \log[Y_1] + \log[N_0 - Y_0] - \log[Y_0] - \log[N_1 - Y_1],$$

and it has an approximate standard error of

$$se[\log(\hat{OR})] \approx \sqrt{\frac{1}{Y_1} + \frac{1}{N_0 - Y_0} + \frac{1}{Y_0} + \frac{1}{N_1 - Y_1}}.$$

An approximate confidence interval for the log-odds ratio appeals to the stable large-sample normality of its ML estimator, producing

$$\log(\hat{OR}) \pm z_{\alpha/2}\sqrt{\frac{1}{Y_1} + \frac{1}{N_0 - Y_0} + \frac{1}{Y_0} + \frac{1}{N_1 - Y_1}}.$$

To obtain an approximate $1 - \alpha$ confidence interval for the odds ratio in its original scale, take the antilog of the limits for $\log(OR)$:

$$\exp\{\log(\hat{OR}) \pm z_{\alpha/2} se[\log(\hat{OR})]\} \,. \tag{4.24}$$

Although use of $\log(\hat{OR})$ helps reduce instability in the statistical inference in small samples, problems may still arise if the number of successes (or failures) is at or very near to zero. In this case, the estimator $\hat{OR}$ may be modified by adding some small number, say 1/2, to the observed counts. This gives an estimator suggested by Gart and Thomas (1982):

$$\tilde{OR} = \frac{\left(Y_1 + \frac{1}{2}\right)\left(N_0 - Y_0 + \frac{1}{2}\right)}{\left(Y_0 + \frac{1}{2}\right)\left(N_1 - Y_1 + \frac{1}{2}\right)} .$$

The natural logarithm of this estimator has standard error

$$se[\log(\tilde{OR})] = \sqrt{\frac{1}{Y_1 + \frac{1}{2}} + \frac{1}{N_0 - Y_0 + \frac{1}{2}} + \frac{1}{Y_0 + \frac{1}{2}} + \frac{1}{N_1 - Y_1 + \frac{1}{2}}}$$

and in similar form to (4.24), it leads to an approximate $1 - \alpha$ confidence interval for OR of the form

$$\exp\left\{\log(\tilde{OR}) \pm z_{\alpha/2} se[\log(\tilde{OR})]\right\}. \tag{4.25}$$

Example 4.1 (continued) Odds ratio for tumorigenicity in mice exposed to 1,3-butadiene

Recall the tumorigenicity study of 1,3-butadiene in $B6C3F_1$ mice, from Table 4.1a. Consider now an estimate of the magnitude of increased tumorigenicity for mice in the 200 ppm concentration group relative to mice in the control group. The cross-product estimator of the odds ratio is $\hat{OR} = 3.128$ with $\log(\hat{OR}) = 0.495$ and $se[\log(\hat{OR})] = 1.173$. Applying (4.24) leads to approximate 95% confidence limits for OR of $\exp\{0.495 \pm 1.96(1.173)\}$, or $0.165 < OR < 16.339$.

Since the data in Table 4.1a show the number of success in the control and exposure groups are somewhat close to zero, calculations with the modified cross-product estimator $\tilde{OR}$ might be considered instead of $\hat{OR}$. This yields a point estimate of $\tilde{OR} = 2.43$ with $\log(\tilde{OR}) = 0.386$ and $se[\log(\tilde{OR})] = 0.997$. From (4.25), approximate 95% confidence limits for OR are $\exp\{0.386 \pm 1.96(0.997)\}$ or $0.208 < OR < 10.378$. In either case, the confidence limits contain 1.0, suggesting that the odds of developing adrenal cortex adenomas after 200 ppm exposure to 1,3-butadiene do not appear to differ from the odds of developing adrenal cortex adenomas in the control group. This corroborates our result from above, where we found no significant difference between the tumorigenicity probabilities. □

Lastly, consider the problem of estimating **relative risk** ratios of response probabilities, $RR = \pi_1/\pi_0$. For small values of π_0 and π_1, $OR \approx RR$, so that the OR often is viewed as an approximation to RR. As with OR, one might be tempted to estimate RR simply by substituting the ML estimates for π_0 and π_1. Here again, however, the small-sample distribution of the raw relative risk is skewed, and it behaves erratically. Instead, we advocate an adjustment to RR that adds a small fraction, again, say 1/2, to the components of RR, and then manipulates the logarithm of this estimator to achieve approximate confidence limits (Santner and Duffy, 1989). Begin with

$$\widetilde{RR} = \frac{\left(\dfrac{Y_1 + \frac{1}{2}}{N_1 + \frac{1}{2}}\right)}{\left(\dfrac{Y_0 + \frac{1}{2}}{N_0 + \frac{1}{2}}\right)} = \frac{\left(Y_1 + \frac{1}{2}\right)\left(N_0 + \frac{1}{2}\right)}{\left(Y_0 + \frac{1}{2}\right)\left(N_1 + \frac{1}{2}\right)}$$

as an estimator of RR. Form $\log(\widetilde{RR})$, and as its standard error take

$$se[\log(\widetilde{RR})] = \sqrt{\frac{1}{Y_1 + 0.5} - \frac{1}{N_1 + 0.5} + \frac{1}{Y_0 + 0.5} - \frac{1}{N_0 + 0.5}}$$

These are employed to form $1 - \alpha$ limits for RR, based on large-sample limits for $\log(RR)$; that is, in similar form to (4.25), RR lies between

$$\exp\left\{\log(\widetilde{RR}) \pm z_{\alpha/2}\, se[\log(\widetilde{RR})]\right\} \tag{4.26}$$

with approximate $1 - \alpha$ confidence.

Example 4.1 (continued) Relative risk for tumorigenicity in mice exposed to 1,3-butadiene

We illustrate calculation of RR using the tumorigenicity data from Example 4.1. Recall that the data are observed proportions of adrenal cortex adenomas in mice after exposure to 200 ppm 1,3-butadiene, compared to a concurrent control group. The modified estimator of RR yields a point estimate of $\widetilde{RR} = 2.33$, with $\log(\widetilde{RR}) = 0.368$ and $se[\log(\widetilde{RR})] = 0.955$. Using (4.26), these lead to a 95% confidence interval for RR of $\exp\{0.368 \pm 1.96(0.955)\}$, or $0.222 < RR < 9.398$. The interpretation of these limits is particularly useful: they give the risk in these mice of developing adrenal cortex adenomas after 200 ppm exposure to 1,3-butadiene as anywhere from about 5 times *less* likely to over 9 times *more* likely than the risk in the control group. Since the interval contains 1.0, this suggests again that the risk of developing adrenal cortex adenomas after 200 ppm 1,3-butadiene exposure does not appear to differ from the background risk of developing adrenal cortex adenomas in these mice. □

4.2.3 Confidence intervals for comparing two Poisson parameters

Confidence limits for the difference between the means of two Poisson distributions may be constructed in a similar fashion as with the binomial case above. Recall that for $Y_{ij} \sim$ (indep.) $\mathcal{Poisson}(\mu_i)$ ($i = 0,1$; $j = 1, \ldots, n_i$), the sample sums Y_{0+} and Y_{1+} are again Poisson-distributed, with means $n_0\mu_0$ and $n_1\mu_1$, respectively. Interest centers on a $1 - \alpha$ confidence interval for the difference in the means, $\Delta = \mu_0 - \mu_1$. (For a method that computes confidence limits on the ratio of means, μ_0/μ_1, using the conditional distribution of Y_{1+} given Y_{++}, see Sahai and Khurshid (1993b).)

The standard method for confidence limits on Δ appeals to a Wald construction: center the interval at some unbiased estimator of Δ, and then add and subtract a multiple of the standard error of this estimator. An unbiased estimator of Δ is the difference in sample means: $\overline{Y}_1 - \overline{Y}_0$. Under a two-sample Poisson model, the standard error of this quantity is

$$se[\overline{Y}_1 - \overline{Y}_0] = \left(\frac{\overline{Y}_1}{n_1} + \frac{\overline{Y}_0}{n_0}\right)^{1/2},$$

which yields Wald-type confidence limits for Δ of the form

$$\overline{Y}_1 - \overline{Y}_0 \pm \mathfrak{z}_{\alpha/2}\sqrt{\frac{\overline{Y}_1}{n_1} + \frac{\overline{Y}_0}{n_0}}\ . \tag{4.27}$$

Use of the standard normal critical point in (4.27) is motivated by the large-sample normality of the sample means, and hence of their difference: as the $n_i \to \infty$ for fixed values of μ_i, the confidence level for the limits in (4.27) approaches $1 - \alpha$.

Unfortunately, in small samples the coverage provided by (4.27) can be less than $1 - \alpha$, especially for very small values of the underlying means μ_i. Thus the Wald interval in (4.27) is recommended for use only with large samples in which the underlying mean rates of response are not expected to be small. A conservative criterion is to require $n_i \geq 10$ and apply (4.27) only if the observed sums, Y_{i+}, exceed 5.0 for both $i = 0$ and $i = 1$.

An improvement to the Wald interval is possible, however, by mimicking Wilson's (1927) quadratic approach. (This approach also led to the JPW confidence limits (4.22) in the two-sample binomial setting.)

Reparameterizing μ_0 and μ_1 into two other values, a nuisance parameter and Δ, leads to a quadratic equation in Δ, the solution of which gives $1 - \alpha$ confidence limits for Δ (Weissfeld *et al.*, 1991). When completed, the computations yield confidence limits of the form

$$\bar{Y}_1 - \bar{Y}_0 + 2(\varphi + 1) \pm z_{\alpha/2}\sqrt{\frac{\bar{Y}_1}{n_1} + \frac{\bar{Y}_0}{n_0} + 2(\varphi + 1)\left(v_1^2 + v_2^2\right)} \quad (4.28)$$

where, similar to the construction in (4.22), $v_1 = \{(1/n_0) + (1/n_1)\}/2$ and $v_2 = \{(1/n_0) - (1/n_1)\}/2$. Notice that the constant φ serves as a selection index for what is essentially a family of confidence intervals in (4.28). At $\varphi = -1$, we recover the Wald intervals in (4.27).

The selection index φ is used in small samples to improve the interval's confidence level. Weissfeld *et al.* (1991, Table II) present a tabular formulation for specifying φ as a function of the sample size ratio n_0/n_1 and of the observed values of Y_{0+} and Y_{1+}. The resulting values for φ lie generally in the range $-1 < \varphi < 0$, with the limiting Wald case of $\varphi = -1$ recommended as the sample sizes and observed sums grow large. If the Weissfeld *et al.* table is unavailable, then $\varphi = -0.25$ is a conservative choice for use in small samples.

Example 4.3 (continued) Genetic toxicity to HI-6 dichloride
Continuing with our example on genetic toxicity to HI-6 dichloride in cultured hamster ovary cells, we consider construction of confidence limits on the difference in mean mutant colony counts between the unexposed and exposed groups. From Table 4.4, the observed group sums are $y_{0+} = 11$ and $y_{1+} = 24$. Since $n_0 = n_1 = 10$ for these data, the corresponding observed group means are $\bar{y}_0 = 1.1$ and $\bar{y}_1 = 2.4$ mutant colonies per plate.

The difference in mean mutant colonies, $\Delta = \mu_1 - \mu_0$, is estimated by the observed difference $2.4 - 1.1 = 1.3$ mutants/plate. A more complete inference, however, includes a set of confidence limits on this difference in means. Equation (4.28) provides such limits. Applying it here, it can be shown that the equi-replicated design ($n_0 = n_1 = 10$) with both observed sums larger than 5.0 gives $\varphi = -1$ in the Weissfeld *et al.* table. This reduces to the Wald interval in (4.27). That is, these data satisfy the large-sample requirements for use of the Wald interval.

Calculation of the Wald interval requires only the group means, $\bar{y}_0 = 1.1$ and $\bar{y}_1 = 2.4$, the group sizes, $n_0 = n_1 = 10$, and a standard normal critical point, $z_{\alpha/2}$. Set $\alpha = 0.05$ so that $z_{0.025} = 1.96$. Then, the approximate 95% confidence limits on Δ become

$$2.40 - 1.10 \pm 1.96\sqrt{0.24 + 0.11} = 1.30 \pm 1.16,$$

that is, $0.14 < \Delta < 2.46$. Thus, with approximate 95% confidence it appears that exposure to HI-6 dichloride can increase the number of mutant colonies by almost 2.5 mutants/plate, but the effect is weak enough that an increase of as low as 0.14 mutants/plate is also possible. ❑

4.3 SUMMARY

In this chapter, basic methods are described for comparing the underlying parameters of two populations, better known as the two-sample comparison problem. Included are both continuous data settings, through consideration of the usual normal distribution model, and also discrete data settings such as the binomial and Poisson parent distribution models. Distribution-free techniques are also noted briefly, including nonparametric rank-based methods (the Mann–Whitney–Wilcoxon test) and permutation/randomization methods. Both significance tests and, for the parametric models, confidence intervals are described.

The following general strategy when comparing two (or more) populations is suggested. First, identify a parent probability model for the populations being sampled. Is the response a continuous measurement? If so, what values of the responses are observable? For instance, if continuous measurements such as weights or failure times are recorded, then a probability model reflecting continuous nonnegative responses would be reasonable. In this situation, the gamma, lognormal, or Weibull distributions might serve as acceptable candidates. On the other hand, if discrete measurements such as counts or proportions are recorded, consider an appropriate discrete distribution, such as the Poisson or binomial, respectively, or perhaps some extension of these simple discrete models, such as the negative binomial or beta-binomial, respectively. (Inferences under some of these distributional models will be discussed in the later chapters; for example, negative binomial counts are considered in Section 6.5.4, beta-binomial proportions in Section 6.5.2, and Weibull time-to-event data in Section 11.2.3.) In any case, if the underlying distribution is sensibly motivated, and if it appears to provide a reasonable fit to the data, then proceed with methods appropriate for those forms of data.

If the choice of parent probability model for the observations is difficult or ambiguous, then consider methods that are distribution-free or otherwise robust to violations of distributional assumptions. For instance, both the rank-based method from Section 4.1.9 and the permutation/randomization testing method noted in Section 4.1.10 may be applied in almost any statistical sampling situation; often, these can provide inferences on the difference between groups that are almost as powerful as the parametric methods, without recourse to any specific distributional model.

EXERCISES

(Exercises with asterisks, e.g. 4.4*, typically are more advanced, or require greater knowledge and understanding of probability distributions and/or calculus.)

4.1. Can you envision a *sensible* setting from your own field of study in which a treatment vs. control experiment or study is considered? Explain.

4.2. Suppose an experiment is conducted to assess the ability of silver iodide to induce rainfall on a coastal ecosystem after the material is used to seed clouds; cf. Inman and Bradley (1994). Rainfall data (in log-acre-feet) are measured from 26 independent seedings. Data are also collected from a group of 26 control clouds. The measurements of log rainfall are

Control clouds: 0.000, 1.589, 1.589, 2.442, 2.851, 3.077, 3.195, 3.262, 3.270, 3.353, 3.367, 3.600, 3.716, 3.857, 4.227, 4.394, 4.397, 4.554, 4.996, 5.096, 5.498, 5.772, 5.845, 5.920, 6.722, 7.092

Seeded clouds: 1.411, 2.041, 2.862, 3.447, 3.487, 3.704, 4.526, 4.748, 4.773, 4.779, 4.864, 5.291, 5.302, 5.491, 5.541, 5.616, 5.616, 5.713, 5.811, 6.064, 6.193, 6.556, 6.886, 7.412, 7.437, 7.918

Assume the data are random samples from normal distributions with unknown means μ_i and common, unknown variance σ^2. Thus, we are assuming that rainfall follows a **lognormal distribution**.

a. Construct a back-to-back stem-and-leaf plot. Does the homogeneous variance assumption on log-rainfall appear valid?

b. Perform a test of the null hypothesis that no difference exists in the means, vs. an alternative that seeding increases rainfall. Set your significance level to $\alpha = 0.01$.

4.3. (Patterson *et al.*, 1994) In a study of the accumulation of polychlorinated biphenyls (PCBs) in humans after chronic environmental exposure, adipose tissue from human subjects was taken and examined for PCB accumulation. Data were collected as parts-per-trillion of PCB (lipid-adjusted) in all subjects. Speculating that African-Americans were at greater risk for PCB exposure, the data were stratified by race. The measurements were

Caucasians:	56.7, 44.5, 48.2, 96.5, 91.0, 34.2, 154.0, 34.5, 41.8, 66.4, 29.5, 49.0, 54.7
African-Americans:	36.7, 174.0, 118.0, 69.9, 62.2, 112.0, 42.0, 67.7, 59.5, 36.4, 62.4, 109.0, 84.0, 35.6, 61.6

Assume the data are random samples from normal distributions with unknown means μ_i and unknown variances σ_i^2.

a. Construct a back-to-back stem-and-leaf plot. Does the heterogeneous variance assumption on the measurements appear reasonable?

b. Perform a test of the null hypothesis that no difference exists in the mean PCB concentrations, vs. an alternative that African-Americans exhibit higher concentrations. Set your significance level to $\alpha = 0.05$.

c. Would your conclusions differ in part (b) if you assumed homogeneous variances?

4.4*. Show that the limiting form as $\lambda \to 0$ of the power transformation $V(\lambda) = (Y^\lambda - 1)/\lambda$ is $V(0) = \log\{Y\}$.

4.5. Apply a power transformation to the PCB exposure data in Exercise 4.3, with $\lambda = 0$; that is, use $V = \log\{Y\}$. Using the transformed data, test the null hypothesis of no difference in transformed means vs. a one-sided alternative. (Assume the transformation stabilized any variance heterogeneity, so that an assumption of homoscedasticity is valid. Do you believe this assumption? How would you check?)

4.6. Apply a power transformation to the PCB exposure data in Exercise 4.3, with $\lambda = 0.5$. Using the transformed data, test the null hypothesis of no difference in transformed means vs. a one-sided alternative. (Assume the transformation stabilized any variance heterogeneity, so that an assumption of homoscedasticity is valid.)

4.7. Consider data-driven selection of the power transformation parameter, λ, for the data in Exercise 4.3. Over the range $\lambda = -1$, -0.5, 0.5, 0, 2, calculate sample variances of the power-transformed data for both the control and treatment groups. What value of λ appears to stabilize the variances?

4.8. (Dimitrov, 1994) In an experiment to compare the effect of storage time on DNA degradation in seeds of hawksbeard (*Crepis capillaris*), the frequency of chromosome aberrations in seeds stored at 22°C for six years was compared with that of a control group of unstored seeds. The data were recorded as proportions of randomly selected seed cells exhibiting chromosome aberrations, out of 400 cells scored per group. Suppose this produced an observed control proportion of 6/400 (1.5%) aberrant cells, and a storage-group proportion of 11/400 (2.75%) aberrant cells. Using the large-sample test statistic in (4.8), test the null hypothesis of no difference between underlying control and storage response probabilities, vs. an alternative that the storage induced an increase in the response probability, at $\alpha = 0.10$.

4.9. In a study of the ability of the toxin methyl methane sulfonate (MMS) to induce *in vivo* genetic damage in bone-marrow cells of laboratory rodents, eight animals were exposed to MMS, and after 24 hours their bone-marrow cells were sampled and scored for micronucleated cells. (Evidence of micronuclei is thought to be associated with chromosomal damage to the blood cells.) For each animal, 1000 cells were scored, producing the following observed proportions of micronucleated cells: 3/1000, 6/1000, 3/1000, 8/1000, 4/1000, 6/1000, 7/1000, 4/1000 (Hothorn, 1994). Assuming a binomial distribution is a valid sampling model for these data, a pooled estimate of the average proportion response to MMS is then 41/8000 (0.525%). For comparison purposes, a control group of 16 animals was also studied, providing the following observed proportions of micronucleated cells: 3/1000, 2/1000, 1/1000, 1/1000, 3/1000, 2/1000, 0/1000, 3/1000, 2/1000, 1/1000, 3/1000, 1/1000, 2/1000, 1/1000, 3/1000, 3/1000. Here, the pooled estimate of responding cells is 31/16 000 (0.194%).

a. Using the pooled proportions and the test statistic in (4.8), test the null hypothesis of no difference between underlying control and MMS-exposure response probabilities, vs. an alternative of some difference in response. In particular, what is the *P*-value? At $\alpha = 0.05$, is this significant?

b. Suppose that you are not justified in pooling samples across animals. How could you evaluate whether MMS induces *in vivo* genetic damage?

4.10. (Motimaya *et al.*, 1994) In an experiment to compare the ability of the DNA chain terminator azidothymidine (AZT) to induce *in vivo* genetic damage in mammalian bone-marrow cells, Swiss Webster mice were given intraperitoneal injections of 14.286 μg AZT per gram body weight over a 48-hour period. At the end of the treatment period, cells were taken from their bone marrow and scored for micronucleation. In addition, a concomitant vehicle control group of mice was given intraperitoneal injections of a saline solution, and their bone marrow cells were also sampled and scored for micronucleation. The two groups of mice produced pooled proportion responses of 19/10 000 (control) and 23/10000 (AZT-treated). Using the large-sample test statistic in (4.8), test the null hypothesis of no difference in response between underlying control and AZT treatment, vs. an alternative of some difference in response. In particular, what is the *P*-value? At $\alpha = 0.10$, is this difference significant? At $\alpha = 0.05$?

4.11. Suppose an experiment is planned to study the toxic effect of exposure to sewage waste on fish, relative to an unexposed control group. The data are to be recorded as proportions of fish dying after acute exposure to the waste. To test the null hypothesis that no difference in probability of lethality exists between the two groups (vs. any difference), it is desired to select appropriate sample sizes for each group. Set the significance level to $\alpha = 0.05$. Suppose a difference between the two proportions of at least $\delta = 0.1$ is to be detected. At the following pre-supposed levels for π_0 and π_1, what is the required sample size for each group when power is set at $\beta' = 0.75$, 0.8, 0.9, or 0.95?

a. $\pi_0 = 0.5$ and $\pi_1 = 0.6$.
b. $\pi_0 = 0.5$ and $\pi_1 = 0.4$.
c. $\pi_0 = 0.1$ and $\pi_1 = 0.2$.
d. $\pi_0 = 0.01$ and $\pi_1 = 0.11$.

4.12. In Exercise 4.11, suppose it is expected that the control response will be close to 0%, so that a greater number of observations will be required to estimate accurately the probability of lethality. If the ratio of treated to control units is changed from 1.0 to 0.5, how does the required sample size change for the control group at $\beta' = 0.75$, 0.8, 0.9, or 0.95? (All other conditions remain unchanged.) What if the ratio is dropped further to 0.25?

4.13. In Exercise 4.11, suppose the desired difference is relaxed to $\delta = 0.25$. (All other conditions remain unchanged.) Now what sample sizes are required at $\beta' = 0.75$, 0.8, 0.9, or 0.95?

4.14. In Exercise 4.11, suppose the desired α level is strengthened to $\alpha = 0.01$. (All other conditions remain unchanged.) Now what sample sizes are required at $\beta' = 0.75$, 0.8, 0.9, or 0.95?

4.15. In Exercise 4.8, apply Fisher's exact test to compare the control and storage groups. Does the result change? How?

4.16. In Exercise 4.9a, apply Yates's continuity correction in your analysis, as given in (4.10). Does the result change? How?

4.17. In Exercise 4.9a, apply Fisher's exact test to compare the control and exposure groups. Does the result change? Is the resulting P-value close to that obtained via Yates's correction in Exercise 4.16?

4.18. An experiment is conducted to study mammalian hepatotoxicity of an agricultural herbicide. Five laboratory rats are exposed to the herbicide daily for 26 weeks, and then their livers are examined for any hepatocellular lesions. An untreated control group of five rats is included for comparison. The data are recorded as the numbers of lesions per animal:

Control animals: 0, 0, 1, 1, 7 Treated animals: 0, 1, 3, 3, 8

a. Assuming the Poisson distribution is a valid sampling model for the counts in both groups, perform a test to assess if the mean treated rate exceeds the mean control rate, using a conditional (on Y_{++}) analysis. Set $\alpha = 0.05$.

b. If the Poisson assumption in part (a) is invalid, a distribution-free test would be an acceptable alternative to assess if the median treated rate exceeds the median control rate. Perform a rank-sum analysis of these data using the Mann–Whitney–Wilcoxon statistic. At $\alpha = 0.05$, you will need the critical value $w_{0.05}(5,5) = 36$. (At $\alpha = 0.01$, $w_{0.01}(5,5) = 39$.) Does your result change substantively from that in part (a)?

4.19. Six months after a spill of toxic contaminants in Chesapeake Bay, ecologists recorded the numbers of different polychaete species observed in samples from two similar test-bed sediment locations. The first location was far distant from the spill and is considered the 'control.' The second is near the spill. The replicate data were:

Control location: 4, 4, 5, 6, 7, 9
Contaminated location: 1, 2, 3, 3, 3, 4, 4.

Assume the Poisson distribution is a valid sampling model for the counts in both groups. Perform a test to assess if the mean species count at the spill is reduced below the mean control count, using the large-sample statistic in (4.15). Set $\alpha = 0.01$.

4.20. In a laboratory experiment to study the carcinogenic potential of the chemical solvent stabilizer 1,2-epoxybutane, 50 male rats were exposed to 200 ppm of the chemical, via inhalation, for 105 weeks (National Toxicology Program, 1988). An independent control group of another 50 rats was included for comparison purposes. As a measure of the tumor burden experienced by each animal, data were recorded as the number of organ sites (per animal) exhibiting tumors. For the two groups, the outcome frequencies may be summarized as follows:

	Dose: 0 ppm	*Dose: 200 ppm*
Number of tumors	*Number of animals*	*Number of animals*
0	0	1
1	8	6
2	12	13
3	17	13
4	11	11
5	1	4
6	1	1
7	0	1

Assume the Poisson distribution is a valid sampling model for the counts in both groups. Perform a test to assess if the mean exposed group rate exceeds the mean control rate, using the statistic in (4.15). Set $\alpha = 0.01$.

4.21. In an experiment on HI-6 dichloride genotoxicity similar to that reported in Example 4.3, Putman *et al.* (1996) give data on counts of mutant colonies per plate when an 'activating' chemical is added to the Petri dishes. (The activating chemical is thought to mimic *in vivo* enzymatic effects as would be seen if a larger organism were exposed to HI-6 dichloride.) At the same levels studied in Example 4.3, the data are:

Experimental group	*Number of offspring*				
Control	0	0	0	3	0
	4	1	4	4	2
5000 μg/ml HI-6 dichloride	0	0	0	2	0
	3	1	4	2	2

Assume the Poisson distribution is a valid sampling model for the counts in both groups. Assess if the mean exposure group count exceeds the mean control group count, using:

a. The (conditional) binomial *P*-value. Set $\alpha = 0.05$.

b. The large-sample statistic in (4.15). Set $\alpha = 0.05$.

4.22. Bailer and Oris (1993) report data from a study of toxic reproductive response in the aquatic organism *Ceriodaphnia dubia* to the herbicide nitrofen. A measure of reproductive stress in *C. dubia* after exposure to the chemical is offspring counts from exposed females. A decrease in mean response in the exposure group suggests a toxic response to the chemical. Data from 20 independent animals were recorded as total number of offspring for three broods per animal, giving:

Experimental group	*Number of offspring*				
Control	27	32	34	33	36
	34	33	30	24	31
160 μg/l nitrofen	29	29	23	27	30
	31	30	26	29	29

Apply the rank-sum statistic (4.16) to these data. Test against a negative shift in median response, H_a:$\boldsymbol{m}_0 > \boldsymbol{m}_1$, at $\alpha = 0.05$. Use the large-sample approximation, adjusting for ties.

4.23. Apply the rank-sum statistic (4.16) to the data in Example 4.3. Test against a positive shift in median response, H_a:$\boldsymbol{m}_0 < \boldsymbol{m}_1$, at $\alpha = 0.05$. Use the large-sample approximation, adjusting for ties.

4.24. Apply the rank-sum statistic (4.16) to the data in Exercise 4.21. Test against a positive shift in median response, H_a:$\boldsymbol{m}_0 < \boldsymbol{m}_1$, at $\alpha = 0.05$. Use the large-sample approximation, adjusting for ties.

4.25. For the data from Exercise 4.2, construct a 99% confidence interval for the difference in mean response between the two groups. Does the interval correspond with the inference you reported from the hypothesis test?

4.26. Repeat Exercise 4.25 at 95% confidence. Does the 95% interval correspond with the inference you reported from the hypothesis test in Exercise 4.2? Should it?

4.27. For the data from Exercise 4.3, assume that there may be some difference between the variances σ_1^2 and σ_2^2, so that the assumption of equal variances is not met. Construct a 95% confidence interval for the difference in mean response between the two groups. Does the interval correspond with the inference you reported from the hypothesis test?

4.28. For the data from Exercise 4.3, apply the logarithmic transform (as in Exercise 4.5). Assume that this transformation stabilized any variance heterogeneity. Construct a 95% confidence interval for the difference in mean transformed response between the two groups. Does the interval correspond with the inference you reported from the hypothesis test in Exercise 4.5? Does it correspond to the inference from the test in Exercise 4.3?

4.29. For the data from Exercise 4.8, construct a 95% confidence interval for the difference in proportion response between the two groups. Use the large-sample interval from (4.20). Does the interval correspond with the inference you reported from the hypothesis test?

4.30*. Verify the assertion in Section 4.2.2 that the variance function $V(\delta,\tau)$ in (4.21) is strictly positive for any $N_0 \geq 1$ and $N_1 \geq 1$. (*Hint*: work with $\delta = p_0 + p_1$ and $\tau = p_0 - p_1$.)

4.31. For the data from Exercise 4.8, construct a 95% confidence interval for the difference in proportion response between the two groups. Use the Wilson-type interval from (4.22) or (4.23). How does this interval compare to the interval from Exercise 4.29?

4.32. For the data from Exercise 4.19, construct a 90% confidence interval for the difference in mean response between the two groups. Use the Wilson-type interval from (4.28), and set $\varphi = -0.25$.

4.33. For the data in Exercise 4.20, construct a 99% confidence interval for the difference in mean response between the two groups using:

a. The simple large-sample interval from (4.27). Does the interval correspond with the inference you reported from the hypothesis test?

b. The Wilson-type interval from (4.28), with $\varphi = -0.25$. How does this compare with the interval from part (a)?

4.34. For the data from Exercise 4.21, construct a 95% confidence interval for the difference in mean response between the two groups. For the Wilson-type interval from (4.28), one has $\varphi = -1$ with these data, thus the interval collapses to the simpler Wald interval in (4.27). Apply it here.

5

Treatment-versus-control multiple comparisons

An important extension of the simple two-group comparisons discussed in Chapter 4 is the multiple-group comparison, that is, comparison of mean or proportion responses among more than two treatment groups. For example:

- Laboratory animals exposed to an environmental toxin at different exposure levels may exhibit differential responses. Can we detect whether the observed toxic response at any of these treatment levels is different from that in a concurrent set of control animals?
- Field studies of a species at different ecological sites may indicate differences in response, compared to a standard or benchmark site/condition. Viewing the established standard as a form of control group (and including a sample of responses from it as part of the experimental design), can we assess which of the sites' responses differ significantly from the control?

As in Chapter 4, 'treatment' is used here in a generic sense, indicating some exposure, intervention, observational condition, etc., being applied to the experimental units. Responses in the concurrent control represent the background state of the experimental units – including possibly a vehicle control or sham exposure. Inclusion of the concurrent control allows for proper comparability among the study groups. Similar to Chapter 4, the presentation in this chapter focuses on identification of significant differences between the treated experimental units, relative to the controls. In the following sections, we will describe a series of methods for making properly adjusted **multiple comparisons** between the treatment groups and the control group, emphasizing their use with environmental data.

5.1 COMPARING MORE THAN TWO POPULATIONS

The methods described in Chapter 4 for assessing differences between a single treatment and a concurrent control group can be extended to the multiple-treatment setting, where more than a single treatment level is included in the experimental design. If some quantitative variable such as dose is associated with each treatment level, then the statistical analyses focus on testing or estimating the nature of the dose response. In order to give inferences for dose response appropriate emphasis, we defer our discussion of them to Chapters 6 and 7. For the more general problem of comparing multiple treatments vs. a control when no dose-response model is involved, a wide variety of applicable methods is available, and we present a selection of these in this chapter.

Suppose there are $T \geq 2$ treatment groups or levels, to be compared to a single control group, where the investigator has allocated the experimental units randomly to the $T + 1$ groups. By way of review, the global null hypothesis of no difference in mean response among all $T + 1$ groups is written as

$$\mathrm{H_0}: \mu_0 = \mu_1 = \cdots = \mu_T \, ,$$

and this may be tested against the global alternative hypothesis of

$$\mathrm{H_a}: \mu_i \neq \mu_j \text{ (for some } i \neq j)$$

via common methods of **analysis of variance** or **ANOVA**. These ANOVA techniques may be viewed as an extension of the two-group pooled variance t-statistic (4.1) that we described in Chapter 4. For the $(T+1)$-group case, we compare variability between group means to variability within groups about their respective means via the F-statistic

$$F_{\text{calc}} = \frac{\frac{1}{T} \sum_{i=0}^{T} n_i (\overline{Y}_{i+} - \overline{Y}_{++})^2}{s^2} \, , \tag{5.1}$$

where $\overline{Y}_{i+}$ is the sample mean for the ith group, $\overline{Y}_{++} = Y_{++}/n_+$ is the overall grand mean, and

$$s^2 = \frac{\sum_{i=0}^{T} (n_i - 1) s_i^2}{\sum_{i=0}^{T} (n_i - 1)} = \frac{1}{n_+ - (T+1)} \sum_{i=0}^{T} (n_i - 1) s_i^2 \, .$$

s^2 is a weighted, or pooled, average of all the individual sample variances. We say that this estimator has $\nu = \sum_{i=0}^{T} (n_i - 1)$ *df*. It corresponds to the usual **mean square for error** (MSE) from a one-factor ANOVA. The test statistic in (5.1) is referenced to an F distribution with T and $\nu = n_+ - (T +$

1) df. If F_{calc} exceeds the upper-α critical point from this reference distribution, then the null hypothesis of equal means is rejected in favor of the alternative that at least two means differ. More complete introductions to these methods are given in a host of texts, including Neter *et al.* (1996 Chapter 16).

Example 5.1 Web Building in House Spiders
In a study of the effects of temperature on the web-building behavior of the common house spider *Achaearanea tepidariorum*, interest centered on whether ambient temperature had any effect on the size of the web constructed. Since bigger spiders tend to spin larger webs, the ratio of the web weight to spider weight was employed as the response variable. (The ratios were transformed via a base-10 logarithm, in order to stabilize variance.) The data appear in Table 5.1. (The data were kindly provided by L. Barghusen of Miami University.)

Table 5.1 Temperature effects on $\log_{10}$(web weight/spider weight) for *A. tepidariorum*

Temp. (°C)	*Sample size*	$log_{10}\left\{\frac{web\ weight}{spider\ weight}\right\}$	*Sample mean (std. dev.)*
10	8	–2.81508, –2.74254, –2.69473, –2.58087 2.30309, –2.82513, –2.63073, –2.10109	–2.5867 (0.2572)
15	8	–2.23378, –2.85680, –2.31367, –2.93023 –2.69093, –2.56165, –2.61604, –2.97926	–2.6478 (0.2743)
20	10	–1.97753, –1.98788, –1.92117, –2.42344 –2.53879, –2.29301, –2.19948, –2.27348 –2.13957, –2.43604	–2.2190 (0.2126)
25	9	–2.02078, –2.54437, –2.41178, –1.95839 –3.02343, –2.63536, –2.63536, –2.75174 –1.99863	–2.4422 (0.3755)

For this study, the null hypothesis was that the average log weight ratio was the same in all four temperature conditions, $H_0{:}\mu_{10} = \mu_{15} = \mu_{20} = \mu_{25}$, vs. the alternative that at least two temperature conditions differed with respect to this mean log ratio. Side-by-side box-and-whisker plots for these data are provided in Fig. 5.1.

From the figure, we see that the log ratio varies among different temperature conditions, although the variances appear roughly homogeneous. Formal assessment of whether the center of the ratio distribution is the same for the four temperature conditions is evaluated via the F-test described above. For the group means and variances in Table 5.1, we have $T = 3$, $n_+ = 35$, and $\overline{Y}_{++} = -2.4585$. Thus,

$$F_{\text{calc}} = \frac{\frac{1}{3}\left\{ (8)[-2.587 - (-2.459)]^2 + \cdots + (9)[-2.442 - (-2.459)]^2 \right\}}{\frac{1}{31}\left\{ (7)(0.257)^2 + \cdots + (8)(0.376)^2 \right\}},$$

or $F_{\text{calc}} = 0.3313/0.0814 = 4.068$. This has an associated P-value of $P[F(3,31) > 4.068] = 0.015$. The research hypothesis that at least two of the temperature conditions differ with respect to log-weight ratios is supported by this result. The question of where the specific temperature differences lie for these data will be explored below.

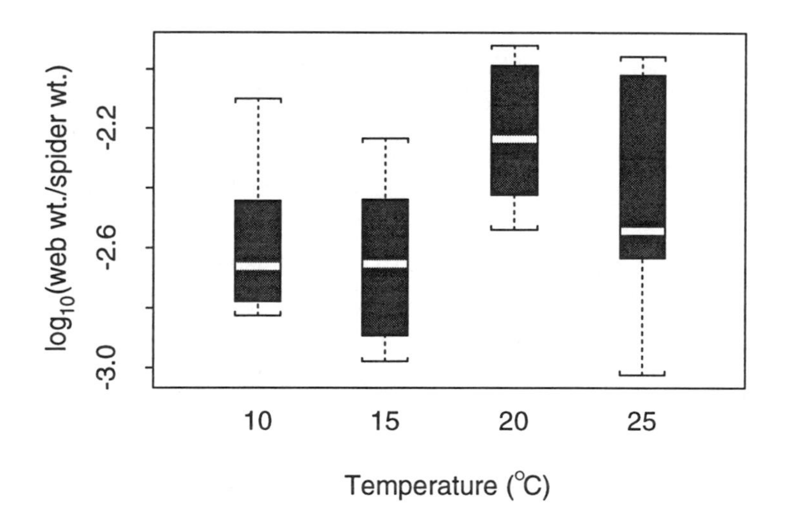

Fig. 5.1 Side-by-side boxplots for weight ratios with common house spiders.

The F-statistic can be calculated via SAS PROC GLM, as illustrated in Fig. 5.2. In the figure, the `class` statement defines the classification variable denoting group membership. In this example, the variable `temp` assumes four values corresponding to 10, 15, 20, and 25°C; see `Class Level Information` in Fig. 5.3.

In the (edited) SAS output from Fig. 5.3, the value of F_{calc} is listed under `F Value`; the associated P-value is given under `Pr > F`.

```
* SAS code for spider data from Table 5.1;

data spiders;
  input temp logratio @@;
  cards;
10 -2.81508  10 -2.69473  10 -2.30309  10 -2.63073
10 -2.74254  10 -2.58087  10 -2.82513  10 -2.10109
15 -2.23378  15 -2.31367  15 -2.69093  15 -2.61604
15 -2.85680  15 -2.93023  15 -2.56165  15 -2.97926
20 -1.97753  20 -1.92117  20 -2.53879  20 -2.19948
20 -2.13957  20 -1.98788  20 -2.42344  20 -2.29301
20 -2.27348  20 -2.43604
25 -2.02078  25 -2.41178  25 -3.02343  25 -2.63536
25 -2.54437  25 -1.95839  25 -2.63536  25 -2.75174
25 -1.99863

proc sort; by temp;
proc univariate plot;  *descriptive summaries;
  by temp;             *incl. side-by-side boxplots;
  var logratio;

proc glm order=data;
  class temp;
  model logratio = temp;
```

Fig. 5.2 SAS PROC GLM code for comparing web-spider weight ratios from Table 5.1.

Alternatively, the *F*-statistic can be calculated using the S-PLUS `aov` function. Assume a data frame, say `spider.df`, has been defined with `temp` as a factor and `spiderwt` and `webwt` as two additional variables. Then, the *F*-test for comparing log-weight ratios among different temperatures is formed using the S-PLUS command

```
summary(aov(log10(webwt/spiderwt)~temp,data=spider.df))
```

where `log10` transforms the weight ratio, `aov` fits the one-factor ANOVA model, and `summary` constructs the ANOVA table. In S-PLUS, the tilde ('~') notation reflects the syntax for model construction, where response variables are placed on the left side of the tilde and explanatory variables (or functions of the explanatory variables) appear on the right side. See Chambers and Hastie (1992a) for details on model formulation in S-PLUS, Chambers (1992a) for data frame specification and construction, and Chambers *et al*. (1992) for discussions on S-PLUS functions for ANOVA. □

```
                    The SAS System
           General Linear Models Procedure

               Class Level Information
              Class     Levels     Values
              TEMP           4     10 15 20 25
        Number of observations in data set = 35

Dependent Variable: LOGRATIO
Source         DF     Sum of Squares     F Value     Pr > F
Model           3         0.99386107        4.07     0.0151
Error          31         2.52462736
C  Total       34         3.51848843

    R-Square              C.V.           LOGRATIO Mean
    0.282468          -11.60796          -2.45845286
```

Fig. 5.3 SAS output (edited) for *F*-test with web-spider weight ratios from Table 5.1.

In many environmental applications, comparisons among multiple groups often extend beyond the global hypotheses $H_0{:}\mu_0 = \mu_1 = \cdots = \mu_T$. Specific questions can include whether any or all of the various treatment groups differ significantly from the concurrent control group, or whether significant differences exist in the responses among adjacent treatment groups, etc. These inferences involve **multiple comparisons** among the $T + 1$ groups, which in and of themselves cannot be assessed via global ANOVA methodology.

For example, suppose interest exists in identifying which specific treatment groups differ from a control group, such as in Chapter 4 (but now with $T \geq 2$ treatment groups). At first glance, differences among the $T + 1$ groups might be assessed by performing T separate analyses corresponding to the T different treatment-vs.-control comparisons, that is, calculating a different test statistic for each comparison. This overlooks an important fact about the experimental error rate, however: assuming each of the separate T comparisons is conducted at significance level α, we know each is subject to a possible false positive (Type I) error with probability α. These **per-comparison (Type I) errors** (also called **comparisonwise errors**) can accumulate across the ensemble of comparisons for the entire experiment. As a result, the probability of making at least one false positive error somewhere among the family of comparisons of interest can be much larger than the desired α level. (It can even approach 1 if T is large!) This is known as the **multiplicity problem**, and the goal in multiple comparison analysis is to adjust the per-comparison error rates so that the broader **familywise error rate** (also called the **experimentwise error rate**) is no greater than α (Hochberg and Tamhane, 1987, Chapter 1). This is accomplished by incorporating **simultaneous inferences** on the various multiple comparisons, as illustrated in the sections below.

Perhaps the most common form of multiple comparison problem in environmental biology is that noted above of testing the response from a series of $T \geq 2$ treatment groups against that from a concurrent control group. In terms of the mean response, the null effect occurs when the differences $\mu_i - \mu_0$ $(i \neq 0)$ are all zero, which implies that no differences in mean response exist between the control and any treatment group. This is a **multiple comparisons with control** (MCC) problem, also referred to as a **many-to-one** comparison problem. Notice that with T treatment groups and one control group under study, there are T comparisons of interest.

Alternatively, interest may include **all-pairwise differences** $\mu_i - \mu_h$, that is, study inter-treatment comparisons as well as control-vs.-treatment comparisons. Assessing all pairwise differences is often denoted as the 'MCA' problem, in contrast to the 'MCC' problem. Notice that there are many more comparisons to be made among all pairs of treatments than among all treatments vs. a control: the MCA problem produces $T(T+1)/2$ possible pairwise differences for study.

Other forms of comparisons are possible as well, such as all comparisons with the unknown best or highest responding treatment ('MCB' – Hsu, 1996, Chapter 4) or comparisons among factor levels with multiple or crossed factors (Genizi and Marcus, 1994). Discussion of these advanced topics is beyond the scope of this chapter, and we will center our presentation on MCC approaches. These are quite similar in form to MCA methods, and we will include brief discussions on MCA methods and other associated issues as well, in Section 5.6.

We assume throughout that true *inferences* are desired on the family of multiple hypotheses under study, leading to the adjustments for multiplicity described herein. If instead an exploratory analysis or other form of 'data snooping' is undertaken – where the research questions are less clearly defined – less stringent adjustments might be envisioned. For discussions of the philosophical and structural aspects of multiple testing, and for other, more advanced multiple comparison methods, see Hochberg and Tamhane (1987), Neter *et al.* (1996, Chapter 17), Hsu (1996), or the review by Shaffer (1995).

5.2 MULTIPLE COMPARISONS VIA BONFERRONI'S INEQUALITY

We begin our discussion with a useful (although generally conservative) catch-all method for performing simultaneous inferences. The method relies upon **Bonferroni's inequality**, which states that the familywise probability of any set of events occurring is always at least as large as 1 minus the sum of the probabilities of each individual event not occurring. This concept may seem complicated at first, but it has a simple symbolic representation: denote a series of events as $\mathcal{A}_j$ $(j = 1, \ldots, J)$, and denote their complement as $\mathcal{A}_j^c$; that is, $\mathcal{A}_j^c$ denotes that $\mathcal{A}_j$ has not occurred. In the context of a set of hypothesis tests, $\mathcal{A}_j$ corresponds to the event that a correct statistical decision was made for the jth comparison in a family of J comparisons. For example, if the jth comparison of interest is a test of the null hypothesis H_0: $\mu_0 = \mu_j$, then $\mathcal{A}_j$ is the event that this null hypothesis is not rejected when it is true, while $\mathcal{A}_j^c$ represents the event that this null hypothesis is rejected when it is true. Thus, the complementary event, $\mathcal{A}_j^c$, corresponds to a false positive error on the jth comparison, and $P[\mathcal{A}_j^c]$ corresponds to the associated false positive (Type I) error rate. Alternatively, in the context of a set of confidence intervals, $\mathcal{A}_j$ corresponds to the event that the parameter of interest was covered correctly by the jth interval. For example, if the jth parameter of interest is $\mu_j - \mu_0$, then $\mathcal{A}_j^c$ represents the event that the confidence interval did not contain the true value for the difference. Notice, both for hypothesis tests and confidence intervals, that the multiple events $\mathcal{A}_j$ often are correlated, since they are derived from the same set of data. This does not affect the Bonferroni method; it is valid for any form of correlation structure among the $\mathcal{A}_j$.

Using this notation, let $P[\mathcal{A}_1\mathcal{A}_2\cdots\mathcal{A}_J]$ be the probability that all J events occur simultaneously. Then the Bonferroni inequality may be written as

$$P[\mathcal{A}_1\mathcal{A}_2\cdots\mathcal{A}_J] \geq 1 - \sum_{j=1}^{J} P[\mathcal{A}_j^c]. \tag{5.2}$$

In the realm of simultaneous statistical inference, $\{\mathcal{A}_1\mathcal{A}_2\cdots\mathcal{A}_J\}$ represents the joint event that correct decisions are made on all J hypothesis tests for a family of tests (or that all of the J confidence intervals cover their respective parameters of interest).

If determination of the complementary probabilities, $P[\mathcal{A}_j^c]$, is difficult, an alternative form for (5.2) is possible:

$$P[\mathcal{A}_1\mathcal{A}_2\cdots\mathcal{A}_J] \geq \sum_{j=1}^{J} P[\mathcal{A}_j] - (J-1)\,. \tag{5.3}$$

Except in very specific cases – some of which are described in the following sections – finding critical points to set the probability $P[\mathcal{A}_1\mathcal{A}_2\cdots\mathcal{A}_J]$ at exactly $1-\alpha$ is not feasible. But, if we had critical points corresponding to the individual events $\mathcal{A}_j^c$ such that their probabilities, $P[\mathcal{A}_j^c]$, were calculable, we could find a lower bound for the simultaneous coverage probability $P[\mathcal{A}_1\mathcal{A}_2\cdots\mathcal{A}_J]$ using (5.2). The simplest way to construct a lower bound from (5.2) is to set the complementary probabilities, $P[\mathcal{A}_j^c]$, equal to one another, say $P[\mathcal{A}_j^c] = \gamma$. The lower bound in (5.2) is then $1 - J\gamma$. If, for example, we desire that the simultaneous probability represent a simultaneous confidence level of $1-\alpha$, it is clear that taking $\gamma = \alpha/J$ produces the desired result. That is, if each complementary event occurs with probability α/J, the simultaneous collection of original events occurs with probability at least $1-\alpha$, using (5.2). We call this **minimal simultaneous coverage**.

Example 5.2 Simultaneous Bonferroni intervals for the spider web data
To illustrate the Bonferroni approach, consider the setting where paired comparisons are desired only with a pre-selected, 'standard' treatment group. For example, in the web-building data from Example 5.1, the 20°C temperature condition corresponds to typical house spider habitats. Viewed as the standard, comparisons of the 20°C group with the other temperature conditions might be of scientific interest.

To compare 20°C vs. all other temperature conditions, $J = 3$ distinct pairwise differences are calculated: $\mu_{20} - \mu_{10}$, $\mu_{20} - \mu_{15}$, and $\mu_{20} - \mu_{25}$ where the μs refer to true mean log-weight ratios. (For purposes of illustration, the indexing has been modified to represent the temperature condition of each group.) In order to obtain simultaneous inferences for the three paired differences, we apply the Bonferroni inequality. We take the 'event' $\mathcal{A}_j$ to be coverage of $\mu_{20} - \mu_j$ ($j = 10, 15, 25$) using the standard t-interval from (4.17):

$$\overline{Y}_{20} - \overline{Y}_j \pm t_{\gamma/2}(\nu)\, s \sqrt{\frac{1}{n_{20}} + \frac{1}{n_j}}$$

where s is the root mean square error, and ν is the pooled *df*, $\nu = \sum_{i=0}^{T}(n_i - 1)$. (We will specify a value for γ below.) The complementary event, $\mathcal{A}_j^c$, is *non*coverage, that is, that $\mu_{20} - \mu_j$ is not contained in this interval. Since we used the tail area subscript $\gamma/2$ in $t_{\gamma/2}(\nu)$, this set of intervals produces noncoverage probabilities for each j of simply $P[\mathcal{A}_j^c] = 1 - (1 - \gamma) = \gamma$, $j = 10, 15, 25$.

Our goal is minimal simultaneous $1 - \alpha$ coverage, so we wish to set $P[\mathcal{A}_{10}\mathcal{A}_{15}\mathcal{A}_{25}] \geq 1 - \alpha$. Since $P[\mathcal{A}_j^c] = \gamma$, (5.2) gives $P[\mathcal{A}_{10}\mathcal{A}_{15}\mathcal{A}_{25}] \geq 1 - 3\gamma$. Hence, we achieve our goal if $\gamma = \alpha/3$:

$$\bar{Y}_{20} - \bar{Y}_j \pm t_{\alpha/6}(\nu)\, s \sqrt{\frac{1}{n_{20}} + \frac{1}{n_j}}\,. \tag{5.4}$$

The Bonferroni inequality ensures that (5.4) represents a set of minimal simultaneous $1 - \alpha$ confidence limits for all three differences $\mu_{20} - \mu_j$, $j = 10, 15, 25$. At $1 - \alpha = 0.95$, we set $\gamma = 0.05/3 = 0.0167$.

Recall that the root mean square error from the spider data was $s = 0.2854$, associated with $\nu = 31$ *df*. The corresponding Bonferroni critical value from a t distribution for these intervals is $t_{\gamma/2}(\nu) = t_{0.0083}(31) = 2.531$. This leads to the following set of simultaneous confidence intervals:

$$\mu_{20} - \mu_{10}: \quad [-2.2190 - (-2.5867)] \pm 2.531(0.2854)\sqrt{\frac{1}{10} + \frac{1}{8}}$$

$$\mu_{20} - \mu_{15}: \quad [-2.2190 - (-2.6478)] \pm 2.531(0.2854)\sqrt{\frac{1}{10} + \frac{1}{8}}$$

$$\mu_{20} - \mu_{25}: \quad [-2.2190 - (-2.4422)] \pm 2.531(0.2854)\sqrt{\frac{1}{10} + \frac{1}{9}}\,.$$

That is, $0.025 < \mu_{20} - \mu_{10} < 0.710$, $0.086 < \mu_{20} - \mu_{15} < 0.771$, and $-0.109 < \mu_{20} - \mu_{25} < 0.555$. This analysis suggests that the web-to-spider weight ratios are significantly larger in the 20°C temperature condition than in either the 10°C or 15°C condition. We do not have evidence of a significant difference between the 20°C and 25°C conditions, however. These inferences are consistent with the expectation that the web yield of spiders are largest under typical temperature conditions. □

Notice that the only effect of the Bonferroni correction on the final intervals in Example 5.2 is to adjust downward the probability level for the two-sided critical point: instead of $t_{0.025}(31)$, we used $t_{0.0083}(31)$ from a standard t-table. This illustrates perhaps the greatest value of the Bonferroni method: its exceeding simplicity. (It is also true that the Bonferroni method excels in its great applicability. It is valid as long as the joint probability of interest exists and if the probabilities that form the lower bound can be evaluated.) Contrastingly, however, since the Bonferroni method is based on an inequality, it can only ensure minimal $1 - \alpha$ coverage. Indeed, in some instances the actual simultaneous coverage can be quite high, for example, 99.9% instead of 95%. This added confidence implies excess width in the simultaneous intervals. Application of the method carries some trade-offs, although the inequality's simplicity and applicability have endowed it with widespread acceptance. In general, the Bonferroni inequality is most useful when a small or unique set of pre-specified comparisons is under study, as in Example 5.2.

It is possible to improve upon the Bonferroni result, and in certain settings a number of sharper inequalities exist that may be employed instead of (5.2) (Hochberg and Tamhane, 1987, Appendix 2). We mention one in particular here, since in the normal sampling case it competes favorably with (5.2), and also since an extension of it is applied in the T3 procedure discussed in Section 5.3.2. Due to Šidák (1967), the inequality in its basic form is multiplicative rather than additive: start with J normally distributed random variables X_j with zero means, arbitrary variances σ_j^2, and arbitrary correlations ρ_{jk} $(j \neq k)$. Denote as $\mathcal{A}_j$ the event $\left\{ |X_j| \leq \xi_j \right\}$ for some arbitrary constants ξ_j. Then, Šidák's inequality states

$$\mathrm{P}[\mathcal{A}_1\mathcal{A}_2\cdots\mathcal{A}_J] \geq \prod_{j=1}^{J} \mathrm{P}[\mathcal{A}_j] .$$

As a consequence of this, minimal simultaneous $1 - \alpha$ confidence limits on any set of $J > 1$ pairwise (or other form of) contrasts among normally distributed sample means $\overline{Y}_j$ and $\overline{Y}_k$ are

$$\overline{Y}_j - \overline{Y}_k \pm t_{\tilde{\alpha}/2}(\nu)\, s \sqrt{\frac{1}{n_j} + \frac{1}{n_k}} \quad .$$

$(j \neq k)$, where $\tilde{\alpha}$ is the Šidák-adjusted value $\tilde{\alpha} = 1 - (1 - \alpha)^{1/J}$. The Šidák adjustment is similar to the Bonferroni adjustment here: it acts only upon the critical point, and then only by changing the upper probability level used to find that point. As J increases, however, application of Šidák's inequality leads to narrower intervals than the corresponding Bonferroni adjustment. (Additional improvement is possible by exploiting the relationship of Šidák's inequality with the Studentized maximum modulus critical points $|\mathcal{M}|_{T,\nu}^{(\alpha)}$; see Hochberg and Tamhane (1987, Section 3.2.3).)

5.3 MULTIPLE COMPARISONS WITH A CONTROL – NORMAL SAMPLING

When the index $i = 0$ corresponds to a control group from an environmental study, it is common to ask if any of the various treatment groups differ significantly from it. This is a problem of multiple comparisons with a control (MCC).

In some applications, it is common to see the multiple comparisons preceded by a preliminary F-test of the global null hypothesis $H_0: \mu_0 = \mu_1 = \cdots = \mu_T$, using (5.1). Ostensibly, *if* the F-test is seen to be insignificant it makes little sense to continue with specific comparisons among the μ_is. Otherwise, one goes on to a second stage and makes individual pairwise comparisons among the μ_is. If the common t-statistic (4.1) between any two means, μ_i and μ_h, is seen to be significant, *each at per-comparison significance level* α, then the pairwise difference is labeled significant as well. This is known as Fisher's **protected least significant difference** (LSD) procedure (Fisher, 1935a, Section 24). It adjusts for multiplicity, in effect, via the first-stage F-test screen.

Although reasonable in the abstract, in practice the protected LSD *fails* to protect fully the familywise false positive error rate. When the global null H_0 is true the familywise false positive (Type I) error rate is held fixed at α, but configurations of the means can exist away from H_0 that drive the false positive rate above the nominal level (Hochberg and Tamhane, 1987, Example 2.4.6; Keselman *et al.*, 1991). Because of this weakness, we will avoid use of protected tests using preliminary F-test screens. Our discussion will be limited primarily to single-stage multiple-comparison procedures applied directly on the unknown means or proportions.

In this section, we present methodologies for performing MCC inferences appropriate for environmental problems. We emphasize the well-established case of sampling from a normal parent distribution, but we will see in future sections that these methods extend to nonnormal data as well.

5.3.1 Homogeneous variances

Suppose continuous data are generated under normal distribution sampling with homogeneous variances: $Y_{ij} \sim N(\mu_i, \sigma^2)$, $i = 0, \ldots, T$; $j = 1, \ldots, n_i$. Consider testing the null hypothesis that no differences exist between any of the treatment groups ($i \neq 0$) and the concurrent control ($i = 0$),

$$H_0: \mu_i - \mu_0 = 0 \quad \text{(for all } i \neq 0\text{)},$$

which is equivalent to the global null hypothesis $H_0: \mu_0 = \mu_1 = \cdots = \mu_T$. The two-sided (multiple) alternative hypothesis is

$$H_a: \mu_i - \mu_0 \neq 0 \quad \text{(for some } i \neq 0\text{)}.$$

This two-sided alternative implies that *at least one* of the treatment levels exhibits a response that differs significantly from the control. Our goal is to identify at which, if any, of the treatment means the difference is significant.

One-sided multiple alternatives are also possible; for example, $H_a: \mu_i - \mu_0 > 0$ (for some $i \neq 0$) suggests that at least one of the treatment levels exhibits an increased mean response relative to the control.

This problem was developed by Dunnett (1955), and we refer to the MCC inferences as **Dunnett comparisons**. Under Dunnett's MCC analysis, the t-statistic from Section 4.1.1 is the basic measure of departure from H_0 for each of the T possible comparisons. At each $i = 1, \ldots, T$, calculate

$$t_{i0} = \frac{\overline{Y}_i - \overline{Y}_0}{s_p \left[\frac{1}{n_i} + \frac{1}{n_0}\right]^{1/2}}. \tag{5.5}$$

As in (5.1), s_p^2 is the pooled estimator of the common variance, with $\nu = n_+ - (T + 1)$ *df*.

For the two-sample (one-comparison) setting, the t-statistic was referred to a t distribution to test H_0. With multiple comparisons, however, the null reference distribution for the set of t_{i0}-statistics extends to a **multivariate *t* distribution**. As might be expected, the parameters for this multivariate reference distribution are more complex than those for the simple univariate case. They include (a) the number of comparisons, T; (b) the *df* associated with the pooled variance estimate, ν; and (c) a correlation matrix, $\mathbf{R} = \{\rho_{ih}\}$, representing the correlations between estimates of treatment effect, $\rho_{ih} = \text{Corr}[(\overline{Y}_i - \overline{Y}_0),(\overline{Y}_h - \overline{Y}_0)]$ $(i \neq h)$. For the homoscedastic normal distribution sampling setting considered here, the correlations take on a fairly simple form:

$$\rho_{ih} = \sqrt{\frac{n_i n_h}{(n_i + n_0)(n_h + n_0)}} \quad . \tag{5.6}$$

The definition of ρ_{ih} is symmetric, so that $\rho_{ih} = \rho_{hi}$. Thus, there are $T(T + 1)/2$ different ρ_{ih}-values in a $(T + 1)$-group design. Notice that if the experiment design is **treatment-balanced**, with $n_i = n$ $(i = 1, \ldots, T)$ but $n_0 \neq n$, then the correlations simplify to a single constant, $\rho = n/(n_0 + n)$. If the design is **fully balanced**, with $n_0 = n_1 = \cdots = n_T$, then $\rho = 1/2$.

Rejection of the global null hypothesis $H_0: \mu_i - \mu_0 = 0$ (for all i) in favor of the two-sided alternative $H_a: \mu_i - \mu_0 \neq 0$ (for some $i \neq 0$) occurs when *any* of the t_{i0}s exceed in absolute value the two-sided upper-α critical point from the appropriate multivariate t distribution, denoted by $|\mathcal{T}|^{(\alpha)}_{T,\nu,\{\rho_{ih}\}}$. This is equivalent to rejecting H_0 when the largest absolute t_{i0} exceeds $|\mathcal{T}|^{(\alpha)}_{T,\nu,\{\rho_{ih}\}}$, that is, when

$$\max\left\{ |t_{10}|, \ldots, |t_{T0}| \right\} > |\mathcal{T}|^{(\alpha)}_{T,\nu,\{\rho_{ih}\}}.$$

For testing against the one-sided alternative that at least one of the μ_is exceeds μ_0, reject H_0 when the largest t_{i0} exceeds the one-sided upper-α critical point $\mathcal{T}^{(\alpha)}_{T,\nu,\{\rho_{ih}\}}$. Against a decreasing one-sided alternative, reject H_0 when the smallest t_{i0} drops below $-\mathcal{T}^{(\alpha)}_{T,\nu,\{\rho_{ih}\}}$.

An important feature of this simultaneous construction is that inferences on individual departures from H_0 can be included at the same familywise significance level as the parent inference on H_0. That is, H_0 and H_a may be viewed as a (finite) family of local hypotheses

$$H_{0i}: \mu_i - \mu_0 = 0 \quad \text{vs.} \quad H_{ai}: \mu_i - \mu_0 \neq 0$$

for each $i = 1, \ldots, T$. Rejection of H_{0i} in favor of any individual H_{ai} implies rejection of the larger, global family of hypotheses H_0. If global rejection of H_0 occurs, it is valid to report not only that H_0 is rejected in favor of H_a, but also at which specific, local treatment levels the departure from H_{0i} occurs in favor of H_{ai}, all at (simultaneous) significance level α.

When calculating the critical point $|\mathcal{T}|^{(\alpha)}_{T,\nu,\{\rho_{ih}\}}$, it may be the case that the ρ_{ih} are all equal to the same value, say, ρ (as in the case of balanced designs). If this is true, we simplify the notation and write $|\mathcal{T}|^{(\alpha)}_{T,\nu,\rho}$ and $\mathcal{T}^{(\alpha)}_{T,\nu,\rho}$ for the two-sided and one-sided critical points, respectively. Also, as $\nu \to \infty$, the T-variate t distribution approaches a T-variate normal distribution with identical correlation matrix. Thus one may simplify the notation to $|\mathcal{Z}|^{(\alpha)}_{T,\rho} = |\mathcal{T}|^{(\alpha)}_{T,\infty,\rho}$; cf. Section 5.4.1, below.

Unfortunately, in the general, unbalanced case, there will be as many possible multivariate t distributions referenced to the t_{i0} – and hence as many possible tables of critical points $|\mathcal{T}|^{(\alpha)}_{T,\nu,\{\rho_{ih}\}}$ – as there are different possible configurations of the ρ_{ih}. Tables for this prohibitive number of possibilities do not exist, although in certain cases one can find selected tables for use. For instance, Dunnett (1964) provides tables of critical points for $\rho = 1/2$, which corresponds to the case where all the treatment levels and the control have equal sample sizes. For the treatment-balanced case where the per-treatment sample sizes are all equal to, say, n (but $n_0 \neq n$), there is only one correlation parameter, $\rho = n/(n_0 + n)$. Tables for $|\mathcal{T}|^{(\alpha)}_{T,\infty,\rho}$ can be found for some common values of ρ; see Hochberg and Tamhane (1987, Tables 4–5) or Bechhofer and Dunnett (1988). In the general unbalanced case, computer calculation of the critical points is necessary; see, for example, the SAS function PROBMC, or computer code by Dunnett (1989) for the limiting normal case as $\nu \to \infty$.

When it is not possible to design the study in a treatment-balanced form, the ρ_{ih}s may differ. If computer resources for finding the exact critical points $|\mathcal{T}|^{(\alpha)}_{T,\nu,\{\rho_{ih}\}}$ are unavailable, a simple approximation may be used instead: calculate a central value for ρ, say $\bar{\rho}$, and then replace $|\mathcal{T}|^{(\alpha)}_{T,\nu,\{\rho_{ih}\}}$ with the approximate critical point $|\mathcal{T}|^{(\alpha)}_{T,\nu,\bar{\rho}}$ found from existing equicorrelated tables. A common recommendation for $\bar{\rho}$ in this case is the average correlation (Hochberg and Tamhane, 1987, Section 5.2) among those parameters being studied. For example, if testing is to be performed only against the control group, find the individual per-control correlations ρ_{i0} via (5.6), and use the average

$$\bar{\rho} = \frac{1}{T}\sum_{i=1}^{T}\rho_{i0}. \tag{5.7}$$

Example 5.3 MCC for the spider web data
We illustrate calculation of the MCC hypothesis tests using the spider web data presented from Example 5.2. As discussed there, 20°C represents the common temperature level experienced by spiders while constructing their webs, thus it represents a 'standard' group against which multiple comparisons are desired.

In this study, the sample sizes are $n_{10} = n_{15} = 8$ and $n_{25} = 9$, while $n_{20} = 10$. (As in Example 5.2, the indexing has been modified to indicate the temperature condition of each group.) Since the 20°C group is the standard/reference condition against which comparisons are desired, we view n_{20} as n_0, that is, the 'control' sample size is $n_{20} = 10$.

Due to the unbalanced nature of the design, we cannot easily find in published tables the critical point upon which to base the simultaneous confidence intervals. A useful approximation is the arithmetic average of the $T = 3$ different correlation parameters associated with the control, as in (5.7). For this, we need to find the values of $\rho_{i,20}$. For example,

$$\rho_{10,20} = \sqrt{\frac{n_{10}n_{20}}{(n_{20} + n_{10})(n_{20} + n_{20})}} = \sqrt{\frac{80}{(18)(20)}} = 0.4714.$$

Similarly, we find $\rho_{15,20} = 0.4714$ and $\rho_{25,20} = 0.4867$. Applying (5.7) gives $\bar{\rho} = (\rho_{10,20} + \rho_{15,20} + \rho_{25,20})/3 = 0.4765$. Thus an approximate critical value available for simultaneous tests of the individual hypotheses

$$\mathrm{H}_{0i}{:}\,\mu_i - \mu_{20} = 0 \quad \text{vs.} \quad \mathrm{H}_{ai}{:}\,\mu_i - \mu_{20} \neq 0$$

(i = 10, 15, 25) at $\alpha = 0.05$ is $|\mathcal{T}|^{(0.05)}_{3,31,0.4765}$. Reject any H_{ai} if the corresponding test statistic, $t_{i,20}$ from (5.5), exceeds this value.

The exact value of this multivariate t critical point is available via the SAS PROBMC function. Figure 5.4 gives the SAS code and (edited) output. From it, we find $|\mathcal{T}|^{(0.05)}_{3,31,0.4765} = 2.474$. If the SAS function is unavailable, an approximation to the critical point $|\mathcal{T}|^{(0.05)}_{3,31,0.4765}$ employs linear interpolation on $1/(1-\rho)$ in established sources such as Hochberg and Tamhane (1987, Table 5). If the exact value for error $df = 31$ is not tabulated, a conservative critical point is the nearest (lower) value for the df: $|\mathcal{T}|^{(0.05)}_{3,31,0.4765} \leq |\mathcal{T}|^{(0.05)}_{3,30,0.4765} \approx 2.476$. This compares well with the value used above of 2.474.

The t-statistics for these comparisons are $t_{10,20} = -2.716$, $t_{15,20} = -3.167$, and $t_{25,20} = -1.702$. In absolute value, the first two exceed $|\mathcal{T}|^{(0.05)}_{3,31,0.4765} = 2.474$, but the third does not. Hence, as with the Bonferroni comparisons illustrated previously, log-weight ratios at the 20°C temperature condition are significantly larger than either the 10°C or 15°C conditions, but there is no evidence of a significant difference between the 20°C and 25°C conditions. Notice that a natural research hypothesis to develop from these results is that maximal web production occurs at the optimal 20°C condition. Future data could help validate this proposition, using larger sample sizes to increase power (see below) and operating under the one-sided alternatives $H_{ai}: \mu_i - \mu_{20} < 0$. □

```
* SAS code to find multivariate t critical points;
* rcon PROBMC params are the square roots of rhos;

data tcrit;
    rcon1 = sqrt(.4765);
     rcon2 = rcon1;
      rcon3 = rcon1;
t331 = probmc('dunnett2', . , .95 , 31 , 3 ,
                                    OF rcon1-rcon3);
cards;
proc print;
```

```
                  The SAS System

   OBS       RCON1       RCON2       RCON3       T331
     1     0.69029     0.69029     0.69029    2.47446
```

Fig. 5.4 SAS program and output (edited) to find multivariate t critical point.

We end this section by noting that prior sample size selection can be considered with MCC-type comparisons in mind. In particular, it is common to employ a treatment-balanced design with $n_i = n$ for $i = 1, \ldots, T$, and to set the control sample size, n_0, larger than the treatment sample size(s), n. This has some intuitive value: since the control group is being used as an anchor for the T multiple comparisons, it seems reasonable to design the study to yield more precise information on the control response. Greater precision is accomplished by increasing the sample size, hence the recommendation that $n_0 > n$. But how much of an increase is needed? One suggestion, dating back to Dunnett's original work on the problem (Dunnett, 1955), is to take the nearest integer to $n_0 = n\sqrt{T}$, where the total sample size, N, is predetermined by a prior power specification/calculation, resource limitations, or other constraints. For $n_0 = n\sqrt{T}$, N satisfies $N = n(T + \sqrt{T})$. Thus, given N, find the nearest integer to $n = (T + \sqrt{T})^{-1}N$, and then $n_0 = N/(1 + \sqrt{T})$. This allocation of sample sizes minimizes the common variance of the observed mean differences, and possesses other asymptotic optimality properties. More complex allocation schemes are discussed by Hochberg and Tamhane (1987, Section 6.1).

Example 5.3 (continued) MCC for the spider web data – sample size allocations

Suppose in the spider web study we were going to design a new experiment with the previous $T + 1 = 4$ temperature conditions 10, 15, 20, and 25°C. Suppose further that we can afford to allocate $N = 40$ spiders to the entire experiment. Assume that the 20°C condition serves as our control group. Using the minimum variance allocation scheme suggested above, we assign to temperature groups 10°C, 15°C, and 25°C

$$n_{10} = n_{15} = n_{25} = \frac{40}{(3 + \sqrt{3})} = 8.45 \approx 8,$$

and

$$n_{20} = \frac{40}{1 + \sqrt{3}} = 14.64 \approx 15$$

for the 20°C group.

Notice that the total sample size under this allocation is $n_{10} + n_{15} + n_{20} + n_{25} = 39$. For simplicity, the remaining available spider could be assigned to the control temperature condition: $n_{20} = 16$. □

5.3.2 Multiple comparisons with a control – Heterogeneous variances

The MCC construction given above relies heavily on the assumption of equal variances. When the homogeneous variance assumption is not met, tests based on use of a pooled estimator of variance in (5.1) are inappropriate. To adjust for possibly heterogeneous variances, one can apply the Bonferroni adjustment described in Section 5.2 directly to modified t-statistics such as (4.2). Alternatively, one can refer these modified t-statistics to corresponding versions of the multivariate t distribution, as follows.

Given data $Y_{ij} \sim \mathrm{N}(\mu_i, \sigma_i^2)$, $i = 0, \ldots, T$; $j = 1, \ldots, n_i$, consider the global null hypothesis of no difference between any of the treatment groups ($i \neq 0$) and the concurrent control ($i = 0$):

$$\mathrm{H}_0\text{:}\, \mu_i - \mu_0 = 0 \quad (\text{for all } i \neq 0).$$

The two-sided multiple alternative hypothesis is

$$\mathrm{H}_a\text{:}\, \mu_i - \mu_0 \neq 0 \quad (\text{for some } i \neq 0).$$

To detect departure from H_0 at any level i, we modify the t-statistic by using the individual sample variances, s_i^2, as in (4.2):

$$t_{i0} = \frac{\overline{Y}_i - \overline{Y}_0}{\sqrt{\dfrac{s_i^2}{n_i} + \dfrac{s_0^2}{n_0}}} .$$

With heterogeneous variances, however, the null reference distribution for t_{i0} is no longer multivariate t. We apply instead a Welch–Satterthwaite correction from (4.3): for the ith comparison, record the modified *df*

$$\nu_{i0} = \frac{\left(u_i + u_0\right)^2}{\dfrac{u_i^2}{n_i - 1} + \dfrac{u_0^2}{n_0 - 1}}, \tag{5.8}$$

where, as above, $u_i = s_i^2/(n_i - 1)$. With this correction, the approximate *df* are ν_{i0} in the multivariate-t reference distribution. The associated values to use for ρ_{ih} in $|\mathcal{T}|^{(\alpha)}_{T,\nu_{i0},\{\rho_{ih}\}}$ are quite complex with heterogeneous variances, however. A useful simplification uses an approximate multivariate-t reference distribution for t_{i0}, with all the correlations set equal to zero. This is, in fact, conservative, since it is known that $|\mathcal{T}|^{(\alpha)}_{T,\nu_{i0},\{\rho_{ih}\}} \leq |\mathcal{T}|^{(\alpha)}_{T,\nu_{i0},0}$ for any collection of correlations $\{\rho_{ih}\}$ (Kimball, 1951).

For the homogeneous variance case the absolute value of a zero-correlated multivariate t has the **Studentized maximum modulus** (SMM) distribution. If we set $\rho_{ij} = 0$, we could denote the two-sided SMM upper-α critical points as $|\mathcal{T}|^{(\alpha)}_{T,\nu,0}$. The SMM distribution is useful in many multiple-comparison settings, however, so in Section 1.5.4 it is accorded a separate notation: its (two-sided) upper-α critical points are denoted by $|\mathcal{M}|^{(\alpha)}_{T,\nu}$. Specialized tables of $|\mathcal{M}|^{(\alpha)}_{T,\nu}$ are available, for example, in Hochberg and Tamhane (1987, Table 7), or via the SAS PROBMC function. For one-sided testing, the reference distribution is the **Studentized**

maximum distribution from Section 1.5.4. Studentized maximum critical points $\mathcal{M}^{(\alpha)}_{T,\nu}$ are found in Hochberg and Tamhane (1987, Table 6); see also Hothorn and Lemacher (1991).

With heterogeneous variances, modifications for the MCC setting were described by Dunnett (1980), as part of what he called the **T3 procedure**. Here, the T3 procedure rejects H_0 in favor of H_a with simultaneous significance α when any $|t_{i0}|$ exceeds the corresponding approximate critical point $|\mathcal{M}|^{(\alpha)}_{T,\nu_{i0}}$.

5.3.3 Confidence intervals for comparisons with a control – Normal distribution sampling

The constructions described above for developing MCC hypothesis tests may be inverted to yield **simultaneous confidence intervals** for the T differences $\mu_i - \mu_0$. Just as the hypothesis tests are said to have simultaneous false positive error equal to α, the corresponding intervals possess simultaneous $1 - \alpha$ confidence that they cover all T differences $\mu_i - \mu_0$. For normal distribution sampling with homogeneous variances, the simultaneous limits are straightforward to compute, using the statistics employed for the hypothesis tests, above. Center the intervals at the point estimate $\overline{Y}_i - \overline{Y}_0$, and then add or subtract its standard error, multiplied by a critical point designed to provide the simultaneous $1 - \alpha$ coverage. The critical point is from the multivariate t, $|\mathcal{T}|^{(\alpha)}_{T,\nu,\{\rho_{ih}\}}$, producing

$$\overline{Y}_i - \overline{Y}_0 \pm |\mathcal{T}|^{(\alpha)}_{T,\nu,\{\rho_{ih}\}}\, s_p \sqrt{\frac{1}{n_i} + \frac{1}{n_0}}. \qquad (5.9)$$

When the correlations ρ_{ih} are constant, the multivariate-t critical point simplifies to $|\mathcal{T}|^{(\alpha)}_{T,\nu,\rho}$. If the correlations are heterogeneous, use (5.7) to find a constant $\bar{\rho}$; from which the approximate critical point is $|\mathcal{T}|^{(\alpha)}_{T,\nu,\bar{\rho}}$.

Example 5.4 Simultaneous MCC intervals for spider web data
Returning to the spider web data from Examples 5.1 and 5.3, it is straightforward to invert the test statistics calculated therein to find a set of simultaneous 95% confidence limits for the three differences of interest. (Recall that the inferences are approximate, since we use an approximate correlation $\bar{\rho} = 0.4765$.) From (5.9), the intervals are

$$\mu_{20} - \mu_{10}: \quad [-2.2190 - (-2.5867)] \pm 2.474(0.2854)\sqrt{\frac{1}{10} + \frac{1}{8}}$$

$$\mu_{20} - \mu_{15}: \quad [-2.2190 - (-2.6478)] \pm 2.474(0.2854)\sqrt{\frac{1}{10} + \frac{1}{8}}$$

$$\mu_{20} - \mu_{25}: \quad [-2.2190 - (-2.4422)] \pm 2.474(0.2854)\sqrt{\frac{1}{10} + \frac{1}{9}}.$$

That is, $0.033 < \mu_{20} - \mu_{10} < 0.703$, $0.094 < \mu_{20} - \mu_{15} < 0.764$, and $-0.101 < \mu_{20} - \mu_{25} < 0.548$.

As with the Bonferroni intervals in Example 5.2, these intervals suggest that the log-weight ratios are significantly larger in the 20°C temperature condition than in either the 10°C or 15°C conditions, but there is no evidence of a significant difference between the 20°C and 25°C conditions. Notice, however, that the Bonferroni procedure is more conservative than the Dunnett MCC method, as reflected in the larger critical value (2.531 for Bonferroni vs. 2.479 for the approximate MCC) and the wider simultaneous confidence intervals. □

If the population variances are heterogeneous, (5.9) is contraindicated. Instead, the T3 procedure from Section 5.3.2 can be inverted to yield approximate $1 - \alpha$ simultaneous limits on the differences $\mu_i - \mu_0$. These are

$$\overline{Y}_i - \overline{Y}_0 \pm |\mathcal{M}|^{(\alpha)}_{T,\nu_{i0}}\sqrt{\frac{s_i^2}{n_i} + \frac{s_0^2}{n_0}}, \tag{5.10}$$

where the approximate *df* for the SMM critical points, ν_{i0}, are derived from Welch–Satterthwaite corrections as in (5.8).

5.4 MULTIPLE COMPARISONS AMONG BINOMIAL POPULATIONS

When data occur in the form of proportions, a binomial sampling model often is valid. In this case, the normal-based MCC approaches are no longer exact and require modification. We assume data are recorded among the $T + 1$ groups as independent binomial counts, $Y_i \sim$ (indep.) $\mathcal{Bin}(N_i,\pi_i)$, producing observed proportions $p_i = Y_i/N_i$, $i = 0, \ldots, T$. Of interest is the difference between the treated and control proportions, parameterized as $\delta_i = \pi_i - \pi_0$, $i = 1, \ldots, T$. As we saw in Chapter 4, however, other relationships are also of potential interest, such as multiple comparisons among the risk ratios, π_i/π_0, or among the odds ratios, $[\pi_i(1 - \pi_0)]/[\pi_0(1 - \pi_i)]$. Coupling the approaches described in Chapter 4 with the Bonferroni adjustment from Section 5.2 can yield methods for multiple comparisons for risk ratios or odds ratios. In this section, however, we center attention on inference for the differences $\delta_i = \pi_i - \pi_0$, by generalizing many of the MCC concepts from previous sections.

5.4.1 Hypothesis tests comparing treatments to a control – Binomial sampling

For hypothesis testing, the null hypothesis of no difference in response probability between the treatment groups ($i \neq 0$) and the concurrent control ($i = 0$) is

$$H_0{:}\pi_i - \pi_0 = 0 \quad (\text{for all } i \neq 0).$$

The corresponding two-sided alternative hypothesis is

$$H_a{:}\pi_i - \pi_0 \neq 0 \quad (\text{for some } i \neq 0).$$

To test H_0 against H_a, a traditional approach has been to transform the observed proportions, and assume the transformation sufficiently normalizes the data to apply the normal-based MCC methods above. The arc sine/square root is a common choice for transforming binomial proportion data; see Levy (1975). As an MCC approach, transformations are most useful when testing hypotheses, but suffer when used to construct simultaneous confidence intervals since they may yield uninterpretable limits or limits that cannot be transformed back to the original scale. As such, we will direct attention to MCC approaches for binomial proportions without use of scale-changing transformations. For the settings discussed below, we focus on large-sample comparisons, appealing to the asymptotic normality of the p_i.

Perhaps the simplest approach to take for testing H_0 against H_a is to extend the Dunnett test from Section 5.3 and reject H_0 when the largest absolute difference $|p_i - p_0|$ divided by its standard error exceeds an appropriate critical point (Passing, 1984). This is a form of Wald test. The test statistic refers in large samples to a multivariate standard normal distribution with two-sided upper-α critical points $|\mathfrak{z}|^{(\alpha)}_{T,\{\rho_{ih}\}}$. Thus, H_0 is rejected when

$$\max\left\{ |z_{10}|, \ldots, |z_{T0}| \right\} > |\mathfrak{z}|^{(\alpha)}_{T,\{\rho_{ih}\}},$$

where

$$z_{i0} = \frac{|p_i - p_0|}{\left[\dfrac{p_i(1 - p_i)}{n_i} + \dfrac{p_0(1 - p_0)}{n_0}\right]^{1/2}} \tag{5.11}$$

are the individual Wald statistics for each treatment-vs.-control comparison.

Unfortunately, under binomial sampling the correlation parameters ρ_{ih} in $|\mathfrak{z}|^{(\alpha)}_{T,\{\rho_{ih}\}}$ are especially complex. They represent the large-sample correlations between $p_i - p_0$ and $p_h - p_0$ $(i \neq h)$, and for binomial data these are difficult to calculate exactly. Instead, consistent estimates

$$r_{ih} = \frac{1}{\left[1 + \dfrac{n_0\, p_i\,(1 - p_i)}{n_i\, p_0\,(1 - p_0)}\right]^{1/2}} \frac{1}{\left[1 + \dfrac{n_0\, p_h\,(1 - p_h)}{n_h\, p_0\,(1 - p_0)}\right]^{1/2}} \qquad (5.12)$$

are available . Of course, once the estimates r_{ih} are calculated from (5.12), one still must find a critical point for each different correlation matrix $\{ r_{ih} \}$. When $T = 2$, there is only one estimated correlation, r_{12}. The approximate critical point is then $|\mathfrak{z}|^{(\alpha)}_{2,r_{12}}$, which is available from tabular sources (Hochberg and Tamhane, 1987, Table 3). When $T = 3$, there are three correlations, r_{12}, r_{13}, and r_{23}. For this case it is possible to find the critical points relatively simply, using, for example, the SAS PROBMC function (as in Fig. 5.4, but with $df_E = \infty$ instead of $df_E = 3$).

When $T > 3$, computing the many estimated correlations can become prohibitive, and we often turn to approximations for the critical point using a single correlation parameter, r. The approximate critical point $|\mathfrak{z}|^{(\alpha)}_{T,r}$ is determined from tabular sources that assume a constant correlation, such as Hochberg and Tamhane (1987, Table 3). To approximate a single correlation, say $\bar{r}$, we use the average estimated correlation from (5.7):

$$\bar{r} = \frac{1}{T} \sum_{i=1}^{T} r_{i0} \ ,$$

where the values of r_{i0} are based on (5.12). Reject H_0 in favor of H_a when

$$\max\left\{ \ |z_{10}|, \ldots, |z_{T0}| \ \right\} \ > \ |\mathfrak{z}|^{(\alpha)}_{T,\bar{r}} \ . \qquad (5.13)$$

An alternative test procedure for the MCC setting has been given by Simes (1986) that avoids determination of multivariate critical points and also improves the global test of H_0 in some cases. This is achieved in a **sequentially rejective** fashion: we modify the per-comparison α levels at which each individual hypothesis is tested, depending on test results from previous hypotheses in the family. Begin by calculating the individual test statistics z_{i0} from (5.11), and then find the corresponding two-sided *P*-values $P_i = 2\{1 - \Phi(z_{i0})\}$, $i = 1, \ldots, T$. Denote the **ordered *P*-values** as $P_{(i)}$; that is, $P_{(1)} \leq P_{(2)} \leq \cdots \leq P_{(T)}$. Then, reject H_0 in favor of the global alternative hypothesis H_a when $P_{(i)} \leq i\alpha/T$ at any $i > 0$.

Unfortunately, Simes's procedure is not guaranteed to control the familywise false positive (Type I) error rate (Samuel-Cahn, 1996), although cases where the familywise error rate rises above α are rare. (The approach does control familywise error if the original, unordered P-values, P_i, are independent.) To improve the inferences from Simes's adjustment and also to assess individual, local departures such as $H_{ai}: \pi_i - \pi_0 \neq 0$, Hommel (1988) described a form of **stagewise rejective test**, based on a form of simultaneous inference known as **closed testing** (Marcus *et al.*, 1976). Closed testing is generally a more powerful way to perform multiple comparisons (Shaffer, 1995; Hothorn, 1997), although it is also more complex than the simpler methods described above.

To use Hommel's test, calculate the index

$$\kappa = \max\left\{ i \in \{1,\ldots,T\}: P_{(T-i+h)} > \frac{h\alpha}{i}, \text{ for all } h = 1,\ldots,i \right\}, \quad (5.14)$$

that is, the largest index i between 1 and T such that $P_{(T-i+h)} > h\alpha/i$ is true for all indexes up to and including that i. Reject H_0 in favor of any local departure H_{ai} when $P_i \leq \alpha/\kappa$. If κ cannot be calculated, that is, the maximum index under (5.14) does not exist, then reject H_0 at all $i = 1, \ldots, T$. Hommel showed that this modification does control the familywise false positive error rate at level α. Hochberg (1988) described similar modifications of Simes's method, although these are slightly less powerful than Hommel's approach (Hommel, 1989; Dunnett and Tamhane, 1993). Rom (1990; 1992) gave a more sensitive modification of Hommel's test that avoids calculation of κ, but requires special tables to determine each stagewise rejection of H_{ai}. Shaffer (1995) reviews other modifications of Simes's method.

Example 5.5 Multiple-dose aneuploidy in hamster cell cultures

Consider again the data on aneuploidy induction in Chinese hamsters from Example 4.2. Those data were part of a larger experiment using three exposures to diethylstilbestrol (DES) and an additional, untreated control. (Recall that the control group used for comparisons in this experiment was a concurrent vehicle control.) As an extension of the two-sample comparison from Example 4.2, we now perform multiple comparisons of the response in the three exposure groups, along with the untreated control, to the vehicle control. Here, $T = 4$.

Data from the entire experiment appear in Table 5.2. The table also presents summary statistics and intermediate calculations associated with the Wald statistics from (5.11) and with Hommel's stagewise test procedure. From these, we find $\bar{r} = 0.369$. Applying this 'common' correlation for

Table 5.2 Multiple-dose aneuploidy induction in Chinese hamster cells

	Vehicle control ($i = 0$)	Untr. control	DES doses (µg/ml) 5	10	15
Aneuploid cells	9	20	36	30	36
Total cells scored	150	300	200	150	175
Obs'd proportions, p_i	0.060	0.067	0.180	0.200	0.206
Obs'd differences, $p_i - p_0$	—	0.007	0.120	0.140	0.146
Standard errors, $se[p_i - p_0]$	—	0.024	0.033	0.038	0.036
Wald statistics, z_{i0}	—	0.29	3.64	3.68	4.06
P-values, P_i	—	0.772	2.7×10^{-4}	2.3×10^{-4}	4.9×10^{-5}
Ordered P-values, $P_{(i)}$	—	4.9×10^{-5}	2.3×10^{-4}	2.7×10^{-4}	0.772

Data from Dulout and Natarajan (1987).

the multivariate normal critical points gives $|\mathfrak{z}|^{(0.05)}_{4,0.369} = 2.39$ at $\alpha = 0.05$, or $|\mathfrak{z}|^{(0.01)}_{4,0.369} = 3.08$ at $\alpha = 0.01$ (Hochberg and Tamhane, 1987, Table 3). The calculated statistics at the DES doses are all well above these points. Hence, the DES-induced aneuploid response differs significantly from the vehicle control at every exposure level. No significant difference is seen, however, between the untreated and vehicle control groups.

Given the ordered P-values $P_{(i)}$, Hommel's stagewise critical index κ from (5.14) is the largest index, i, such that $P_{(T-i+h)} > h\alpha/i$, for all $h = 1, \ldots, i$. To find κ, the relevant comparisons at $\alpha = 0.01$ and for $i = 4$ are:

h	1	2	3	4
$\frac{h\alpha}{i}$	$\frac{\alpha}{4} = 0.0025$	$\frac{\alpha}{2} = 0.005$	$\frac{3\alpha}{4} = 0.0075$	$\alpha = 0.01$
$P_{(T-i+h)}$	4.9×10^{-5}	2.3×10^{-4}	2.7×10^{-4}	0.772

This shows that $\kappa \neq 4$, since, for example, $P_{(1)} < \alpha/4$. The same effect holds at $i = 3$ and $i = 2$. At $i = 1$, however, the single pertinent comparison is $P_{(4)} > \alpha$; that is, is $0.772 > 0.01$? Since this holds, one sets $\kappa = 1$, and therefore rejects H_0 in favor of H_{ai} when $P_i \leq \alpha/\kappa = \alpha$. Thus at $\alpha = 0.01$, rejection occurs at $i = 2,3,4$, corroborating the Dunnett test results. □

5.4.2 Confidence intervals for comparisons with a control – Binomial sampling

If interest in the treatment-vs.-control comparisons centers on the actual changes in the response probabilities above or below background, $1 - \alpha$

simultaneous confidence intervals are required. Thus, as in the previous section, assume the binomial counts $Y_i \sim \mathcal{Bin}(N_i,\pi_i)$ yield observed proportions $p_i = Y_i/N_i$. The simultaneous confidence intervals are to be constructed for the T differences $\pi_i - \pi_0$, $i = 1, \ldots, T$.

At first, a natural approach is to invert the Dunnett-adjusted Wald-type tests in Section 5.3.1. (It is not clear how to invert Hommel's stagewise test based on (5.14) to achieve simultaneous confidence limits for $\pi_i - \pi_0$.) Unfortunately, inverting the Wald test by pivoting on the test statistics in (5.11) produces confidence intervals than can fail to achieve $1 - \alpha$ coverage in small to moderate samples (Piegorsch, 1991). In moderate to large samples, a possible alternative involves extending the JPW limits in (4.22) to the T differences $\pi_i - \pi_0$. The procedure is straightforward: perform all the calculations leading to the limits in (4.22), but replace the critical point $\mathfrak{z}_{\alpha/2}$ with the multiplicity-adjusted critical point from (5.13): $|\mathfrak{z}|^{(\alpha)}_{T,\bar{r}}$.

5.5 MULTIPLE COMPARISONS WITH A CONTROL – POISSON SAMPLING

When the data follow a Poisson distribution, MCC calculations differ from those seen for the normal and binomial cases, although many similarities exist in style and approach. We assume data Y_{ij} are recorded as unbounded counts, with $Y_{ij} \sim$ (indep.) $\mathcal{Poisson}(\mu_i)$, $i = 0, \ldots, T$; $j = 1, \ldots, n_i$. For comparing the treatment means, μ_i, to the control mean, μ_0, the global hypotheses are again $\mathrm{H}_0{:}\mu_i - \mu_0 = 0$ (for all $i \neq 0$) vs. $\mathrm{H}_a{:}\mu_i - \mu_0 \neq 0$ (for some $i \neq 0$).

5.5.1* Exact conditional simultaneous inferences

As with the two-sample case in Section 4.1.8, for maximum power we condition the analysis on the grand sum Y_{++}. With more than two groups, however, conditioning on Y_{++} now induces a conditional multinomial distribution on the array of group sums $[Y_{0+}\ Y_{1+} \ldots\ Y_{T+}]$, where, as above, $Y_{i+} = \sum_{j=1}^{n_i} Y_{ij}$, $i = 0, \ldots, T$. We write

$$[Y_{0+}\ Y_{1+} \cdots\ Y_{T+}] \mid Y_{++} \sim \mathcal{Multinom}(Y_{++}, [\Psi_0\ \Psi_1 \cdots \Psi_T])\ ,$$

where the conditional probability parameters are $\Psi_i = n_i\mu_i / \sum_{k=0}^{T} n_k\mu_k$, $i = 0, \ldots, T$. Under H_0, these simplify to n_i/n_+, making it possible to find conditional Wald statistics for each individual comparison:

$$z_{i0} = \frac{n_+(Y_{i+} - Y_{0+}) - Y_{++}(n_i - n_0)}{\left[Y_{++}\left\{n_+(n_i + n_0) - (n_i - n_0)^2\right\}\right]^{1/2}}, \tag{5.15}$$

$i = 1, \ldots, T$; see Suissa and Salmi (1989) for details. To test H_0 vs. H_a, calculate $z_{calc} = \max\{\,|z_{10}|, \ldots, |z_{T0}|\,\}$ and reject H_0 when z_{calc} exceeds its corresponding upper-α critical point. To find these values, note that the conditional multinomial sampling distribution for the Y_{i+}s is discrete, so it is possible to perform exact calculations via full enumeration of the sample space. In particular, exact P-values can be computed from the multinomial p.m.f.: given an observed value of z_{calc}, the two-sided P-value is

$$P = \sum_{\mathcal{Y}} (Y_{++})! \prod_{i=0}^{T} \frac{(n_i/n_+)^{Y_{i+}}}{(Y_{i+})!}. \tag{5.16}$$

In (5.16), the index set $\mathcal{Y}$ is the collection of all possible values of observed sums y_{i+} ($i = 0, \ldots, T$) such that

$$\max_{i=1,\ldots,T} \left\{ \left| \frac{n_+(y_{i+} - y_{0+}) - Y_{++}(n_i - n_0)}{\left[Y_{++}\left\{n_+(n_i + n_0) - (n_i - n_0)^2\right\}\right]^{1/2}} \right| \right\} > z_{calc}.$$

The upper-α critical points are those values of z such that (5.16) is no larger than α.

Notice that (5.16) depends upon the design of the study through the n_is. Thus, the P-value calculation differs for each possible design configuration. Except perhaps for the simple cases $T = 1$ or $T = 2$, this requires computer calculation, particularly as T and the y_{i+} grow large. To facilitate implementation, Suissa and Salmi (1989) provide tables of exact one-sided P-values and critical points under a balanced design, $n_0 = n_1 = \cdots = n_T$, for the cases $T = 1,\ldots,4$. (Their statistic $Z^{(1)}$ is $z_{calc} = \max\{z_{10}, \ldots, z_{T0}\}$; enter their tables with $M = Y_{++}$ and $k = T$ to find the appropriate critical point. Reject $H_0{:}\mu_i - \mu_0 = 0$ (for all $i \neq 0$) vs. $H_a{:}\mu_i - \mu_0 > 0$ (for some $i \neq 0$) when $z_{calc} = \max\{z_{10}, \ldots, z_{T0}\}$ exceeds the tabled critical point.)

5.5.2 Large-sample simultaneous inferences

In moderate to large samples – say, where $Y_{i+} \geq 10$ for all i – an approximation for the exact enumeration in (5.16) is available via the conditional asymptotic distribution of (5.15): as $Y_{++} \to \infty$ (and where the n_i/n_+ all remain far from zero), $z_{i0} \mid Y_{++} \stackrel{\cdot}{\sim} \mathrm{N}(0,1)$. Thus an approximation to the conditional sampling distribution of z_{calc} is based on the multivariate standard normal distribution, with two-sided upper-α critical points $|\mathfrak{z}|^{(\alpha)}_{T,\{\rho_{ih}\}}$ The conditional correlations are based on the sample sizes:

$$\rho_{ih} = \frac{n_0(n_i + n_h - n_0 + n_+) - n_i n_h}{\left[\left\{n_+(n_i + n_0) - (n_i - n_0)^2\right\}\left\{n_+(n_h + n_0) - (n_h - n_0)^2\right\}\right]^{1/2}} .$$

From these, a 'common' correlation, $\bar{\rho}$, may be calculated using (5.7). For the case of a fully balanced design with constant sample sizes, the correlations are constant: $\rho_{ih} = 1/(2T + 2)$. This may be used in $|\mathfrak{z}|^{(\alpha)}_{T,\rho}$, *viz*. reject H_0 in favor of H_a when $z_{calc} > |\mathfrak{z}|^{(\alpha)}_{T,\bar{\rho}}$. In addition, local departures from H_0 are evidenced if any of the individual statistics $|z_{i0}|$ exceed this critical point, and these inferences all hold with simultaneous significance level α.

Extensions of this approximate construction to the one-sided case are straightforward: reject H_0 in favor of H_a: $\mu_i - \mu_0 > 0$ (for some $i \neq 0$) when $z_{calc} = \max\{z_{10}, \ldots, z_{T0}\} > \mathfrak{z}^{(\alpha)}_{T,\bar{\rho}}$. If testing against H_a: $\mu_i - \mu_0 < 0$ (for some i), reject when $z_{calc} = \min\{z_{10}, \ldots, z_{T0}\} < -\mathfrak{z}^{(\alpha)}_{T,\bar{\rho}}$.

Example 5.6 Multiple-dose aquatic toxicity of azinphosmethyl

As an illustration of these methods for testing Poisson populations, we consider a study of aquatic toxicity to the pesticide azinphosmethyl in the infaunal copepod *Amphiascus tenuiremis*. (The data were kindly provided by Dr Andrew Green of the US Army Corps of Engineers.) The pesticide is an organophosphate that runs off into benthic layers of aquatic ecosystems. To determine aquatic toxicity, attention centered on the capacity of *A. tenuiremis* to produce young that grow to the adult stage. Presented in Table 5.3 is one such capacity measure, the number of female offspring that survive to adulthood after a 26-day maturation period. The data are reported for $T = 2$ exposure levels and include an untreated control group. The copepods were contained in three replicate tanks per exposure group, so that the design is balanced: $n_i = 3$ for all $i = 0,1,2$.

Table 5.3 Multiple-dose reproductive toxicity to azinphosmethyl in *Amphiascus tenuiremis*

Experimental group	*Number of offspring, Y_{ij}*			*Total, Y_{i+}*
Control	103	118	142	363
10 μg/l azinphosmethyl	90	39	28	157
40 μg/l azinphosmethyl	62	66	91	219

Denote the observed counts in Table 5.3 as Y_{ij}, and assume these are independent Poisson random variables, with means μ_i (i = 0,1,2). Since interest is in whether chemical exposure decreases mean female counts, we consider testing H_0:$\mu_i - \mu_0 = 0$ (for all $i \neq 0$) against the global one-sided alternative H_a:$\mu_i - \mu_0 < 0$ (for some $i \neq 0$). (As an aside, it would be of additional interest to assess the trend in response with these data as a function of increasing concentration; this topic is discussed in detail in the next chapter.)

The group sums, Y_{i+}, are given in the far right column of Table 5.3. From these, we find $Y_{++} = 739$. These values are large enough to preclude convenient use of the exact conditional test. Instead, we will turn to the large-sample approximation for z_{calc} discussed above.

We calculate the individual values of z_{i0} from (5.15) as $z_{10} = -1854/\sqrt{39906} = -9.281$, and $z_{20} = -1296/\sqrt{39906} = -6.488$. Notice that the balanced design simplifies the calculations greatly since, for example, quantities such as $n_i - n_0$ collapse to zero. The balance also induces a constant correlation, which for this design is $\rho = 1/6$. At α = 0.01, the corresponding critical point is $-\mathfrak{z}^{(0.01)}_{2,0.167}$, which is found from Table 2 of Hochberg and Tamhane (1987) to be –2.57.

The test statistic z_{calc} is the smaller (that is, larger in a negative direction) of z_{10} and z_{20}. If this test statistic is more extreme (here, smaller) than –2.57, we reject H_0 in favor of the one-sided decrease. Both test statistics are smaller than the critical point, leading to clear rejection of H_0 at α = 0.01. Indeed, the simultaneous nature of the test's constructions allows us to conclude that a significant decrease is evidenced at each level. □

5.6 ALL-PAIRWISE MULTIPLE COMPARISONS

In selected instances, it may be of interest to include comparisons among the actual treatment levels, not just to the control group. In the most extreme case, one would be comparing all possible pairs of treatment levels, a multiple-comparison problem which is known as all-pairwise comparisons, or 'MCA.' Although our emphasis in this chapter is on comparisons with the control group, we include here a brief description of the MCA calculations to illustrate their similarity with the MCC paradigm.

For testing hypotheses, the MCA setting asks whether

$$H_0: \mu_i - \mu_h = 0 \quad (\text{for all } i \neq h)$$

vs. the multiple alternative hypothesis that

$$H_a: \mu_i - \mu_h \neq 0 \text{ (for some } i \neq h).$$

Notice that this encompasses the MCC construction. Indeed, since all treatment levels (including the control) are now contained within the set of global hypotheses, there will be many more comparisons to perform. Specifically, for $T + 1$ groups, the number of all-pairwise comparisons is $T(T + 1)/2$. This number is always larger than the T MCC comparisons, for any $T \geq 2$.

It is possible to perform the MCA analysis by applying the Bonferroni approach from Section 5.2 to the $T(T + 1)/2$ all-pairwise comparisons. Unfortunately, for most values of T (say, $T \geq 5$) the large number of comparisons weakens the Bonferroni method. For example, if an experiment contains five treated groups and one control group, $T + 1 = 6$, so $T(T + 1)/2 = 21$ pairwise comparisons are considered. If a familywise Type I error rate of $\alpha = 0.05$ is employed, the Bonferroni adjustment induces a per-comparison error rate of $0.05/21 = 0.0024$. This error rate is quite small, and its use requires very strong departures from H_0 to achieve significance. More precise methods for MCA-type comparisons are needed.

We will focus on constructions to test MCA alternative hypotheses that are similar in nature to the MCC constructions presented in the preceding sections, but that do *not* require an explicit Bonferroni adjustment for multiplicity. Generally, all that is required is a change in the reference distribution from which the critical points are derived. Similarly, for $1 - \alpha$ simultaneous confidence intervals on the pairwise differences $\mu_i - \mu_h$ ($i \neq h$), we modify only the critical point used to multiply the standard error. Here are the relevant changes for the cases described above (parentheses indicate the original sections in which the sampling scenario was discussed):

(From Section 5.3.1.) If $Y_{ij} \sim N(\mu_i, \sigma^2)$, $i = 0, \ldots, T$; $j = 1, \ldots, n_i$: continue to use the t-statistic with a pooled standard deviation as in (5.1),

$$t_{ih} = \frac{\overline{Y}_i - \overline{Y}_h}{s_p \sqrt{\dfrac{1}{n_i} + \dfrac{1}{n_h}}}$$

($i \neq h$), but now reject H_0 in favor of the two-sided alternative when

$$\max_{i \neq h} \left\{ \, |t_{ih}| \, \right\} > \frac{Q_{T+1,\nu}^{(\alpha)}}{\sqrt{2}}, \qquad (5.17)$$

where $Q_{T+1,\nu}^{(\alpha)}$ is the upper-α critical point from the **Studentized range distribution**. This is available in Hochberg and Tamhane (1987, Table 7) or Neter *et al.* (1996, Table B.9). This approach is known as the **Tukey–Kramer procedure** for all-pairwise comparisons.

When the design is fully balanced, so that $n_0 = n_1 = \cdots = n_T$, Tukey (1953) showed that the rejection region in (5.17) ensures exact α (simultaneous) significance for the test of H_0. When the design is unbalanced, however, the actual significance level is lower than α (Hayter, 1984). In this case, exact critical points are difficult to achieve, especially for large T, and the Tukey–Kramer points are employed despite their conservative nature. For the special case of $T = 2$, however, Spurrier and Isham (1985) gave exact two-sided critical points, denoted as c^*_α; also see Uusipaikka (1985). With these, reject H_0 in favor of H_a when

$$\max\left\{ |t_{10}|, |t_{20}|, |t_{12}| \right\} > c^*_\alpha;$$

(From Section 5.3.2.) If $Y_{ij} \sim N(\mu_i, \sigma_i^2)$, $i = 0, \ldots, T; j = 1, \ldots, n_i$: with heterogeneous variances, continue to use the modified t-statistic from (4.2),

$$t_{ih} = \frac{\overline{Y}_i - \overline{Y}_h}{\sqrt{\dfrac{s_i^2}{n_i} + \dfrac{s_h^2}{n_h}}},$$

and estimate the df via the Welch–Satterthwaite correction, as in (5.8):

$$\nu_{ih} = \frac{\left(u_i + u_h\right)^2}{\dfrac{u_i^2}{n_i - 1} + \dfrac{u_h^2}{n_h - 1}},$$

where $u_i = s_i^2/(n_i - 1)$. Then, the only change to the T3 procedure from Section 5.3.2 is to increase the dimension of the SMM critical point from T to $T(T+1)/2$; that is, reject H_0 in favor of H_a with simultaneous significance α when any $|t_{ih}|$ exceeds its corresponding approximate critical point $|\mathcal{M}|^{(\alpha)}_{T(T+1)/2,\, \nu_{ih}}$ (Dunnett, 1980).

(From Section 5.3.3.) Simultaneous MCA confidence intervals under normal sampling are modified similarly. For homogeneous variances, conservative $1 - \alpha$ simultaneous limits on all pairwise differences $\mu_i - \mu_h$ are

$$\overline{Y}_i - \overline{Y}_h \ \pm\ Q^{(\alpha)}_{T+1,\nu}\, s_p \left\{\frac{1}{2}\left(\frac{1}{n_i} + \frac{1}{n_h}\right)\right\}^{1/2};$$

cf. (5.9). If $T = 2$, use the Spurrier and Isham (1985) critical points c^*_α to achieve exact $1 - \alpha$ coverage. When the normal population variances are heterogeneous, modify (5.10) to account for the additional comparisons across all pairs of treatments:

$$\overline{Y}_i - \overline{Y}_h \ \pm\ |\mathcal{M}|^{(\alpha)}_{T(T+1)/2,\, \nu_{ih}} \left(\frac{s_i^2}{n_i} + \frac{s_h^2}{n_h}\right)^{1/2}.$$

(From Section 5.4.1.) If $Y_i \sim \mathcal{B}in(N_i, \pi_i)$, with observed proportions $p_i = Y_i/N_i$ $(i = 0, \ldots, T)$: calculate the pairwise statistics similar to (5.11),

$$z_{ih} = \frac{|p_i - p_h|}{\left[\frac{p_i(1-p_i)}{n_i} + \frac{p_h(1-p_h)}{n_h}\right]^{1/2}},$$

and employ Simes's (1986) method. First find the $T(T+1)/2$ P-values $P_i = 2\{1 - \Phi(z_{ih})\}$ $(i \neq h)$, then order the P-values into $P_{(1)} \leq P_{(2)} \leq \cdots \leq P_{(T[T+1]/2)}$. Reject $H_0: \pi_i = \pi_h$ (for all i,h) in favor of the global alternative $H_a: \pi_i \neq \pi_h$ (for some $i \neq h$) when $P_{(k)} \leq 2k\alpha/[T(T+1)]$ for any $k = 1, \ldots, T(T+1)/2$. A Hommel-type stagewise rejective test is similar.

(From Section 5.4.2.) Simultaneous MCA confidence intervals under binomial sampling are calculated via the JPW approach in (4.22) by replacing the two-sample critical point $\mathfrak{z}_{\alpha/2}$ with the critical point from (5.17), $Q^{(\alpha)}_{T+1,\nu}/\sqrt{2}$ (Piegorsch, 1991).

(From Section 5.5.) If $Y_{ij} \sim$ (indep.) $\mathcal{P}oisson(\mu_i)$, $i = 0, \ldots, T$; $j = 1, \ldots, n_i$: for moderate to large samples, continue to use z_{calc}, which in the two-sided case is the maximum in absolute value of the individual z-statistics

$$z_{ih} = \frac{n_+(Y_{i+} - Y_{h+}) - Y_{++}(n_i - n_h)}{\left[Y_{++}\left\{n_+(n_i + n_h) - (n_i - n_h)^2\right\}\right]^{1/2}}.$$

Reject H_0 in favor of some difference among at least one pair of means when z_{calc} exceeds the MCA critical point $Q^{(\alpha)}_{T+1,\infty}/\sqrt{2}$, as in (5.17).

5.7 SUMMARY

In this chapter, extensions of the basic, two-sample, treatment-vs.-control comparison are described. These apply when more than one treatment group is studied in an experimental or observational study. Pairwise comparisons of any single treatment with the control remain the basis for assessing differences from the control group, but adjustments and corrections for multiplicity are presented that allow the investigator to control the familywise false positive rate within the family of multiple comparisons being performed. Emphasis is placed on multiple comparisons with a single, concurrent control group, but extensions to other forms of multiple comparisons also are noted. Specific methodologies for both hypothesis testing and for simultaneous confidence interval constructions are given, for both the normal distribution and binomial sampling scenarios. Methodology for testing multiple comparisons among rates of occurrence under Poisson sampling also is discussed.

As noted in Section 4.1.9, it is also possible to derive statistical inferences when the underlying parent distributions are unknown. Methods for this case are based on the ranks of the data, and it is possible to construct rank-based multiple comparison procedures in similar form to the parametric forms discussed above. Interested readers may refer to Miller (1981, Chapter 4), and also to selected modern references, including Chakraborti and Gibbons (1991; 1993), Hayter and Stone (1991), and Spurrier (1993).

EXERCISES

(Exercises with asterisks, e.g. 5.17*, typically are more advanced, or require greater knowledge and understanding of probability distributions and/or calculus.)

5.1. The data on serum levels of PCB after environmental exposures used in Exercise 4.3 are part of a larger study of serum levels after PCB exposure (Patterson *et al.*, 1994). For example, the data from Exercise 4.3 were on 3,3′,4,4′,5,5′-PCB. For the related chemical 3,3′,4,4′,5-PCB, replicate data were taken on human serum levels after storing the sera to study reliability of the storage process. The data are:

Year	*n*	*Sample mean*	*Sample std. deviation*
1982	3	0.281	0.0577
1988	3	0.183	0.0203
1989	6	0.135	0.0233

Assume these values represent summary statistics from independent random samples from normal distributions with unknown means μ_i and unknown common variance σ^2. View the most recent information (1989) as a control group and test the null hypothesis that no difference exists between μ_{1989} and the other μ_is, vs. the alternative that there is some difference. Set $\alpha = 0.05$.

5.2. How would you proceed in Exercise 5.1 if you believed that the assumption of homogeneous variance was suspect? Perform such an analysis, and compare it to your results in Exercise 5.1.

5.3. For the PCB exposure concentrations in Exercise 5.1, construct simultaneous 95% confidence intervals on the difference in means between the 1989 data and the other two years, if the variances are assumed homogeneous. Do your results change if you assume that the variances are heterogeneous?

5.4. Using the data from Example 5.5, perform a simple two-group comparison of whether the vehicle and untreated control groups differ in their proportion response. What differences, if any, do you achieve in your result relative to that given in the example? In particular, how does your P-value compare to the Dunnett-based P of 0.772? Is this difference expected?

5.5. Yanagawa *et al.* (1994) report data on liver tissue damage after rats were exposed to various doses of an environmental toxin. The proportions of animals in which no liver damage was evidenced, stratified by dose, are given as follows:

Dose	*Number with no damage*	*Total number of animals*
0	16	17
0.5	15	19
2	10	19
10	11	22

Use the Wald test statistics in (5.11) to assess whether any significant difference is observed between the dose groups and the control. Set $\alpha = 0.05$.

5.6. Use Simes's adjustment to the Dunnett procedure – discussed after (5.9) – to compare differences between the control and nonzero doses for the liver toxicity data in Exercise 5.5. Set $\alpha = 0.05$. Is there a difference between this inference and that achieved in Exercise 5.5?

5.7. Use the stagewise rejective test based on the index in (5.14) to identify which, if any, of the dose groups in Exercise 5.5 appear significantly different from the control. Set $\alpha = 0.05$.

5.8. The data on cell nucleus damage after exposure to methyl methane sulfonate (MMS) used in Exercise 4.9 are part of a larger study of the chemical's genotoxic potential. MMS was administered over a number of doses, and produced the following pooled proportions of micronucleated (MN) cells as a function of dose:

Dose	*MN*	*Cells scored*
0	31	16 000
5	9	8 000
10	24	8 000
20	42	8 000
40	204	8 000

Use Simes's adjustment to the Dunnett procedure – discussed after (5.13) – to compare differences between the control and nonzero doses with these data. Set $\alpha = 0.01$.

5.9. Use the stagewise rejective test based on the index in (5.14) to identify which, if any, of the dose groups in Exercise 5.8 appear significantly different from the control. Set $\alpha = 0.01$.

5.10. Using the JPW confidence limits based on (4.22), but replacing the two-sample critical point $\mathfrak{z}_{\alpha/2}$ with the multiplicity-adjusted critical point from (5.13), construct simultaneous 99% confidence limits on the differences in MN proportions between the control group and each dose group in Exercise 5.8.

5.11. In a laboratory experiment to study the inflammatory effects of asbestos exposure to mammalian lung tissue, rodent lung cells were exposed *in vitro* to asbestos dust. Prior to exposure, the cells were pre-treated with a benign radioactive marker that is released when a cell generates an inflammatory response. Cells in culture were then exposed to asbestos dust for 0 (control), 19, 24, or 31 hours. Suppose the results showed the following proportions of responding cell cultures, as a function of time of exposure:

Exposure time (h)	*Responding cultures*	*Cultures examined*
0 (control)	5	32
19	5	20
24	15	20
31	13	20

Using the JPW confidence limits based on (4.22), but replacing the two-sample critical point $\mathfrak{z}_{\alpha/2}$ with the multiplicity-adjusted critical point from (5.13), construct simultaneous 95% confidence limits on the differences in response probabilities between the control group and each exposure group.

5.12. Reporting on aquatic sediment contamination of Vancouver Harbor, Green *et al.* (1993) present infaunal abundance data for three different benthic species. A portion of the data follows:

Benthic species	*Numbers recorded*
Arthropods	1, 2
Mollusks	3, 7
Polychaetes	6, 0

View arthropods as the standard or control group for comparison in these data. Using the conditional statistics for Poisson data in

(5.15), test the null hypothesis that abundance rates are similar across species, vs. an alternative that rates for polychaetes and/or mollusks exceed the arthropod rate. (You will need the one-sided critical point from the tables in Suissa and Salmi (1989). For these data at $\alpha = 0.05$, it is 2.24.)

5.13. For the aquatic toxicity study of the pesticide azinphosmethyl described in Example 5.6, data were also recorded on numbers of male *A. tenuiremis* offspring that survived to adulthood after 26-day maturation. These were:

Azinphosmethyl dose	*Number of offspring*
0 (control)	32, 39, 32
10 μg/l	38, 20, 41
40 μg/l	37, 25, 21

Apply the MCC analysis from Example 5.6 to these male offspring data, testing for any decrease in offspring viability after pesticide exposure. Set $\alpha = 0.01$.

5.14. Reanalyze the PCB exposure data in Exercise 5.1 by performing a comparison across all pairs of years, that is, an MCA comparison. Set $\alpha = 0.05$.

5.15. For the benthic abundance data in Exercise 5.12, it might be considered arbitrary to set arthropods as the standard group for comparison. As such, reanalyze these data by performing an MCA comparison across all pairs of species. Use the following approaches, and discuss any differences in the inferences they produce:

a. Transform the counts using a Freeman–Tukey transformation: $V_{ij} = (Y_{ij} + 1)^{1/2} + (Y_{ij})^{1/2}$, $i = 1,2,3$; $j = 1,2,3$. Assume this has stabilized any variance heterogeneity and produced approximately normal random variables with unit variance. (The transformation assures that in large samples $\text{Var}[V_{ij}] \approx 1$ (Freeman and Tukey, 1950). Use this value instead of any estimate(s) of variance, such as a pooled sample variance.) Apply the Tukey–Kramer method for MCA comparisons. Set $\alpha = 0.10$. (The appropriate critical point is based on the Studentized range point $Q^{(0.1)}_{3,\infty} = 2.90$.)

b. Perform every possible pairwise comparison between species using the unadjusted statistic in (4.15), but apply a Bonferroni correction from Section 5.2 to adjust the familywise error rate to $\alpha = 0.10$.

5.16*. Show that the Bonferroni inequality in (5.2) is the same as that given in (5.3), by showing that the lower bounds in each inequality are identical. (*Hint*: $P[\mathcal{A}_j^c] = 1 - P[\mathcal{A}_j]$.)

5.17*. An interesting application of the Bonferroni approach involves comparisons on only the pairs of **successive differences**: $(\mu_1 - \mu_0)$, $(\mu_2 - \mu_1)$, ..., $(\mu_T - \mu_{T-1})$. Apply the Bonferroni inequality to obtain simultaneous confidence limits for the successive differences, as follows:

a. Assume normally distributed data with homogeneous variances, and construct each individual comparison based on the standard t-interval from (4.17). If each individual interval is to have per-comparison coverage of $1 - \gamma$, show that this yields simultaneous confidence limits for each difference $\mu_j - \mu_{j-1}$ of the form

$$\overline{Y}_j - \overline{Y}_{j-1} \pm t_{\gamma/2}(\nu)\, s_p \sqrt{\frac{1}{n_j} + \frac{1}{n_{j-1}}} ,$$

where $\nu = \sum_{i=0}^{T}(n_i - 1)$ are the pooled *df*, and s_p is the root mean square error.

b. To apply the Bonferroni method, view the individual events $\mathcal{A}_j$ as coverage of $\mu_j - \mu_{j-1}$. The complementary events, $\mathcal{A}_j^c$, are *non*coverage, that is, $\mu_j - \mu_{j-1}$ is not contained in this interval. Show that the noncoverage probability for each $\mathcal{A}_j$ is simply $P[\mathcal{A}_j^c] = 1 - (1 - \gamma) = \gamma$, for all $j = 1,...,T$.

c. Our goal is minimal simultaneous $1 - \alpha$ coverage, so we wish to set $P[\mathcal{A}_1\mathcal{A}_2 \cdots \mathcal{A}_T] \geq 1 - \alpha$. Since $P[\mathcal{A}_j^c] = \gamma$, the inequality in (5.2) implies that $P[\mathcal{A}_1\mathcal{A}_2 \cdots \mathcal{A}_T] \geq 1 - T\gamma$. Give a value for γ that will achieve this goal. What is the final form of the simultaneous $1 - \alpha$ intervals?

d. How would you adjust this construction to perform Bonferroni-adjusted tests of the multiple hypotheses $H_{0j}: \mu_j = \mu_{j-1}$ vs. $H_{aj}: \mu_j \neq \mu_{j-1}$?

5.18. How would you modify the simultaneous confidence intervals for successive differences in Exercise 5.17 if the data were proportions rather than normally distributed measurements? (Continue to use the Bonferroni adjustment; explain what other changes you would make to the simultaneous limits.) Apply your method over the successive time levels of the asbestos exposure data in Exercise 5.11.

6

Trend testing

An important extension of the simple two-group and multi-group comparisons discussed in Chapters 4 and 5 occurs when some ordered variable, x_i, is recorded independently of the observed responses Y_i. When it is felt that the x-variable plays a role in explaining or predicting the nature of the responses in Y_i the information in x should be included in the statistical analysis. This is a more complex modeling scenario than the multiple-group comparisons discussed in Chapter 5: now we observe an independent variable that may act to predict the mean response over its ordered scale. For example:

- Laboratory animals exposed to a suspected environmental toxin at different exposure levels may exhibit differential responses across the different levels. If the exposure levels are doses of the toxin, we question whether the observed toxic response varies in a deterministic manner as the doses increase, for example, is there a significant **dose response** or **concentration response**?
- Increasing levels of attention paid by a predator to a prey may be recorded, along with the predator's rate of success. Although the attention variable cannot be quantified, it can be ordered: none, some, much, etc. In this sense, we may still wish to **test the trend** of the success rate over the increasing levels of attention to see whether it has value as a predictor of predatory success.

As in Chapters 4 and 5, we use the term 'treatment' in a generic sense, indicating some exposure, intervention, observational condition, etc., being applied to the experimental units. We assume that nonrandom, numerical values, x_i, are recorded, representing the ordered levels of the treatment. In most cases, x_i is referred to as an **independent variable** or **concomitant variable**, since it is ostensibly independent of the response variable Y_i. If x_i is in fact a quantitative variable, we often refer to it also as a **predictor variable**, highlighting the opportunity to use it for predictive purposes. As in Chapter 5, it is common to include a concurrent control level in the study design; from a dose-response perspective, the control level often represents a zero dose, $x_0 = 0$. For our discussions, we use 'dose' or 'concentration' interchangeably, even though some toxicologists use 'dose' to quantify the amount of toxin present at a target site, and 'concentration' to quantify the environmental exposure.

The presentation in this chapter concerns methods for testing the significance of any observed, increasing (or decreasing) trend in the response over the ordered levels of the independent variable, x. (If x is quantitative, then we may be interested also in modeling the dose response as a function of x, and using this model to infer qualities or predict features of the mean dose response. We study this aspect of statistical modeling further in Chapter 7.)

6.1 SIMPLE LINEAR REGRESSION FOR NORMAL DATA

The multiple-comparison methods described in Chapter 5 for assessing differences among multiple treatments and/or vs. a concurrent control group require no assumptions about the levels of the treatment. Indeed, to employ those methods for data analysis, the treatment could be any form of indicator or classification variable. Thus, multiple comparisons focus on detecting *qualitative* differences between experimental groups. When the treatment indicator is ordered in some way, however, it is possible to refine the inferences from Chapter 5 to take advantage of this ordering, that is, to exploit the quantitative differences between experimental groups. Suppose a study is conducted where some quantifiable measure such as an exposure dose is associated with each treatment level. Then, statistical emphasis concerns testing or estimating the nature of the dose response over this ordered variable. Different methods exist for different forms of response, however. Perhaps the simplest involves the normal distribution model used in **simple linear regression**. Specifically, given quantitative values of a nonrandom predictor variable, x_i, suppose that $Y_i \sim$ (indep.) $N(\mu_i, \sigma^2)$, $i = 1, \ldots, n$, where we link the relationship between the mean response function and the independent variable in a simple linear fashion:

$$\mu_i = \mu(x_i) = \beta_0 + \beta_1 x_i. \tag{6.1}$$

In effect, (6.1) models each observation as a random departure from the central value $\mu(x_i)$, say, $Y_i = \mu(x_i) + \varepsilon_i$, where $\varepsilon_i \sim$ i.i.d. $N(0, \sigma^2)$.

In (6.1), the **regression parameters**, β_0 and β_1, represent the Y-intercept and slope, respectively, of the straight-line response. The units of β_0 are the same as the units of Y, while the units of β_1 correspond to units of Y per unit of x.

In trend testing, β_1 is the parameter of interest. If $\beta_1 = 0$, then the mean response does not vary in a linear fashion with the predictor variable, and no linear trend with x exists. In this sense, this model can be used for simple trend analysis, by testing $H_0{:}\,\beta_1 = 0$. (Of course, complicated nonlinear relationships may exist between x and Y even when $\beta_1 = 0$, so model (6.1) is often only a simple approximation of some underlying trend or effect.)

The simple linear regression model is described in many statistical textbooks, such as Moore and McCabe (1993, Chapter 9) or Neter *et al.* (1996, Chapters 1–3). Here, we review only those features useful for trend testing. For instance, the least squares (LS) estimators of the regression parameters take the closed forms

$$\hat{\beta}_1 = \frac{\sum_{i=1}^{n}(x_i - \bar{x})(Y_i - \bar{Y})}{\sum_{i=1}^{n}(x_i - \bar{x})^2} \qquad \text{and} \qquad \hat{\beta}_0 = \bar{Y} - \hat{\beta}_1\bar{x},$$

where $\bar{x} = \sum_{i=1}^{n}x_i/n$ and $\bar{Y} = \sum_{i=1}^{n}Y_i/n$ are the sample means of the **predictor variable** or **independent variable** and **response variable** or **dependent variable**, respectively. Also, the estimated standard errors of the intercept and slope estimators are, respectively,

$$se[\hat{\beta}_0] = \hat{\sigma}\left(\frac{1}{n} + \frac{\bar{x}^2}{\sum_{i=1}^{n}(x_i - \bar{x})^2}\right)^{1/2},$$

and

$$se[\hat{\beta}_1] = \frac{\hat{\sigma}}{\left\{\sum_{i=1}^{n}(x_i - \bar{x})^2\right\}^{1/2}},$$

where $\hat{\sigma}$ is the root mean square error,

$$\hat{\sigma} = \left\{\frac{1}{n-2}\sum_{i=1}^{n}(Y_i - \hat{\beta}_0 - \hat{\beta}_1 x_i)^2\right\}^{1/2}.$$

To conduct a test for trend under this simple linear model, we calculate the t-statistic $t_{calc} = \hat{\beta}_1/se[\hat{\beta}_1]$. Under the normal distribution model, these LS estimators coincide with the maximum likelihood (ML) estimators for β_0 and β_1, and the null reference distribution of t_{calc} is $t(n-2)$. We reject H_0 in favor of an increasing trend, $H_a{:}\,\beta_1 > 0$, if t_{calc} exceeds the upper-α

critical point $t_\alpha(n-2)$. Against a decreasing trend, $H_a{:}\beta_1 < 0$, reject H_0 when $t_{calc} < -t_\alpha(n-2)$. Alternatively, the P-value for an increasing trend is $P = P[t(n-2) \geq t_{calc}]$; for a decreasing trend it is $P = P[t(n-2) \leq t_{calc}]$. In either case, reject H_0 when $P < \alpha$. Used in this sense, the simple linear regression model is an effective method for testing trend when the relationship between x_i and μ_i can be assumed linear, as in (6.1).

If the relationship between x_i and μ_i is not linear, but is at least approximately linear over the range of the recorded values of x_i, then this model often performs adequately for testing trend.

Example 6.1 Linear regression test for trend over PCB exposure levels
Lahvis *et al.* (1995) report blood concentration data in wild bottlenose dolphins (*Tursiops truncatus*) exposed to polychlorinated biphenyls (PCBs). For $n = 5$ of the dolphins, data were collected on blood levels of various PCBs, to study whether PCB concentrations increase as a function of age of the animal. The data appear in Table 6.1.

Table 6.1 Blood PCB concentrations in wild dolphins (*T. truncatus*)

Animal, i:	*1*	*2*	*3*	*4*	*5*
Age in years, x_i:	3	9	13	21	32
Tetrachloro-PCB conc. (ng/g)	1.5	9.1	5.8	17.6	15.1

Data from Lahvis *et al.* (1995).

To consider whether an increasing trend in PCB concentration is evident for these data, we assume that the simple linear regression model (6.1) adequately represents the mean PCB concentration response as a function of age. From the LS equations, we find $\hat{\beta}_0 = 2.1726$, $\hat{\beta}_1 = 0.4902$, $se[\hat{\beta}_0] = 3.4465$, and $se[\hat{\beta}_1] = 0.1856$. The t-statistic for testing increasing trend is the ratio $t_{calc} = \hat{\beta}_1/se[\hat{\beta}_1] = 2.6412$, on $n - 2 = 3$ df. The associated P-value is $P = P[t(3) > 2.6412]$; Fig. 6.1 displays SAS code that finds this to be $P = 0.0388$. Thus, for example, at significance level $\alpha = 0.05$, we conclude there a significant increasing trend in blood PCB concentration as the animals age.

As an alternative to direct calculation, SAS PROC REG can be used to find the LS estimates. SAS code for fitting the linear regression model is given in Fig. 6.2. The `model` statement in PROC REG is required: it relates the response variable `pcb` to the predictor variable `age`. The associated output appears in Fig. 6.3.

```
* SAS code to find t(3) P-value;
data pval;
  cdf = probt(2.6412, 3);
  pval = 1 - cdf;
cards;
proc print;
```

```
                The SAS System
        OBS          CDF            PVAL
         1        0.96121        0.038788
```

Fig. 6.1 SAS program and output (edited) to find $t(3)$ P-value.

```
* SAS code for LS regression;
data PCBlevel;
input age pcb @@;
cards;
3 1.5   9 9.1   13 5.8   21 17.6   32 15.1
proc reg;
 model pcb = age;
```

Fig. 6.2 SAS PROC REG code to fit LS regression line and obtain a test of linear trend for data in Table 6.1.

The LS estimates (and standard errors) for β_0 and β_1 are given in Fig. 6.3 under `Parameter Estimates`; see the columns labeled `Parameter Estimate` and `Standard Error`. Each associated t_{calc} statistic is presented alongside, under `T for H0:Parameter=0`. The P-value for testing H_0: $\beta_1 = 0$ is found in the `AGE` row under `Prob > |T|`. (This P-value is associated with the two-sided alternative H_a: $\beta_1 \neq 0$.) When testing for increasing trend, however, we are interested in the one-sided P-value $P = P[t(3) > t_{calc}] = P[t(3) > 2.641]$. As we saw in Fig. 6.1, this is $P[t(3) > 2.641] = 0.0388$.

This analysis can be performed in S-PLUS via the functions `summary` and `lm`. To do so, place the data in Table 6.1 into a data frame, say `data.df`, that contains the variables `pcb` and `age`. Then LS estimates, standard errors, and hypothesis tests are generated using the command

```
summary(lm(pcb~age,data=data.df))
```

□

```
                          The SAS System
Model: MODEL1
Dependent Variable: PCB
                    Analysis of Variance
                     Sum of           Mean
Source     DF       Squares         Square     F Value    Prob>F
Model       1     121.88850      121.88850       6.976    0.0776
Error       3      52.41950       17.47317
C Total     4     174.30800
   Root MSE          4.18009       R-square          0.6993
   Dep Mean          9.82000       Adj R-sq          0.5990
   C.V.             42.56712
                       Parameter Estimates
                 Parameter Standard    T for H0:
Variable DF      Estimate    Error   Parameter=0   Prob > |T|
INTERCEP  1      2.17256   3.44651      0.630         0.5732
AGE       1      0.49022   0.18561      2.641         0.0776
```

Fig. 6.3 SAS PROC REG output (edited) for a regression test of linear trend with data from Table 6.1.

We caution that the t-test can suffer great losses in stability when the data depart from the normal distribution assumptions (Hyrenius, 1950). For example, an important constraint in the use of the simple linear regression model is that the variance must be homogeneous over the levels of the independent variable; that is, $\mathrm{Var}[Y_i] = \sigma^2$ is constant with respect to x_i. If this is not the case, the method is not appropriate for use. One can modify the LS estimator to account for heterogeneous variance in select instances, however. Suppose it is known that the variance changes in a proportional manner with respect to some function of x_i, say $\mathrm{Var}[Y_i] \propto x_i\sigma^2$ (assuming $x_i > 0$) or $\mathrm{Var}[Y_i] \propto x_i^2\sigma^2$, etc. ($\sigma^2$ can be known or unknown). Then, one can apply a form of **weighted least squares** estimation, by weighting each observation inversely to its variance. Assume the known weights, w_i, are given as the reciprocals of the varying portions of $\mathrm{Var}[Y_i]$; for example, $w_i = x_i^{-2}$ if $\mathrm{Var}[Y_i] \propto x_i^2\sigma^2$, $i = 1, \ldots, n$. (If these weights are used with a control group coded as $x_0 = 0$, a linear transformation can be applied to the predictor variables prior to the construction of the weights. For example, if $x_i' = a + bx_i$, where a and b are constants so that $x_i' \neq 0$ for all i, take as weights $w_i' = (x_i')^{-2}$.) More generally, if r_i replicate observations are available at each x_i, one can calculate the per-group sample variances s_i^2, and then take as inverse-variance weights $w_i = 1/s_i^2$.

As with regular least squares, the goal in weighted least squares is to find estimators $\hat{\beta}_0$ and $\hat{\beta}_1$ that minimize the (weighted) objective function $Q_w = \sum_{i=1}^{n} w_i(Y_i - \hat{\beta}_0 - \hat{\beta}_1 x_i)^2$. The result for the simple linear regression model is weighted LS estimators of the form

$$\hat{\beta}_{0,w} = \frac{\sum_{i=1}^{n} w_i Y_i - \hat{\beta}_1 \sum_{i=1}^{n} w_i x_i}{w_+}$$

and

$$\hat{\beta}_{1,w} = \frac{\sum_{i=1}^{n} w_i x_i Y_i - \frac{1}{w_+} \sum_{i=1}^{n} w_i x_i \sum_{i=1}^{n} w_i Y_i}{\sum_{i=1}^{n} w_i x_i^2 - \frac{1}{w_+}\left(\sum_{i=1}^{n} w_i x_i\right)^2},$$

with $w_+ = \sum_{i=1}^{n} w_i$. The standard error of $\hat{\beta}_{1,w}$ is

$$se[\hat{\beta}_{1,w}] = \frac{\hat{\sigma}_w}{\left\{\sum_{i=1}^{n} w_i x_i^2 - w_+^{-1}\left(\sum_{i=1}^{n} w_i x_i\right)^2\right\}^{1/2}},$$

where the proportionality constant σ^2 is estimated by the weighted mean square error: $\hat{\sigma}_w^2 = \sum_{i=1}^{n} w_i(Y_i - \hat{\beta}_{0,w} - \hat{\beta}_{1,w} x_i)^2/(n-2)$. From these, the weighted t-statistic is $t_{calc} = \hat{\beta}_{1,w}/se[\hat{\beta}_{1,w}]$, which tests $H_0: \beta_1 = 0$ vs. an increasing trend, $H_a: \beta_1 > 0$. Under H_0, t_{calc} is distributed as $t(n-2)$, so we reject H_0 in favor of H_a if t_{calc} exceeds $t_\alpha(n-2)$. Tests for decreasing trend are similar. Further details on weighted LS methods for regression models are available, for example, in Neter *et al.* (1996, Section 10.1).

Example 6.2 Weighted linear regression test for trend over nitrofen exposure levels

In Exercise 4.22 we examined reproductive toxicity in the aquatic organism *C. dubia* after exposure to the chemical toxin nitrofen. The measure of reproductive stress used in that exercise was offspring counts in exposed females (Bailer and Oris, 1993).

The data from Exercise 4.22 were part of a larger dose-response study of nitrofen: in total, 50 independent animals were studied over one control and four exposure groups. The response was total number of offspring for three broods per animal, Y_{ij}, where $i = 1, \ldots, 5$, and $j = 1, \ldots, 10$ indexes the replicate animals producing young in each exposure group. Notice that the

Table 6.2 Offspring counts for *C. dubia* exposed to nitrofen

Experimental group	*Number of offspring, Y_{ij}*					*Mean, $\overline{Y}_{i+}$*	*Variance, s_i^2*
Control	27	32	34	33	36		
	34	33	30	24	31	31.4	12.933
80 µg/l nitrofen	33	33	35	33	36		
	26	27	31	32	29	31.5	10.722
160 µg/l nitrofen	29	29	23	27	30		
	31	30	26	29	29	28.3	5.567
235 µg/l nitrofen	23	21	7	12	27		
	16	13	15	21	17	17.2	34.844
310 µg/l nitrofen	6	6	7	0	15		
	5	6	4	6	5	6.0	13.778

Data from Bailer and Oris (1993).

design here is balanced, with $r_i = 10$ in each group. The complete data, along with per-exposure summary statistics $\overline{Y}_{i+}$ and s_i^2, appear in Table 6.2.

The concern here is whether nitrofen induces a significant decrease in the reproductive capabilities of *C. dubia*, as measured by the number of offspring. Over a series of increasing doses, this concern translates into a question of whether the mean response decreases with exposure dose.

Since these data are counts, it is natural to consider a Poisson parent assumption. As we know from the population features of the Poisson distribution, the population variance equals the population mean. Therefore, if the mean is assumed to vary with some exposure variable, the variance will as well. More generally, it is often the case that count data exhibit heterogeneous variances across experimental groups. For the reproductive toxicity counts in Table 6.2, this is illustrated by the sample variances in the table's last column: the variances vary from 5.567 to 34.844. In this case, a weighted LS regression trend test employs inverse-variance weights, $w_i = 1/s_i^2$, at each exposure level $i = 1, \ldots, 5$.

As noted in Section 4.1.2, formal tests of significance for assessing homogeneous variances are possible, although in general it is ill advised to precede significance tests on the mean response with a test of variance homogeneity, due to the sensitivity of the variance tests to the underlying distributional assumptions (Bailer, 1989).

For the data in Table 6.2, the weighted LS estimates for the slope and intercept are $\hat{\beta}_{0,w} = 37.0851$ and $\hat{\beta}_{1,w} = -0.0819$, while the weighted root mean square error is $\hat{\sigma}_w = 1.633$. The standard error the slope estimate is $se[\hat{\beta}_{1,w}] = 0.0079$. To test decreasing trend based upon the weighted LS estimates, we calculate $t_{calc} = \hat{\beta}_{1,w}/se[\hat{\beta}_{1,w}] = -0.0819/0.0079 = -10.37$, and refer it to a t distribution with 48 *df*. The associated P-value is $P = P[t(48) < -10.37] < 0.0001$. This suggests strong evidence of a decreasing trend in offspring count with increasing nitrofen exposure. □

In general, the adequacy of the various assumptions made to implement the model should be examined whenever possible. These **regression diagnostics** include checks (a) of linearity, (b) for influential values of x_i, (c) of constant variance, (d) of independence and normality of the Y_is, and (e) for any other outlying values of Y_i (Neter *et al.*, 1996, Chapter 9).

Example 6.3 Adequacy of the linearity assumption in simple linear regression

Implicit in the trend analysis for the aquatic toxicity data in Example 6.2 is the assumption that a linear relationship exists (or, at least provides a reasonable approximation) between the number of offspring and nitrofen concentration. To assess the adequacy of the linearity assumption, a simple yet effective tool is a scatterplot of the data: number of offspring, Y, vs. nitrofen concentration, x. Such a scatterplot is displayed in Fig. 6.4. The weighted LS regression line from Example 6.2, $\hat{Y}_{ij} = 37.085 - 0.082x_i$, is superimposed on the figure.

We can see from Fig. 6.4 that the estimated intercept is overpredicting observed control group response; the simple linear model cannot capture the **curvature** in the concentration-response pattern. Thus, the test of trend, while significant, is based upon a model in which the functional relationship between the response variable, number of offspring, and the predictor variable, nitrofen concentration, is misspecified.

The model misspecification is corroborated by a scatterplot of the (raw) **residuals** from the model fit, that is, a plot of $R_{ij} = Y_{ij} - \hat{Y}_{ij} = Y_{ij} - (37.085 - 0.082x_i)$ vs. the predicted values $\hat{Y}_{ij} = 37.085 - 0.082x_i$. If the linearity assumption is valid, such a **residual plot** should exhibit a random scatter about a zero reference line. If, however, the data correspond better to some form of nonlinear curvature, the residual plot should help display this. Figure 6.5 illustrates the effect for the nitrofen data, where it is clear that the residuals from the linear fit exhibit a strongly nonrandom pattern. The residuals tend to be negative for low and high predicted responses and positive for mid-range responses. This is a strong indication of the need for greater curvature to be built into the model. (We will explore this further in Example 7.5.) As illustrated here, scatterplots and residual plots should be fundamental components of any standard analysis of trend. (Scatterplots and residual plots for the data in Example 6.1 are part of Exercise 6.5.) □

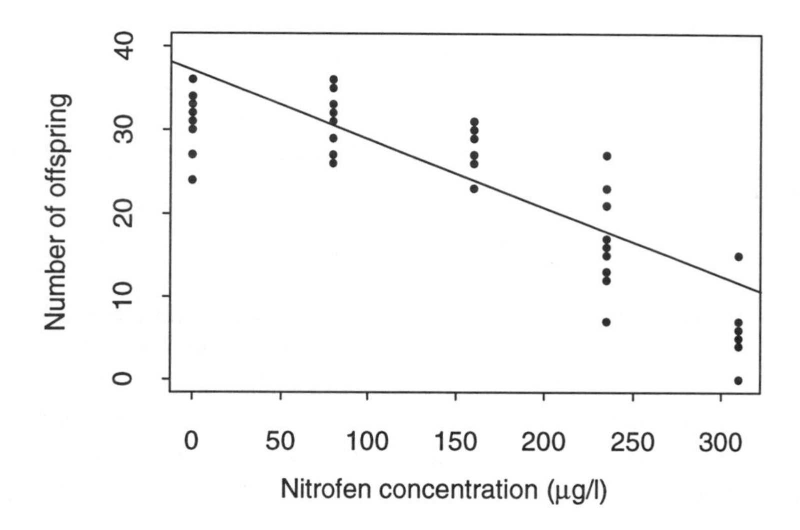

Fig. 6.4 Scatterplot of offspring counts in *C. dubia* vs. levels of nitrofen; data from Table 6.2. The weighted least squares prediction line is superimposed on the plot.

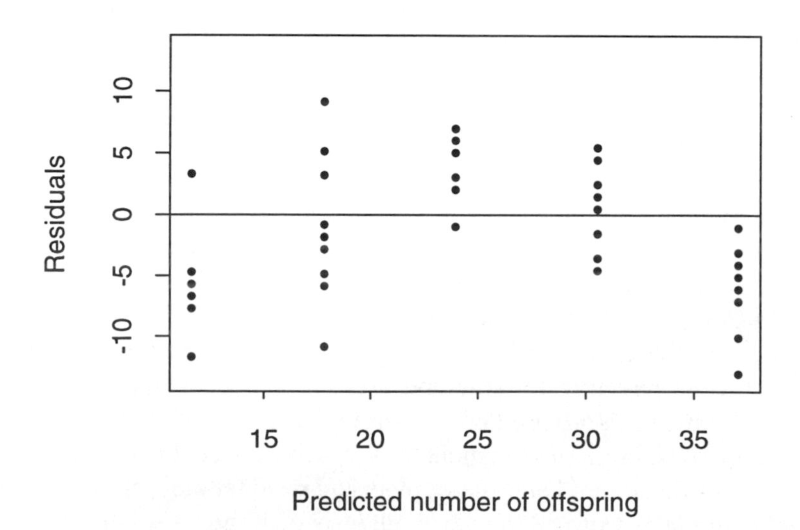

Fig. 6.5 Scatterplot of the residuals from a simple linear model for offspring counts in *C. dubia* after exposure to nitrofen vs. the predicted values; data from Table 6.2. A horizontal reference line at zero is superimposed on the plot.

For modeling departures from linearity, we describe extensions of the simple linear model to other forms of curvilinear dose response in Chapter 7. We will also discuss in detail generalizations of the linear model to both nonlinear response functions and different models for the parent distributions in Chapter 8, when we present generalized linear models. More advanced forms of departure from the simple model, and ways to address them, are beyond the scope of this chapter, however. Interested readers may consult the discussion in Neter *et al.* (1996, Chapters 13–14) and the references therein.

6.2 WILLIAMS' TEST FOR NORMAL DATA

When it is felt that the normal distribution is still a valid parent distribution for the Y_is, but that the simple linear model (6.1) is inappropriate, or when the independent variable is ordered but not fully quantified, it is still possible to perform a test for trend in the mean response. The method is due to Williams (1971; 1972), and it is straightforward to implement. Using Williams' notation, begin by supposing that there are $k + 1$ treatment groups or levels, where the lowest level of the ordered, independent variable corresponds to a zero-dose control, say $x_0 = 0$, while the other ordered levels of x_i are strictly increasing. At each treatment level, $i = 0, \ldots, k$, suppose we take r replicate observations, Y_{ij}, with $j = 1, \ldots, r$. Under a normal parent model, we assume $Y_{ij} \sim N(\mu_i, \sigma^2)$. Then, no trend among the mean responses corresponds to the usual no-effect null hypothesis $H_0: \mu_0 = \mu_1 = \cdots = \mu_k$. The alternative hypothesis of a **monotone increasing trend** is

$$H_a: \mu_0 \leq \mu_1 \leq \cdots \leq \mu_k, \tag{6.2}$$

with some strict inequality assumed to exist somewhere among the μ_is. This is known as an **ordered alternative hypothesis**, and it has numerous applications in environmental biology (Boyd, 1982; Schoenfeld, 1986; Hothorn and Lemacher, 1991).

Williams' trend test requires calculation of the maximum likelihood estimate of each μ_i, say $\hat{\mu}_i$, *under the ordered alternative*, that is, under $\mu_0 \leq \mu_1 \leq \cdots \leq \mu_k$. To calculate these values, we first find the observed sample means $\overline{Y}_{i+} = \sum_{j=1}^{r} Y_{ij}/r$, and assess whether they satisfy the ordered alternative. If so, then we take $\hat{\mu}_i = \overline{Y}_{i+}$, $i = 0, \ldots, k$. If not, however, we pool any adjacent violators into the same value, beginning at $i = 0$ and proceeding up until $i = k$. For example, suppose $k = 3$, and the sample means are $\overline{Y}_{0+} = 1.4$, $\overline{Y}_{1+} = 2.3$, $\overline{Y}_{2+} = 1.9$, and $\overline{Y}_{3+} = 3.2$. Notice that

$\overline{Y}_{1+} = 2.3 > \overline{Y}_{2+} = 1.9$, so the observed pattern of the sample means does not satisfy the ordered alternative. To make it so, we apply the **pool-adjacent-violators algorithm**: retain $\hat{\mu}_0 = \overline{Y}_{0+} = 1.4$, but pool $\overline{Y}_{1+}$ and $\overline{Y}_{2+}$: $\hat{\mu}_1 = \hat{\mu}_2 = \{\overline{Y}_{1+} + \overline{Y}_{2+}\}/2 = (2.3 + 1.9)/2 = 2.1$. Now $\hat{\mu}_0 < \hat{\mu}_1 = \hat{\mu}_2$, so the ML estimates are in compliance with the ordered alternative. Lastly, since $\overline{Y}_{3+} = 3.2$ is greater than $\hat{\mu}_2 = 2.1$, set $\hat{\mu}_3 = \overline{Y}_{3+}$ to achieve the properly ordered estimators. The pool-adjacent-violators algorithm under an increasing ordered alternative may be written compactly as

$$\hat{\mu}_i = \max_{a=1,\ldots,i} \min_{b=i,\ldots,k} \left\{ \frac{1}{b - a + 1} \sum_{h=a}^{b} \overline{Y}_{h+} \right\}. \tag{6.3}$$

The algorithm is discussed more fully in numerous sources, including Ayer *et al.* (1955), Williams' original article (1971), and the compendium by Robertson *et al.* (1988, Section 1.2).

Given the order-restricted ML estimates $\hat{\mu}_i$, Williams' test rejects H_0 in favor of H_a when the modified t-statistic,

$$\bar{t}_{\text{calc}} = \frac{\hat{\mu}_k - \overline{Y}_{0+}}{\sqrt{2\hat{\sigma}/r}}, \tag{6.4}$$

exceeds an appropriate upper-α critical point, denoted by $\bar{t}_\alpha(\nu)$. In this construction, $\hat{\sigma}$ is the root mean square error,

$$\hat{\sigma} = \left\{ \frac{1}{\nu} \sum_{i=1}^{n} \sum_{j=1}^{r} (Y_{ij} - \overline{Y}_{i+})^2 \right\}^{1/2},$$

and $\nu = (k+1)(r-1)$ are the error *df*. Notice how (6.4) is being used: after ordering the ML estimates at each treatment level, the control sample mean is subtracted from the largest order-restricted estimate, and an increasing trend is indicated if this difference is too large.

Williams' test may be applied also to test against a decreasing order, H_a: $\mu_0 \geq \mu_1 \geq \cdots \geq \mu_k$. An expedient way to perform this is to multiply all the observations by –1, then apply the test against an increasing order as described above. One can also multiply by –1 after completing the pool-adjacent-violators operations; see Example 6.4.

Williams gave tables of the $\bar{t}$ critical points, $\bar{t}_\alpha(\nu)$, for $\alpha = 0.01, 0.05$, and for a variety of values of ν and k (Williams, 1971, Tables 1–2). Note that these tables assume a balanced design, that is, $r_i = r$ for all $i = 0, \ldots, k$. Unfortunately, for designs where the r_is are unbalanced, the associated critical points are much more difficult to calculate. For the special case of treatment balance – where $r_i = r$ over $i = 1, \ldots, k$, but r_0 may differ from r – Williams (1972) gave extended tables. (Williams assumed that $r_0 > r$, which would be reasonable if interest existed in comparing the various treatment levels to the control; cf. Section 5.3.1.) In this case, (6.4) becomes

$$\bar{t}_{\text{calc}} = \frac{\hat{\mu}_k - \bar{Y}_{0+}}{\hat{\sigma}\left(\frac{1}{r_0} + \frac{1}{r}\right)}, \tag{6.5}$$

and one rejects H_0 when $\bar{t}_{\text{calc}}$ exceeds the treatment-balanced critical point $\bar{t}_\alpha(\nu, \omega)$, with $\nu = \sum_{i=0}^{k}(r_i - 1)$ and $\omega = r_0/r$. Williams noted that a simple approximation exists between $\bar{t}_\alpha(\nu, \omega)$ and $\bar{t}_\alpha(\nu,1)$, which he wrote as $\bar{t}_\alpha(\nu, \omega) \approx \bar{t}_\alpha(\nu,1) - (0.01)\beta(1 - \omega^{-1})$. Notice that the critical point $\bar{t}_\alpha(\nu,1)$ corresponds to $\bar{t}_\alpha(\nu)$, the critical value under a balanced design. $\bar{t}_\alpha(\nu,1)$ and the extrapolation coefficient β are found in Williams (1972, Tables 1–4), where table values are given in the form $\bar{t}_\alpha(\nu,1)^\beta$, that is, as a tabulated value for the balanced critical point and as a superscript on the tabulated value for the extrapolation constant. Williams (1972) also noted that if some slight imbalance exists throughout the design, say $0.8 \le r_i/r_k \le 1.25$ for all $i = 1, \ldots, k - 1$, use of the treatment-balanced critical points $\bar{t}_\alpha(\nu, \omega)$ still serves as a good approximation.

Example 6.4 Body weight changes in rats after exposure to 1,4-dichlorobenzene

Consider the problem of testing for a decreasing trend in mean body weights of laboratory animals after they have been exposed to an environmental carcinogen. For example, the animal repellent 1,4-dichlorobenzene was studied by the US National Toxicology Program (1987b) to assess its toxicological potential in female mice. Animals were exposed to either a zero-dose control ($i = 0$), or to one of $k = 3$ levels of the chemical (600, 900, or 1000 mg/kg in feed). Forty mice were used, with $r = 10$ mice assigned randomly to each group. The outcome variable was body weight change, that is, how much each mouse's body weight changed over the course of the 13-week study. Summary statistics are given in Table 6.3.

Exposure to 1,4-dichlorobenzene may inhibit growth in the mice, as exhibited by a decrease in their weight gains. We can assess this formally using Williams' test, with the ordered alternative here of the form H_a: $\mu_0 \ge \mu_1 \ge \mu_2 \ge \mu_3$. A normal parent distribution is considered reasonable for these data (at least approximately), so we assume each animal's weight change, Y_{ij}, is normal, with mean μ_i and (constant) variance σ^2 ($i = 0, \ldots, 3; j = 1, \ldots, 10$). To apply Williams' test in this setting, we first calculate the ML estimates of each μ_i under the decreasing-order alternative hypothesis: begin with $\hat{\mu}_0 = \overline{Y}_{0+} = 8.9$ and note that at $i = 1$, $\overline{Y}_{1+} = 5.4 < \hat{\mu}_0$, so, initially, the decreasing order is preserved. Now, however, move to $i = 2$, where $\overline{Y}_{2+} = 6.5 > 5.4 = \overline{Y}_{1+}$. Thus an order violation occurs, forcing us to impose the pool-adjacent-violators algorithm via (6.3): set $\hat{\mu}_1 = \hat{\mu}_2 = \{\overline{Y}_{1+} + \overline{Y}_{2+}\}/2 = (5.4 + 6.5)/2 = 5.95$. Since this now satisfies the decreasing-order restriction $\hat{\mu}_0 \geq \hat{\mu}_1 \geq \hat{\mu}_2$ for the first three indices, we move on to $i = 3$.

Table 6.3 Body weight changes (mg) in female mice

	Control	*1,4-dichlorobenzene (mg/kg)*		
	(i = 0)	*600*	*900*	*1000*
Number of animals, r	10	10	10	10
Sample means, $\overline{Y}_{i+}$	8.9	5.4	6.5	6.3
Sample std. deviations	0.8	0.3	0.7	0.3
Ordered ML estimates	8.9	6.067	6.067	6.067

Data from NTP (1987b). Two additional high-dose groups from the original study exhibited excessive animal mortality and provided only limited information; they are not reported here.

Another violation of the decreasing-order restriction occurs at $i = 3$, where $\overline{Y}_{3+} = 6.3 > \hat{\mu}_1 = \hat{\mu}_2 = 5.95$. So, we apply the pool-adjacent-violators algorithm at $i = 1,2,3$: set $\hat{\mu}_1 = \hat{\mu}_2 = \hat{\mu}_3 = \{\overline{Y}_{1+} + \overline{Y}_{2+} + \overline{Y}_{3+}\}/3 = (5.4 + 6.5 + 6.3)/3 = 6.0667$. This now satisfies fully the decreasing-order restriction $\hat{\mu}_0 \geq \hat{\mu}_1 \geq \hat{\mu}_2 \geq \hat{\mu}_3$. (These final, order-restricted estimates are given in the last row of Table 6.3.)

To construct the test statistic, we multiply the ML estimates by –1 (in order to force the problem into one of assessing an increasing trend). Using (6.4), the calculated trend test statistic becomes

$$\bar{t}_{\text{calc}} = \frac{-6.0667 - (-8.9)}{(0.2)\sqrt{0.3275}}$$

where the mean square error, $\hat{\sigma}^2 = 0.3275$, is a weighted average of the individual sample variances, s_i^2: $\hat{\sigma}^2 = \sum_{i=0}^{k}(r_i - 1)s_i^2/\sum_{i=0}^{k}(r_i - 1)$, as in Section 5.1. The final calculations give $\bar{t}_{\text{calc}} = 24.76$; the *df* for this trend statistic are $\nu = (k + 1)(r - 1) = (4)(9) = 36$.

Rejection of the no-trend null hypothesis in favor of an increasing trend (in the *negative* values) occurs when $\bar{t}_{\text{calc}}$ exceeds $\bar{t}_\alpha(36)$. Suppose $\alpha = 0.01$. From Williams' (1971) tables, one finds $\bar{t}_{0.01}(35) = 2.50$, while $\bar{t}_{0.01}(40) = 2.48$. Interpolating, we find $\bar{t}_\alpha(36) \approx 2.49$. Since $\bar{t}_{\text{calc}}$ is well above this critical point, we conclude that a strongly increasing trend exists for the negative values, which is equivalent to a strongly decreasing trend in the original body weight changes. ❑

When it is unreasonable to assume a normal parent distribution for the observations, it is possible to conduct a distribution-free trend test. Given by Shirley (1977), the test is essentially a rank-based version of Williams' test, modified to take advantage of the information in the rank ordering. Williams (1986) provides a modification of Shirley's procedure that ensures a proper familywise Type I error rate. In a comparison of both approaches, Hothorn (1989) suggests that Williams' test is fairly robust to departures from the normal sampling assumption, although he notes that with nonnormal data, Shirley's nonparametric test is generally more powerful when the samples sizes are moderately large, say, $r_i \geq 10$ at all $i = 0, \ldots, k$. Hothorn also describes an adjustment to Williams' test that allows the investigator to make sequential rejections of the ordered alternative hypothesis, in effect identifying at which levels of x_i the departure from H_0 is insignificant. This is a form of **sequentially rejective test** (as in Section 5.4.1), constructed to control the familywise false positive rate, α. The methods can be employed to formulate extensions of Williams' original approach to trend testing (Antonello *et al.*, 1993; Rom *et al.*, 1994).

Although designed strictly as a significance testing methodology, it is possible to invert Williams' approach to provide simultaneous confidence limits on various functions of the μ_is under an ordered alternative. Schoenfeld (1986) describes methods that include a form of inverted Williams test, with applications in biology and toxicology, while Hayter (1990) gives an approach based on multiple comparisons among the μ_is; see also Hayter and Hsu (1994).

6.3 TREND TESTS FOR PROPORTIONS

The use of Williams' test is limited to continuous data such as concentrations, growth measurement, changes in structure, etc., where an assumption of normal parent distributions is appropriate. When the original data are recorded as discrete counts or proportions, however, the normal distribution assumptions necessary to apply Williams' test are violated. A natural alternative for trend testing in these cases is to employ some form of distribution-free method. For instance, one can apply methods based on ranks, such as Shirley's method (Shirley, 1977; Williams, 1986) noted above, or a more general test known as the **Jonckheere–Terpstra trend test**, which we discuss in Section 6.6. When the parent distribution of the data is known, however, it is still appropriate to use this information in constructing tests for trend.

With proportion data, it is common to use a binomial parent distribution: $Y_i \sim$ (indep.) $\mathcal{Bin}(N_i,\pi_i)$, where π_i represents the response probability at the ith treatment level ($i = 0, \ldots, k$). Of interest is testing the null hypothesis $H_0{:}\pi_0 = \pi_1 = \cdots = \pi_k$ vs. the increasing-trend alternative $H_a{:}\pi_0 \leq \pi_1 \leq \cdots \leq \pi_k$, with strict inequality somewhere among the π_is.

6.3.1 The Cochran–Armitage trend test

A powerful approach for trend testing with discrete data was proposed by Cochran (1954) and Armitage (1955) – and, in a more general form, by Yates (1948) – and is known as the **Cochran–Armitage trend test**. The basic principle behind the test is simple: begin with some set of independent measures, x_i, that quantify the ordered levels, such as doses or log-doses of an exposure regimen. In cases where no such values exist, a common recommendation is to assign a set of reasonable measures, such as some form of equal spacing, for the x_is. (The trend test is generally insensitive to minor distortions in the underlying scale of the ordered measures; see Snedecor and Cochran (1980, Section 11.8).)

Denote the ordered measures associated with the individual levels of the treatment factor as $x_0 < x_1 < \cdots < x_k$. At each of these values, the maximum likelihood estimate of π_i is the sample proportion $p_i = Y_i/N_i$. If the no-trend hypothesis, H_0, were true, we would expect these values to equal roughly the pooled proportion $\overline{p} = \sum_{i=0}^{k} Y_i / \sum_{i=0}^{k} N_i$. The Cochran–Armitage statistic quantifies the departure from this pooled value, by assessing the amount of any trend in the p_is over the ordered levels of x_i. The statistic is essentially a weighted regression of p_i on x_i, with weights taken as the reciprocals of the estimated variances of p_i assuming H_0 is true: $w_i = N_i/\{\overline{p}\,\overline{q}\}$, where $\overline{q} = 1 - \overline{p}$ (Cochran, 1954; Margolin, 1988). A regression slope estimator is calculated, which is divided by its standard error to achieve the test statistic. (Details of the construction are left for Exercise 6.12.) The result is

$$z_{\mathrm{calc}} = \frac{\sum_{i=0}^{k} (x_i - \overline{x}) Y_i}{\left\{ \overline{p}\,\overline{q} \sum_{i=0}^{k} N_i (x_i - \overline{x})^2 \right\}^{1/2}} \tag{6.6a}$$

or equivalently,

$$z_{\text{calc}} = \frac{\sum_{i=0}^{k} (x_i - \bar{x}) N_i p_i}{\left\{ \bar{p}\bar{q} \sum_{i=0}^{k} N_i (x_i - \bar{x})^2 \right\}^{1/2}} = \frac{\sum_{i=0}^{k} (x_i - \bar{x}) N_i (p_i - \bar{p})}{\left\{ \bar{p}\bar{q} \sum_{i=0}^{k} N_i (x_i - \bar{x})^2 \right\}^{1/2}}. \quad (6.6b)$$

In (6.6), $\bar{x}$ is the sample-weighted average independent measure, $\bar{p}$ is the sample-weighted average proportion exhibiting the response of interest, and p_i is the sample proportion in the ith experimental group; that is, $\bar{x} = \sum_{i=0}^{k} N_i x_i / \sum_{i=0}^{k} N_i$, $\bar{p} = \sum_{i=0}^{k} N_i p_i / \sum_{i=0}^{k} N_i$, and $p_i = Y_i/N_i$.

In large samples – say, when the N_i all exceed 10, and the total sample size exceeds 50 – the Cochran–Armitage statistic is referred to a standard normal distribution. One rejects the no-trend hypothesis H_0 in favor of an increasing trend in H_a when $z_{\text{calc}} \geq z_\alpha$, where z_α is the upper-α standard normal critical point. Alternatively, the large-sample P-value of the test is $P \approx 1 - \Phi(z_{\text{calc}})$. To test against a decreasing trend in the π_is, calculate (6.6), and reject if $z_{\text{calc}} \leq -z_\alpha$. The corresponding large-sample P-value is $P \approx \Phi(z_{\text{calc}})$.

The Cochran–Armitage statistic in (6.6) possesses many optimal statistical properties; for example, it is invariant to linear transformations of the x_i, so that a change in the stimulus scale from x_i to $A + Bx_i$ does not change the value of z_{calc}. More importantly, the test is uniformly most powerful (UMP; see Section 2.4) against any alternative hypothesis in which the π_is increase (or decrease) via the form

$$\pi(x_i) = \frac{1}{1 + \exp\{-\beta_0 - \beta_1 x_i\}} \quad (6.7)$$

(Cox, 1958). (This particular pattern is known as a **logistic regression model**, and we discuss it in more detail in Section 7.2.1.) From the UMP feature of the test, one expects to detect departures from homogeneity in this logistic form more often than with any other trend test for proportions operating at the same α level.

The Cochran–Armitage trend test corresponds also to the score test of $H_0\!:\beta_1 = 0$; that is, it is the score statistic for testing H_0 vs. $H_a\!:\beta_1 \neq 0$ under (6.7). It is also locally most powerful against any form of twice-differentiable, monotone function that creates a trend in the π_is (Tarone and Gart, 1980), giving it a form of omnibus optimality against a variety of smoothly increasing (or decreasing) ordered alternatives. Thus, even though the quantity in (6.6) can be viewed as a form of weighted linear regression test statistic, it has application to any smoothly increasing (or decreasing) trend in the proportion response.

It is important to note, however, that computation of z_{calc} as given in (6.6) assumes that the x_is are reasonably symmetric about $\bar{x}$. When this is not the case, and if a simple transformation does not reduce the skew, a skewness correction is advocated (Tarone, 1986). Also, for the case of just two groups (for example, control vs. treatment, where $k = 1$), the Cochran–Armitage statistic collapses to the statistic for comparing two proportions based on (4.8), and thus may be viewed as an extension of this simple two-sample test (Margolin *et al.*, 1983).

Example 6.5 Chromosome damage after exposure to hydroquinones
Dobo and Eastmond (1994) report data on micronucleation in Chinese hamster cells, induced by the compound *tert*-butylhydroquinone (*t*BHQ). A micronucleus is a small portion of the original cell nucleus that has been detached, typically due to some intracellular genetic damage. By studying micronucleus formation, investigators learn more about the nature and origin of genotoxic effects in cells exposed to environmental agents.

In the *t*BHQ study, each cell was treated with an antibody after chemical exposure. The antibody generates a protein label indicative of breakage along the chromosome. If the label is seen in the micronucleus, we infer that the original event leading to micronucleus formation was chromosome breakage. Table 6.4 and Fig. 6.6 illustrate data on this phenomenon, with cells exposed to *t*BHQ *in vitro* for 1 hour over a series of $k = 5$ exposure concentrations (and including a zero-dose solvent control).

Notice in both Table 6.4 and Fig. 6.6 that the observed proportions p_i appear to increase as the exposure levels increase. To identify whether this increasing trend is significant, we apply the Cochran–Armitage trend test, using (6.6). The pooled response proportion is $\bar{p} = 127/12000 = 0.0106$, with $\bar{q} = 1 - \bar{p} = 0.9894$. The sample-weighted average exposure level, $\bar{x}$, is just the simple average of all the exposure levels (since the design sets the N_is equal): $\bar{x} = 360.0$. Then, the trend statistic in (6.6) is based on

$$\sum_{i=0}^{5}(x_i - \bar{x})Y_i = (-360)(8) + (-240)(9) + (-120)(9) + (120)(18) + (240)(39) + (360)(44)$$

which calculates to 21 240, and

$$\sum_{i=0}^{5}N_i(x_i - \bar{x})^2 = (2000)\{(-360)^2 + (-240)^2 + (-120)^2 + (120)^2 + (240)^2 + (360)^2\}$$

which calculates to 8.064×10^8.

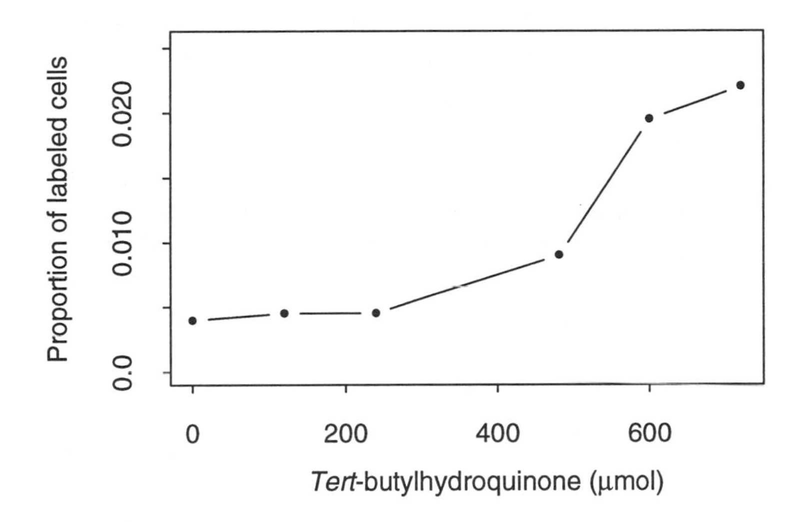

Fig. 6.6 Scatterplot of proportions of labeled cells vs. concentration of *tert*-butylhydroquinone. The observed proportions are connected with solid lines.

Table 6.4 Induction of micronucleated erythrocytes in mammalian cell cultures by *tert*-butylhydroquinone

	Control	*tBHQ concentration (μmol)*				
	(i = 0)	*120*	*240*	*480*	*600*	*720*
Number of labeled cells, Y_i	8	9	9	18	39	44
Number of total cells, N_i	2000	2000	2000	2000	2000	2000
Proportions, $p_i = Y_i/N_i$	0.0040	0.0045	0.0045	0.0090	0.0195	0.0220

Data from Dobo and Eastmond (1994).

Employing these values in (6.6) yields $z_{\text{calc}} = 21240.0/\sqrt{8444077.6} = 7.309$. Compared to the upper-0.005 critical point of a standard normal distribution, $z_{0.005} = 2.576$, we see the trend in Table 6.4 is strongly significant. (The one-sided P-value is well below 0.001.) It appears that the probability of tBHQ-induced chromosome breakage trends upward in these hamster cells.

The Cochran–Armitage trend statistic can be calculated by computer, using SAS's PROC FREQ. The SAS code in Fig. 6.7 first illustrates a SAS `data` step that inputs the data from Table 6.4, constructs new variables that indicate whether cell labeling occurred (`label`), and counts of the number of cells in each concentration-label condition (`count`). The code then invokes PROC FREQ to calculate the trend statistic. The `order=data` option in PROC FREQ is used to specify that the rows and columns of data be presented in the order in which the levels of the variables are defined during data entry (in contrast to the SAS default of sorting the levels alphabetically). Also, the statement

```
table conc*label / chisq;
```

outputs a table of counts cross-classified by `conc` (row factor) and `label` (column factor). (We discuss the analysis of cross-classified tabular data in greater detail in Chapter 9.) The `chisq` option in the `table` statement produces the Cochran–Armitage trend statistic, along with other tests. Also, the separate `weight` statement labels the variable containing the count of observations in each concentration–label condition.

```
* SAS code for C-A trend statistic;
data tBHQ;
 input conc nlabel @@;
 total = 2000;
 notlabel = total - nlabel;
 pi = nlabel/total;
    label="Yes"; count=nlabel; output;
    label=" No"; count=notlabel; output;
 cards;
0 8 120 9 240 9 480 18 600 39 720 44
proc freq order=data;
 table conc*label / chisq;
 weight count;
```

Fig. 6.7 SAS program to obtain Cochran–Armitage trend statistic, applied to data from Table 6.4.

The SAS output appears in Fig. 6.8. It includes a table of the number of labeled cells (`Yes` column) and unlabeled cells (`No` column) by *t*BHQ concentration, along with various tests of association between the column and row variables. The row labeled `Mantel-Haenszel Chi-Square` contains the square of the Cochran–Armitage trend statistic. (Indeed, $z_{calc} = \sqrt{53.422} = 7.309$, as calculated above.) The Mantel-Haenszel method is a general approach for testing homogeneity in data with categorical structure (Agresti, 1990, Section 7.4).

Suppose for purposes of illustration that an alternate coding of exposure levels was considered with these data. For instance, assign $u_0 = -3$, $u_1 =$

-2, $u_2 = -1$, $u_3 = 1$, $u_4 = 2$, and $u_5 = 3$, that is, $u_i = (x_i - 360)/120$ with $\bar{u} = 0$. The trend test now employs the values $\sum_{i=0}^{5}(u_i - \bar{u})Y_i = 177$ and $\sum_{i=0}^{5}N_i(u_i - \bar{u})^2 = 56\,000$. This leads to $z_{\text{calc}} = 177/\sqrt{586.3943} = 7.309$, which is the same result as with the previous coding. This illustrates the trend statistic's invariance to linear transformations of the x_is. ❑

```
                    The SAS System
                  The FREQ Procedure

               TABLE OF CONC BY LABEL

               CONC        LABEL
           Frequency|
           Row Pct  |Yes     | No     |  Total
           ---------+--------+--------+
                  0 |      8 |   1992 |   2000
                    |   0.40 |  99.60 |
           ---------+--------+--------+
                120 |      9 |   1991 |   2000
                    |   0.45 |  99.55 |
           ---------+--------+--------+
                240 |      9 |   1991 |   2000
                    |   0.45 |  99.55 |
           ---------+--------+--------+
                480 |     18 |   1982 |   2000
                    |   0.90 |  99.10 |
           ---------+--------+--------+
                600 |     39 |   1961 |   2000
                    |   1.95 |  98.05 |
           ---------+--------+--------+
                720 |     44 |   1956 |   2000
                    |   2.20 |  97.80 |
           ---------+--------+--------+

           STATISTICS FOR TABLE OF CONC BY LABEL
 Statistic                         DF      Value         Prob
 ------------------------------------------------------------
 Chi-Square                         5     62.974        0.000
 Likelihood Ratio Chi-Square        5     60.542        0.000
 Mantel-Haenszel Chi-Square         1     53.422        0.000

 Sample Size = 12000
```

Fig. 6.8 SAS output (edited) for calculating the Cochran–Armitage trend test statistic (`Mantel-Haenszel Chi-Square`) with data in Table 6.4.

We should note that many other methods are available for testing trend in proportion data, based primarily on manipulations with the ordered inequalities making up H_a (Robertson *et al.*, 1988, Section 1.5). Typically, these require extensive computer iteration to calculate the test statistics, and they have been shown to offer only limited improvement in either increased stability or sensitivity over use of the Cochran–Armitage trend test (Collings *et al.*, 1981). It is also possible to employ parametric regression models to test for trend in binomial response, as we will see in Chapter 7.

6.3.2 Trend tests incorporating differing time at risk

In some instances, binary outcomes used to assess or measure environmental toxicity may not be observed directly, due to some competing or alternate risk factor that masks the actual event of interest. For example, in **environmental carcinogenicity testing**, laboratory studies with rodents are undertaken to study the carcinogenic effects of various environmental stimuli. A common experimental protocol for such a study exposes N_i animals to increasing doses, x_i, of some environmental agent over an extended period, usually 104–105 weeks. A concurrent set of N_0 control animals is used to provide a measure of the spontaneous or background carcinogenic response, as noted in Section 3.1.2. The data are recorded as simple binary observations, where each animal's response is coded as 1 if it exhibits the tumor, or as 0 if it does not. Pooled within each dose group, the independent binary observations sum to a binomial random variable, with parameters N = {number of animals at each exposure level}, and π = {probability of carcinogenic response at that exposure}. The study's goal is identification of increasing carcinogenic dose response to the agent.

At first glance, the statistical analysis of such data is straightforward: employ the Cochran–Armitage trend test for proportions from Section 6.3.1. In such an experiment, however, the analyst must recognize that both tumor onset and animal mortality may increase in response to high levels of the toxin. Typically, tumor onset occurs later in the life of an animal than exposure-related (noncancer) mortality, so the binary observation of whether an animal is or is not tumor-bearing may be inhibited at higher doses, due to the competing mortality. (Other potential complications, such as whether the tumor incidence rates vary with the age of the animal, lead to additional adjustments in the simple analysis; see Example 8.7.) In such a case, we say the observations are **censored** by the effects of exposure-related, noncancer mortality.

To account for effects of differential tumor onset, an adjustment is required in the Cochran–Armitage trend statistic. (In Chapter 11, we address more general questions associated with censored observations when we present methods for survival data analysis.) For testing trend in proportions over some independent measure x_i, the statistic from (6.6) is

$$z_{\text{calc}} = \frac{\sum_{i=0}^{k}(x_i - \bar{x})N_i p_i}{\left\{\bar{p}\bar{q}\sum_{i=0}^{k} N_i(x_i - \bar{x})^2\right\}^{1/2}},$$

where $p_i = Y_i/N_i$, $\bar{x} = \sum_{i=0}^{k} N_i x_i / \sum_{i=0}^{k} N_i$, and $\bar{p} = \sum_{i=0}^{k} N_i p_i / \sum_{i=0}^{k} N_i$. The N_is in this formula are the number of experimental units at risk for the event of interest, for example, tumor onset. If each organism is 'on test' for the same length of time, then N_i measures the time at risk in the ith group, with each of the N_i organisms contributing weight $\omega_{ij} = 1$ to the total time at risk; that is, $\sum_{j=1}^{N_i} \omega_{ij} = N_i$. Intuitively, an organism that dies without the tumor prior to the scheduled termination of the study has not contributed a full quantity of time at risk, say, $\omega_{ij} < 1$, while an organism developing the tumor receives a full quantity of time-at-risk: $\omega_{ij} = 1$.

To quantify this, assume that the jth organism in the ith experimental group is exposed to a dose x_i and let $Y_{ij} = 0$ if the event of interest is not present, and $Y_{ij} = 1$ if the event of interest is present. Notice that $Y_i = \sum_{j=1}^{N_i} Y_{ij}$. In addition to dose and an event indicator, assume also the length of time on study is recorded, say t_{ij}. Finally, let T represent the total length of time for an experiment, implying $t_{ij} \leq T$ for all organisms in a study. (This approach can be extended to incorporate terminations that are less than T, arising as a result of scheduled **interim sacrifices**.)

Gart *et al.* (1979) recommend that trend tests be applied to adjusted proportions in which the data associated with deaths prior to the first observed event of interest are eliminated. If t^* is the time of the first observed event of interest, then the time-at-risk weight for the jth organism in the ith experimental group is $\omega_{ij} = 0$ if $t_{ij} < t^*$ and $\omega_{ij} = 1$ if $t_{ij} \geq t^*$. The adjusted proportions are $p_i^* = \sum_{j=1}^{N_i} Y_{ij}/N_i^*$, where $N_i^* = \sum_{j=1}^{N_i} \omega_{ij}$ is the adjusted time at risk. These adjusted quantities are used in place of their unadjusted counterparts in z_{calc} above.

To increase sensitivity to detect increasing trends in response hidden by time-related censoring, Bailer and Portier (1988) modified the time adjustment to reflect dose-related differences in the underlying tumor onset distribution. They found that time-at-risk weights $\tilde{\omega}_{ij}$ constructed as powers of the survival time t_{ij}, divided by the total study time T, provided a stable adjustment to the trend test. That is, take

$$\tilde{\omega}_{ij} = \begin{cases} 1 & \text{if } Y_{ij} = 1 \\ \left(\frac{t_{ij}}{T}\right)^{\lambda} & \text{if } Y_{ij} = 0 \end{cases}$$

for some $\lambda > 0$. The value $\lambda = 3$ is recommended as a useful rule of thumb (Portier and Bailer, 1989), resulting in what is called a Poly-3 trend test. The corresponding adjusted times at risk $\tilde{N}_i = \sum_{j=1}^{N_i} \tilde{\omega}_{ij}$ and adjusted proportions $\tilde{p}_i = \sum_{j=1}^{N_i} Y_{ij}/\tilde{N}_i$ are used in z_{calc} above. Bieler and Williams (1993) further modified the Bailer–Portier adjustment by incorporating a large-sample approximation to the variance of the adjusted proportions. Their statistic is

$$z_{\text{adj}} = \frac{a_+ \sum_{i=0}^{k} a_i \tilde{p}_i x_i - \left(\sum_{i=0}^{k} a_i x_i\right)\left(\sum_{j=0}^{k} a_j \tilde{p}_j\right)}{\left\{\left(\frac{a_+}{N_+ - k - 1}\right)\left[a_+ \sum_{i=0}^{k} a_i x_i^2 - \left(\sum_{i=0}^{k} a_i x_i\right)^2\right] \sum_{i=0}^{k} \sum_{j=1}^{N_i} (R_{ij} - \bar{R}_{i+})^2\right\}^{1/2}},$$

where $R_{ij} = Y_{ij} - \tilde{p}_+ \tilde{\omega}_{ij}$, $\tilde{p}_+ = \sum_{i=1}^{k} Y_i / \sum_{i=1}^{k} \tilde{N}_i$, $a_i = (\tilde{N}_i)^2/N_i$, $a_+ = \sum_{i=0}^{k} a_i$, and $N_+ = \sum_{i=0}^{k} N_i$.

In all these cases, we reject the null hypothesis of no trend in the tumorigenic response against an alternative hypothesis of increasing trend when the test statistic, z_{calc} or z_{adj}, exceeds $\mathfrak{z}_{\alpha}$ from (1.14). The respective approximate P-values are $1 - \Phi(z_{\text{calc}})$ or $1 - \Phi(z_{\text{adj}})$.

Example 6.6 Trends in alveolar/bronchiolar tumors after vinylcyclohexene diepoxide exposure

Portier and Bailer (1989) present an analysis of the tumorigenicity of the chemical vinylcyclohexene diepoxide in female B6C3F$_1$ mice. Of interest were the effects of a 105-week chronic exposure on the formation of murine alveolar/bronchiolar tumors. Four groups were examined (at exposure levels 0, 25, 50, and 100 mg/ml), with 50 animals per group. Data from this experiment, along with intermediate calculations for the Poly-3 adjusted weights $\tilde{\omega}_{ij} = (t_{ij}/T)^3$ and the adjusted differences $R_{ij} = Y_{ij} - \tilde{p}_+ \tilde{\omega}_{ij}$, are given in Table 6.5. Notice in the table that in each exposure group except the highest (100 mg/ml) some mice survived until the 105-week termination time. (In the highest exposure group, all animals had died by week 85.)

Also, all of the tumors recorded in the control and lowest-exposure groups were detected after 85 weeks, and a majority (eight of 11) of the tumors recorded in the 50 mg/ml exposure group were detected after 85 weeks. This suggests that the accelerated mortality pattern seen in the 100 mg/ml exposure group may have acted to censor from possible observation later-developing tumors, and that the observed count of only seven tumors at this exposure level may require statistical adjustment. Hence these data provide a natural context in which to apply the adjusted trend tests.

Here, we compare adjusted Cochran–Armitage-type trend statistics for the vinylcyclohexene diepoxide data in Table 6.5, using the Gart *et al.* and the Bailer–Portier adjusted proportions, p_i^* and $\tilde{p}_i$, respectively. (We also include calculation of the Bieler–Williams modification to the Bailer–Portier method.) For all of these adjustments, we first calculate the adjusted time-at-risk values, N_i^* or $\tilde{N}_i$. These quantities are given in Table 6.6. Figure 6.9 displays the raw and adjusted proportions, plotted as a function of dose, x_i. Notice from the figure that the raw proportions p_i and the Gart *et al.* adjusted proportions p_i^* do not increase monotonically with dose. This would be expected if exposure-related mortality were acting to censor late-developing tumors. The Bailer–Portier adjusted proportions $\tilde{p}_i$ are constructed to overcome such an effect, however; as seen in Fig. 6.9, the $\tilde{p}_i$ at $\lambda = 3$ do exhibit a strictly increasing pattern.

Applying the unadjusted Cochran–Armitage statistic (6.6) to the raw proportions, p_i, in Table 6.6 yields $z_{calc} = 0.6275$ ($P \approx 0.53$); no significant increasing trend is evidenced. Similarly, substituting the Gart *et al.* adjusted values N_i^* and p_i^* into (6.6) yields $z_{calc} = 0.8224$ ($P \approx 0.41$); again, no significant increase is evidenced.

In both these cases, however, the lack of significance may be affected by the downturn in the proportion response at the 100 mg/ml exposure level. (The unadjusted Cochran–Armitage statistic is known to suffer losses in sensitivity when the response function expresses a nonmonotone, unimodal form (Collings *et al.*, 1981), such as seen in Fig. 6.9.) The adjusted proportions $\tilde{p}_i$ are designed to minimize this nonmonotone effect, however, and applying them with $\lambda = 3$ in z_{calc} does appear to isolate a tumorigenic effect: the trend statistic is $z_{calc} = 2.128$ ($P \approx 0.033$). Similarly, using $\tilde{N}_i$ and $\tilde{p}_i$ in the Bieler–Williams modified statistic yields

$$z_{adj} = \frac{927 - \dfrac{(3361.532)(22.383)}{108.610}}{\left\{(0.1168)\left(\dfrac{199144.612 - (3361.532)^2}{108.610}\right)\right\}^{1/2}} = 2.231,$$

Table 6.5 Alveolar/bronchiolar adenomas or carcinomas in female $B6C3F_1$ mice after exposure to vinylcyclohexene diepoxide: animal number (j), tumor indicator (Y_{ij}), adjusted time-at-risk weight $(\tilde{\omega}_{ij})$, and adjusted Bieler–Williams difference (R_{ij}) at $i = 0$ (0 mg/ml), $i = 1$ (25 mg/ml), $i = 2$ (50 mg/ml); for $i = 3$ (100 mg/ml) see page 239

j	Y_{0j}	t_{0j}	$\tilde{\omega}_{0j}$	R_{0j}	Y_{1j}	t_{1j}	$\tilde{\omega}_{1j}$	R_{1j}	Y_{2j}	t_{2j}	$\tilde{\omega}_{2j}$	R_{2j}
1	0	10	0.00	0.00	0	5	0.00	0.00	0	1	0.00	0.00
2	0	56	0.15	−0.03	0	28	0.02	0.00	0	1	0.00	0.00
3	0	63	0.22	−0.05	0	31	0.03	−0.01	0	7	0.00	0.00
4	0	65	0.24	−0.05	0	60	0.19	−0.04	0	58	0.17	−0.04
5	0	65	0.24	−0.05	0	62	0.21	−0.04	0	63	0.22	−0.05
6	0	72	0.32	−0.07	0	66	0.25	−0.05	0	65	0.24	−0.05
7	0	77	0.39	−0.08	0	69	0.28	−0.06	0	67	0.26	−0.06
8	0	82	0.48	−0.10	0	74	0.35	−0.08	1	71	1.00	0.78
9	0	85	0.53	−0.11	0	79	0.43	−0.09	1	71	1.00	0.78
10	0	85	0.53	−0.11	0	80	0.44	−0.10	0	81	0.46	−0.10
11	0	95	0.74	−0.16	0	83	0.49	−0.11	0	82	0.48	−0.10
12	0	97	0.79	−0.17	0	92	0.67	−0.14	0	82	0.48	−0.10
13	0	97	0.79	−0.17	0	101	0.89	−0.19	0	83	0.49	−0.11
14	0	100	0.86	−0.19	0	102	0.92	−0.20	1	83	1.00	0.78
15	0	100	0.86	−0.19	0	102	0.92	−0.20	0	84	0.51	−0.11
16	0	101	0.89	−0.19	0	102	0.92	−0.20	0	85	0.53	−0.11
17	0	102	0.92	−0.20	1	102	1.00	0.78	0	86	0.55	−0.12
18	0	103	0.94	−0.20	1	102	1.00	0.78	1	87	1.00	0.78
19	1	103	1.00	0.78	0	104	0.97	−0.21	0	93	0.69	−0.15
20	0	103	0.94	−0.20	1	105	1.00	0.78	0	96	0.76	−0.16
21	1	105	1.00	0.78	1	105	1.00	0.78	1	96	1.00	0.78
22	1	105	1.00	0.78	1	105	1.00	0.78	0	97	0.79	−0.17
23	1	105	1.00	0.78	1	105	1.00	0.78	0	97	0.79	−0.17
24	0	105	1.00	−0.22	1	105	1.00	0.78	0	98	0.81	−0.17
25	0	105	1.00	−0.22	1	105	1.00	0.78	0	98	0.81	−0.17
26	0	105	1.00	−0.22	1	105	1.00	0.78	0	99	0.84	−0.18
27	0	105	1.00	−0.22	0	105	1.00	−0.22	1	100	1.00	0.78
28	0	105	1.00	−0.22	0	105	1.00	−0.22	0	101	0.89	−0.19
29	0	105	1.00	−0.22	0	105	1.00	−0.22	0	101	0.89	−0.19
30	0	105	1.00	−0.22	0	105	1.00	−0.22	0	103	0.94	−0.20
31	0	105	1.00	−0.22	0	105	1.00	−0.22	0	104	0.97	−0.21
32	0	105	1.00	−0.22	0	105	1.00	−0.22	0	104	0.97	−0.21
33	0	105	1.00	−0.22	0	105	1.00	−0.22	0	104	0.97	−0.21
34	0	105	1.00	−0.22	0	105	1.00	−0.22	0	104	0.97	−0.21
35	0	105	1.00	−0.22	0	105	1.00	−0.22	1	104	1.00	0.78
36	0	105	1.00	−0.22	0	105	1.00	−0.22	1	105	1.00	0.78
37	0	105	1.00	−0.22	0	105	1.00	−0.22	1	105	1.00	0.78
38	0	105	1.00	−0.22	0	105	1.00	−0.22	1	105	1.00	0.78
39	0	105	1.00	−0.22	0	105	1.00	−0.22	1	105	1.00	0.78
40	0	105	1.00	−0.22	0	105	1.00	−0.22	0	105	1.00	−0.22
41	0	105	1.00	−0.22	0	105	1.00	−0.22	0	105	1.00	−0.22
42	0	105	1.00	−0.22	0	105	1.00	−0.22	0	105	1.00	−0.22
43	0	105	1.00	−0.22	0	105	1.00	−0.22	0	105	1.00	−0.22
44	0	105	1.00	−0.22	0	105	1.00	−0.22	0	105	1.00	−0.22
45	0	105	1.00	−0.22	0	105	1.00	−0.22	0	105	1.00	−0.22
46	0	105	1.00	−0.22	0	105	1.00	−0.22	0	105	1.00	−0.22
47	0	105	1.00	−0.22	0	105	1.00	−0.22	0	105	1.00	−0.22
48	0	105	1.00	−0.22	0	105	1.00	−0.22	0	105	1.00	−0.22
49	0	105	1.00	−0.22	0	105	1.00	−0.22	0	105	1.00	−0.22
50	0	105	1.00	−0.22	0	105	1.00	−0.22	0	105	1.00	−0.22

continued

Table 6.5 (continued)

j	Y_{3j}	t_{3j}	$\tilde{\omega}_{3j}$	R_{3j}	j	Y_{3j}	t_{3j}	$\tilde{\omega}_{3j}$	R_{3j}
1	0	1	0.00	0.00	26	0	78	0.41	-0.09
2	0	24	0.01	0.00	27	0	79	0.43	-0.09
3	0	27	0.02	0.00	28	0	79	0.43	-0.09
4	0	54	0.14	-0.03	29	0	79	0.43	-0.09
5	0	56	0.15	-0.03	30	0	79	0.43	-0.09
6	0	59	0.18	-0.04	31	0	80	0.44	-0.10
7	0	62	0.21	-0.04	32	0	81	0.46	-0.10
8	1	64	1.00	0.78	33	0	83	0.49	-0.11
9	0	65	0.24	-0.05	34	0	83	0.49	-0.11
10	0	67	0.26	-0.06	35	0	83	0.49	-0.11
11	0	68	0.27	-0.06	36	1	83	1.00	0.78
12	0	68	0.27	-0.06	37	0	83	0.49	-0.11
13	0	70	0.30	-0.06	38	0	83	0.49	-0.11
14	1	71	1.00	0.78	39	0	85	0.53	-0.11
15	0	72	0.32	-0.07	40	0	85	0.53	-0.11
16	0	72	0.32	-0.07	41	1	85	1.00	0.78
17	0	72	0.32	-0.07	42	1	85	1.00	0.78
18	0	75	0.36	-0.08	43	0	85	0.53	-0.11
19	0	75	0.36	-0.08	44	1	85	1.00	0.78
20	0	75	0.36	-0.08	45	1	85	1.00	0.78
21	0	76	0.38	-0.08	46	0	85	0.53	-0.11
22	0	77	0.39	-0.08	47	0	85	0.53	-0.11
23	0	77	0.39	-0.08	48	0	85	0.53	-0.11
24	0	78	0.41	-0.09	49	0	85	0.53	-0.11
25	0	78	0.41	-0.09	50	0	85	0.53	-0.11

Data from Portier and Bailer (1989).

Table 6.6 Observed numbers at risk and proportions (N_i, p_i), Gart *et al.* adjusted numbers at risk and proportions, and Bailer–Portier adjusted numbers at risk and adjusted proportions for data from Table 6.5

Exposure				*Adjusted values*			
level	*dose*			*Gart et al.*		*Bailer–Portier*	
i	x_i	N_i	p_i	N_i^*	p_i^*	$\tilde{N}_i$	$\tilde{p}_i$
0	0	50	0.08	47	0.0851	41.8370	0.0956
1	25	50	0.18	45	0.2000	40.9661	0.2197
2	50	50	0.22	45	0.2444	38.4904	0.2858
3	100	50	0.14	43	0.1628	22.8135	0.3068

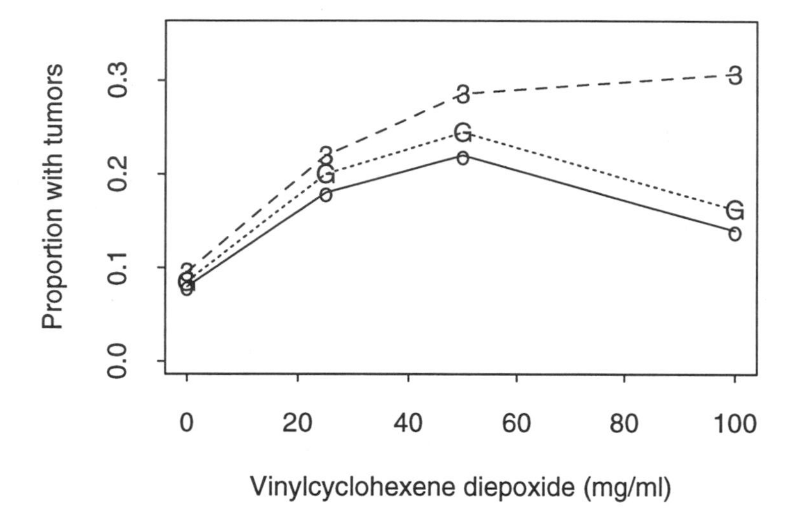

Fig. 6.9 Observed proportions (o) of alveolar/bronchiolar tumors in female B6C3F$_1$ mice from Table 6.5, Gart *et al.* adjusted proportions (G), and Bailer–Portier adjusted proportions (at $\lambda = 3$) plotted against vinylcyclohexene diepoxide concentration.

with $P \approx 0.0257$. In both cases, use of $\lambda = 3$ to adjust the proportions suggests a significant increasing response. We conclude from this analysis that increasing concentrations of vinylcyclohexene diepoxide may indeed be associated with significantly increasing alveolar/bronchiolar tumors.

The Bailer–Portier method can be calculated in SAS using PROC FREQ, with the response variable defined to reflect the adjusted time-at-risk values. SAS code for calculating this is given in Fig. 6.10. The code is similar to that displayed in Fig. 6.7, except that now the `nocol` and `nopercent` options are invoked in the `table` statement to suppress the calculation of the column marginal and joint table relative frequencies. The associated output appears in Fig. 6.11.

As noted with the PROC FREQ output in Fig. 6.8, the trend statistic is obtained in Fig. 6.11 from the square root of the `Mantel-Haenszel Chi-Square` statistic. In this example, the resulting Bailer–Portier adjusted statistic is $z_{calc} = \sqrt{4.497} = 2.121$ (a slight discrepancy with the correct value of 2.128 occurs here, since the adjusted time-at-risk values were entered to only four decimal places). Again, a significantly increasing trend is evidenced. □

```
* SAS code for adjusted CA trend statistic;
data vcdp;
  input conc x nrisk nriskBP @@;
  nsurv = nrisk - x;
  nsurvBP = nriskBP - x;      cards;
   0  4 50 41.8370   25  9 50 40.9661
  50 11 50 38.4904  100  7 50 22.8135
data BPtrend; set vcdp;
  tumor="yes"; count=x; output;
  tumor=" no"; count=nsurvBP; output;
  keep conc tumor count;
proc freq order=data;
 table conc*tumor / chisq nocol nopercent;
 weight count;
```

Fig. 6.10 SAS program to obtain Bailer–Portier time-at-risk adjusted trend test statistic, applied to data from Table 6.6.

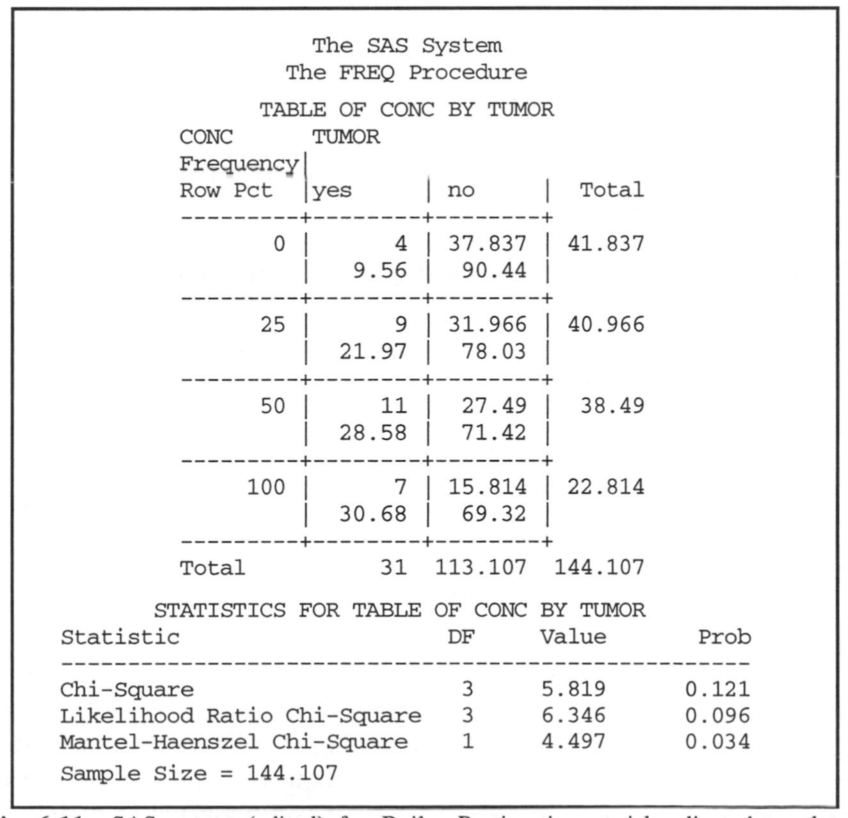

```
                    The SAS System
                  The FREQ Procedure
                TABLE OF CONC BY TUMOR
          CONC       TUMOR
          Frequency|
          Row Pct  |yes     | no     |  Total
          ---------+--------+--------+
                 0 |      4 | 37.837 | 41.837
                   |   9.56 |  90.44 |
          ---------+--------+--------+
                25 |      9 | 31.966 | 40.966
                   |  21.97 |  78.03 |
          ---------+--------+--------+
                50 |     11 |  27.49 |  38.49
                   |  28.58 |  71.42 |
          ---------+--------+--------+
               100 |      7 | 15.814 | 22.814
                   |  30.68 |  69.32 |
          ---------+--------+--------+
          Total          31  113.107  144.107

          STATISTICS FOR TABLE OF CONC BY TUMOR
Statistic                      DF     Value      Prob
------------------------------------------------------
Chi-Square                      3     5.819     0.121
Likelihood Ratio Chi-Square     3     6.346     0.096
Mantel-Haenszel Chi-Square      1     4.497     0.034
Sample Size = 144.107
```

Fig. 6.11 SAS output (edited) for Bailer–Portier time-at-risk adjusted trend test statistic, applied to data from Table 6.6.

6.4 COCHRAN–ARMITAGE TREND TEST FOR COUNTS

When the response is a discrete count that is not in the form of a proportion, a natural form to employ for the parent distribution is the Poisson model from Section 1.2.5. Here, the data are nonnegative integers: $Y_{ij} \sim$ (indep.) $\mathcal{Poisson}(\mu_i)$, where μ_i is the mean response rate at the ith exposure level ($i = 0, \ldots, k$), and $j = 1, \ldots, r_i$ indexes replicate observations at each i. Of interest is testing the homogeneity null hypothesis H_0:$\mu_0 = \mu_1 = \cdots = \mu_k$ vs. the increasing-trend alternative (6.2).

To construct a trend test of H_0 vs. H_a, both Cochran (1954) and Armitage (1955) noted that the basic formulation of their trend test for proportions may be applied also to count data. That is, assume that some ordered measures, $x_0 < x_1 < \cdots < x_k$, quantify the ordered levels of the treatment. (As in Section 6.3.1, if no such set of measures exists, a common recommendation is to assign a set of reasonable values, such as some form of equal spacing, for the x_is. Here again, the trend test is generally insensitive to minor distortions in the underlying scale that quantifies the ordered levels of the treatment.) At each x_i, the maximum likelihood estimate of μ_i is the sample mean $\overline{Y}_{i+} = \sum_{j=1}^{r_i} Y_{ij}/r_i$. If H_0 were true, we would expect these values to approximate the pooled mean $\overline{Y}_{++} = \sum_{i=0}^{k}\sum_{j=1}^{r_i} Y_{ij}/\sum_{i=0}^{k} r_i$. The Cochran–Armitage statistic quantifies departure from this pooled value, by assessing the amount of any trend in the $\overline{Y}_{i+}$s over the ordered levels of x_i. Once again, the statistic is a weighted regression of $\overline{Y}_{i+}$ on x_i, with weights taken as the reciprocals of the estimated variances of $\overline{Y}_{i+}$ assuming H_0 is true: $w_i = r_i/\overline{Y}_{++}$. A weighted LS slope estimator is calculated, which is then divided by its standard error (Exercise 6.19). The result is

$$z_{\text{calc}} = \frac{\sum_{i=0}^{k} r_i (x_i - \overline{x})\overline{Y}_{i+}}{\left\{ \overline{Y}_{++} \sum_{i=0}^{k} r_i\,(x_i - \overline{x})^2 \right\}^{1/2}}, \tag{6.8}$$

where $\overline{x} = \sum_{i=0}^{k} r_i x_i / \sum_{i=0}^{k} r_i$.

Significant departure from H_0 in favor of H_a is suggested when (6.8) is larger than an upper-α critical point from the standard normal distribution. As with its proportion-based cousin, the test based on (6.8) possesses many optimal properties for testing a monotone trend under a Poisson parent distribution for the Y_{ij}s (Tarone, 1982a). It is a preferred approach for trend analysis of count data when the Poisson parent assumption is correct.

Example 6.7 Cochran–Armitage test for trends after nitrofen exposure
Recall from Example 6.2 that significantly decreased trends in offspring counts were observed in the aquatic organism *C. dubia* after exposure to increasing levels of the toxin nitrofen. We continue our examination of these data, by illustrating here trend assessment via the Cochran–Armitage statistic for count data.

For the data in Table 6.2, we assume the Y_{ij}s are distributed as independent Poisson random variables with means μ_i (i = 0, ..., 4). Of interest is whether nitrofen induces a significant decrease in the reproductive capabilities of *C. dubia*, as measured by the number of offspring. This is a question of assessing decreasing trend, and we set the hypotheses under study to $H_0{:}\mu_0 = \mu_1 = \cdots = \mu_k$ vs. $H_a{:}\mu_0 \geq \mu_1 \geq \cdots \geq \mu_k$, with strict inequality in H_a somewhere among the μ_is.

The calculations for (6.8) require an independent measure, x_i, at each level of the exposure factor. Here, we use the recorded dose levels: $x_0 = 0$, $x_1 = 80, \ldots, x_4 = 310$. The associated exposure mean is $\bar{x} = 785/5 = 157.0$, while the sum of squares $\sum_{i=0}^{k} r_i(x_i - \bar{x})^2$ is $(10)(0 - 57)^2 + (10)(80 - 157)^2 + \cdots + (10)(310 - 157)^2 = 600\,800$.

To complete the Cochran–Armitage computations, we use the sample means $\overline{Y}_{i+}$, given also in Table 6.2, and the grand mean $\overline{Y}_{++} = 1144/50 = 22.88$. The numerator of (6.8) is $(10)(0 - 157)(31.4) + (10)(80 - 157)(31.5) + \cdots + (10)(310 - 157)(6.0) = -50\,108$, while the denominator of (6.8) is the square root of $(22.88)(600\,800) = 13\,746\,304$. The resulting test statistic is $z_{\text{calc}} = -50\,108/3707.6 = -13.515$. At $\alpha = 0.001$, $-z_{0.001} = -3.09$, and we see the decreasing trend in Table 6.2 is strongly significant. (The one-sided P-value is well below 0.001.) □

Example 6.8 Cochran–Armitage test for trends in soybean leaf spot toxicity to azaserine
In soybeans (*Glycine max*), mutagenicity of an environmental agent may be measured via the nature and appearance of colored spots on a grown plant's first compound leaf after the plant was exposed to the agent as a seed. Increases in the number of leaf spots are attributed to chemical interference with nucleic acid synthesis, leading to structural changes in the leaf cells' chromosomes. Thus, an increasing trend in the number of leaf spots suggests an increasing mutagenic effect of the agent. The Cochran–Armitage trend statistic (6.8) is useful for assessing such an effect.

Katoh *et al.* (1995) use this system to investigate mutagenicity of the diazo compound *O*-diazoacetyl-L-sterine (azaserine). Azaserine induces mutations in bacteria, and it is of interest to study the compound in a more

Table 6.7 Mutagenicity of azaserine, assessed via leaf spot counts in *G. max*

Experimental group	*Number of leaves,* r_i	*Total spots,* Y_{i+}	*Mean spot frequency,* $\overline{Y}_{i+}$
Control	196	237	1.209
0.05 mg/ml azaserine	127	218	1.716
0.10 mg/ml azaserine	130	282	2.169
0.25 mg/ml azaserine	119	269	2.261
0.50 mg/ml azaserine	113	426	3.770

Data from Katoh *et al.* (1995).

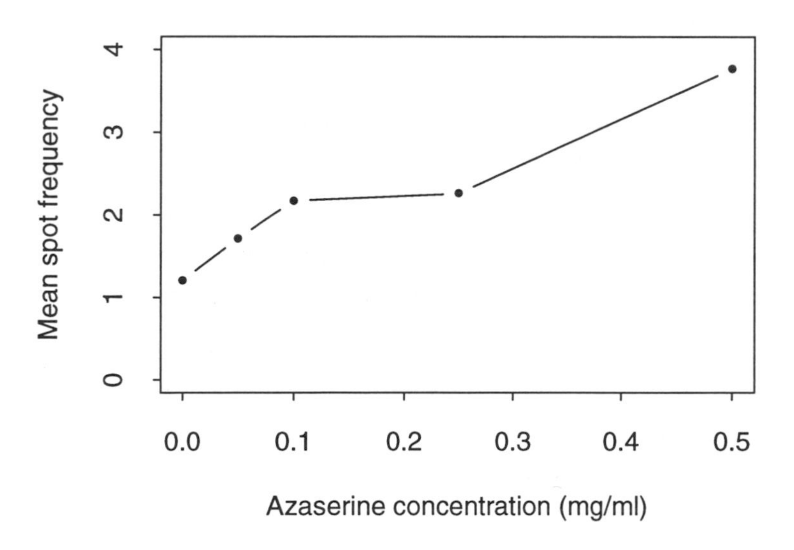

Fig. 6.12 Scatterplot of mean leaf spot frequency as a function of azaserine concentration for data from Table 6.7.

complex organism. Summary data from an experiment using a control group and $k = 4$ exposure groups are given in Table 6.7 and Fig. 6.12. Notice that the design here is unbalanced, with the number of leaves studied per group, r_i, varying between $r_4 = 113$ and $r_0 = 196$.

We assume the original leaf spot counts, Y_{ij}, leading to the summary values in Table 6.7 are distributed as independent Poisson random variables, with means μ_i ($i = 0, \ldots, 4$). Of interest is whether azaserine induces a significant increase in the counts of leaf spots in *G. max*, indicating an increase in the mutagenic effect over the control group. We set the hypotheses under study to $H_0{:}\mu_0 = \mu_1 = \cdots = \mu_k$ and the increasing-order alternative (6.2). Under a Poisson assumption on the observed counts, application of the Cochran–Armitage trend statistic (6.8) is appropriate.

The calculations require an independent measure, x_i, at each level of the exposure factor. Here, in order to use a measure that is approximately equi-spaced, we use the natural logarithms of the recorded dose levels. Unfortunately, at the control level the logarithm of zero is undefined. To overcome this, we employ a formula that averages the other log-doses to approximate an equally spaced value at $i = 0$. Known as **consecutive-dose average spacing** (Margolin *et al.*, 1986), the formula for x_0 is

$$x_0 = x_1 - \frac{x_k - x_1}{k - 1}, \tag{6.9}$$

where x_i is the logarithm of each original dose ($i = 1, \ldots, k$). Since for these data this leads to negative x_is at every exposure level, we simplify the presentation by adding a constant, +4, to every log-dose. For the doses in Table 6.7, we obtain $x_0 = 0.236$, $x_1 = 1.004$, $x_2 = 1.697$, $x_3 = 2.614$, and $x_4 = 3.307$. The associated mean (weighted for imbalance among the r_is) is $\bar{x} = 1079.131/685 = 1.575$, while the sum of squares $\sum_{i=0}^{k} r_i(x_i - \bar{x})^2$ is $(196)(0.236 - 1.575)^2 + (127)(1.004 - 1.575)^2 + \cdots + (113)(3.307 - 1.575)^2 = 848.169$.

To complete the Cochran–Armitage computations for these data, we use the sample means $\bar{Y}_{i+}$, as given in Table 6.7, and also find the grand mean $\bar{Y}_{++} = 1432/685 = 2.091$. Then, the numerator of (6.8) is $(96)(0.236 - 1.575)(1.209) + (127)(1.004 - 1.575)(1.716) + \cdots + (113)(3.307 - 1.575)(3.770) = 610.068$, while the denominator of (6.8) is the square root of $(2.091)(848.169) = 1773.521$. The test statistic is $z_{calc} = 14.486$. Compared to, say, the upper-0.001 critical point of a standard normal distribution, $\mathfrak{z}_{0.001} = 3.09$, we see the trend in Table 6.7 and Fig. 6.12 is strongly significant. From this we conclude that azaserine induces significantly increasing leaf spot mutagenesis in *G. max*. ❑

6.5 OVERDISPERSED DISCRETE DATA

The Cochran–Armitage trend tests described in Sections 6.3.1 and 6.4 are powerful techniques for assessing monotone trends in discrete data. A critical assumption for their implementation, however, is that the binomial or Poisson assumptions on the parent distributions are valid. As discussed in Sections 1.2.2 and 1.2.5, environmental data often are subject to sources of excess variability that disturb the basic distributions for discrete data. With data in the form of proportions, we called this **extra-binomial**

variability; with counts, it is **extra-Poisson variability**. A single term we use to describe both cases is **overdispersion**.

In the presence of overdispersion, the Cochran–Armitage trend tests based on (6.6) and (6.8) can misrepresent the significance of the true trend (Lockhart *et al.*, 1992; Piegorsch, 1993), since they do not take the overdispersion into account. (In effect, they underestimate the variability in the data, and consequently reject the null hypothesis of no trend with a frequency much greater than $100\alpha\%$ of the time when, in fact, it is true.) In such cases, (6.6) and (6.8) are contraindicated for use, and some adjustment in their construction is required.

6.5.1 Assessing overdispersion in proportion data

To determine whether a set of proportions exhibits overdispersion, we assess the excess variability statistically. We require that the data consist of replicate proportions at each exposure level, say $p_{ij} = Y_{ij}/N_{ij}$ ($i = 0, \ldots, k$ groups; $j = 1, \ldots, r_i$ replicates per group), and test the significance of any observed excess variability in the proportions at the ith exposure level via the score statistic

$$T_i^2 = \frac{\left(\sum_{j=1}^{r_i} \frac{[Y_{ij} - N_{ij}\hat{p}_i]^2}{\hat{p}_i(1 - \hat{p}_i)} - \sum_{j=1}^{r_i} N_{ij} \right)^2}{2\sum_{j=1}^{r_i} N_{ij}(N_{ij} - 1)} \tag{6.10}$$

(Tarone, 1979), where $\hat{p}_i = \sum_{j=1}^{r_i} Y_{ij} / \sum_{j=1}^{r_i} N_{ij}$ is the pooled proportion response at each $i = 0, \ldots, k$. In large samples, $T_i^2 \stackrel{\cdot}{\sim} \chi^2(1)$, so to test for significant extra-binomial variability, calculate (6.10) and reject the null hypothesis of no overdispersion at each exposure level if $T_i^2 > \chi^2_\alpha(1)$. The associated P-value is, approximately, $\mathrm{P}[\chi^2(1) > T_i^2]$.

Tarone (1979) showed that the test based on (6.10) is locally most powerful against overdispersion that includes the beta-binomial distribution from Section 1.2.2. The specific form of local optimality is known as $C(\alpha)$ optimality (Neyman and Scott, 1966), hence the test based on (6.10) is referred to as the **C(α)-test for extra-binomial variability**. Alternative forms of testing for extra-binomial variability are also possible (Exercise 6.26).

To extend the $C(\alpha)$-test to assess excess variability across all the exposure levels, aggregate the individual $C(\alpha)$-statistics over the independent exposures: $T_+^2 = \sum_{i=0}^{k} T_i^2$. In large samples, $T_+^2 \stackrel{\cdot}{\sim} \chi^2(k+1)$, so one rejects the global null hypothesis of no extra-binomial variability in favor of overdispersion at some level when $T_+^2 > \chi_\alpha^2(k+1)$. Further extensions to more complex regression settings, where a concomitant predictor variable is recorded with each observation, are discussed by Dean (1992).

Example 6.9 Extra-binomial variability in litter proportions
We noted in Example 3.6 that extra-binomial variability is common with *in utero* teratogenesis experiments generating correlated binary data. There, laboratory animals are exposed to a toxic environmental agent and their offspring are examined for toxic effects incurred during the embryonic or fetal stage of development. If the endpoint under study represents a simple dichotomy such as malformed/not malformed, or stillborn/alive, the parameter of interest is the probability, π_i, of malformation (or death, etc.), $i = 0, \ldots, k$. Each female contributes an observed proportion, $p_{ij} = Y_{ij}/N_{ij}$, where Y_{ij} is the number of malformed (or dead, etc.) fetuses and N_{ij} is the litter size for the jth animal in the ith exposure group, $j = 1, \ldots, r_i$. As we saw in Example 3.6, fetuses sampled from the same female represent multiple observations on a single experimental unit, inducing an **intralitter correlation** in the data; we called this a **litter effect**.

Statistically, intralitter correlations generate variability in the Y_{ij}s in excess of that assumed under a simple binomial model. To assess this in a set of per-litter proportions, we apply the C(α)-test. Consider the per-litter proportions given in Table 6.8. These data represent results from a US National Toxicology Program (NTP) study of developmental toxicity in CD-1 Swiss mice (NTP, 1989a), after females were exposed to the insecticide boric acid at $k = 3$ dose levels: $x_1 = 0.1\%$ boric acid in feed, $x_2 = 0.2\%$, $x_3 = 0.4\%$, and a control group with $x_0 = 0\%$. Recorded were the number of dead fetuses, Y_{ij}, and the litter size, N_{ij}, for each animal at each exposure level.

An initial concern with the data in Table 6.8 is whether they exhibit any extra-binomial variability. To assess this, we calculate the C(α)-statistic from (6.10) at each dose level; the results are summarized in Table 6.9.

From the P-values in Table 6.9, we infer that there is no apparent departure from standard binomial variability at the control and two lowest doses, but there is strong departure at the high dose ($P < 0.001$). Thus, any procedure designed to employ the binomial parent distribution for these data would be inappropriate. We will return to this experiment below, and assess whether a significantly increasing trend is exhibited in the overdispersed proportions.

Table 6.8 Fetal deaths after maternal exposure to boric acid; number of dead fetuses Y_{ij}, total implants N_{ij}

			Dose (% in feed)				
Control		*Low dose: 0.1*		*Middle dose: 0.2*		*High dose: 0.4*	
Y_{ij}	N_{ij}	Y_{ij}	N_{ij}	Y_{ij}	N_{ij}	Y_{ij}	N_{ij}
0	15	0	6	1	12	12	12
0	3	1	14	0	12	1	12
1	9	1	12	0	11	0	13
1	12	0	10	0	13	2	8
1	13	2	14	0	12	2	12
2	13	0	12	0	14	4	13
0	16	0	14	4	15	0	13
0	11	3	14	0	14	1	13
1	11	0	10	0	12	0	12
2	8	2	12	1	6	1	9
0	14	3	13	2	13	3	9
0	13	1	11	0	10	0	11
3	14	1	11	1	14	1	14
1	13	0	11	1	12	0	10
0	8	0	13	0	10	3	12
0	13	0	10	0	9	2	21
2	14	1	12	1	12	3	10
3	14	0	11	0	13	3	11
0	11	2	10	1	14	1	11
2	12	2	12	0	13	1	11
0	15	2	15	0	14	8	14
0	15	3	12	1	13	0	15
2	14	1	12	2	12	2	13
1	11	0	12	1	14	8	11
1	16	1	12	0	13	4	12
0	12	1	13	0	12	2	12
0	14	1	15	1	7		

Data from NTP (1989a).

Table 6.9 Overdispersion analyses for boric acid data from Table 6.8

		Dose (% in feed)	
Control	*Low dose: 0.1*	*Middle dose: 0.2*	*High dose: 0.4*
Litters examined:			
$r_0 = 27$	$r_1 = 27$	$r_2 = 27$	$r_3 = 26$
Responding fetuses:			
$\sum Y_{0j} = 23$	$\sum Y_{1j} = 28$	$\sum Y_{2j} = 17$	$\sum Y_{3j} = 64$
Total implantations:			
$\sum N_{0j} = 334$	$\sum N_{1j} = 323$	$\sum N_{2j} = 326$	$\sum N_{3j} = 314$
Pooled proportions:			
$\hat{p}_0 = 0.069$	$\hat{p}_1 = 0.087$	$\hat{p}_2 = 0.052$	$\hat{p}_3 = 0.204$
C(α)-statistic, T_i^2:			
$T_0^2 = 0.509$	$T_1^2 = 0.042$	$T_2^2 = 1.716$	$T_3^2 = 140.94$
C(α) P-value, $P[\chi^2(1) > T_i^2]$:			
0.476	0.837	0.190	<0.001

```
# Detecting overdispersion in Binomial data

T2.overdisp.bin<-function(yy,rr) {

# yy = vector of observed events
# rr = vector of replicate counts (same length as yy)
   ppi<- sum(yy)/sum(rr)
   denom <- 2*sum(rr*(rr-1))
   top1<- sum((yy-rr*ppi)^2)/(ppi*(1-ppi))
   top2<- sum(rr)
   (top1 - top2)^2/denom
}

# input the boric acid teratogenicity data
dead.0<-c(0,0,1,1,1,2,0,0,1,2,0,0,3,1,0,0,2,3,0,2,0,0,2,1,1,0,0)
implant.0<-c(15,3,9,12,13,13,16,11,11,8,14,13,14,13,
          8,13,14,14,11,12,15,15,14,11,16,12,14)
dead.1<-c(0,1,1,0,2,0,0,3,0,2,3,1,1,0,0,0,1,0,2,2,2,3,1,0,1,1,1)
implant.1<-c(6,14,12,10,14,12,14,14,10,12,13,11,11,
          11,13,10,12,11,10,12,15,12,12,12,12,13,15)
dead.2<-c(1,0,0,0,0,0,4,0,0,1,2,0,1,1,0,0,1,0,1,0,0,1,2,1,0,0,1)
implant.2<-c(12,12,11,13,12,14,15,14,12,6,13,10,14,
          12,10,9,12,13,14,13,14,13,12,14,13,12,7)
dead.4<-c(12,1,0,2,2,4,0,1,0,1,3,0,1,0,3,2,3,3,1,1,8,0,2,8,4,2)
implant.4<-c(12,12,13,8,12,13,13,13,12,9,9,11,14,10,
          12,21,10,11,11,11,14,15,13,11,12,12)

# calculate the test of overdispersion
T2.0<-T2.overdisp.bin(dead.0,implant.0)
T2.1<-T2.overdisp.bin(dead.1,implant.1)
T2.2<-T2.overdisp.bin(dead.2,implant.2)
T2.4<-T2.overdisp.bin(dead.4,implant.4)

conc<-c(0,.1,.2,.4)
all.T2<-c(T2.0,T2.1,T2.2,T2.4)
for (jj in 1:4)
 cat("conc = ",conc[jj],": T2 = ",signif(all.T2[jj]),
  "; P-value = ",signif(1-pchisq(all.T2[jj],1)),"\n")
 cat("\nOVERALL:  sum(T2) = ",signif(sum(all.T2)),
  "; P-value = ",signif(1-pchisq(sum(all.T2),4)),"\n")
```

```
conc =  0   :  T2 =  0.509 ; P-value =  0.475565
conc =  0.1 :  T2 =  0.042 ; P-value =  0.837406
conc =  0.2 :  T2 =  1.716 ; P-value =  0.19015
conc =  0.4 :  T2 =140.941 ; P-value =  0
OVERALL:   sum(T2)  =  143.209 ; P-value =  0
```

Fig. 6.13 S-PLUS program to detect overdispersion in teratology data from Table 6.8.

The C(α)-test of extra-binomial variability can be implemented in S-PLUS, as illustrated in Fig. 6.13. Our S-PLUS function, `T2.overdisp.bin`, takes as input a vector of observed events (`dead.*`) and a vector of the number of trials (`implants.*`). The function outputs the value of T_i^2. P-values can be found using the S-PLUS χ^2 probability function `pchisq`. These are displayed at the end of Fig. 6.13, as applied to the data from Table 6.8. The results confirm the values reported in Table 6.9. □

6.5.2 Trend tests under extra-binomial variability

To assess a monotone trend with overdispersed proportions, the simple Cochran–Armitage statistic in (6.6) is not appropriate: as noted above, the statistic fails to take into account the overdispersion, and in effect underestimates the excess variability. The associated test suffers from instability in its false positive rate, with two- or even fourfold increases possible over the nominal α level (Lockhart *et al.*, 1992; Piegorsch, 1993; Fung *et al.*, 1994). Adjustments to account for overdispersion are many and varied, however, and it is not possible to recommend unequivocally one particular adjustment over another for testing trend in extra-binomial proportions. Nonetheless, two general guidelines may be proffered. First, in the special case where it is reasonable to model the extra-binomial variability by assuming a hierarchical model with a beta distribution for π, the distribution of Y can be taken as beta-binomial. Under the beta-binomial model, a general test for trend can be developed. Second, when the specific extra-binomial distribution for the data is not known, we recommend using some form of distribution-free trend test.

In the first case, when the beta-binomial model is valid, a trend statistic may be constructed similar to the Cochran–Armitage statistic in (6.6). For simplicity, we impose the additional assumption that the overdispersion is constant across all exposure levels (including at the control group level). That is, the dispersion parameters, φ_i, are assumed identical, and we write $\varphi_i = \varphi$ for all $i = 0, \ldots, k$. To test $H_0: \pi_0 = \pi_1 = \cdots = \pi_k$ vs. the increasing-order alternative $H_a: \pi_0 \leq \pi_1 \leq \cdots \leq \pi_k$, suppose we are given an ordered measure, x_i, associated with each observed proportion, Y_{ij}/N_{ij}. Take as the sample-weighted average of x_i the quantity

$$\overline{x} = \left(\sum_{i=0}^{k}\sum_{j=1}^{r_i} N_{ij}x_i\right) / \left(\sum_{i=0}^{k}\sum_{j=1}^{r_i} N_{ij}\right).$$

Also, as a summary estimator of the proportion response under H_0 calculate

$$\overline{p} = \left(\sum_{i=0}^{k}\sum_{j=1}^{r_i} Y_{ij}\right) / \left(\sum_{i=0}^{k}\sum_{j=1}^{r_i} N_{ij}\right).$$

With these, a Cochran–Armitage-type test statistic is

$$z_{\text{calc}} = \frac{\sum_{i=0}^{k} \sum_{j=1}^{r_i} (x_i - \bar{x}) Y_{ij}}{\left\{ \bar{p}(1-\bar{p}) \sum_{i=0}^{k} \sum_{j=1}^{r_i} N_{ij} (x_i - \bar{x})^2 \left[\frac{1 + N_{ij}\tilde{\varphi}}{1 + \tilde{\varphi}} \right] \right\}^{1/2}} \tag{6.11}$$

where $\tilde{\varphi}$ is a method of moments (MoM) estimator of the constant dispersion parameter. The dispersion estimator is found by equating the pooled sample variance of the data to its expected value under H_0. It may be expressed in terms of the **intralevel** or **intraclass correlation**, ρ, via the relation $\rho = \varphi/(1+\varphi)$: for $\nu_m = \sum_{i=0}^{k}\sum_{j=1}^{r_i} N_{ij}^m$ $(m = 1, 2, 3)$, the MoM estimator of ρ is

$$\tilde{\rho} = \frac{\nu_2 - \nu_1 \left\{ \nu_1 - \sum_{i=0}^{k} \sum_{j=1}^{r_i} \frac{[Y_{ij} - N_{ij}\bar{p}]^2}{\bar{p}(1-\bar{p})} \right\}}{(\nu_2 - \nu_1)\left(\nu_1 + \frac{\nu_2}{\nu_1}\right) - 2(\nu_3 - \nu_2)}$$

so that

$$\tilde{\varphi} = \begin{cases} 0 & \text{if } \tilde{\rho} \le 0 \\ \dfrac{\tilde{\rho}}{1 - \tilde{\rho}} & \text{if } 0 < \tilde{\rho} < 1 \end{cases} .$$

(If $\tilde{\rho} \ge 1$, $\tilde{\varphi}$ is taken as ∞.) This translates into the following adjustment for the denominator of (6.11):

$$\left[\frac{1 + N_{ij}\tilde{\varphi}}{1 + \tilde{\varphi}} \right] = \begin{cases} 1 & \text{if } \tilde{\rho} \le 0 \\ 1 + \tilde{\rho}(N_{ij} - 1) & \text{if } 0 < \tilde{\rho} < 1 \\ N_{ij} & \text{if } \tilde{\rho} \ge 1 \end{cases} .$$

In large samples, $z_{\text{calc}} \mathrel{\dot\sim} N(0,1)$, so one rejects H_0 in favor of H_a when $z_{\text{calc}} > z_\alpha$. The corresponding approximate P-value is $1 - \Phi(z_{\text{calc}})$.

More sophisticated models for trend analysis are possible in the presence of extra-binomial variability, but they exceed the scope of this chapter. For further reading, see Lefkopoulou *et al.* (1996), Pendergast *et al.* (1996), Williams (1991; 1996), and the references therein.

Our second guideline concerns the case where the specific extra-binomial parent distribution for the data is not known, or where the extra-binomial variability is not constant across all exposure levels. In this case, we recommend some form of distribution-free trend test. One possibility is the rank-based Jonckheere–Terpstra trend test, discussed below in Section 6.6. Another distribution-free trend test that is slightly more compact than the Jonckheere–Terpstra test involves extending the Cochran–Armitage statistic via the method of generalized estimating equations (GEEs) in Section 2.7.2. Applied to overdispersed proportions, the method in effect groups the individual dichotomous outcomes into a longitudinal array of observations, and applies the GEEs to achieve inferences on any trend associated with some independent measure, x_i. The method may be viewed as distribution-free, since no assumptions are required on nature of the parent distribution for the overdispersed proportions. (There are limited assumptions required on the mathematical structure of the mean and variance, but these are not obtrusive. For instance, the test is appropriate for use when the proportion response follows the commonly accepted logistic form in (6.7); see Lefkopoulou *et al.* (1996).)

The construction underlying the GEE statistic is beyond the scope of this text; however, it is possible to write the trend test statistic in its closed form:

$$z_{\text{GEE}} = \frac{\sum_{i=0}^{k}\sum_{j=1}^{r_i}(x_i - \bar{x})Y_{ij}}{\left\{\sum_{i=0}^{k}\left[(x_i - \bar{x})^2\sum_{j=1}^{r_i}(Y_{ij} - N_{ij}\bar{p})^2\right]\right\}^{1/2}}. \tag{6.12}$$

Notice that (6.12) is very similar to the basic Cochran–Armitage statistic in (6.6); the major difference is that the term estimating the binomial variance, $N_i\bar{p}\,\bar{q}$, is replaced by the more robust empirical variance estimate $\sum_{j=1}^{r_i}(Y_{ij} - N_{ij}\bar{p})^2$ (Exercise 6.27). Indeed, the statistic in (6.12) is a form of generalized score statistic (Boos, 1992; Lefkopoulou and Ryan, 1993; Carr and Gorelick, 1995), mimicking the Cochran–Armitage construction.

In large samples, $z_{\text{GEE}} \dot{\sim} N(0,1)$, and one rejects H_0 in favor of an increasing trend when $z_{\text{GEE}} > \mathfrak{z}_\alpha$. The corresponding approximate *P*-value is $1 - \Phi(z_{\text{GEE}})$.

Trend testing with the GEE statistic is known to exhibit very stable false positive error, and also good power to detect true increases in trend with overdispersed proportions, at least in large samples (Piegorsch, 1993; Carr and Gorelick, 1995). Its combination of simplicity and good robustness without great losses in sensitivity makes it highly recommended for use with overdispersed proportions. (There can be stability problems with the test when both the individual denominators, N_{ij}, are large and the sample sizes,

r_i, are small. In this case, extensions to (6.12) are discussed by Rotnitzky and Jewell (1990).]

The issue of overdispersion with proportion data is a common concern in **cluster sampling**, where experimental units are studied for some binary response within related *clusters*. In our notation, the N_{ij}s are the cluster sizes while the r_is represent the number of clusters sampled from a particular population. See Bieler and Williams (1995) for further discussion on this cluster sampling perspective.

Example 6.9 (continued) Extra-binomial variability in litter proportions – Trend testing

Continuing with our analysis of the developmental toxicity data in Table 6.8, we now assess whether the litter proportions exhibit an increasing dose response to the insecticide boric acid. Recall from Table 6.9 that significant extra-binomial variability was seen at the high dose. Thus use of the simple Cochran–Armitage statistic (6.6) in testing for a dose response or trend is inappropriate. Instead, we illustrate use of the GEE statistic in (6.12).

Note that it is inappropriate to consider use of the beta-binomial trend statistic (6.11) for these data. Recall that the statistic in (6.11) was constructed under the assumption of constant overdispersion across all dose levels. With the boric acid data, however, we saw in Table 6.9 that the overdispersion was not evident at all doses: only the high dose exhibits significant extra-binomial variability.

The GEE computations for the boric acid data in Table 6.8 involve the quantities $\bar{x} = \{(0)(334) + (0.1)(323) + (0.2)(326) + (0.4)(314)\}/1297 = 0.172$, and $\bar{p} = (23 + 28 + 17 + 64)/1297 = 0.102$. The numerator of (6.12) is 9.094; the denominator is the square root of 14.041. Thus the test statistic is $z_{GEE} = 9.094/\sqrt{14.041} = 2.4270$. The corresponding upper-tail P-value from a standard normal reference distribution is $P \approx 0.0076$, suggesting a significant increasing trend in the overdispersed proportions. It appears that boric acid can act as a significant developmental toxin in these mice.

The GEE trend test can be implemented in S-PLUS, as illustrated in Fig. 6.14. Our S-PLUS function, `Z.GEE.bin`, takes as input vectors of concentrations (`conc`), observed events (`yy`), and the number of trials (`rr`). The function reports z_{GEE}, with $\bar{x}$ and $\bar{p}$ generated as part of the output. P-values can then be calculated using the S-PLUS N(0,1) probability function `pnorm`. These are displayed at the end of Fig. 6.14, as applied to the data from Table 6.8. As found above, $z_{GEE} = 2.427$, with associated P-value approximately 0.0076. □

```
# Trend tests under extra-binomial variability

Z.GEE.bin<-function(conc,yy,rr)   {

# formula (6.12)
# conc = vector of concentrations
# yy = vector of observed events
# rr = vector of replicate counts
         ngrp <- length(unique(conc))
     conc.lst <- split(conc,conc)
       yy.lst <- split(yy,conc)
       rr.lst <- split(rr,conc)

         xbar <- sum(rr*conc)/sum(rr)
         pbar <- sum(yy)/sum(rr)

   top <- sum( (conc-xbar)*dead )
   bot <- 0
   for (ii in 1:ngrp) {
      bot <- bot + sum( ((conc.lst[[ii]]-xbar)^2)*
             ((yy.lst[[ii]] - rr.lst[[ii]]*pbar)^2))
  }
   cat("xbar = ",signif(xbar),"\n")
   cat("pbar = ",signif(pbar),"\n")
   cat("Z.GEE= ",signif(top/sqrt(bot)),"\n")
   top/sqrt(bot)
}

# place the teratogenicity data into a data frame
dead    <-c(dead.0,dead.1,dead.2,dead.4)
implant<-c(implant.0,implant.1,implant.2,implant.4)
conc    <-rep(c(0,.1,.2,.4),c(27,27,27,26))

terat.df<-data.frame(conc=conc,
                     dead=dead,implant=implant)
Z.GEE<-Z.GEE.bin(terat.df$conc,
                 terat.df$dead,terat.df$implant)
```

```
xbar =   0.172
pbar =   0.102
Z.GEE=   2.427

1-pnorm(Z.GEE)
[1] 0.007611753
```

Fig. 6.14 S-PLUS function for calculating the overdispersion GEE trend test statistic for data in Table 6.8.

6.5.3 Assessing overdispersion with count data

Concerns with overdispersion are just as critical when the data are counts as when they are proportions. Suppose, then, that r_i independent counts Y_{ij} are observed, each with mean rate of response μ_i at each exposure level ($i = 0, \ldots, k; j = 1, \ldots, r_i$). If the Poisson assumption is valid, $\mathrm{Var}[Y_{ij}]$ equals μ_i, so we expect the sample mean to equal roughly the sample variance.

As with extra-binomial data, the Cochran–Armitage test using (6.8) applied to extra-Poisson data fails to take the excess variability into account, and consequently misrepresents the significance of the true trend. With overdispersed counts, the test statistic underestimates the variability in the data, and rejects the null hypothesis of no trend with a frequency much greater than $100\alpha\%$ of the time when this null hypothesis is true. In such cases, use of (6.8) is contraindicated; some adjustment is required that accounts for the extra-Poisson variability.

To assess whether count data exhibit overdispersion, we take advantage of the constant mean-to-variance ratio characteristic of Poisson data. In the presence of extra-Poisson variability, the per-group sample variances should exceed the per-group sample means, from which a dispersion statistic may be calculated (Fisher *et al.*, 1922; Fisher, 1950):

$$X_i^2 = \frac{\sum_{j=1}^{r_i} (Y_{ij} - \overline{Y}_{i+})^2}{\overline{Y}_{i+}}, \tag{6.13}$$

where the sample means are $\overline{Y}_{i+}$ ($i = 0, \ldots, k$). In large samples, $X_i^2 \stackrel{\cdot}{\sim} \chi^2(r_i - 1)$. To test for significant extra-Poisson variability, calculate (6.13) and reject the null hypothesis of no overdispersion at any i if $X_i^2 > \chi^2_\alpha(r_i - 1)$. The associated P-value is, approximately, $\mathrm{P}[\chi^2(r_i - 1) > X_i^2]$.

We extend the dispersion test to assess excess variability across all exposures by aggregating over the $k + 1$ independent levels: $X_+^2 = \sum_{i=0}^{k} X_i^2$. The aggregate statistic is distributed asymptotically as χ^2 with *df* equal to $\nu_+ = \sum_{i=0}^{k}(r_i - 1)$ (Margolin, 1985). In large samples, one rejects the global null hypothesis of no extra-Poisson variability at all exposure levels in favor of overdispersion at some level when $X_+^2 > \chi^2_\alpha(\nu_+)$. (A slightly more powerful test of extra-Poisson variability may be constructed using the similar statistic $C^2 = \sum_{i=0}^{k}\sum_{j=1}^{r_i}(Y_{ij} - \overline{Y}_{i+})^2/\overline{Y}_{++}$ (Collings and Margolin, 1985), but its null reference distribution is complex, and hence somewhat difficult to apply in practice.] Dispersion tests for more complex regression settings are discussed by Dean (1992).

Example 6.10 Extra-Poisson variability in mutant plate counts
A common application of tests of overdispersion is to large, controlled data-bases where similar control trials and replicate plates are available. Margolin *et al.* (1981) reported on such data for the *Salmonella* mutagenesis assay introduced in Example 3.7. They were able to demonstrate that the Poisson model is generally inadequate to describe statistical variability when analyzing *Salmonella* mutagenicity experiments. Further research has confirmed this extra-Poisson variability for *Salmonella* data (Margolin *et al.*, 1989), although the overdispersion can be mitigated by imposing strict protocol adherence, and in particular by harvesting all the data on the same day (Hamada *et al.*, 1994).

To illustrate the use of Fisher's dispersion test for data of this sort, consider the counts of *Salmonella* mutant colonies given in Table 3.2. Recall that these data represent the mutagenic response of *S. typhimurium* bacteria to the toxin 1,3-butadiene after gaseous exposure. The dispersion test requires the per-dose sample means, $\overline{Y}_{i+}$ (i = 0, ..., 5), given in the last column of Table 3.2 and also in Table 6.10. With these, the calculations are straightforward: the test statistics at each dose and the associated, approximate P-values from reference to χ^2 distributions with $r_i - 1$ *df* are given in Table 6.10. (Notice that the design for this experiment was balanced, with $r_i = 3$ for all $i = 0, \ldots, 5$.)

Table 6.10 Summary statistics and overdispersion analyses for 1,3-butadiene data from Table 3.2

			Dose (ppm)		
Control	*0.002*	*0.007*	*0.014*	*0.020*	*0.030*
Plates providing data:					
$r_0 = 3$	$r_1 = 3$	$r_2 = 3$	$r_3 = 3$	$r_4 = 3$	$r_5 = 3$
Average plate count, $\overline{Y}_{i+}$:					
26.0	93.7	149.3	189.0	171.0	152.0
Dispersion statistic, X_i^2:					
2.385	0.690	10.183	6.106	2.678	0.092
Approximate P-value, $\mathrm{P}[\chi^2(r_i - 1) > X_i^2]$:					
0.303	0.708	0.006	0.047	0.262	0.955

From Table 6.10 we see that some extra-Poisson variability may exist with these data: the overdispersion P-value at $x_2 = 0.007$ ppm is strongly significant, while the P-value at $x_3 = 0.014$ ppm may imply overdispersion there as well. The low numbers of replicate plates at each dose produce very small *df* for each dispersion test, however, and this may lead to low power to detect excess variability at the other doses. Aggregating the individual X_i^2-values into the single dispersion statistic X_+^2 increases the number of *df* and the sensitivity. The test statistic is $X_+^2 = 22.134$, and the aggregate *df* are the sum of the individual *df*, $\nu_+ = 12$. The approximate *P*-value is $\text{P}[\chi^2(12) > 22.134] \approx 0.004$. Significant extra-Poisson variability is evidenced in these data. □

Example 6.11 Overdispersion in aquatic toxicity data
As another example, consider application of the Fisher variance test using (6.13) to the aquatic toxicity data from Table 6.2. Recall that these data are numbers of offspring in *C. dubia* after exposure to the herbicide nitrofen (Bailer and Oris, 1993). We saw, using the Cochran–Armitage trend statistic (6.8), that a significant decrease ($P < 0.001$) was evidenced in the mean counts for animals exposed to the chemical, suggesting a toxic exposure-related response. To perform this analysis, however, we assumed that the data were distributed as Poisson counts, and that no overdispersion existed in the data. Was this a valid assumption?

Table 6.11 Summary statistics and overdispersion analyses for data from Table 6.2

		Nitrofen exposure (μg/l)		
Control	*80*	*160*	*235*	*310*
Animals providing data:				
$r_0 = 10$	$r_1 = 10$	$r_2 = 10$	$r_3 = 10$	$r_4 = 10$
Dispersion statistic, X_i^2:				
3.707	3.063	1.770	18.233	20.667
Approximate *P*-value, $\text{P}[\chi^2(r_i - 1) > X_i^2]$:				
0.930	0.962	0.995	0.033	0.014

Application of (6.13) is again straightforward for these data. The results appear in Table 6.11. Therein, we see that no apparent departure from Poisson variability is evidenced at the control level, nor at the two lowest exposure levels. At thc two highest exposures, however, both *P*-values are low, and significant extra-Poisson variability is suggested. To corroborate this, we also calculate the aggregate dispersion statistic $X_+^2 = 47.440$, with $\nu_4 = 45$ *df*. The associated *P*-value is, approximately, $\text{P}[\chi^2(45) > 47.44] \approx 0.373$. Thus in the aggregate, no significant overdispersion is evidenced with these data, although at the higher doses, some limited extra-Poisson variability may be lurking.

Fisher's test of overdispersion can be coded in S-PLUS, as illustrated in Fig. 6.15. Our S-PLUS function, `X2.overdisp.Poisson`, takes as

```
# Detecting overdispersion in Poisson data
X2.overdisp.Poisson<-function(yy)  {
# yy = vector of observed counts
  css <- var(yy)*(length(yy)-1)
  ybar<- mean(yy)
  X2 <- css/ybar
  Pvalue <- 1-pchisq(X2,length(yy)-1)
  list(X2=X2,Pvalue=Pvalue)
}
offsprg.0  <-c(27,32,34,33,36,34,33,30,24,31)
offsprg.80 <-c(33,33,35,33,36,26,27,31,32,29)
offsprg.160<-c(29,29,23,27,30,31,30,26,29,29)
offsprg.235<-c(23,21, 7,12,27,16,13,15,21,17)
offsprg.310<-c( 6, 6, 7, 0,15, 5, 6, 4, 6, 5)
nitro.conc<-rep(c(0,80,160,235,310),c(10,10,10,10,10))
offsprg<-c(offsprg.0,offsprg.80,offsprg.160,
           offsprg.235,offsprg.310)
nitro.df<-data.frame(conc=nitro.conc,
                     offspring=offsprg)
nitro.ngrp<-length(unique(nitro.df$conc))
nitro.lst<-split(nitro.df$offspring,nitro.df$conc)
for (iii in 1:nitro.ngrp) {
    cat("Concentration = ",
        signif(unique(nitro.df$conc)[iii]),"\n")
  junk<-unlist(X2.overdisp.Poisson(nitro.lst[[iii]]))
  cat("\t","X2 = ",signif(junk[1]),
      "; P-value = ",signif(junk[2]),"\n")
}
```

```
Concentration =  0
 X2 =  3.70701 ; P-value =  0.929617
Concentration =  80
 X2 =  3.06349 ; P-value =  0.961724
Concentration =  160
 X2 =  1.77032 ; P-value =  0.994601
Concentration =  235
 X2 =  18.2326 ; P-value =  0.0325693
Concentration =  310
 X2 =  20.6667 ; P-value =  0.0142156
```

Fig. 6.15 S-PLUS program to detect overdispersion in count data from Table 6.2.

input a vector of observed counts (yy), and reports the value of each individual X_i^2-statistic, along with its approximate P-value. These are displayed at the end of Fig. 6.15, as applied to the data from Table 6.2.

Notice in Table 6.2 that at $x = 310$ μg/l, the overdispersion appears to be driven almost entirely by a single large observation: $Y_{45} = 15$. Whether this is an unusual value is unclear, but further analysis using the Bonferroni inequality from Section 5.2 corroborates that the excess variability may not be important when viewed in the aggregate (Exercise 6.32.) ☐

6.5.4 Trend tests under extra-Poisson variability

To assess a monotone trend with overdispersed counts, methods are available similar in form to those used with overdispersed proportions. From Section 1.2.3, a specific parent distribution for the counts that incorporates excess variability is the negative binomial: if $Y \sim \mathcal{NB}(\mu, \delta)$ then Y has mean $E[Y] = \mu$ and $Var[Y] = \mu + \delta\mu^2$. The dispersion parameter $\delta > 0$ represents the extra-Poisson variability: in the limit as $\delta \rightarrow 0$, the negative binomial distribution approaches the Poisson distribution. Recall that this can be formulated as a form of hierarchical model, where a hierarchical gamma assumption on the mean of a Poisson variable induces overdispersion that leads to a negative binomial distribution for the count data.

Assuming $Y_{ij} \sim$ (indep.) $\mathcal{NB}(\mu_i, \delta)$, one tests the null hypothesis $H_0: \mu_0 = \mu_1 = \cdots = \mu_k$ against (6.2) via modification of the Cochran–Armitage trend statistic: simply replace the denominator in (6.8) with an estimator based on the negative binomial variance (Margolin, 1985). That is, replace $\overline{Y}_{++}$ under the square root with $\overline{Y}_{++}(1 + \hat{\delta}\overline{Y}_{++})$, where $\hat{\delta}$ is some consistent estimator of δ (see below). The test statistic becomes

$$z_{calc} = \frac{\sum_{i=0}^{k} r_i(x_i - \overline{x})\overline{Y}_{i+}}{\left\{\overline{Y}_{++}(1 + \hat{\delta}\overline{Y}_{++})\sum_{i=0}^{k} r_i(x_i - \overline{x})^2\right\}^{1/2}} . \tag{6.14}$$

In large samples, $z_{calc} \stackrel{\cdot}{\sim} N(0,1)$, and one rejects H_0 in favor of an increasing trend when $z_{calc} > z_\alpha$. The approximate P-value is $1 - \Phi(z_{calc})$.

The dispersion estimator, $\hat{\delta}$, in (6.14) may be found via many different estimation methods. Two simpler possibilities are the method of moments

(MoM) and maximum likelihood (ML), both described in Section 2.3. The MoM estimator is

$$\hat{\delta}_{\text{MOM}} = \frac{1}{k+1}\sum_{i=0}^{k}\frac{s_i^2 - \overline{Y}_{i+}}{\overline{Y}_{i+}^2}, \tag{6.15}$$

where the s_i^2-values are the individual sample variances, $s_i^2 = \sum_{j=1}^{r_i}(Y_{ij} - \overline{Y}_{i+})^2\}/(r_i - 1)$. Although simple to calculate, one concern with use of $\hat{\delta}_{\text{MOM}}$ is that it is not guaranteed to be positive. Indeed, there are settings where it is valid to interpret $\delta < 0$ as evidence of **underdispersion**, for example, environmental toxicity testing with litter implant counts (Lockhart *et al.*, 1992). More generally, however, negative estimates of δ are either uninterpretable, or they suggest that there is no departure from Poisson sampling (for example, when they are not significantly different from $\delta = 0$).

The ML estimator of δ is based on differentiating the log-likelihood function with respect to δ. This achieves a single estimating equation

$$\psi(\delta^{-1})\sum_{i=0}^{k} r_i + \sum_{i=0}^{k}\sum_{j=1}^{r_i}\left\{\log\left(1 + \delta\overline{Y}_{i+}\right) - \psi(Y_{ij} + \delta^{-1})\right\} = 0, \tag{6.16}$$

the solution of which provides $\hat{\delta}_{\text{MLE}}$. In (6.16), $\psi(\cdot)$ is the **digamma function**, which is available via tables (Abramowitz and Stegun, 1972) or computer algorithms (Bernardo, 1976). Once calculated, either $\hat{\delta}_{\text{MOM}}$ or $\hat{\delta}_{\text{MLE}}$ may be substituted for $\hat{\delta}$ into (6.14). The ML estimator can provide somewhat more powerful inferences while retaining stable Type I error (Piegorsch, 1993), and when computational resources are available for solving (6.16), it is recommended for use under the negative binomial assumption.

When the parent distribution of the data is not known or indeterminate, it is possible to find a generalized score statistic for testing trend in $\text{E}[Y_{ij}] = \mu_i$, via a construction similar to (6.12). The trend statistic is

$$z_{\text{QL}} = \frac{\sum_{i=0}^{k} r_i(x_i - \overline{x})\overline{Y}_{i+}}{\left\{\sum_{i=0}^{k} r_i(x_i - \overline{x})^2 \sum_{j=1}^{r_i}(Y_{ij} - \overline{Y}_{i+})^2\right\}^{1/2}}. \tag{6.17}$$

Here again (6.17) is very similar to the basic trend statistics in (6.8) or (6.14); the major difference is that the terms estimating the Poisson or negative binomial variances – for example, $\overline{Y}_{++}$ in (6.8) – are replaced by the more robust empirical variance estimate $\sum_{j=1}^{r_i}(Y_{ij} - \overline{Y}_{i+})^2$. We write QL in the subscript of (6.17), since it can be motivated from a maximum quasi-likelihood perspective (Breslow, 1990). In large samples, $z_{QL} \dot{\sim} N(0,1)$; one rejects H_0 in favor of an increasing trend when $z_{QL} > \mathfrak{z}_{\alpha}$. The approximate P-value is $1 - \Phi(z_{QL})$.

Example 6.11 (continued) Extra-Poisson variability in offspring counts – Trend testing

Continuing with our aquatic toxicity example on *C. dubia* exposure to nitrofen, we saw in Table 6.11 an indication of extra-Poisson variability with these data, although the effect across the entire data was not compelling. Nonetheless, the potential overdispersion may lead a conservative analyst to apply the generalized score statistic (6.17) for testing trend in the mean offspring counts, rather than the usual Cochran–Armitage test from (6.8).

Recall that increased toxicity to the chemical is indicated by decreases in the mean offspring counts. Applying (6.17) to the data in Table 6.2, we saw previously that $\bar{x} = 785/5 = 157.0$, so the numerator of (6.17) computes to –50 108. The denominator is the square root of $\sum_{i=0}^{k}\{ r_i(x_i - \bar{x})^2\sum_{j=1}^{r_i}(Y_{ij} - \bar{Y}_{i+})^2\} = (10)(0 - 157)^2(116.4) + (10)(80 - 157)^2(96.5) + \cdots + (10)(310 - 157)^2(124.0) = 82524014.0$. Thus the test statistic is $z_{QL} = -50108/9084.273 = -5.516$, in contrast to $z_{calc} = -13.515$ from the Cochran–Armitage test.

Compared to the lower-0.001 critical point of a standard normal distribution, $-\mathfrak{z}_{0.001} = -3.09$, we see again that the trend in Table 6.2 is strongly significant. (The one-sided P-value is still below 0.001.) Even when correcting for potential overdispersion, nitrofen induces significant downward trends in *C. dubia* reproductive response.

The QL trend test can be implemented in S-PLUS, as illustrated in Fig. 6.16. Our S-PLUS function, `Z.QL.Poisson`, takes as input a vector of concentrations (`conc`) and a vector of observed events (`yy`). The function reports z_{QL} with $\bar{x}$ generated as part of the output. P-values can then be calculated using the S-PLUS N(0,1) probability function `pnorm`. These are displayed at the end of Fig. 6.16, as applied to the data from Table 6.2. □

6.6 DISTRIBUTION-FREE TREND TESTING

As noted above, the parent distribution of a set of environmental data can be difficult or even impossible to identify in select instances: excess variability can induce differential patterns of overdispersion in counts or proportions, or continuous data may possess skew or asymmetries of unknown origin.

```
# QL Trend test for count data

Z.QL.Poisson<-function(conc,yy) {

# formula (6.17)
# conc = vector of concentrations
# yy = vector of observed counts
     nreps <- unlist(lapply(split(yy,conc),length))
     means <- unlist(lapply(split(yy,conc),mean))
      vars <- unlist(lapply(split(yy,conc),var))
      xbar <- sum(conc)/length(conc)
     uconc <- unique(conc)
   top <- sum( nreps*(uconc-xbar)*means )
   bot <- sum( nreps*(uconc-xbar)^2*(nreps-1)*vars)
   cat("xbar = ",signif(xbar),"\n")
   cat("Z.QL= ",signif(top/sqrt(bot)),"\n")
   top/sqrt(bot)
}

Z.QL<-Z.QL.Poisson(nitro.df$conc,nitro.df$offspring)
```

```
xbar =  157
Z.QL=  -5.51591
```

```
pnorm(Z.QL)
```

```
[1] 1.734933e-08
```

Fig. 6.16 S-PLUS program to calculate the QL trend statistic for data in Table 6.2.

In these cases, **distribution-free methods** often possess omnibus applicability for testing trend in the data. Of these, **rank-based methods** (also known as **nonparametric methods**) are some of the most widely applicable.

There are numerous rank-based technologies available for significance testing, applicable to a variety of data-analytic settings (Hollander and Wolfe, 1973; Hettmansperger, 1984; Gibbons, 1993). We discussed one simple case in Section 4.1.9: the Mann–Whitney–Wilcoxon U-statistic for two-sample testing. This method is, in fact, the basis of a common rank-based test for trend: the **Jonckheere–Terpstra trend test** (Jonckheere, 1954). For the Jonckheere–Terpstra analysis, one assumes that the observed data, Y_{ij}, have medians $\boldsymbol{m}_i$ ($i = 0, \ldots, k; j = 1, \ldots, r_i$); the goal is to test the null hypothesis H_0:$\boldsymbol{m}_1 = \cdots = \boldsymbol{m}_T$ against the ordered alternative H_a:$\boldsymbol{m}_1 \le \cdots \le \boldsymbol{m}_T$, with strict inequality somewhere among the $\boldsymbol{m}_i$.

We warn that use of the Jonckheere–Terpstra test requires that the shapes of the distributions of the Y_{ij}s be equivalent. (In many cases, this translates to requiring that the Y_{ij}s have a common variance.) This homogeneity assumption may not hold in all cases; when large disparities occur, the applicability of the Jonckheere–Terpstra test is brought into question.

To perform the Jonckheere–Terpstra test for trend, calculate over $i = 0, \ldots, k$, every possible pair of comparisons among treatment levels, and compute from (4.16) the Mann–Whitney–Wilcoxon U-statistic for each pairing. For $h < i$, this is

$$U_{hi} = \sum_{u=1}^{r_h} \sum_{v=1}^{r_i} I(Y_{hu} < Y_{iv}).$$

where $I(x < y)$ is the ordering indicator

$$I(x < y) = \begin{cases} 1 & \text{if } x < y \\ \frac{1}{2} & \text{if } x = y \\ 0 & \text{if } x > y \end{cases} \quad . \tag{6.18}$$

Essentially, each U_{hi} counts the number of times data values at treatment level i exceed those in a preceding treatment level, $h < i$, where ties count 1/2. Notice that over $i = 0, \ldots, k$, there are $k(k+1)/2$ possible U-statistics that can be so calculated.

Next, aggregate the $k(k+1)/2$ U-statistics into a single sum:

$$J_{\text{calc}} = \sum_{h=0}^{k-1} \sum_{i=h+1}^{k} U_{hi} \,. \tag{6.19}$$

J_{calc} is the Jonckheere–Terpstra trend statistic. Reject H_0 when J_{calc} exceeds an upper-α critical point, $J_{\alpha,k+1}(r_0, r_1, \ldots, r_k)$, from the null distribution of this statistic. Tables of $J_{\alpha,k+1}(r_0, r_1, \ldots, r_k)$ are common; see, for example, Hollander and Wolfe (1973, Table A.8). Lin and Haseman (1976) give corrections for the J-statistic if there are many tied observations at a single rank value, as might occur with dose-response data whose control mean is near a lower bound of zero, or, in the case of a negative trend, whose responses at high doses are near a lower bound of zero.

In large samples, it is possible to construct a standardized J-statistic whose null reference distribution is approximately standard normal:

$$z_{\text{calc}} = \frac{J_{\text{calc}} - \frac{1}{4}\left\{(r_+)^2 - \sum_{i=0}^{k} r_i^2\right\}}{\left\{\frac{1}{72}\left[(r_+)^2(2r_+ + 3) - \sum_{i=0}^{k} r_i^2(2r_i + 3)\right]\right\}^{1/2}} , \tag{6.20}$$

where $r_+ = \sum_{i=0}^{k} r_i$. The large-sample distribution of (6.20) is $z_{calc} \dot{\sim} N(0,1)$, so reject H_0 in favor of an increasing-trend when $z_{calc} > z_\alpha$; the approximate P-value for the increasing trend alternative is $1 - \Phi(z_{calc})$. For a decreasing trend, reject H_0 when $z_{calc} < -z_\alpha$; the P-value is $\Phi(z_{calc})$.

Example 6.12 Nonparametric test for trends in offspring counts after nitrofen exposure

We illustrate calculation of the Jonckheere–Terpstra trend statistic with the *C. dubia* aquatic toxicity data from Example 6.2. Recall that this study involved $k = 4$ nitrofen exposures (80, 160, 235, and 310 mg/ml), along with a zero-exposure control. Ten female organisms were studied in each experimental group, so that $r_0 = r_1 = r_2 = r_3 = r_4 = 10$. The response variable was number of offspring produced by each organism. If nitrofen exposure induces reproductive toxicity in these aquatic organisms, then a trend of decreasing number of offspring with increasing nitrofen concentration would be expected.

Table 6.12 Example of calculation of $I(y_{0u} < y_{1v})$ from (6.18) for aquatic toxicity data from Table 6.2. y_{0u} represents the response of an organism in the control group and y_{1v} represents the response of an organism in the 80 mg/ml exposure group

					y_{1v}					
y_{0u}	26	27	29	31	32	33	33	33	35	36
24	1	1	1	1	1	1	1	1	1	1
27	0	$\frac{1}{2}$	1	1	1	1	1	1	1	1
30	0	0	0	1	1	1	1	1	1	1
31	0	0	0	$\frac{1}{2}$	1	1	1	1	1	1
32	0	0	0	0	$\frac{1}{2}$	1	1	1	1	1
33	0	0	0	0	0	$\frac{1}{2}$	$\frac{1}{2}$	$\frac{1}{2}$	1	1
33	0	0	0	0	0	$\frac{1}{2}$	$\frac{1}{2}$	$\frac{1}{2}$	1	1
34	0	0	0	0	0	0	0	0	1	1
34	0	0	0	0	0	0	0	0	1	1
36	0	0	0	0	0	0	0	0	0	$\frac{1}{2}$

For these data, the Jonckheere–Terpstra statistic requires the calculation of $k(k + 1)/2 = 10$ individual U-statistics. We illustrate this calculation for the first U-statistic, U_{01}, in Table 6.12. There, we compare the $(h = 0)$ control group with the $(i = 1)$ 80 mg/ml exposure group. The table lists values of the indicator function $I(y_{0u} < y_{1v})$ from (6.18), where y_{0u} represents the observed response of the uth organism in the control group and y_{1v} represents the observed response of the vth organism in the control group, $u,v = 1,\ldots, 10$.

Summing up all of the entries in Table 6.12 yields $U_{01} = 49$. All the other intermediate U-statistics can be determined in a similar fashion, leading to the Jonckheere–Terpstra test statistic

$$J_{\text{calc}} = \sum_{h=0}^{k-1} \sum_{i=h+1}^{k} U_{hi}$$

$$= (U_{01} + U_{02} + U_{03} + U_{04}) + (U_{12} + U_{13} + U_{14}) + (U_{23} + U_{24}) + U_{34}$$

$$= (49 + 19 + 1.5 + 0) + (21.5 + 1.5 + 0) + (1.5 + 0) + 4,$$

or simply 98. Standardized, this is

$$z_{\text{calc}} = \frac{J_{\text{calc}} - \frac{1}{4}\left\{(r_+)^2 - \sum_{i=0}^{4} r_i^2\right\}}{\left\{\frac{1}{72}\left[(r_+)^2(2r_+ + 3) - \sum_{i=0}^{4} r_i^2(2r_i + 3)\right]\right\}^{1/2}}$$

$$= \frac{98 - \frac{1}{4}\left\{(50)^2 - 500\right\}}{\left\{\frac{1}{72}\left[(50)^2(103) - 11\,500\right]\right\}^{1/2}} = -6.88,$$

with approximate P-value $\Phi(-6.88) < 0.0001$. Once again, this suggests a strongly decreasing trend with increasing nitrofen exposure in the number of offspring produced. ❑

6.7 NONPARAMETRIC TESTS FOR NONMONOTONE ('UMBRELLA') TRENDS

An interesting application of trend analysis occurs when environmental data exhibit a **nonmonotone dose response**. For example, the data may respond in an increasing fashion over low doses, but then a **downturn in dose response** occurs at higher doses. In terms of the mean responses, μ_i, at each dose x_i, this is

$$\mu_0 \leq \mu_1 \leq \cdots \leq \mu_v \geq \mu_{v+1} \geq \cdots \geq \mu_k, \qquad (6.21)$$

with some strict inequality among the μ_is. This form of dose response is known as an **umbrella model** (Mack and Wolfe, 1981). (The nonmonotonicity may also be decreasing then increasing, but this is less common with environmental trend data.) As above, the null hypothesis of no dose response is represented as H_0:$\mu_0 = \mu_1 = \cdots = \mu_k$, while the nonmonotone alternative, H_a, is given by (6.21). Testing for significance in the increasing-trend portion of (6.21) is essentially a problem of trend testing under an **umbrella alternative hypothesis**.

When the nonmonotone dose response is known to have a specific functional form, LS and ML methods are appropriate to estimate features of the response, including increasing trend. (Selected methods for analysis of nonmonotone dose response are discussed in Chapter 7.) When there is no strong motivation for a parametric model to describe the nonmonotone response, however, we turn to nonparametric statistical methods. A number of nonparametric approaches are available for testing against umbrella alternatives (Mack and Wolfe, 1981; Hettmansperger and Norton, 1987; Shi, 1988; Chen and Wolfe, 1990; Schumacher and Schmoor, 1991). One that is particularly useful for trend testing employs the Jonckheere–Terpstra trend statistic from Section 6.6 (Simpson and Margolin, 1986): estimate the **umbrella index** υ at which the response reaches its peak, via some form of recursive statistical testing; then, perform the Jonckheere–Terpstra test for increasing trend up to, *but not past* the estimated umbrella point. P-values and significance levels in the trend analysis step are adjusted for the recursive estimation step, so that the entire process operates at a pre-specified experimentwise false positive rate, α.

One begins the recursive procedure by calculating a two-sample Mann–Whitney statistic that compares the ith dose group with the responses from all the preceding $i - 1$ dose groups after pooling the latter into one large 'group.' That is, assuming data of the form Y_{ij} ($i = 0, \ldots, k; j = 1, \ldots, r_i$), take as in (4.16) the k pairwise Mann–Whitney statistics

$$U_{c-1,c} = \sum_{a=1}^{r_{c-1}} \sum_{b=1}^{r_c} I(Y_{c-1,a} < Y_{cb})$$

where $c = 1, \ldots, k$, and $I(x < y)$ is given in (6.18). With these, let

$$V_i = \sum_{c=1}^{i} U_{c-1,i} \tag{6.22}$$

be the pooled Mann–Whitney statistic that compares the response at level i with all previous levels ($i = 1, \ldots, k$). Thus,

$$\begin{aligned}
V_1 &= U_{0,1} \\
V_2 &= U_{0,2} + U_{1,2} \\
V_3 &= U_{0,3} + U_{1,3} + U_{2,3} \\
&\vdots \\
V_k &= U_{0,k} + U_{1,k} + U_{2,k} + \cdots + U_{k-1,k}.
\end{aligned}$$

To estimate the umbrella index, start at $i = k$ and work backwards: use V_k to test whether the response information at $i = k$ deviates significantly above the pooled response information at $i = 1, \ldots, k-1$. If so, an estimate, u, of the umbrella index at maximal response is $u = k$. If not, drop the information at $i = k$, and repeat the process at $i = k - 1$. In this way, the test employs the values of V_i $(i = k, k-1, \ldots, 1)$ to estimate the umbrella index via recursive pre-testing.

For the recursive step, find the largest dose index, u, such that V_u is larger than a specified critical value $c_{u,(1-q)}$; that is,

$$u = \max\left\{i = 1, \ldots, k:\ V_i > c_{i,(1-q)}\right\},$$

where $c_{1,(1-q)} = 0$ and $c_{i,(1-q)}$ is the upper-$(1-q)$ critical point from the distribution of V_i under H_0; $i = 2, \ldots, k$. For example, we can use tables for the two-sample Mann–Whitney–Wilcoxon rank-sum statistic, W from Section 4.1.9, with sample sizes $R_{i-1} = \sum_{c=0}^{i-1} r_i$ and r_i. Recall that W is linearly related to the simpler Mann–Whitney statistic; here, $W_i = V_i + [(r_i + 1)r_i]/2$. Thus we relate $c_{i,(1-q)}$ to the rank-sum critical point, $w_{1-q}(R_{i-1}, r_i)$:

$$c_{i,(1-q)} = w_{1-q}(R_{i-1}, r_i) - \frac{r_i(r_i+1)}{2}. \tag{6.23}$$

(In (6.23), find the value of $w_{1-q}(R_{i-1}, r_i)$ that yields no more than $1 - q$ upper tail probability.) The tuning parameter q must be pre-specified between 0 and 1, and is usually set at or near the middle of this range: $q = 0.50$ (Simpson and Margolin, 1990; Piegorsch, 1992).

Given the estimated umbrella index u, the recursive procedure calculates a Jonckheere–Terpstra statistic up to that estimated index; here this is

$$J_u = \sum_{h=0}^{u-1} \sum_{i=h+1}^{u} U_{hi},$$

as in (6.19), or, equivalently, $J_u = V_1 + \cdots + V_u$. Reject H_0 in favor of an increasing dose response (up to $i = u$) when J_u exceeds the upper-γ critical point, $J_{\gamma,u+1}(r_0, r_1, \ldots, r_u)$. Given q and the experimentwise significance level α, a conservative choice for γ is $\gamma = \alpha(1-q)/(1-q^k)$ (Simpson and Margolin, 1986). Schumacher and Schmoor (1991) report on an adaptive procedure that can improve the conservatism in this choice for γ.

When the r_is are large, both V_u and J_u are approximately normal, allowing for use of approximate critical points in the recursive procedure. Let $R_i = \sum_{c=0}^{i} r_c$ be the cumulative sample sizes ($i = 0, \ldots, k$). For large r_i, approximate values for $c_{i,(1-q)}$ are

$$c_{i,(1-q)} \approx \frac{1}{2} r_i R_{i-1} + z_{(1-q)} \left\{ \frac{r_i R_{i-1}(R_i + 1)}{12} \right\}^{1/2} \quad (6.24)$$

(note the dependency on q). Approximate values for $J_{\gamma,u+1}(r_0, r_1, \ldots, r_u)$ are slightly more complex:

$$J_{\gamma,u+1}(r_0, r_1, \ldots, r_u) \approx \sum_{c=1}^{u} \frac{r_c R_{c-1}}{2} + z_\gamma \left\{ \sum_{c=1}^{u} \frac{r_c R_{c-1}(R_c + 1)}{12} \right\}^{1/2} \quad (6.25)$$

(Exercise 6.41). With large samples (say, $k \geq 5$ and $\min\{r_i\} \geq 3$; or, $\min\{r_i\} \geq 6$ for any k), one can use the approximate values in place of tabulated values wherever called for in the recursive procedure. A computing algorithm for these calculations is given by Simpson and Dallal (1989).

Example 6.13 Distribution-free umbrella analysis of Salmonella *data*
Recall the *Salmonella* mutagenesis data from Table 3.2, representing the mutagenic response of *S. typhimurium* strain TA1535 to $k = 5$ gaseous exposures of the airborne toxin 1,3-butadiene. As discussed in Example 3.7, an increase in the number of observed colonies suggests a mutagenic effect, even in the presence of a downturn at higher doses caused by the competing toxicity of the chemical exposure. Figure 6.17 illustrates both effects with the *Salmonella* data from Table 3.2.

To assess whether the dose response in Table 3.2 increases significantly, we consider testing $H_0: \mu_0 = \mu_1 = \cdots = \mu_5$ against the umbrella alternative (6.21). To do so, we apply the Simpson–Margolin recursive procedure described above.

We set the experimentwise error rate to $\alpha = 0.01$ and the tuning parameter to $q = 0.5$. To estimate the umbrella point, we calculate the individual Mann–Whitney U-values, and sum them into the pooled statistics, V_i from (6.22), starting at $i = 5$ and working backwards to $i = 1$. We stop the recursive step at $i = u$ when $V_u > c_{u,0.5}$, where the critical point is built from the relationship in (6.23). Table 6.13 displays the results of these computations. As seen therein, at $i = 5$ we find $c_{5,0.5} = 23$ is smaller than $V_5 = 24.5$. Thus we stop the recursive calculations immediately, and conclude that the estimated umbrella index is $u = 5$. (We also include the approximate critical points using (6.24) in Table 6.13. For this design at $q = 0.5$, the approximation appears to be reasonable, although slight discrepancies are noted at every level of i.)

Table 6.13 Recursive umbrella analysis at $q = 0.5$ for 1,3-butadiene data from Table 3.2

Level, i	*5*	*4*	*3*	*2*	*1*	*0*
Sample size, r_i	3	3	3	3	3	3
Cumulative sample size, R_i	18	15	12	9	6	3
Rank-sum critical point*, $w_{0.5}(R_{i-1}, r_i)$	29	25	20	16	11	—
Critical point from (6.23), $c_{i,0.5}$	23	19	14	10	5	—
Approximate critical point from (6.24)	22.5	18	13.5	9	4.5	—
Recursive statistic, V_i	24.5	28	26	18	9	—

* Rank-sum critical points from Hollander and Wolfe (1973, Table A.5). Value listed here is largest tabled value of w that gives no more than 50% upper tail probability.

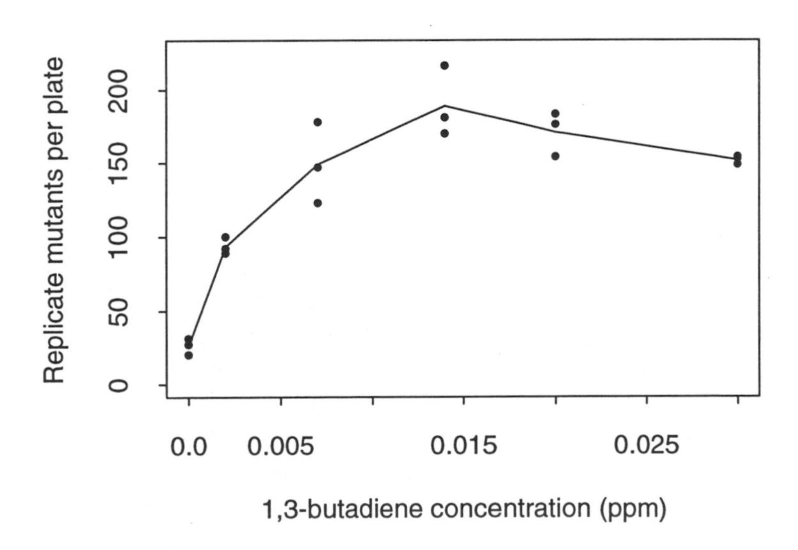

Fig. 6.17 Scatterplot of mutant plate counts as a function of 1,3-butadiene levels for data in Table 3.2. A solid line connects the mean plate counts for each concentration.

For an umbrella index of $u = 5$, the corresponding Jonckheere–Terpstra trend statistic is the sum of the five individual V_is: $J_5 = (24.5 + 28 + \cdots + 9) = 105.5$. To determine if this suggests a significant increase, we calculate the recursion-adjusted false-positive rate $\gamma = \alpha(1 - q)/(1 - q^k) = (0.01)(0.5)/(1 - 2^{-5}) = (0.01)(0.5)/(0.9688) = 0.0052$. Unfortunately, from tables of the Jonckheere–Terpstra statistic (Hollander and Wolfe, 1973, Table A.8), we find that there is no critical point $J_{0.0052,6}(3, 3, \ldots, 3)$; the closest point with upper tail probability at or below 0.0052 is $J_{0.0047,6}(3, 3, \ldots, 3) = 101$. Since the observed value $J_5 = 105.5$ exceeds this critical point, we reject H_0 and conclude there is a significant, increasing portion in the dose response to 1,3-butadiene at $\alpha = 0.01$.

If we employ instead the approximate critical point (6.25) for these data, we find $J_{0.0052,6}(3, 3, \ldots, 3) \approx 67.5 + z_{0.0052}\sqrt{168.75}$. The standard normal critical value here is $z_{0.0052} = 2.562$. This leads to an approximate critical point of 100.784, obviously very close to the tabled value of 101. Again, a significant increase in the trend is evidenced.

We close by noting that the tuning parameter q can be used to increase sensitivity to downturns early in the dose response by raising its value to, say, $q = 0.6$ or $q = 0.7$. If such sensitivity were specified a priori, the analysis for the 1,3-butadiene data would show an even more significant result: at either $q = 0.6$ or $q = 0.7$, the umbrella index is estimated as $u = 4$, given by $J_4 = 81.0$. This is significant at levels as low as $\alpha = 0.005$. □

6.8 SUMMARY

In this chapter, statistical methods are presented for assessing an increasing (or decreasing) trend in environmental data. Emphasizing simple tests for trend, consideration is given to both continuous and discrete data structures, including overdispersed settings for the discrete case. A simple, nonparametric method based on ranks, the Jonckheere–Terpstra trend test, is given as a further alternative when the underlying distribution of the data is unknown or not consistent with the models presented. Also illustrated is an extension of the rank-based method to testing for increasing trend when a possible downturn in the response is evidenced. The method employs a recursive estimation scheme which first estimates the point at which the response peaks (at least, relative to the final observed levels of the response), and then applies the Jonckheere–Terpstra statistic to test for increasing trend up to that point. Adjustments are included to compensate for the recursive pre-testing that identifies the point of maximal departure.

When the dose response is known to have a specific parametric form, it may be more powerful to use that form to estimate features of the response, including increasing trend. Methods for this type of analysis are discussed in Chapter 7, under the general topic of dose-response analysis.

EXERCISES

(Exercises with asterisks, e.g. 6.12*, typically are more advanced, or require greater knowledge and understanding of probability distributions and/or calculus.)

6.1. When testing for trend, must a dose-response experiment always have a control group? If so, must it always be represented as a zero-dose control? Explain.

6.2. Describe three applications (from your own field of study, as appropriate) where an independent variable/predictor variable, x_i, is ordered and

a. quantitative;

b. not quantitative.

Explain.

6.3. For the body weight data in Example 6.4, assume that a simple linear regression model (6.1) adequately represents the mean response. View the sample means in Table 6.3 as individual observations and find the LS estimators for β_0 and β_1, and conduct a test for decreasing trend. Set $\alpha = 0.01$. Does your result differ from the result using Williams' test in the example?

6.4. Repeat Exercise 6.3, but now find weighted LS estimators for β_0 and β_1, using as weights the reciprocals of the squared standard deviations in Table 6.3. Set $\alpha = 0.01$.

6.5. Return to the bottlenose dolphins data from Example 6.1. Assume that a simple linear regression model (6.1) adequately represents the mean concentration response as a function of age for these data.

a. Graph the scatterplot displaying mean concentration (Y) vs. age (x). Does the scatterplot support the linear regression assumption?

b. Verify the calculations for the LS estimators for β_0 and β_1.

c. Graph the residuals $R_{ij} = Y_{ij} - \hat{Y}_{ij}$ vs. the predicted values $\hat{Y}_{ij}$ for these data, as in Example 6.3. Do the residuals appear randomly scattered? What problems, if any, does the plot suggest?

6.6. Return to the nitrofen exposure data from Example 6.2. Verify the calculations for the weighted LS estimators $\hat{\beta}_{0,w}$ and $\hat{\beta}_{1,w}$ using the reciprocal sample variances, $w_i = 1/s_i^2$, $i = 1, \ldots, 5$.

6.7. Facemire *et al.* (1995) report data from a study of hormone concentrations in wild Florida panthers (*Felis concolor coryi*) as a function of the animals' ages. The response variable is the ratio of blood-level estriadol to blood-level testosterone (each in pg/ml). For male panthers, consider the following data:

Animal, i:	*1*	*2*	*3*	*4*	*5*	*6*	*7*	*8*
Age in years, x_i:	1	2	2	3	3	9	9	10
Hormone ratio	0.42	0.39	0.54	0.65	0.75	0.56	0.53	0.48

Ratio responses often are lognormally distributed, suggesting that a logarithmic transformation of the data will yield data for which a normal parent distribution is valid.

a. Plot hormone ratio (Y) versus age (x) and also plot log(Y) vs. age. Which plot is better described by a linear regression model?

b. Assuming a simple linear regression model (6.1), find the LS estimators for β_0 and β_1, and conduct a test for increasing trend using hormone ratio as the dependent variable/response variable. Set $\alpha = 0.10$.

c. Repeat part (b) with Y = log(hormone ratio).

d. Comment on the relationship between the two tests in parts (b) and (c) and your observations in part (a).

6.8. For the Florida panther study in Exercise 6.7, data were collected also on female panthers. These were:

Animal, i:	*1*	*2*	*3*	*4*	*5*	*6*
Age in years, x_i:	1	2	4	7	7	11
Hormone ratio	1.58	1.47	3.16	0.77	2.26	4.03

a. Repeat the analyses from Exercise 6.7 for these data.

b. Is the trend more pronounced for the male panthers or the female panthers? Explain.

6.9. Exercise 5.1 gives data on reliability of human serum PCB levels. The explanatory variable, x, is time in storage. The data are:

Year	*x*	*n*	*Sample mean*	*Sample std. deviation*
1989	1	6	0.135	0.0233
1988	2	3	0.183	0.0203
1982	8	3	0.281	0.0577

Assume that a simple linear regression model (6.1) adequately represents the mean response as a function of x. View the sample means as individual observations Y_i, and find weighted LS estimators for β_0 and β_1, with weights taken as the reciprocals of the standard errors of the means. That is, square the sample standard deviations, divide them by the sample sizes, and take the reciprocals of these values as the weights ($w_i = n_i/s_i^2$). Use these values to conduct a test for increasing trend. Set $\alpha = 0.01$.

6.10. Apply Williams' test to assess if there is a significantly increasing trend in the blood PCB storage data from Exercise 6.9. Set $\alpha = 0.01$. (Notice that by viewing the 1989 data as the control, the design is treatment-balanced. You will require the balanced critical point $\bar{t}_{0.01}(9,1) = 2.903$ and the extrapolation coefficient $\beta = 4$.) How does your result compare with the result in Exercise 6.9?

6.11. The study by Facemire *et al.* (1995) reported in Exercise 6.7 also provided data on the ratio of blood-level estriadol to blood-level testosterone (each in pg/ml) in male panthers. A portion of the data follows:

Level, i	*0*	*0*	*0*	*1*	*1*	*2*	*2*	*3*	*3*
Age (yrs), x_i	1	1	1	2	2	3	3	5	5
Hormone ratio	0.27	0.29	0.35	0.70	0.77	0.60	0.91	0.50	0.69

As in Exercise 6.7, apply a logarithmic transform to the hormone ratios before analyzing the data. Use Williams' test to assess if there is a significantly increasing trend in the log-transformed data. Set $\alpha = 0.05$. (Use the treatment-balanced critical point to perform the test, with $\omega = 5/3$. You will need the critical point $\bar{t}_{0.05}(6,1) = 2.098$ and the extrapolation coefficient $\beta = 4$.)

6.12*. (Margolin, 1988) Construct the Cochran–Armitage trend statistic (6.6) for proportions via the following steps:

a. Begin by defining independent measures x_i, and observed proportions $p_i = Y_i/N_i$, for $i = 0, \ldots, k$. We will perform a simple linear weighted regression of p_i on x_i. Let $\bar{x} = \sum_{i=0}^{k} N_i x_i / \sum_{i=0}^{k} N_i$.

b. We choose for weights the reciprocals of the estimated variances of p_i. Since we assume $Y_i \sim$ (indep.) $\mathcal{Bin}(N_i, \pi_i)$, we know the variance of p_i is $\text{Var}[p_i] = \pi_i(1 - \pi_i)/N_i$. The weights are then estimates of $1/\text{Var}[p_i]$, *under the no-effect null hypothesis* H_0:$\pi_0 = \cdots = \pi_k$. Show that this leads to $w_i = N_i/\{\bar{p}(1 - \bar{p})\}$, where $\bar{p} = \sum_{i=0}^{k} Y_i / \sum_{i=0}^{k} N_i$.

c. Employ the equations for weighted linear regression from Section 6.1 to show that the weighted LS estimator of slope can take any of the following four forms:

$$\hat{\beta}_{1,w} = \frac{\sum_{i=0}^{k}(x_i - \bar{x})N_i(p_i - \bar{p})}{SS_X} = \frac{\sum_{i=0}^{k}x_iN_i(p_i - \bar{p})}{SS_X} = \frac{\sum_{i=0}^{k}(x_i - \bar{x})N_ip_i}{SS_X} = \frac{\sum_{i=0}^{k}(x_i - \bar{x})Y_i}{SS_X},$$

where $SS_X = \sum_{i=0}^{k}N_i(x_i - \bar{x})^2$.

d. The standard error of $\hat{\beta}_{1,w}$ is the square root of the estimated variance. To find this, first find the population variance of $\hat{\beta}_{1,w}$, using the last form for $\hat{\beta}_{1,w}$ from part (c); that is, find

$$\mathrm{Var}\left[\frac{\sum_{i=0}^{k}(x_i - \bar{x})Y_i}{SS_X}\right].$$

(*Hint*: recall that for independent random variables Y_i, and constants a_i and c, $\mathrm{Var}\left[\sum_{i=0}^{k}a_iY_i/c\right] = c^{-2}\sum_{i=0}^{k}a_i^2\mathrm{Var}[Y_i]$.)

e. Now, replace π_i wherever it appears in the variance term from part (d) with its estimate under H_0, $\bar{p}$. Simplify; this is the estimated variance. The standard error is the square root of this estimated variance.

f. Complete the construction by dividing $\hat{\beta}_{1,w}$ from part (c) by its standard error from part (e). Show that this ratio is equivalent to (6.6).

6.13*. Show that the Cochran–Armitage test statistic (6.6) is invariant to linear transformations of the independent measures (x_i).

6.14. Witt *et al.* (1995) give data on induction of chromosome aberrations (CAs) in cultured Chinese hamster ovary cells by the chemical phenolphthalein. The data are proportions of cells with CAs at each of $k = 3$ exposure levels (including a control group):

Dose (μg/ml):	*0*	*30*	*40*	*50*
Cells with CAs	2	3	9	58
Cells scored	200	200	200	200

a. Graph the relationship between proportion of cells scored (Y/N) and dose (x).

b. Perform a Cochran–Armitage trend test with x_is assigned as the observed exposure levels to assess if there is an increasing trend in CA proportion response with dose. Set $\alpha = 0.01$.

c. Repeat part (b) using $x_0 = 1$, $x_1 = 2$, $x_2 = 3$, and $x_3 = 4$. Does this different set of x_is lead to a different conclusion regarding the presence or absence of trend in these data?

6.15. White and Hoffman (1995) report on reproductive productivity in nesting wood ducks (*Aix sponsa*) after dioxin exposure. Given as proportions of nests in which at least one egg hatched, the data are:

Dose (ppt):	≤5	5–20	20–50	>50
Nests with hatched eggs, Y	29	27	10	9
Nests, N	29	30	14	13

a. Define measures, x_i, to represent the ordered categories of dioxin exposure. Graph the relationship between proportion of nests with hatched eggs (Y/N) and exposure (x).

b. With the measures you defined in part (a), perform a Cochran–Armitage trend test to assess if there is a decreasing trend in hatching proportion with exposure. Set $\alpha = 0.05$.

6.16. Exercise 5.5 gives data on liver tissue damage after rats were exposed to an environmental toxin. The data may be viewed as:

Dose	*Animals with damage*	*Total animals*
0	1	17
0.5	4	19
2	9	19
10	11	22

Take as measures, x_i, the natural log of dose. (Use consecutive-dose average spacing from (6.9) to define the control value.)

a. Graph the relationship between proportion of animals exhibiting damage (Y/N) and log-dose (x).

b. Perform a Cochran–Armitage test to assess if there is an increasing trend in proportion of liver damage over log-dose. Set $\alpha = 0.05$.

6.17. Kerckaert *et al.* (1996) report on cell transformation experiments with Syrian hamster embryo cells that are used as predictors of carcinogenicity of environmental agents. The data are proportions, Y_i/N_i, of cell colonies that exhibit preneoplastic morphological transformations (MTs) after exposure to the agent at various exposures x_i ($i = 0, \ldots, k$). For each of the following data sets, perform a Cochran–Armitage test to assess if there is an increasing trend over dose in proportion of MT colonies. Set $\alpha = 0.10$.

a. Agent: Primaclone.

Dose (μg/ml):	*0*	*200*	*250*	*300*	*350*	*400*
Number of MTs	8	9	11	8	10	11
Total colonies	1374	1435	1394	1431	1445	1499

b. Agent: Phenolphthalein (cf. Exercise 6.14).

Dose (μg/ml):	*0*	*15*	*17.5*	*20*	*22.5*	*25*
Number of MTs	6	13	13	17	18	10
Total colonies	1323	1337	1276	1054	959	802

6.18. Suppose data are recorded on number of hunting forays a bird of prey makes during a 12-hour daylight period, as a function of ambient temperature (°C), over the span of a week. The results are:

Temp.	n_i	*Forays*
< 10	2	1, 3
10–20	3	3, 5, 7
> 20	3	5, 7, 9

Define measures, x_i, to represent the ordered categories of ambient temperature. Perform a Cochran–Armitage trend test to assess if there is an increasing trend in number of hunting forays over this exposure measure. Set $\alpha = 0.01$.

6.19*. Consider a set of measures x_i, and observed counts Y_{ij} ($i = 0, \ldots, k; j = 1, \ldots, r_i$). Suppose the counts are distributed as independent Poisson random variables, each with mean μ_i. Let

$$\bar{x} = \sum_{i=0}^{k} r_i x_i / \sum_{i=0}^{k} r_i$$

be the sample-weighted average measure, and

$$\bar{Y}_{++} = \sum_{i=0}^{k} \sum_{j=1}^{r_i} Y_{ij} / \sum_{i=0}^{k} r_i$$

be a pooled estimate of the mean response under the null hypothesis $H_0: \mu_0 = \cdots = \mu_k$. Mimic the steps in Exercise 6.12 to show that the Cochran–Armitage trend statistic (6.8) for testing H_0 with count data may be derived from a weighted LS regression slope estimator.

6.20. Hakulinen and Dyba (1994) report data on malignant skin cancer counts in a cohort of Swedish women over a period of years, in order to ascertain if cancer rates are increasing over time, perhaps due to increasing exposure to ultraviolet radiation. The data are:

Index, i:	*1*	*2*	*3*	*4*	*5*
Time range:	1960–64	1965–69	1970–74	1975–79	1980–84
New cancer cases:	122	197	202	292	352

a. Chose an appropriate measure, x_i, to represent time, and graph the relationship between the number of new cancer cases (Y) and time (x).

b. Using the measures you chose in part (a), calculate the Cochran–Armitage trend statistic (6.8) for these data and test if there is an increasing trend in rate of new cases over time. Set $\alpha = 0.10$.

6.21. The data in Exercise 4.20 on rodent tumorigenicity to the chemical solvent stabilizer 1,2-epoxybutane are actually part of a larger experiment that included an additional exposure group at 400 ppm (National Toxicology Program, 1988). Recall that the data are numbers of organ sites (per animal) exhibiting tumors. The entire data may be written as follows:

	Number of animals		
Number of tumors	*Dose: 0 ppm*	*Dose: 200 ppm*	*Dose: 400 ppm*
0	0	1	2
1	8	6	4
2	12	13	11
3	17	13	11
4	11	11	12
5	1	4	4
6	1	1	4
7	0	1	2

Assume the Poisson parent model is valid for these data.

a. Provide an appropriate graphical display illustrating the relationship between tumor onset and dose.

b. Use the Cochran–Armitage trend statistic (6.8) to assess whether there is an increasing trend in number of tumors (per animal) over the exposure dose. Set $\alpha = 0.01$.

6.22. Can a statistic for assessing extra-binomial variability in proportions such as (6.10) be calculated if there is only one replicate proportion at each dose level (that is, $r_i = 1$ at every $i = 0, \ldots, k$)? Explain. If not, what condition(s) on the sample sizes must hold for a test to be possible?

6.23. Exercise 5.8 gives data on proportions of micronucleated (MN) cells in laboratory rodents after exposure to methyl methane sulfonate. For that exercise, the data are pooled over replicate animals to give single proportions of MN response at each exposure level. The unpooled, per-animal data are available, however (Hothorn, 1994). For each animal, 1000 cells were examined, from which the following numbers of MN cells were scored:

Dose	*Anim. ID*	*MN*	*Dose*	*Anim. ID*	*MN*	*Dose*	*Anim. ID*	*MN*
0	1a	0	5	1b	0	20	1d	3
	2a	1		2b	0		2d	3
	3a	1		3b	1		3d	4
	4a	1		4b	1		4d	4
	5a	1		5b	1		5d	6
	6a	1		6b	2		6d	7
	7a	2		7b	2		7d	7
	8a	2		8b	2		8d	8
	9a	2	10	1c	1	40	1e	19
	10a	2		2c	1		2e	23
	11a	3		3c	3		3e	23
	12a	3		4c	3		4e	25
	13a	3		5c	3		5e	25
	14a	3		6c	3		6e	26
	15a	3		7c	4		7e	29
	16a	3		8c	6		8e	34

a. Verify that pooling these data at each dose produces the proportions given in Exercise 5.8.

b. Test for extra-binomial variability. Set $\alpha = 0.10$.

c. Perform a trend test to assess if there is an increasing trend in MN response with dose. Use an appropriate test, based on your results in part (b); set $\alpha = 0.05$. How do your results compare qualitatively with those from Exercise 5.8?

6.24. In a developmental toxicity study similar to that described in Example 6.9, *in utero* lethality in rat fetuses was studied after maternal exposure to the airborne toxin 1,3-butadiene over a series of exposures (Piegorsch, 1994, Section 3.5). The per-litter replicate proportions (number dead/total implants) were:

Dose = 0 ppm: $\frac{0}{11}, \frac{1}{9}, \frac{0}{13}, \frac{0}{15}, \frac{0}{15}, \frac{0}{11}, \frac{0}{12}, \frac{0}{14}, \frac{2}{14}, \frac{1}{14}, \frac{1}{17}, \frac{1}{14}, \frac{0}{14}, \frac{0}{15}, \frac{0}{13}, \frac{0}{12}, \frac{0}{12}, \frac{0}{16}, \frac{0}{14}, \frac{3}{15}, \frac{3}{17}, \frac{1}{13}, \frac{0}{14}, \frac{1}{16}, \frac{0}{14}, \frac{1}{16}, \frac{0}{9}, \frac{1}{10}, \frac{0}{2}, \frac{0}{9}, \frac{0}{14}, \frac{0}{14}, \frac{1}{14}, \frac{0}{4}, \frac{0}{15}, \frac{0}{16}$

Dose = 200 ppm: $\frac{0}{13}, \frac{0}{14}, \frac{0}{15}, \frac{0}{3}, \frac{2}{14}, \frac{0}{12}, \frac{1}{9}, \frac{4}{14}, \frac{0}{10}, \frac{0}{12}, \frac{0}{17}, \frac{0}{14}, \frac{1}{10}, \frac{0}{16}, \frac{5}{16}, \frac{1}{13}, \frac{0}{9}, \frac{1}{10}, \frac{0}{15}, \frac{0}{16}, \frac{2}{15}, \frac{0}{15}$

Dose = 1000 ppm: $\frac{0}{13}, \frac{1}{14}, \frac{1}{12}, \frac{0}{13}, \frac{0}{13}, \frac{0}{15}, \frac{0}{10}, \frac{2}{13}, \frac{0}{13}, \frac{1}{15}, \frac{0}{15}, \frac{0}{11}, \frac{0}{17}, \frac{1}{17}, \frac{0}{10}, \frac{2}{17}, \frac{2}{15}, \frac{0}{13}, \frac{3}{14}, \frac{0}{16}, \frac{1}{15}, \frac{2}{16}, \frac{0}{17}$

Dose = 8000 ppm: $\frac{0}{12}, \frac{1}{13}, \frac{5}{13}, \frac{0}{17}, \frac{1}{14}, \frac{0}{15}, \frac{1}{11}, \frac{0}{14}, \frac{1}{12}, \frac{1}{13}, \frac{9}{16}, \frac{0}{16},$
$\frac{0}{14}, \frac{2}{16}, \frac{0}{15}, \frac{1}{13}, \frac{0}{8}, \frac{0}{14}, \frac{0}{15}, \frac{0}{16}, \frac{1}{13}, \frac{0}{14}, \frac{0}{13}$

$r_0 = 36$, $r_1 = 22$, $r_2 = 23$, and $r_3 = 23$ litters were examined.

a. Mimic the display in Table 6.9 and verify that there is extra-binomial variability in these proportions using the C(α)-test from (6.10). Set $\alpha = 0.10$.

b. From your result(s) in part (a), does the extra-binomial variability appear constant across dose? If so, employ the trend statistic (6.11) to test for an increasing trend in per-litter lethality over log-dose. (Use consecutive-dose average spacing from (6.9) to define the control value.) If not, employ the GEE trend statistic (6.12) to test for an increasing trend in per-litter lethality over log-dose. Set $\alpha = 0.05$.

6.25. In a developmental toxicity study similar to that described in Example 6.9, the US National Toxicology Program examined *in utero* lethality in rat fetuses after maternal exposure to the narcotic analgesic codeine over a series of gavage exposures (NTP, 1987a). The per-litter replicate proportions (number dead/total implants) were:

Dose = 0 ppm: $\frac{1}{17}, \frac{0}{11}, \frac{1}{13}, \frac{0}{11}, \frac{2}{15}, \frac{0}{12}, \frac{2}{12}, \frac{3}{10}$

Dose = 75 ppm: $\frac{0}{14}, \frac{1}{13}, \frac{1}{13}, \frac{0}{14}, \frac{0}{12}, \frac{0}{14}, \frac{2}{10}, \frac{1}{15}, \frac{2}{10}, \frac{2}{11}, \frac{1}{12}, \frac{0}{6},$
$\frac{0}{13}, \frac{2}{9}, \frac{3}{8}, \frac{2}{14}, \frac{0}{11}, \frac{0}{12}, \frac{0}{14}$

Dose = 150 ppm: $\frac{2}{16}, \frac{0}{12}, \frac{0}{9}, \frac{0}{13}, \frac{2}{14}, \frac{0}{14}, \frac{0}{14}, \frac{0}{11}, \frac{1}{6}, \frac{1}{8}, \frac{2}{13}, \frac{0}{10},$
$\frac{2}{14}, \frac{1}{15}, \frac{0}{13}, \frac{1}{14}$

Dose = 300 ppm: $\frac{2}{13}, \frac{1}{10}, \frac{3}{14}, \frac{1}{17}, \frac{2}{12}, \frac{0}{12}, \frac{1}{12}, \frac{0}{9}, \frac{0}{11}, \frac{2}{14}, \frac{1}{14}, \frac{4}{12},$
$\frac{0}{2}, \frac{1}{12}, \frac{1}{11}$

$r_0 = 8$, $r_1 = 19$, $r_2 = 16$, and $r_3 = 15$ litters were examined.

a. Mimic the display in Table 6.9 and test for extra-binomial variability in these proportions using the C(α)-statistic from (6.10). Set $\alpha = 0.10$.

b. Perform a trend test to assess if there is an increasing trend in per-litter lethality over dose. Use an appropriate test, based on your results in part (a); set $\alpha = 0.10$.

6.26. A number of test statistics are available for assessing extra-binomial variability (Risko and Margolin, 1996). Similar to the C(α)-statistic in (6.10) is the binomial variance statistic (Cochran, 1954)

$$X_i^2 = \sum_{j=1}^{r_i} \frac{(Y_{ij} - N_{ij}\hat{p}_i)^2}{N_{ij}\,\hat{p}_i\,(1 - \hat{p}_i)},$$

where the notation is identical to that in (6.10). In large samples, $X_i^2 \mathrel{\dot\sim} \chi^2(r_i - 1)$. Reject the null hypothesis of no extra-binomial variability when $X_i^2 > \chi^2_\alpha(r_i - 1)$. Aggregation over independent exposure groups is also possible: $X_+^2 \mathrel{\dot\sim} \chi^2(r_+ - k - 1)$. Apply the overdispersion test statistic X_i^2 to the data from the following exercises. Set $\alpha = 0.10$. Compare your outcomes here with those achieved previously.

a. Example 6.9.
b. Exercise 6.23.
c. Exercise 6.24.
d. Exercise 6.25.

6.27*. The GEE statistic in (6.12) may be written in many equivalent forms. For example, show that the numerator in (6.12), $\sum_{i=0}^{k}\sum_{j=1}^{r_i}(x_i - \bar{x})Y_{ij}$, may be written as

$$\sum_{i=0}^{k}\sum_{j=1}^{r_i} x_i(Y_{ij} - N_{ij}\bar{p}),$$

where

$$\bar{x} = \frac{\sum_{i=0}^{k}\sum_{j=1}^{r_i} N_{ij}x_i}{\sum_{i=0}^{k}\sum_{j=1}^{r_i} N_{ij}} \quad \text{and} \quad \bar{p} = \frac{\sum_{i=0}^{k}\sum_{j=1}^{r_i} Y_{ij}}{\sum_{i=0}^{k}\sum_{j=1}^{r_i} N_{ij}}.$$

Notice that this mimics the numerator in the Cochran–Armitage trend statistic (6.6); cf. Exercise 6.12c.

6.28. Return to the counts in Table 4.4 on mutant cell colonies after *in vitro* exposure to HI-6 dichloride. Use the dispersion statistic (6.13) to test for overdispersion in those data. Set $\alpha = 0.05$.

6.29. For the aquatic toxicity study of the pesticide azinphosmethyl described in Example 5.6, count data also were recorded on numbers of male *A. tenuiremis* offspring that survived to adulthood after 26-day maturation. From Exercise 5.13 these are:

Azinphosmethyl dose	*Number of offspring*
0 (control)	32, 39, 32
10 μg/l	38, 20, 41
40 μg/l	37, 25, 21

a. Graph the relationship between the number of offspring (Y) and azinphosmethyl dose (x).

b. Test for overdispersion in these data. Set $\alpha = 0.05$.

c. Based on your results, conduct an appropriate test for decreasing trend in offspring viability after pesticide exposure. Set $\alpha = 0.01$.

6.30. Muhle *et al.* (1994) report data on mineral fiber persistence in rodent lung tissue over time, after the animals have been exposed to the fibers via inhalation. For glass fibers, the data are:

Time since exposure	*Fibers/animal*
1 day	137, 180, 220
180 days	102, 109, 120
365 days	49, 84, 110

a. Graph the relationship between fibers/animal (Y) and time since exposure (x).

b. Test for overdispersion in these data. Set $\alpha = 0.05$.

c. Based on your results, conduct an appropriate test for decreasing trend in fibers/animal over time. Set $\alpha = 0.025$.

6.31. Frome *et al.* (1996) present data on hypersensitivity of blood cells to beryllium, where counts of responding lymphocytes are recorded as a function of *in vitro* beryllium stimulus. The observed counts after 5 day culturing are:

$BeSO_4$ (μmol)	*Number of responding cells*
0 (control)	965, 1173, 828, 862
	1474, 1237, 1021, 976
	1500, 1729, 1672, 1992
1	1050, 706, 1434, 687
10	1551, 1466, 1661, 2301
100	3571, 5780, 4011, 5229

Take for x_i the base-10 logarithm of beryllium stimulus. (Use consecutive-dose average spacing from (6.9) to define the control value.)

a. Graph the relationship between the number of responding cells (Y) and $\log_{10}$(beryllium stimulus) (x).

b. At each stimulus level ($i = 1, \ldots, 4$), calculate the mean number of responding cells, $\overline{Y}_i$, and the sample variance of responding cells, s_i^2. Is any relationship apparent?

c. Test for overdispersion in these data. Set $\alpha = 0.10$.

d. Conduct an appropriate test for increasing trend in responding cells after beryllium stimulation. Set $\alpha = 0.05$.

6.32. Return to the nitrofen toxicity data in Examples 6.2 and 6.11. Table 6.11 gives the individual dispersion test P-values for testing extra-Poisson variability in these count data; it is seen that no significant excess variability is evidenced in the lower three exposure levels (including the control group). At the two higher exposure levels, however, the P-values are below 0.05, suggesting possibly that significant overdispersion exists. To assess this more fully:

a. Corroborate the calculation of the aggregate dispersion statistic, X_+^2, given in the text; that is, show that $X_+^2 = 47.440$. Corroborate also the calculation of the aggregate *df*, $\nu_4 = 45$. At $\alpha = 0.05$, what is the corresponding χ^2 critical value? Is the result significant?

b. Consider application of the Bonferroni inequality from Section 5.2 to *adjust* the individual P-values, in order to correct the individual inferences for experimentwise multiplicity (Rosenthal and Rubin, 1983). In this setting, the inequality in (5.2) applies via an argument similar to that described in Section 5.2: if an effect is significant when its individual P-value, say P_i, is less than α, then a group of J multiple effects are *simultaneously* significant when $P_i \leq \alpha/J$, or, equivalently, when $JP_i \leq \alpha$. Apply this to the $J = 5$ P-values in Table 6.11. At $\alpha = 0.05$, are any of the P-values significant after this correction for multiplicity? (More advanced multiplicity-adjusted P-values are discussed by Westfall and Wolfinger (1997).)

c. Can you suggest how these data might be evaluated for the presence of *underdispersion* relative to a Poisson model?

6.33. Oehlert *et al.* (1995) present data on counts, Y_i, of asbestos fiber counts over pre-set scanned volumes, x_i in cm^3, in a school building. For scanned volumes between 200 cm^3 and 300 cm^3, the data are as follows:

Scanned vol., x_i	225.1	235.5	242.6	251.2	256.2	256.7
Asbestos count, Y_i	15	0	3	1	0	4
Scanned vol., x_i	262.9	267.6	269.0	270.0	270.7	270.8
Asbestos count, Y_i	3	8	0	1	1	2

continued

Scanned vol., x_i	263.1	266.1	266.1	271.6	272.0	273.9
Asbestos count, Y_i	3	6	14	3	4	10
Scanned vol., x_i	274.6	277.3	278.9	281.9	281.9	282.3
Asbestos count, Y_i	4	0	0	1	1	0
Scanned vol., x_i	282.5	284.4	285.3	286.7	287.7	289.4
Asbestos count, Y_i	0	0	0	3	11	2
Scanned vol., x_i	292.4	292.6	293.7	296.1	298.4	
Asbestos count, Y_i	0	3	0	0	1	

The authors note that extra-Poisson variability is apparent in these counts, and suggest that a negative binomial parent distribution may be more appropriate than the Poisson distribution for the counts. Assume that $Y_i \sim$ (indep.) $\mathcal{NB}(\mu_i, \delta)$, where μ_i may be a function of the scanned volumes, x_i.

a. Graph the relationship between asbestos count (Y) and scanned volume (x).

b. Perform a test for increasing trend on these data, using the NB-modified Cochran–Armitage statistic (6.14). Estimate δ via the method of moments in (6.15). Set $\alpha = 0.05$.

c. Perform a test for increasing trend on these data, using the generalized score statistic, z_{QL} in (6.17). Set $\alpha = 0.05$. How does your result compare to that achieved in part (b)?

6.34*. Repeat Exercise 6.33b, using the ML estimating equation (6.16) to estimate δ. At $\alpha = 0.05$, do your results change?

6.35. Ramakrishnan and Meeter (1993) report data from a study of aquatic species abundance after the environmental toxin ethyl benzene was accidentally discharged into a river. They suggest that the counts of same-species fish at various stations downstream from the spill may be subject to extra-Poisson variability due to various ecological factors (clustering of fish in schools, abundance heterogeneity among different fish species, shape and effect of the sampling device on fish recovery, etc.), and propose the negative binomial distribution as a parent distribution for the counts. The data for three different species of shiner are:

Station, i:		*1*	*2*	*3*	*4*	*5*	*6*	*7*
Distance from spill, x_i (m)		−1.5	−0.4	0.5	2.0	3.5	4.5	6.0
No. of fish	(species A)	86	151	3	15	8	144	142
	(species B)	126	77	19	47	15	64	29
	(species C)	125	8	4	46	109	50	76

a. Ignore the different species designations within the family of shiners, and test for overdispersion at each station. Set α = 0.10. Also, aggregate the dispersion statistics and test for overall overdispersion in the data.

b. Ignore the different species designations within the family of shiners, and test for an increasing trend in the animal abundance over only stations 3 through 7 using the NB-modified Cochran–Armitage trend statistic (6.14). Estimate δ via the method of moments in (6.15). Set $\alpha = 0.05$.

c. Ignore the different species designations within the family of shiners, and test for an increasing trend in the animal abundance over only stations 3 through 7 using the generalized score statistic, z_{QL} in (6.17). Set $\alpha = 0.05$. How does your result compare to that achieved in part (b)?

d. Provide a graphical display of fish abundance as a function of distance from the spill.

e. Stratify your analysis over the different species designations, and test each species separately for a *decreasing* trend in the animal abundance over all stations using the generalized score statistic, z_{QL} in (6.17). Set $\alpha = 0.05$.

f. Adjust your analysis in part (e) for multiplicity, using a Bonferroni correction to the critical point or the P-value (Exercise 6.32). Does this affect your conclusions?

6.36. In Exercise 6.33b, the MoM estimator of δ provides a measure of the extra-Poisson variability. Does it appear substantive? If so, we would expect that a simple, Poisson-based Cochran–Armitage trend test based on (6.8) would be too sensitive to any trend in the data, and would reject more often than an overdispersion-adjusted test such as (6.17). Calculate the simple trend statistic (6.8) and compare it to (6.17). Are they substantially different? What do you conclude about these data?

6.37. Perform a Jonckheere–Terpstra rank-based test for increasing trend on the data from the following exercises. Compare your outcomes here with those achieved previously.

a. Exercise 6.18. Set $\alpha = 0.01$. (You will require the critical point $J_{0.01,3}(2,3,3) = 20$.)

b. Exercise 6.29. Set $\alpha = 0.01$. (You will require the critical point $J_{0.01,3}(3,3,3) = 25$.)

6.38. Perform a Jonckheere–Terpstra rank-based trend test for increasing trend on the data from the following exercises. Set $\alpha = 0.05$. Use the large-sample approximation in (6.20). Compare your outcomes here with those achieved previously.

a. Exercise 6.11.

b. Exercise 6.23.

c. Exercise 6.24.

d. Exercise 6.25.

6.39. In an environmental mutagenesis study similar to that described in Examples 3.7 and 6.13, data were taken on the mutagenic response of *S. typhimurium* strain TA100 to $k = 7$ exposures of the chemical intermediate *ortho*-nitroanisole (Piegorsch, 1992, Section 3). The counts of mutant colonies were:

Dose	*Replicate mutants per plate*		
0	132	158	126
33	137	166	150
100	227	222	215
333	273	343	338
666	385	418	401
1000	297	288	357
1200	46	108	139
1500	46	53	–

a. Assess if extra-Poisson variability is evidenced in these data. Set $\alpha = 0.10$.

b. Graph the relationship between mutant counts (Y) and dose (x). Should downturns be considered in any trend analysis with these data?

c. Perform a simple Jonckheere–Terpstra trend test on these data, ignoring any possible downturn in dose response that may be present. Use the large-sample approximation in (6.20), and set $\alpha = 0.05$.

d. Recognize that the *Salmonella* assay is subject to downturns in the dose response, and repeat your analysis in part (b), but now adjust for downturns using the recursive procedure from Section 6.7. Set $\alpha = 0.05$ and $q = 0.5$. (You may require the following critical points: $J_{0.0220,6}(3,\ldots,3) = 94$

$J_{0.0234,5}(3,\ldots,3) = 65$ $\quad J_{0.0183,4}(3,\ldots,3) = 42$

$w_{0.478}(21,2) = 25$ $\quad w_{0.481}(18,3) = 34$ $\quad w_{0.5}(15,3) = 29$

$w_{0.473}(12,3) = 25$ $\quad w_{0.5}(9,3) = 20$ $\quad w_{0.452}(6,3) = 16$

For critical points not listed, use the appropriate large-sample approximation.)

6.40. In an environmental mutagenesis study similar to that described in Examples 3.7 and 6.13, the US National Toxicology Program examined the mutagenic response of *S. typhimurium* strain TA98 to $k = 5$ exposures of the chemical nitrofurantoin (NTP, 1989b). The counts of mutant colonies were:

Dose	*Replicate mutants per plate*		
0.00	46	43	44
0.33	52	38	48
1.0	67	77	78
3.3	108	99	100
10.0	46	19	26
33.0	30	34	18

a. Check to see if any extra-Poisson variability is evidenced in these data. Set $\alpha = 0.10$.

b. Graph the relationship between mutant counts (Y) and dose (x).

c. Perform a simple Jonckheere–Terpstra trend test on these data, ignoring any possible downturn in dose response that may be present. Set $\alpha = 0.05$. (You will require the critical point $J_{0.0452,6}(3, \ldots, 3) = 90$.)

d. Recognizing that the *Salmonella* assay is subject to downturns in the response, repeat your analysis in part (c), but now incorporate an adjustment for possible downturns using the recursive procedure from Section 6.7. Set $\alpha = 0.05$ and $q = 0.5$. (Note that the experimental design here is identical to that in Example 6.13, so the critical points $w_{0.5}(R_{i-1}, r_i)$ from Table 6.13 may prove useful. You may also require the following critical points: $J_{0.0220,6}(3, \ldots, 3) = 94$, $J_{0.0234,5}(3, \ldots, 3) = 65$, or $J_{0.0183,4}(3, \ldots, 3) = 42$.)

6.41*. Establish a relationship between the large-sample approximation for the Jonckheere–Terpstra statistic in (6.20), and components of the approximate critical points described by (6.25). In particular:

a. Show that the numerator term in (6.20),

$$\frac{1}{4}\left\{(r_+)^2 - \sum_{i=0}^{k} r_i^2\right\},$$

is equivalent to

$$\frac{1}{2}\sum_{i=1}^{k} r_i R_{i-1}$$

where $R_i = r_0 + r_1 + \ldots + r_i$. (Note that $R_k = r_+$.)

b. Show that the denominator from (6.20),

$$\frac{1}{72}\left[(r_+)^2(2r_+ + 3) - \sum_{i=0}^{k} r_i^2(2r_i + 3)\right],$$

is equivalent to

$$\frac{1}{12}\sum_{i=1}^{k} r_i R_{i-1}(R_i + 1).$$

7

Dose-response modeling and analysis

In Chapter 6 we discussed methods for assessing trends in environmental data when some ordered variable, x_i, is recorded in conjunction with the observed responses Y_i. There, we noted that if the x_is play a role in explaining or predicting the nature of the responses in Y_i, this information should be included in the statistical analysis. Trend tests from that chapter are most appropriate, however, when the actual parametric form of the response is not modeled directly, or when predictions of response at particular levels of the ordered variable are not of interest. This is common when uncertainty about the specific features of the dose response hinders its functional specification. If, on the other hand, it is known or assumed what precise parametric form the response function takes, more powerful statistical methods may be applied for estimation and testing of dose effects. For example:

- If a treatment is known to affect an organism in a curvilinear fashion, that is, to cause the dose response to increase differently than predicted by a straight-line relationship, that knowledge should be incorporated into some nonlinear model for the dose response.
- If basic biological principles suggest specific nonlinear forms for the dose response, then it is of interest to model the dose response via these forms, and assess how well the biomathematical model fits the observed data.

As above, 'treatment' is used here in a generic sense, indicating some exposure, intervention, observational condition, etc., being applied to the experimental units. We assume that nonrandom, numerical values, x_i, are recorded, representing the quantitative levels of the treatment. As in Chapter 6, we refer to x_i as the **predictor variable**.

In this chapter we extend and enhance the trend-testing approaches from Chapter 6. We use a series of common parametric models of environmental dose response. The goal will be to incorporate functional aspects of the dose response into the statistical analysis in order to (a) estimate parameters of the dose-response function, (b) graph or otherwise illustrate qualities of the response to the dose stimulus, (c) predict features of the underlying mean dose response, and/or (d) invert the estimated dose-response relationship in order to predict a dose associated with the particular level of adverse response.

7.1 DOSE-RESPONSE MODELS ON A CONTINUOUS SCALE

Suppose we observe continuous data, Y_i, such as tissue concentrations, times, body weights, etc. Associated with these data is some quantitative dose variable, x_i ($i = 1, \ldots, n$), producing a pair of measurements for each experimental unit: (x_i, Y_i). Our goal is to model the population mean of Y_i, $E[Y_i] = \mu_i$, as a function of x_i, and also of $P + 1$ unknown parameters, β_j ($j = 0, \ldots, P$), that relate μ and x. The unknown parameters are written in a compact **vector form**: $\boldsymbol{\beta} = [\beta_0\ \beta_1\ \cdots\ \beta_P]$.

We assume the data are independently distributed, and take a normal distribution as their parent distribution: $Y_i \sim$ (indep.) $N(\mu_i, \sigma^2)$. (Notice that the population variances, σ^2, are assumed constant at all $i = 1, \ldots, n$. If this is thought to be inappropriate, one can extend the dose-response model to account for the heterogeneous variances in a functional manner, say $\sigma_i^2 = \sigma^2(x_i)$. As we will see in Chapter 8, generalized linear models provide one mechanism for accommodating heterogeneous variances in select instances.) Other continuous distributions also are possible, such as the gamma or lognormal for positive-valued random variables, or the beta distribution for data restricted to the unit interval.

For example, the simple linear regression model (6.1) treats the mean as a linear function of the predictor variable, x. Thus $\mu_i = \beta_0 + \beta_1 x_i$, and $P = 1$ with two parameters: β_0 and β_1. As we saw earlier, β_0 corresponds to the Y-intercept, which is the value of the mean response when the predictor variable is zero, and β_1 corresponds to the slope, which is the unit increase in response associated with a single unit increase in x. In many instances the simple linear model provides a reasonable approximation to the true dose–response relationship between μ and x, at least for the dose range under study. Testing for increasing (or decreasing) dose response is then tantamount to testing $H_0: \beta_1 = 0$ vs. $H_a: \beta_1 > 0$ (or vs. $H_a: \beta_1 < 0$), and this is accomplished using the methods described in Section 6.1.

When it is felt that the simple linear model in (6.1) is not a good approximation to the true mean response, more flexible, **nonlinear models** must be considered. In most cases, these are curvilinear extensions or generalizations of (6.1). An additional set of $P - 1$ parameters, $\beta_2, \ldots,$

β_P, is then employed as part of the model relationship. In this more general form, we say $\mu_i = h(x_i;\boldsymbol{\beta})$, where $h(x;\boldsymbol{\beta})$ is a pre-specified function thought to represent adequately the relationship between μ and x.

7.1.1 Polynomial response models

Perhaps the most natural extension of the simple linear model (6.1) is to add a series of higher-order terms to the mean response function. This produces a **polynomial function** of dose for μ_i:

$$\mu_i = \beta_0 + \beta_1(x_i - \bar{x}) + \beta_2(x_i - \bar{x})^2 + \cdots + \beta_P(x_i - \bar{x})^P, \qquad (7.1)$$

$i = 1, \ldots, n$. Special cases of (7.1) are the quadratic model with $P = 2$,

$$\mu_i = \beta_0 + \beta_1(x_i - \bar{x}) + \beta_2(x_i - \bar{x})^2;$$

and the cubic model with $P = 3$,

$$\mu_i = \beta_0 + \beta_1(x_i - \bar{x}) + \beta_2(x_i - \bar{x})^2 + \beta_3(x_i - \bar{x})^3.$$

The quadratic model ($P = 2$) is useful when the dose-response pattern changes from strictly increasing (decreasing) over one dose region to strictly decreasing (increasing) over a subsequent dose region: a simple **nonmonotone curvature**. The cubic model ($P = 3$) is useful if the dose–response relationship exhibits a pattern such as increasing-decreasing-increasing over three contiguous, nonoverlapping intervals of doses; see, for example, Neter *et al.* (1996, Fig. 7.5). In this latter case, we say the dose response exhibits an **inflection point** at which its *rate of change* shifts from strictly decreasing to strictly increasing.

It is often unlikely that the true mean response function is exactly quadratic or cubic in x_i. (Exceptions occur when some underlying biomathematical mechanism induces the polynomial forms, such as solutions to differential equations, etc.) Nonetheless, these polynomial functions may approximate well the curvature for the dose range under study. As such, it is often best to view polynomial models as tools to provide useful approximations of the data at hand, rather than exact models whose coefficients have strict biological interpretations.

Notice the use of the centered doses, $x_i - \bar{x}$, in (7.1). This affects the interpretation of the intercept parameter: β_0 now represents the mean response at the average level of the predictor variables, $x = \bar{x}$. Nonetheless, centering is a common and important adjustment, since it reduces instabilities that occur when the values of x_i provide overlapping information contained in their higher powers. That is, when values of x_i are very similar to their corresponding values of x_i^2, etc., one encounters large statistical correlations among the least squares estimators. This effect is known as **multicollinearity**, and it can create computational instabilities in the LS estimators for $\boldsymbol{\beta}$ (Bradley and Srivastava, 1979; Neter *et al.*, 1996, Section 7.7). The effect is most simply corrected by centering the predictor variable via $x_i - \bar{x}$ (although see Exercise 7.4). In any polynomial regression, we recommend always centering the x-variable about it mean, $\bar{x}$.

Note that multicollinearity is not a concern under the simple linear model from (6.1): $\mu_i = \beta_0 + \beta_1 x_i$. When $P = 1$ there is, in effect, only one x-term in the model, and so overlapping information from another x-term does not exist. Thus, for example, in (6.1) we did not chose to center the x-variable. (If we had, the maximum likelihood estimator for β_1 would not have changed, although the estimator for β_0 simplifies to $\bar{Y}$. Hence, inferences on the significance of any trend in x are unaffected. Either form of the model, with or without centering, is acceptable when $P = 1$.)

Given a set of observations, Y_i, and of associated doses, x_i, we take $Y_i \sim$ (indep.) $N(\mu_i, \sigma^2)$, where μ_i is given in (7.1). This is a **polynomial regression model**. The process of fitting this model to the data produces estimators, $\hat{\boldsymbol{\beta}}$, which are used to predict values of the mean response at any x_i, or to test for a significant dose response, etc. The most common fitting algorithm used is least squares (LS) from Section 2.3.2, and under a normal parent assumption the LS estimators correspond to ML estimators as well. Unfortunately, for most values of P it is difficult to express the LS/ML estimators for (7.1) in simple closed forms. Their calculation is facilitated in this case by use of common statistical computer programs, such as SAS PROC REG or PROC GLM, the S-PLUS `lm` function (Chambers, 1992b; Venables and Ripley, 1997) or the GLIM statistical system (Francis *et al.*, 1993). This is not unusual: as we move into the later chapters, we will see that advanced parametric models for environmental data often require some sort of computer calculation in order to estimate their unknown parameters.

Under the normal parent assumption for the data, estimators for $\boldsymbol{\beta}$ in (7.1) from an LS fit also correspond to ML estimators, and most computer outputs assume this feature to produce standard errors of the LS/ML estimators. This allows for construction of confidence intervals or simple tests of hypotheses on $\boldsymbol{\beta}$. For example, with $P = 2$ in (7.1), the case of no dose response corresponds to $H_0: \beta_1 = \beta_2 = 0$. Using standard regression methodology (Neter *et al.*, 1996, Chapters 6–7), H_0 may be tested via an F-test. This corresponds to a comparison of **discrepancy measures** from two competing models for the mean response. Notice that the null hypothesis $H_0: \beta_1 = \beta_2 = 0$ implies that the **reduced model** $\mu_i = \beta_0$ is sufficient for describing the mean response, while the alternative hypothesis implies that the full model

$$\mu_i = \beta_0 + \beta_1(x_i - \bar{x}) + \beta_2(x_i - \bar{x})^2$$

is a better descriptor of the mean response as a function of the predictor variable x_i. Let SSE(RM) represent the sum of squared errors after fitting the reduced model when the null hypothesis is assumed true, that is, $\text{SSE}(RM) = \sum_{i=1}^{n}(Y_i - \tilde{\beta}_0)^2$, where $\tilde{\beta}_0$ is the LS estimator of β_0 under H_0. (For this simple null hypothesis, this is $\tilde{\beta}_0 = \bar{Y}$.) In addition, let SSE($FM$) represent the sum of squared errors after fitting the full model when the alternative hypothesis is assumed true, that is,

$$\text{SSE}(FM) = \sum_{i=1}^{n}\left\{Y_i - \left[\hat{\beta}_0 - \hat{\beta}_1(x_i - \bar{x}) - \hat{\beta}_2(x_i - \bar{x})^2\right]\right\}^2,$$

where $\hat{\beta}_0$, $\hat{\beta}_1$, and $\hat{\beta}_2$ are the unrestricted LS estimators of β_0, β_1, and β_2, respectively. Notice that the reduced model is hierarchically **nested** within the full model in that it is a special case of the latter, with $\beta_1 = \beta_2 = 0$. The comparison of such a hierarchy of models is a critically important concept in the analysis of linear and generalized linear models. To test the null hypothesis of no dose response, $H_0: \beta_1 = \beta_2 = 0$, we compare SSE($RM$) and SSE($FM$), using the F-ratio

$$F_{\text{calc}} = \frac{\frac{1}{Q}[\text{SSE}(RM) - \text{SSE}(FM)]}{\frac{1}{n-3}\text{SSE}(FM)},$$

where $1/Q$ in the numerator corresponds to the Q parameters set equal to zero under the reduced model; here $Q = 2$. Reject H_0 if $F_{\text{calc}} > F_\alpha(Q, n - 3)$. If this occurs, we conclude that either $\beta_1 \neq 0$, or $\beta_2 \neq 0$, or both $\beta_1 \neq 0$ and $\beta_2 \neq 0$.

In general, suppose a full model contains $P + 1$ unknown parameters (including the intercept). To compare it to a reduced model with Q fewer parameters, use the F-ratio

$$F_{\text{calc}} = \frac{\frac{1}{Q}[\text{SSE}(RM) - \text{SSE}(FM)]}{\frac{1}{n-(P+1)}\text{SSE}(FM)}.$$

Reject the null hypothesis that the Q regression coefficients all equal zero if $F_{\text{calc}} > F_\alpha(Q, n - [P + 1])$. We discuss F-tests for regression models in further detail in Section 8.1.3, below.

Other null hypotheses could be tested in this manner; for example,

$$\mu_i = \beta_0 + \beta_1(x_i - \bar{x})$$

is another reduced model that could be compared to the full model described above. This reduced model suggests that the quadratic term is not needed, given the presence of a linear term in the model. Or,

$$\mu_i = \beta_0 + \beta_2(x_i - \bar{x})^2$$

is a third reduced model that could be compared to the full model. In this case, the relationship between mean response and the predictor variable is purely quadratic with no linear component. (Such a reduced model would be unusual, however; many modeling strategies force retention of lower-order terms when any higher-order terms are included.)

We caution the reader that use of polynomial models can be taken to extremes: as higher-order terms are added to (7.1), the polynomial function becomes more and more flexible, but also less and less valid as an approximation to some simple underlying curvilinearity. It is unusual and typically unwise to employ (7.1) past the cubic ($P = 3$) order. (If, of course, there is strong environmental or biomathematical motivation for moving to higher-order polynomials, then these should be included in the modeling process.)

Example 7.1 Quadratic regression – Panther hormone concentrations
In Exercise 6.7, we presented data associated with a study of hormone concentrations in wild Florida panthers (*Felis concolor coryi*) as a function of the animals' ages, x. In this example, we view x as the 'dose.' The response variable was the ratio of blood-level estriadol to blood-level testosterone (each in pg/ml). For male panthers, $n = 8$ observations give a sample mean age of $\bar{x} = 4.875$ years. (Refer to Exercise 6.7 for the data, or see the SAS code in Fig. 7.1.)

Ratio responses often are lognormally distributed, so that a logarithmic transformation yields normally distributed data. Thus we take the natural logarithms of the hormone ratios as our Y-variable. A quick examination of the transformed data shows that the log ratios are increasing, but that they may depart from simple linearity, reaching a peak response at early ages, then dropping back. As a simple approximation, then, consider a quadratic model, $P = 2$ in (7.1), for these data. A short SAS program to perform the fit using SAS PROC REG is given in Fig. 7.1. Notice use of the mean-adjusted predictor variable `age-agebar`.

The SAS output (edited for presentation) corresponding to the code in Fig. 7.1 appears in Fig. 7.2. From the figure, we can construct a test of the need for a quadratic term, *given the presence of the linear term*, by comparing the SSEs from appropriate full and reduced models. From Fig. 7.2, the discrepancy measures are SSE(FM) = 0.12376 (using `MODEL1`) and

SSE(RM) = 0.32183 (using MODEL2). The difference in SSEs has $Q = 1$ df, while SSE(FM) has $n - 3$ df. The corresponding F-ratio is

$$F_{\text{calc}} = \frac{\text{SSE}(RM) - \text{SSE}(FM)}{\frac{1}{n-3}\text{SSE}(FM)} = \frac{0.32183 - 0.12376}{\frac{1}{5}(0.12376)} = \frac{0.19807}{0.02475},$$

or $F_{\text{calc}} = 8.0028$. The associated P-value is $\text{P}[F(1,5) > 8.0028] = 0.0367$. This is significant at $\alpha = 0.05$, suggesting that a quadratic term is needed above and beyond a linear term.

```
* SAS code to fit Panther data from Exercise 6.7;
data mpanther;
input age ratio @@;
y = log(ratio);
    agebar = 4.875 ;
          x = age - agebar;
          x2 = x*x;
cards;
 1  0.42   2  0.39   2  0.54  3  0.65   3  0.75
 9  0.56   9  0.53  10  0.48
* Use 2 MODEL statements;
* 1st for full model, and 2nd for reduced model;
proc reg;
   model y = x x2;      * Fit quadratic model;
   model y = x;         * Fit reduced linear model;
```

Fig. 7.1 SAS PROC REG program for quadratic regression fit and linear regression fit to data in Exercise 6.7.

A number of other features in Fig. 7.2 correspond to this same inference on the value of the quadratic term. For example, in the first `Parameter Estimates` section, the t-test result associated with the X2 component is the signed square root of F_{calc}, with the same P-value. Equivalently, a 95% confidence interval for the β_2 parameter associated with the quadratic term is $\hat{\beta}_2 \pm t_{0.025}(5)se[\hat{\beta}_2]$. The estimated LS regression coefficient for the quadratic term is $\hat{\beta}_2 = -0.0325$, with standard error $se[\hat{\beta}_2] = 0.0115$, so the interval for β_2 is $-0.0325 \pm (2.571)(0.0115)$, or $-0.062 < \beta_2 < -0.003$. Since the interval does not contain zero, we infer again that a significant quadratic effect due to age appears in these data.

The hormone data are plotted in Fig. 7.3 along with the fitted quadratic regression model $\hat{\mu}_i = -0.235 + 0.056(x_i - \bar{x}) - 0.033(x_i - \bar{x})^2$; $\bar{x}$ is 4.875.

```
                       The SAS System
                      The REG Procedure
Model: MODEL1
Dependent Variable: Y
                     Analysis of Variance
                      Sum of          Mean
Source     DF       Squares         Square    F Value     Prob>F
Model       2       0.19982        0.09991      4.037     0.0905
Error       5       0.12376        0.02475
C Total     7       0.32357
                      Parameter Estimates
                 Parameter Standard   T for H0:
Variable DF  Estimate    Error  Parameter=0   Prob > |T|
INTERCEP  1  -0.23465  0.15257    -1.538        0.1847
X         1   0.05554  0.02408     2.307        0.0692
X2        1  -0.03252  0.01150    -2.829        0.0367

Model: MODEL2
Dependent Variable: Y
                     Analysis of Variance
                      Sum of          Mean
Source     DF       Squares         Square    F Value     Prob>F
Model       1       0.00174        0.00174      0.032     0.8629
Error       6       0.32183        0.05364
C Total     7       0.32357
                      Parameter Estimates
                 Parameter Standard   T for H0:
Variable DF  Estimate    Error  Parameter=0   Prob > |T|
INTERCEP  1 -0.636553  0.08188    -7.774        0.0002
 X        1  0.004199  0.02329     0.180        0.8629
```

Fig. 7.2 Output (edited) from SAS PROC REG quadratic regression fit (MODEL1) and linear regression fit (MODEL2) to data in Exercise 6.7.

Graphically, this model fit appears adequate; however, since there is a substantial gap in age values in the region of greatest curvature, the fitted response pattern between ages 3 and 9 should be viewed with caution. A residual plot for this model fit is generally reasonable, although the residual for the lower response ($y_2 = 0.39$) at age 2 is much larger in absolute value than any other residual, and that point may require additional consideration (Exercise 7.4c). In fact, the overall F-test of $H_0: \beta_1 = \beta_2 = 0$ shows only weak evidence of any effect: from the SAS output in Fig. 7.2, $F_{calc} = 4.037$. The associated P-value is only 0.0905, significant at $\alpha = 0.10$ but not at $\alpha = 0.05$. The evidence in favor of the simple quadratic as a good

Fig. 7.3 A plot of the ratio of blood-level estriadol to blood-level testosterone (each in pg/ml) versus panther age with superimposed quadratic regression model fit.

approximation to the log-response is marginal here, and other model machinations may provide a better fit to the data (Exercise 7.4). Indeed, collection and incorporation of data from 4–8-year-old panthers would provide greater clarity for this modeling exercise, and perhaps the best conclusion to draw is that more data are needed. □

It is possible to extended the basic polynomial model in (7.1) to more than one input variable. For instance, in Example 7.1, besides x_{i1} = age's effect on hormone levels in *F. concolor coryi*, interest may also involve the effects of an environmental variable, such as x_{i2} = temperature, on the animals' hormone levels. Or, in a laboratory toxicity experiment, two different environmental chemicals may be studied for their joint toxic effects on an organism; the dose regimen may involve exposure to a dose, x_{i1}, of the first chemical, and another dose, x_{i2}, of the second. In both cases, the mean response can be modeled as a polynomial function of both x_{i1} and of x_{i2}. Commonly, this is limited to powers involving squares, for example,

$$\begin{aligned}\mu_i = {} & \beta_0 + \beta_1(x_{i1} - \bar{x}_1) + \beta_2(x_{i1} - \bar{x}_1)^2 + \\ & \gamma_1(x_{i2} - \bar{x}_2) + \gamma_2(x_{i2} - \bar{x}_2)^2 + \delta_{12}(x_{i1} - \bar{x}_1)(x_{i2} - \bar{x}_2).\end{aligned}$$

(Notice the inclusion of an **interaction term**, $(x_{i1} - \bar{x}_1)(x_{i2} - \bar{x}_2)$, in the model.) Since this function graphs as a three-dimensional surface over the joint range of x_1 and x_2, the model is referred to as a **response surface** (Neter *et al.*, 1996, Section 7.7). It can have useful applications in modeling curvature for environmental data with two input variables.

We will find it useful to refer back to the polynomial response model in (7.1), primarily as a tool for incorporating curvilinearity in different model scenarios. The general topic of polynomial regression is covered in great depth throughout the statistical and applied sciences literature, however, and we will not pursue it further. Additional details are available in regression texts such as Draper and Smith (1981, Section 5.1) or Neter *et al.* (1996, Section 7.7).

7.1.2 Threshold models and other nonlinear forms

The polynomial response model in (7.1) possesses an important characteristic: it is linear in its unknown parameters. That is, although the model contains polynomial terms in the predictor variable, x, the unknown parameters enter into the model only as simple products ('linearly') of the polynomial predictors. This linear feature allows us to apply broad regression programs such as SAS PROC GLM or PROC REG to fit (7.1).

There are, however, many other mathematical functions that can be used in $\mu_i = \hbar(x_i;\boldsymbol{\beta})$ for which this parameter linearity is not valid. When combined with the distributional specification $Y_i \sim$ (indep.) $N(\mu_i, \sigma^2)$, $i = 1,\ldots, n$, this produces a **nonlinear regression model**. Here again, the variances σ^2 are assumed constant, although further extension to function-specific variances, $\sigma^2(x)$, is also possible (Neter *et al.*, 1996, Section 10.1).

Common examples of nonlinear regression models include exponential forms such as $\hbar(x_i;\boldsymbol{\beta}) = \beta_0 + \exp\{\beta_1 x_i\}$, or $\hbar(x_i;\boldsymbol{\beta}) = \beta_0\exp\{\beta_1 x_i\} + \beta_2\exp\{\beta_3 x_i\}$, which arise in the analysis of pharmacokinetic or toxicokinetic data (Bailer and Piegorsch, 1990; Becka *et al.*, 1992). Others include power-function forms $\hbar(x_i;\boldsymbol{\beta}) = \beta_0 + \beta_1 x_i^{\beta_2}$, with β_2 unknown, which can arise in studies of allometric scaling (Hertzberg and Miller, 1985). When all the β-parameters are unknown, models such as these cannot be written as linear in the βs, so a nonlinear regression analysis is required. To perform a model fit, iterative computer calculation is necessary. Numerous computer packages are available to fit nonlinear models, such SAS PROC NLIN and S-PLUS `nls` (Bates and Chambers, 1992; Venables and Ripley, 1997). The programs produce, typically, LS estimators for $\boldsymbol{\beta}$, and, under normal parent assumptions on Y_i, also give standard errors useful for constructing confidence intervals, significance tests, etc. One note of caution is warranted, however: with some extreme nonlinear forms, large correlations are induced among the LS parameter estimators, making them somewhat difficult to interpret or to use for certain predictive inferences. Caution is advised when fitting complex nonlinear models.

One particular nonlinear model that has received much attention in the analysis of environmental and biological data is the **threshold response model** (Cox, 1987; Gaylor *et al.*, 1988; Ulm, 1991) As the name suggests, the model is based on the concept of a response threshold dose: a dose or other measurable stimulus below which none of the experimental units respond. (If different thresholds are assumed to exist for each organism and this attribute has a distribution across a population of organisms, then a consideration of models for **tolerance distributions** becomes relevant; see Section 7.2.1.) After the dose exceeds the threshold, the response changes as a function of the dose, x. For example, an organism may be able to withstand a small toxic insult with no discernible damage – due to efficient metabolism or repair mechanisms, etc. – but, after a build-up of or substantial exposure to the toxin occurs, the organism may respond to the toxic insult in some linear or nonlinear fashion.

In its general form, the threshold model is

$$\mu_i = \begin{cases} \gamma & \text{if } x \leq \tau \\ \hbar(x_i - \tau; \boldsymbol{\beta}) & \text{if } x > \tau, \end{cases} \tag{7.2}$$

where τ is the **threshold point** or **threshold dose**, γ is the pre-threshold mean response, and $\hbar(\cdot;\boldsymbol{\beta})$ is the post-threshold mean response, modeled to vary as a function of the difference $x - \tau$, for $x > \tau$. For instance, a widely seen form for (7.2) employs a simple linear post-threshold function:

$$\mu_i = \begin{cases} \gamma_0 & \text{if } x \leq \tau \\ \gamma_1 + \beta_1(x_i - \tau) & \text{if } x > \tau. \end{cases}$$

For simplicity, we can absorb terms not including x_i into the post-threshold intercept parameter, and write the model as

$$\mu_i = \begin{cases} \gamma_0 & \text{if } x \leq \tau \\ \beta_0 + \beta_1 x_i & \text{if } x > \tau. \end{cases} \tag{7.3}$$

This relationship is a special case of a **segmented linear regression** (Hudson, 1966; Gallant and Fuller, 1973; Lerman, 1980), since the threshold point τ splits the mean response into two linear segments. If the model is constrained to force $\beta_0 = \gamma_0 - \beta_1\tau$, the mean response is continuous at all values of x. This parameterization is somewhat colorfully termed a **hockey stick regression model** (Takeshima and Yanagimoto, 1978; Yanagimoto and Yamamoto, 1979). Generalizations of (7.3) include **piecewise linear regression** (Tishler and Zang, 1981) or **bilinear segmented regression** (Guegan and Pham, 1989) models.

In general, the threshold model (7.2) and even its special piecewise linear form (7.3) represent true nonlinear models, since for unknown τ it is impossible to write the models as linear functions of the unknown parameters. (In some cases, the pre-threshold response γ_0 may be known; a common example is $\gamma_0 = 0$, where no response is expected below $x = \tau$. This simplification does not alleviate the nonlinear nature of the model, however.) Thus, fitting these models to dose-response data requires computer calculation (Nakamura, 1986). Under the distributional assumption that $Y_i \sim$ (indep.) $N(\mu_i, \sigma^2)$, the LS estimators are also ML estimators, for which large-sample standard errors are calculable. Most computer programs report these values. With them, one can perform inferences on the dose response; for example, to test for no dose response past τ, test $H_0{:}\beta_1 = 0$. From the computer output, find the point estimate $\hat{\beta}_1$ and the standard error $se[\hat{\beta}_1]$, and calculate the Wald statistic $W_{calc} = \hat{\beta}_1/se[\hat{\beta}_1]$. Reject H_0 in favor of an increasing dose response past τ if $W_{calc} > \mathfrak{z}_\alpha$.

Example 7.2 Hockey-stick regression – Mercury toxicity data

Exposure to phenylmercuric compounds used in fungicides and pesticides can lead to adverse toxic effects in mammals, including humans. For example, consider the data displayed in Table 7.1 (Piegorsch, 1987a), from a study of accidental phenylmercury exposure in infants. The toxic outcome studied was increased activity of the renal tubular enzyme γ-glutamyl transpeptidase (γ-GT) as a function of mercury (Hg) concentration in urine. A segmented linear model is motivated here by the supposition that low levels of mercury may be tolerated with little or no toxicity, but after some point the system cannot detoxify the metal properly and a toxic response (indicated by increased γ-GT activity) ensues.

Figure 7.4 illustrates a segmented toxic effect with these data over the Hg concentrations. We employ the segmented threshold model in (7.3) with the constraint that the dose response is continuous, so $\beta_0 = \gamma_0 - \beta_1\tau$. Interest in these data concerns statistical estimation of the three unknown parameters: the plateau value γ_0, the post-threshold slope β_1, and the unknown threshold dose τ. (Notice that the predictor variable, x_i, is taken as the log-Hg concentrations, in order to make the predictor scale more equi-spaced.)

Table 7.1 Neonatal toxic response as renal γ-GT activity (U/l) after Hg exposure (μg/l)

Hg concentration, c_i	$x_i = log(c_i)$	γ-*GT*, Y_i
2.5	0.916	13.57
22	3.09	15.33
60	4.09	16.92
90	4.50	16.08
105	4.65	17.14
144	4.97	18.85
178	5.18	18.63
210	5.35	19.30
233	5.45	25.77
256	5.55	28.44
300	5.70	35.46
400	5.99	45.26

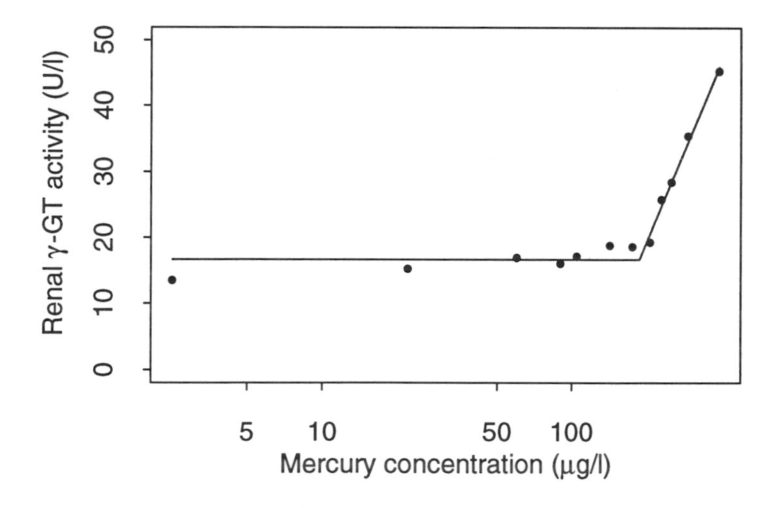

Fig. 7.4 Scatterplot of renal γ-GT activity after Hg exposure, with threshold model fit superimposed.

To analyze these data, we can employ nonlinear regression routines such as SAS PROC NLIN or S-PLUS `nls`. Commonly, nonlinear regression programs require specification of the dose-response function, here (7.3), and also its first derivatives with respect to each unknown parameter. To find these expressions we use indicator functions, as described in Section

1.3.1. Under thc continuity constraint $\beta_0 = \gamma_0 - \beta_1\tau$, (7.3) may be expressed as $\mu = \gamma_0 + \beta_1(x - \tau)I_{(\tau,\infty)}(x)$, where $I_{(\tau,\infty)}(x)$ equals 1 when $x > \tau$, and 0 when $x \le \tau$ (Exercise 7.8). From this, the first derivatives with respect to γ_0 and β_1 are $\partial\mu/\partial\gamma_0 = 1$ and $\partial\mu/\partial\beta_1 = (x - \tau)I_{(\tau,\infty)}(x)$, respectively. The first derivative with respect to τ is not as obvious, however. To find it, recognize that $\mu = \gamma_0$ for $x \le \tau$, and $\mu = \gamma_0 + \beta_1(x - \tau) = \gamma_0 + \beta_1 x - \beta_1\tau$ for $x > \tau$. Since the indicator $I_{(\tau,\infty)}(x)$ is identical to the indicator $I_{(-\infty,x)}(\tau)$, we can write μ as a function of τ: $\mu(\tau) = \gamma_0 + \beta_1(x - \tau)I_{(-\infty,x)}(\tau)$. Thus, the slope of this function with respect to τ is $-\beta_1$ over $x > \tau$, and 0 over $x \le \tau$. This can be expressed compactly as $\partial\mu/\partial\tau = -\beta_1 I_{(\tau,\infty)}(x)$.

SAS code to fit this model to the data from Table 7.1 is given in Fig. 7.5. As given therein, the program requires specification of initial parameter estimates; to form these, we initialized the threshold dose as the log-concentration corresponding to $x_7 = 178$ ($\tau_{\text{start}} = \log\{178\} = 5.818$), the plateau response as the observed response at that point ($y_7 = 18.63$), and the right-segment slope as Piegorsch's (1987a) estimate of 38.88. We chose to employ the nonlinear fitting method discussed by Marquardt (Levenberg, 1944; Marquardt, 1963) – an option available in the PROC NLIN statement.

```
* SAS code to fit mercury data from Table 7.1;
  data mercury;
  input hgconc y @@;
     x = log(hgconc);
  cards;
   2.5  13.57     22 15.33     60  4.09     90  4.50
   105  4.65     144  4.97    178  5.18    210  5.35
   233  5.45     256  5.55    300 35.46    400 45.26

  proc nlin method=marquardt;
  parameters  g = 18.63
              b = 38.88
              t = 5.1818;
  if x <= t then do;                       *left plateau;
              model y = g ;
              der.g = 1;    der.b = 0;
              der.t = 0;    end;
  else do;                                 *post-threshold;
          model y = g + b*(x-t);
          der.g = 1;        der.b = x-t;
          der.t = -b;       end;
```

Fig. 7.5 SAS PROC NLIN program for threshold segmented regression fit to Hg toxicity data in Table 7.1.

```
                     The SAS System
                   The NLIN Procedure

        Non-Linear Least Squares Iterative Phase
Dependent Variable Y    Method: Marquardt
Iter          G            B          T    Sum of Squares
 0    18.630000    38.880000   5.181800        146.579343
 1    16.662122    39.054715   5.244192         24.662222
 2    16.645690    39.265732   5.246076         24.648048
 3    16.645714    39.267023   5.246080         24.648047
 4    16.645714    39.267023   5.246080         24.648047
NOTE: Convergence criterion met.

Non-Linear Least Squares Summary Statistics
Dependent Variable Y
Source                DF Sum of Squares      Mean Square
Regression             3   7086.8332528     2362.2777509
Residual               9     24.6480472        2.7386719
Uncorrected Total     12   7111.4813000
(Corrected Total)     11   1002.6844250

Parameter  Estimate  Asymptotic        Asymptotic 95 %
                     Std. Error   Confidence Interval
                                  Lower         Upper
        G  16.64571     0.62549   15.23074   18.06069
        B  39.26702     3.29097   31.82227   46.71177
        T   5.24608     0.03908    5.15766    5.33450

           Asymptotic Correlation Matrix
Corr                  G              B               T
---------------------------------------------------------
  G                   1   1.668956E-16     0.407559752
  B        1.668956E-16              1     0.775466386
  T         0.407559752    0.775466386               1
```

Fig. 7.6 Output (edited) from SAS PROC NLIN threshold segmented regression fit to Hg toxicity data in Table 7.1.

The SAS output (edited for presentation) appears in Fig. 7.6, where the LS parameter estimates appear under the heading `Estimate`. We find $\hat{\gamma}_0 = 16.646$, $\hat{\beta}_1 = 39.267$, and $\hat{\tau} = 5.246$. (In terms of the original Hg scale, the latter quantity gives an estimated threshold point of $\exp\{5.246\} = 189.806$ μg/l.) The mean prediction equation is then $\hat{y}_i = 16.646 + 39.267(x_i - 5.246)I_{(5.246,\infty)}(x_i)$, with x_i on a log-Hg scale.

Inferences associated with the point estimates are based on the assumption that the parent distribution of the original data is normal, with mean $\mu_i = \gamma_0 + \beta_1(x_i - \tau)I_{(\tau,\infty)}(x_i)$, and with constant variance σ^2. If this is correct, then the intervals given under the heading `Asymptotic 95 % Confidence Interval` provide approximate inferences on the unknown parameters. For example, a 95% confidence interval for the slope of the right linear segment is $31.822 < \beta_1 < 46.712$. Since these limits do not contain zero, we infer that the post-threshold response is significantly increasing. Also, confidence limits for the threshold point in units of the original Hg concentration can be found by taking the antilog of the limits given for τ; the simple computation gives an interval for the threshold between $\exp\{5.158\} = 173.758$ and $\exp\{5.334\} = 207.268$ μg/l.

Note that PROC NLIN also allows for nonlinear LS fits without the need to determine the derivatives of the dose-response function: simply use

```
proc nlin method = dud;
```

and do not specify the `der.*` terms (Ralston and Jennrich, 1979). This analysis can also be performed in S-PLUS via the `nls` function, for example,

```
Hg.conc<-c(2.5,22,60,90,105,144,178,210,233,256,
           300,400)
Y.gammaGT<-c(13.57,15.33,16.92,16.08,17.14,18.85,
             18.63,19.30,25.7,28.44,35.46,45.26)
mercury.df<-data.frame(Hg.conc=Hg.conc,
                 log.conc=log(Hg.conc),resp=Y.gammaGT)
Hg.nls.fit<-nls(resp~g0+b1*(log.conc-tau)*
                  (log.conc>tau),data=mercury.df,
start=list(g0=18.63,b1=38.88,tau=5.1818),trace=T)
summary(Hg.nls.fit)
```

□

More generally, a two-line segmented regression with no constraints on the slope of either linear segment can be modeled as

$$\mu_i = \begin{cases} \gamma_0 + \gamma_1 x_i & \text{if } x_i \leq \tau \\ \beta_0 + \beta_1 x_i & \text{if } x_i > \tau \,. \end{cases} \tag{7.4}$$

If model (7.4) is constrained to force $\beta_0 = \gamma_0 + \tau(\gamma_1 - \beta_1)$, the mean response is continuous at all values of x.

Example 7.2 (continued) Two-line segmented regression

Returning to the mercury toxicity data in Table 7.1, it is instructive to examine the residuals, $y_i - \hat{y}_i$, from the fit in Fig. 7.4, using $\hat{y}_i = 16.646 + 39.267(x_i - 5.246)I_{(5.246,\infty)}(x_i)$. Plotted against x_i, these values are scattered evenly and with no apparent pattern, but only after the threshold, that is, for $x > 5.246$. To the left of the threshold, the residuals are almost always below zero, and increase slowly to zero as x increases (Exercise 7.13). This suggests that there may be an increasing linear response in the data below the threshold point τ, requiring fit of the two-line model (7.4). We include the continuity restriction, forcing $\beta_0 = \gamma_0 + \tau(\gamma_1 - \beta_1)$.

Fitting model (7.4) to the data requires again a nonlinear LS regression program. As above, the program may require derivatives of μ with respect to the unknown parameters. Applying similar indicator functions to those used with model (7.3), we find that model (7.4) with $\beta_0 = \gamma_0 + \tau(\gamma_1 - \beta_1)$ may be expressed as

$$\mu = \gamma_0 + \gamma_1 x I_{(-\infty,\tau)}(x) + \{\tau\gamma_1 + \beta_1(x-\tau)\}I_{(\tau,\infty)}(x).$$

The first derivatives are $\partial\mu/\partial\gamma_0 = 1$, $\partial\mu/\partial\gamma_1 = xI_{(-\infty,\tau)}(x) + \tau I_{(\tau,\infty)}(x)$, $\partial\mu/\partial\beta_1 = (x - \tau)I_{(\tau,\infty)}(x)$, and, by manipulating indicator functions as above, $\partial\mu/\partial\tau = (\gamma_1 - \beta_1)I_{(\tau,\infty)}(x)$.

An example of SAS PROC NLIN code using these expressions is given in Fig. 7.7. The associated output (edited for presentation) appears in Fig. 7.8. The results in Fig. 7.8 are similar to those in Fig. 7.6, particularly as concerns the point estimates and confidence limits on β_1 and τ. Note, however, that the estimated slope of the left linear segment, $\hat{\gamma}_1 = 1.144$, appears to be significant: the large-sample 95% confidence limits, $0.544 < \gamma_1 < 1.745$, do not contain the value $\gamma_1 = 0$. Hence it appears that applying the more complex model is important for these data.

```
* SAS code to fit mercury data from Table 7.1;

data mercury;
input hgconc y @@;
   x = log(hgconc);
cards;
  2.5  13.57     22 15.33    60  4.09     90  4.50
  105  4.65     144  4.97   178  5.18    210  5.35
  233  5.45     256  5.55   300 35.46    400 45.26

proc nlin method=marquardt ;
parameters  g0 = 12.73
            g1 =  1.06
            b1 = 38.88
            t = 5.1818 ;

if x <= t then do;            * left linear segment;
            model y = g0 + g1*x ;
            der.g0 = 1;   der.g1 = x;
            der.b1 = 0;   der.t  = 0;
        end;
else do;                      * right linear segment;
        model y = g0 + g1*t + b1*(x-t);
        der.g0 = 1;           der.g1 = t;
        der.b1 = x-t;         der.t  = g1 - b1;
        end;
```

Fig. 7.7 SAS PROC NLIN program for bilinear segmented regression fit to Hg toxicity data in Table 7.1.

```
                    The SAS System
                  The NLIN Procedure

       Non-Linear Least Squares Iterative Phase
Dependent Variable Y     Method: Marquardt
Iter        G0        G1        B1         T    Sum of Squares
 0      12.730     1.060    38.880     5.182         85.207943
 1      12.306     1.103    38.950     5.280          7.276739
 2      12.170     1.143    39.265     5.286          7.220299
 3      12.165     1.144    39.267     5.286          7.220264
 4      12.165     1.144    39.267     5.286          7.220264
NOTE: Convergence criterion met.

Non-Linear Least Squares Summary Statistics
Dependent Variable Y
Source                DF  Sum of Squares      Mean Square
Regression             4    7104.2610355     1776.0652589
Residual               8       7.2202645        0.9025331
Uncorrected Total     12    7111.4813000
(Corrected Total)     11    1002.6844250

Parameter  Estimate   Asymptotic         Asymptotic 95 %
                      Std. Error    Confidence Interval
                                    Lower            Upper
      G0   12.16532      1.08097    9.67257     14.65807
      G1    1.14433      0.26041    0.54381      1.74485
      B1   39.26702      1.88923   34.91041     43.62364
       T    5.28603      0.02355    5.23171      5.34034

            Asymptotic Correlation Matrix
Corr            G0            G1            B1             T
-----------------------------------------------------------
  G0             1      -0.94322   5.0243E-16      -0.24214
  G1      -0.94322             1  -4.7390E-16       0.39755
  B1   5.0243E-16   -4.7390E-16            1        0.67685
   T      -0.24214       0.39755      0.67685             1
```

Fig. 7.8 Output (edited) from SAS PROC NLIN bilinear segmented regression fit to Hg toxicity data in Table 7.1.

This analysis can be replicated in S-PLUS (without specifying derivatives) by using the following commands:

```
Hg.2seg.fit<-nls(resp~ g0+g1*log.conc*(log.conc<=tau)+
  (g1*tau + b1*(log.conc-tau))*(log.conc>tau),
   data=mercury.df,start=list(g0=12.73,g1=1.006,
   b1=38.88,tau=5.1818),trace=T)
```

In closing, we note that if the join point is known to occur *at* one of the observed dose levels ($\tau = x_i$ for some i), then Draper and Smith (1981, Section 5.4) suggest a simple methodology for fitting linear segment regressions. Their approach assumes the join point is *known*, and uses

indicator variables in any standard linear regression program (such as SAS PROC REG). They point out that this approach can be extended to the case of unknown join points by applying their method to all possible join points (all possible values of x_i, $i = 1, \ldots, n$), and choosing the result with the lowest residual sum of squares. This would, by definition, yield LS estimators for the unknown parameters. This method can be extended to produce confidence interval estimates for the threshold dose τ (Venzon and Moolgavkar, 1988; Ulm, 1991). □

It is possible also to extend the threshold model (7.2) to include multiple segmentation points. For instance, there may be a threshold dose, τ_0, below which the mean response is constant, and also a second threshold, τ_1, above which the mean response plateaus. We call τ_1 in this case the **plateau point**, and the response model represents a **plateau regression**: $Y_i \sim$ (indep.) $N(\mu_i, \sigma^2)$, where

$$\mu_i = \begin{cases} \gamma_0 & \text{if } x \leq \tau_0 \\ \hbar(x_i - \tau; \boldsymbol{\beta}) & \text{if } \tau_0 < x < \tau_1 \\ \gamma_1 & \text{if } x \geq \tau_1 \end{cases},$$

$i = 1, \ldots, n$. Here again, for unknown τ_0 and τ_1, the model cannot be written as linear in the unknown parameters, and computer calculation is necessary to find LS or ML estimators.

7.1.3* Nonmonotone ('umbrella') models: A case study

An interesting example of curvilinear, and possibly nonlinear, response behavior occurs when environmental data exhibit a **nonmonotone dose response**. For example, the data may respond in an increasing fashion over low doses, but then a **downturn in dose response** occurs at higher doses. In terms of the mean responses, μ_i, across ordered doses x_i, this is

$$\mu_1 \leq \mu_2 \leq \cdots \leq \mu_v \geq \mu_{v+1} \geq \cdots \geq \mu_n \tag{7.5}$$

with some strict inequality among the μ_is. This form of dose response is known as an **umbrella model** (Mack and Wolfe, 1981). (The non-monotonicity may also decrease, then increase.) We discussed nonparametric trend analysis of umbrella models in Section 6.7. Here, we note some parametric models that can describe an umbrella response.

Perhaps the simplest parametric model that satisfies the umbrella form (7.5) is the quadratic version ($P = 2$) of (7.1): $\mu_i = \beta_0 + \beta_1(x_i - \bar{x}) +$

$\beta_2(x_i - \bar{x})^2$ (cf. Example 7.1). In many cases, the nonmonotone dose response will be smooth enough so that the quadratic model provides at least a good approximation to the true dose response over the dose range studied. If so, use of (7.1) with $P = 2$ is warranted.

The quadratic model is not always the best form to use in approximating a nonmonotone dose response, however. Indeed, many curvilinear shapes can satisfy (7.5), yet depart strongly from the limited curvature of a parabola. In this case, more complex, highly nonlinear response functions may be considered (possibly in conjunction with some form of segmented model, as in Section 7.1.2). As an example, we present a case study of a complex nonmonotone model associated with the *Salmonella* mutagenicity data from Example 3.7.

Example 7.3 Nonlinear dose-response models for Salmonella *mutagenesis*

Consider again the *Salmonella* mutagenicity assay described in Examples 3.7 and 6.13. Recall that the *S. typhimurium* bacteria used in this mutagenesis assay are affected by both mutation and nonmutagenic toxicity after exposure to an environmental toxin. That is, the assay is structured so that bacteria will grow after expressing mutation, but they remain susceptible to cellular toxicity, causing some to die even if they have mutated. At low doses, the toxicity is not strong enough to affect the mutagenic response (so an increasing dose response occurs), but at higher doses mortality in the microbes is sufficient to reduce the response and create a downturn. We saw an example of this effect in Table 3.2.

A number of authors have discussed parametric models to describe the unusual downturn in dose response for this assay; Broekhoven and Nestmann (1991) and Edler (1992) give comprehensive reviews. Fundamental to the modeling formulation is the recognition that only those microbes that survive the cellular toxicity have the opportunity to mutate. For instance, suppose that out of N_0 plated microbes, the proportion that succumb to overt toxicity and die is π_D. Suppose also that of the surviving proportion, $(1 - \pi_D)$, a further proportion, π_M, mutate. Assuming independence between the toxicity and mutagenicity mechanisms, the probability of observing a mutation is a product of these probabilities: the probability of first *not* dying, $(1 - \pi_D)$, times the probability of mutating given that the microbe does not die, π_M. The corresponding mean response is the number of plated microbes times this probability, $\mu_i = N_0(1 - \pi_D)\pi_M$. Additional adjustments for long-term, residual effects of chemical toxicity lead to slightly more complex forms for μ_i (Margolin *et al.*, 1981; Krewski *et al.*, 1993), but the basic principles do not change. The result is a class of nonlinear product models describing the competing risks of mutagenicity and mortality on the mean response μ_i at dose x_i. A simple form was given by Stead *et al.* (1981) and Myers *et al.* (1981):

$$\mu_i = N_0 \exp\{-\beta_2 x_i\}(1 - \exp\{-[\beta_0 + \beta_1 x_i]\}) , \tag{7.6}$$

where the first exponential term is the probability of first not dying (1 minus the probability of dying), and the second exponential term is the probability of mutation. (For the *Salmonella* assay, N_0 is taken to be 10^8.)

The parameters in model (7.6) must satisfy the constraint that the exponential terms are contained in the interval between 0 and 1, since they represent probabilities. Both forms are based on **single-hit kinetics** to explain the manner in which a microbe is exposed and then responds to a toxic insult; that is, a single 'hit' of the toxin produces some form of result, be it death or mutation. This concept is standard in radiation mutagenesis (Haynes *et al.*, 1984; Kruglikov *et al.*, 1993), and it often leads to response functions of the form $1 - e^{-h(x)}$ (Hoel, 1985).

Margolin *et al.* (1981) updated the mortality term in (7.6) to account for long-term, residual effects of chemical toxicity in the microbes, and gave a modification of (7.6) as

$$\mu_i = N_0 [2 - \exp\{-\beta_2 x_i\}]_+ (1 - \exp\{-[\beta_0 + \beta_1 x_i]\}) \tag{7.7}$$

where $[a]_+$ is the **positive part** of a, $[a]_+ = \max\{0,a\}$, and implicit in the model is the constraint $\beta_1 \geq 0$.

The critical parameters in the competing-risk product models (7.6) and (7.7) are β_1, which is the 'slope' parameter due to mutagenicity, and β_2, which is the 'slope' parameter due to cellular toxicity. (β_0 is related to the background mutagenicity rate.) Thus, $\beta_2 = 0$ implies $\exp\{-\beta_2 x_i\} = 1$ which, in turn, implies no overt toxicity. To test for increasing mutagenicity, set the null hypothesis to $H_0: \beta_1 = 0$ and the alternative to $H_a: \beta_1 > 0$.

Under the parametric response model in (7.7), Margolin *et al.* (1989) proposed a negative binomial parent distribution for the mutation counts generated by the assay, in order to allow for response variability in excess of that predicted by a Poisson parent model. They gave details on a likelihood ratio methodology for testing H_0 vs. H_a. (They noted, however, that the LR test suffers some small-sample instabilities, and suggest caution in its use.)

Other authors have noted from previous studies that a power transformation of the form

$$V_i = \frac{Y_i^{0.2} - 1}{0.2} = 5\left(Y_i^{1/5} - 1\right) \tag{7.8}$$

stabilizes the variance in the Y_is and brings their distribution closer to the normal (Snee and Irr, 1984; Eastwood, 1993). Of interest is LS/ML estimation of the unknown parameters in the transformed mean function $5(\mu_i^{1/5} - 1)$, with μ_i taken from (7.6) or (7.7). A standard nonlinear LS regression program can then estimate the unknown parameters in μ_i.

For example, recall the data in Table 3.2 on *S. typhimurium* mutation counts after exposure to the airborne toxin 1,3-butadiene. Applying the power transform (7.8) to the original data leads to the transformed variates V_i, as given in Table 7.2.

Table 7.2 Power-transformed *Salmonella* mutagenicity data from Table 3.2

Dose (ppm)	*Transformed replicate observations*			*Average*
0.000	4.10	4.93	4.67	4.57
0.002	7.56	7.35	7.27	7.39
0.007	8.57	8.09	9.09	8.58
0.014	9.65	8.97	9.14	9.25
0.020	9.06	8.69	9.17	8.98
0.030	8.69	8.67	8.60	8.66

We model the mean of the transformed variates in Table 7.2 via the function (7.6) under the power transform (7.8), that is, as

$$\mu_i = 5N_0^{1/5}\exp\{-\beta_3 x_i\}(1 - \exp\{-\beta_0 - \beta_1 x_i\})^{1/5} - 5, \qquad (7.9)$$

where for simplicity we have absorbed one of the powers of 0.2 into the first exponent, yielding the new parameter $\beta_3 = \beta_2/5$. Figure 7.9 presents SAS PROC NLIN code for performing the nonlinear LS fit. Notice our use of the fitting option `method=dud`, which does not require specification of partial derivatives for the response function. The nonlinearity of (7.9) is so severe that we chose to simplify the presentation (but in return accept greater computing time) by moving to this fitting option.

The model in (7.9) also demands care when applying a nonlinear fitting routine. In particular, poor initial estimates can lead to lack of convergence, or even to a stationary point that does not represent the true global LS minimum. For (7.9), we suggest the following steps for finding initial estimates of β_0, β_1, and β_3 (although other formulations are possible):

1. Begin with the original nonlinear model in (7.6), and recognize that near $x \approx 0$, the function is $\mu_i \approx N_0\exp\{-\beta_2(0)\}(1 - \exp\{-[\beta_0 + \beta_1(0)]\})$, or simply $N_0(1 - \exp\{-\beta_0\})$. Thus, use Y_1 (or, if there are replicate observations at $i = 1$, $\overline{Y}_{1+}$) to estimate μ at $x = 0$. Then, $Y_1 \approx N_0(1 - \exp\{-\beta_0\})$. This leads to the initial estimate

$$b_0 = -\log\left\{1 - \frac{Y_1}{N_0}\right\},$$

where N_0 is 10^8 for the *Salmonella* assay. Notice that since we are operating on the original scale, we use the original observations, Y_i, and not the transformed values, V_i, in the calculation for b_0.

2. Note also that near $x \approx 0$, the mutagenicity term dominates; that is, when $\exp\{-\beta_2 x\} \approx 1$, $\mu \approx N_0(1 - \exp\{-\beta_0 - \beta_1 x\})$. The derivative with respect to x is then $\partial\mu/\partial x \approx N_0\beta_1\exp\{-\beta_0 - \beta_1 x\}$. Near $x \approx 0$, this is approximately $\partial\mu/\partial x \approx N_0\beta_1\exp\{-\beta_0\}$. Given an estimate of the initial slope of the dose response near zero, say $m = (Y_2 - Y_1)/(x_2 - x_1)$, we solve for β_1 and substitute $b_0 = -\log\{1 - (Y_1/N_0)\}$ from step (a) for β_0. This achieves the initial estimate

$$b_1 = \frac{m}{N_0 - N_0 Y_1}.$$

(Again, if replicate observations are available at $i = 1,2$, use $\overline{Y}_{1+}$ and $\overline{Y}_{2+}$ where appropriate.)

3. Lastly, consider the downturn portion of (7.6), where we expect the mortality term to dominate. Indeed, for large x, the mutagenicity term $1 - \exp\{-\beta_0 - \beta_1 x\}$ is approximately 1, so the dose response at large values of x is approximately $\mu \approx N_0\exp\{-\beta_2 x\}$. Thus if the highest dose studied is x_n, take Y_n as an estimate of μ_n and solve for β_2: $b_2 = -\log\{Y_n/N_0\}/x_n$. For use in (7.9), we have $\beta_3 = \beta_2/5$, so take $b_3 = b_2/5$ in

$$b_3 = \frac{-\log\{Y_n/N_0\}}{5x_n}.$$

These operations lead to values for the data in Table 7.2 of $b_0 = 2.6\times10^{-7}$, $b_1 = 3.4\times10^{-4}$, and $b_3 = 89.312$, and these are the initial estimates used in Fig. 7.9. The resulting SAS output (edited for presentation) appears in Fig. 7.10, where the LS estimate of the mutagenicity 'slope' is $\hat{\beta}_1 = 3.18\times10^{-7}$. If the transformed variates are normally distributed with constant variance (at least approximately), then approximate 95% confidence limits for β_1 are given as $2.57\times10^{-4} < \beta_1 < 3.78\times10^{-4}$. (One way to assess the normality assumption is to graph a histogram or boxplot of the residuals, $V_i - \hat{\mu}_i$, and check that they appear roughly symmetric and unimodal; see Neter *et al.* (1996, Section 3.3).) Since the 95% confidence limits on β_1 do not contain zero, a significant mutagenic effect is evidenced.

Before closing, we note that for small values of a, a first-order Taylor approximation (Edwards and Penney, 1990, Section 11.3) for $1 - e^{-a}$ gives

```
* SAS code to fit transformed data from Table 7.2;
 data salm;
 input x y @@;
    v = 5*((y**0.2) - 1);      * transform Y observations;
    N0tr = 10**1.6;            * transform N0 = 10^8;
 cards;
  0.000    20    0.000   31    0.000   27
  0.002   100    0.002   92    0.002   89
  0.007   147    0.007  123    0.007  178
  0.014   216    0.014  170    0.014  181
  0.020   176    0.020  154    0.020  183
  0.030   154    0.030  153    0.030  149

 proc nlin method=dud ;
 parameters    b0 =   0.00000026
               b1 =   0.00034
               b3 =   89.312 ;
 model v=(5*N0tr*exp(-b3*x)*((1-exp(-b0-b1*x))**.2))-5;
```

Fig. 7.9 SAS PROC NLIN program for power-transformed regression fit to data in Table 7.2.

$1 - e^{-a} \approx a$, so that both (7.6) and (7.7) may be approximated via simpler nonlinear forms; for example, (7.6) becomes $\mu_i \approx N_0 \exp\{-\beta_2 x_i\}(\beta_0 + \beta_1 x_i)$. This approximation may be useful when testing H_0 vs. H_a. Many authors have noted this effect; Leroux and Krewski (1993) review these and other aspects of nonlinear modeling for *Salmonella* mutagenicity data. □

As the preceding example illustrates, a wealth of mathematical forms is available to model nonmonotone dose response. (We note an additional example below in (7.18), as part of our presentation on potency estimation.) These dose-response models are typically highly nonlinear in their unknown parameters, and fitting them to downturns requires computer calculation. Programs such as SAS PROC NLIN or S-PLUS `nls` can be useful, particularly if the parent distribution of the data is assumed normal (or if some transformation of the data drives them towards normality, as in Example 7.3). We caution, however, that the extreme nonlinearity of these forms can lead to instabilities in the model fit, including difficulties in achieving convergence of the estimators, sensitivity to initial estimates that start the fitting process, and, possibly, high correlations and large standard errors among the parameter estimates. To prevent being misled by complex statistical artifacts of the nonlinear fit, data analysts must proceed through these nonlinear fitting routines carefully and prudently.

```
                     The SAS System
                   The NLIN Procedure

Non-Linear Least Squares DUD Initialization/Dependent Variable V
   -4   0.0000002600   0.000340  89.312000  1166.762717
   -3   0.0000002860   0.000340  89.312000  1165.208946
   -2   0.0000002600   0.000374  89.312000  1149.802276
   -1   0.0000002600   0.000340  98.243200  1277.592492
    0   0.0000002600   0.000374  89.312000  1149.802276
    1   0.0000002680   0.000227  61.789486   826.536432
    2   0.0000002731   0.000163  44.777204   580.981312
    3   0.0000001942   0.000244  17.790242    47.074521
    4   0.0000001502   0.000370  18.864635    23.488673
    5   0.0000001515   0.000378  19.063252    23.386129
    6   0.0000001947   0.000343  12.950872     3.491695
    7   0.0000002518   0.000342  13.221994     2.314741
    8   0.0000002677   0.000322  13.067580     2.133764
    9   0.0000002688   0.000319  12.924342     2.126951
   10   0.0000002693   0.000318  12.877414     2.126307
   11   0.0000002721   0.000318  12.889703     2.126224
   12   0.0000002716   0.000317  12.874236     2.125761
   13   0.0000002712   0.000318  12.877459     2.125701
   14   0.0000002712   0.000318  12.874923     2.125700
   15   0.0000002713   0.000318  12.875080     2.125700
   16   0.0000002713   0.000318  12.875061     2.125700
 NOTE: Convergence criterion met.

  Non-Linear Least Squares Summary Statistics
  Dependent Variable V
  Source                DF  Sum of Squares      Mean Square
  Regression             3    1170.1513196      390.0504399
  Residual              15       2.1256998        0.1417133
  Uncorrected Total     18    1172.2770194
  (Corrected Total)     17    47.4415257
                                             Asymptotic 95 %
  Param. Estimate       Asymptotic       Confidence Interval
                        Std. Error       Lower          Upper
   B0  0.00000027 0.00000002975  0.00000021  0.00000034
   B1  0.00031756 0.00002834760  0.00025714  0.00037798
   B3 12.87506094 0.88585229863 10.98691863 14.76320324

            Asymptotic Correlation Matrix
  Corr                     B0             B1            B3
  -----------------------------------------------------------
   B0                       1    -0.30426235   -0.21635434
   B1             -0.30426235              1    0.88981229
   B3             -0.21635434     0.88981229             1
```

Fig. 7.10 Output (edited) from SAS PROC NLIN power-transformed regression fit to data in Table 7.2.

7.2 DOSE-RESPONSE MODELS ON A DISCRETE SCALE

When the response data are discrete, restrictions on the dose-response model must be imposed in order to preserve interpretability in the statistical outcomes. Two such instances are models for response probabilities with dichotomous data, and models for mean rates of response for count data. In the former case, the dose-response model must lead to probabilities that lie between 0 and 1, while in the later case the mean response rate must be positive. We explore some common dose-response models for these two settings in this section.

7.2.1 Dichotomous outcomes: Logistic regression

As in Section 6.3.1, suppose we observe a proportion, $p_i = Y_i/N_i$, across ordered doses, x_i $(i = 1, \ldots, n)$. We assume a binomial parent distribution for the observations: $Y_i \sim$ (indep.) $\mathcal{Bin}(N_i,\pi_i)$. Suppose interest centers on estimation and testing of dose-response features for π_i when it is represented as a function of x_i. This is referred to as the **quantal response** scenario (Morgan, 1993). For example, how can we test the homogeneity null hypothesis $H_0:\pi_1 = \cdots = \pi_n$ vs. the increasing trend alternative $H_a:\pi_1 \leq \cdots \leq \pi_n$ (with some strict inequality among the π_is under H_0)?

Obviously, any parametric dose-response model for π_i must incorporate the feature that π_i is a probability between 0 and 1. Many authors recognize this, and turn to some form of readily available (continuous) parametric c.d.f. from Section 1.3, since these functions are also restricted to lie between 0 and 1. In addition, since c.d.f.s must be monotone nondecreasing functions, their double duty as parametric quantal response functions allows for construction of straightforward tests of H_0 vs. H_a. Thus, one often models π_i as

$$\pi_i = F(\beta_0 + \beta_1 x_i), \tag{7.10}$$

where $F(\cdot)$ is some c.d.f., x_i is the ith dose, and the β-parameters are unknown. The null hypothesis of no dose response corresponds to $H_0: \beta_1 = 0$, while the choice of $F(\cdot)$ is made typically to incorporate $H_a: \beta_1 > 0$.

Before continuing, we note that by modeling the dose response via a c.d.f, a **tolerance interpretation** may be ascribed to the response probabilities, π_i. That is, each organism possesses some intrinsic tolerance to the toxic stimulus, which, if exceeded, leads the organism to respond. If this tolerance varies across the population of organisms, we can associate the **tolerance distribution** with some c.d.f. $F(\cdot)$ (Morgan, 1993, Section 1.5; Jorgensen, 1994). The proportion, π_i, of organisms expected to respond to dose x_i is the proportion of organisms in the population with tolerances less than some function of that dose. If the function of x_i is taken as a simple linear form, $\beta_0 + \beta_1 x_i$, then $\pi_i = F(\beta_0 + \beta_1 x_i)$.

In some cases, mechanistic or other environmental factors may drive the selection of a specific tolerance distribution. If so, this simplifies use of the dose-response model in (7.10). Specification of tolerance distributions is not required in quantal response modeling, however, and we employ them only when the subject matter suggests sensible forms for their use.

Whether or not driven by tolerance distribution considerations, a common choice of a parametric model used in (7.10) is the simple normal c.d.f. from Section 1.3.4: $F(z) = \Phi(z)$. The choice of a normal tolerance distribution leads to $\pi_i = \Phi(\beta_0 + \beta_1 x_i)$, which is the simple linear **probit regression model**. If the normal tolerance distribution has mean μ and variance σ^2, then the linear probit regression parameters β_0 and β_1 are related to μ and σ^2 via $\beta_0 = -\mu/\sigma$ and $\beta_1 = 1/\sigma$.

Alternatively, recall the simple logistic c.d.f. from Section 1.3.6: $F(u) = 1/(1 + e^{-u})$, valid over $-\infty < u < \infty$. Employed in (7.10) with a simple linear predictor $\beta_0 + \beta_1 x_i$, this recovers the linear logistic model in (6.7):

$$\pi_i = \frac{1}{1 + \exp\{-\beta_0 - \beta_1 x_i\}}, \qquad (7.11)$$

$i = 1, \ldots, n$. We say this represents a simple linear **logistic regression model** for π_i. Using the simple linear predictor $\beta_0 + \beta_1 x_i$, the hypothesis H_0: $\beta_1 = 0$ is equivalent to H_0: $\pi_1 = \cdots = \pi_n$, while H_a: $\beta_1 > 0$ corresponds to an increasing dose response. (Obviously, additional terms can be included in the term for the linear predictor; for example, a cubic polynomial logistic regression model is $\pi_i = 1/(1 + \exp\{-\beta_0 - \beta_1 x_i - \beta_2 x_i^2 - \beta_3 x_i^3\})$ (Morgan, 1993, Section 4.3).) The logit and probit models are very similar, often producing comparable point estimates for π_i. However, logistic regression corresponds to a form of **natural parameterization** – also called a **canonical link** (see Table 8.1) – used with quantal response, and we will center attention on the logistic model in (7.11).

In general, to test H_0: $\beta_1 = 0$ vs. H_a: $\beta_1 \neq 0$ under (7.11), a number of options are available. We recommend two: first, as discussed in Section 6.3.1, the score test of H_0 leads to use of the Cochran–Armitage statistic in (6.6). Since we know that this test statistic possesses many optimality properties, it is the method of choice in this simple linear logistic setting. The Cochran–Armitage test is not easily extended (for example, to Pth-order polynomial predictors), however. As such, we also recommend the more flexible likelihood ratio test from Section 2.4.1: recall that the likelihood function under the binomial sampling model is

$$\mathcal{L}(\pi_1, \ldots, \pi_n \,|\, \mathbf{y}) = \prod_{i=1}^{n} \binom{N_i}{y_i} \pi_i^{y_i}(1-\pi)_i^{N_i - y_i}, \tag{7.12}$$

(cf. Example 2.1). By modeling π_i as $1/(1 + \exp\{-\beta_0 - \beta_1 x_i\})$, we reduce the number of unknown parameters from n to 2, and the log-likelihood associated with (7.12), $\ell(\beta_0, \beta_1)$, becomes

$$\sum_{i=1}^{n} \left\{ \log\binom{N_i}{y_i} + y_i(\beta_0 + \beta_1 x_i) - (N_i)\log[1 + \exp\{\beta_0 + \beta_1 x_i\}] \right\}. \tag{7.13}$$

Thus, to test $H_0{:}\beta_1 = 0$ vs. $H_a{:}\beta_1 \neq 0$, we form the LR statistic $G^2_{\text{calc}} = -2\{\ell(\tilde{\beta}_0, 0) - \ell(\hat{\beta}_0, \hat{\beta}_1)\}$, where $\hat{\beta}_0$ and $\hat{\beta}_1$ are the (unrestricted) ML estimates of the β-parameters, and $\tilde{\beta}_0$ is the ML estimate of β_0 under the null restriction that $\beta_1 = 0$. For (7.11), $\tilde{\beta}_0 = \log\{\bar{p}/(1 - \bar{p})\}$, where $\bar{p} = Y_+/N_+$ estimates the underlying (constant) response probability under H_0, as in Section 6.3.1. Unfortunately, the unrestricted ML estimates do not possess closed-forms, and computer calculation is required for their solution. Useful programs for this purpose include SAS PROC LOGISTIC, PROC PROBIT (with `d=logistic` specified) and PROC GENMOD (under `distribution=binomial` and `link=logit`), S-PLUS `glm` (Hastie and Pregibon, 1992; Venables and Ripley, 1997), or the GLIM system (Francis *et al.*, 1993).

In passing, recall that G^2_{calc} is a difference between two log-likelihoods (times 2), say $G^2_{\text{calc}} = 2\log\{\mathcal{L}(\hat{\boldsymbol{\beta}} \,|\, \mathbf{Y})\} - 2\log\{\mathcal{L}(\tilde{\boldsymbol{\beta}} \,|\, \mathbf{Y})\}$, where $\tilde{\boldsymbol{\beta}} = [\tilde{\beta}_0\ 0]$ is the ML estimator under the reduced model, and $\hat{\boldsymbol{\beta}} = [\hat{\beta}_0\ \hat{\beta}_1]$ is the ML estimator under the full model. Suppose we define a **discrepancy measure** by how much any model's likelihood deviates from the fullest possible model fit. This latter quantity is achieved by fitting each datum to its own separate parameter, giving a log-likelihood of the form $\log\{\mathcal{L}(\mathbf{Y} \,|\, \mathbf{Y})\}$. (With proportion data, the fullest possible model would have n separate probability parameters, $\pi_1, \ldots, \pi_n$. It is, in a sense, the best model fit possible, but also the least informative for summarizing the data.) The corresponding model discrepancy is

$$D(\hat{\boldsymbol{\beta}}) = 2\log\{\mathcal{L}(\mathbf{Y} \,|\, \mathbf{Y})\} - 2\log\{\mathcal{L}(\hat{\boldsymbol{\beta}} \,|\, \mathbf{Y})\};$$

we call this a **deviance function**. With this, suppose we compare two different model fits by taking the difference of their deviances: say $D(\tilde{\boldsymbol{\beta}})$ –

$D(\hat{\boldsymbol{\beta}})$, using the ML estimators from the reduced and full models, above. Since the common term $2\log\{\mathcal{L}(\mathbf{Y}\,|\,\mathbf{Y})\}$ cancels,

$$D(\tilde{\boldsymbol{\beta}}) - D(\hat{\boldsymbol{\beta}}) = 2\log\{\mathcal{L}(\hat{\boldsymbol{\beta}}\,|\,\mathbf{Y})\} - 2\log\{\mathcal{L}(\tilde{\boldsymbol{\beta}}\,|\,\mathbf{Y})\}$$

is simply G^2_{calc}. Thus a difference in deviances can be viewed as an LR statistic, as noted earlier in Section 2.4.1. For computational purposes, it is common to see deviances $D(\hat{\boldsymbol{\beta}})$ or some portions thereof, such as $-2\log\{\mathcal{L}(\hat{\boldsymbol{\beta}}\,|\,\mathbf{Y})\}$, presented as part of computer outputs. From these, LR statistics may be calculated. We explore this construction in greater detail in the next chapter.

Once calculated, $\hat{\beta}_0$ and $\hat{\beta}_1$ are substituted into (7.13), and the LR/deviance statistic G^2_{calc} is computed. In large samples G^2_{calc} is approximately χ^2, with *df* equal to the number of parameters specified in H_0. For testing $\beta_1 = 0$, this is $df = 1$. Reject H_0 when $G^2_{\text{calc}} > \chi^2_{\alpha}(1)$. Notice that this is, in effect, a two-sided rejection region: the LR/deviance statistic ignores the sign of $\hat{\beta}_1$ and rejects H_0 in favor of any departure, $H_a\colon \beta_1 \neq 0$.

As noted above, the LR/deviance statistic G^2_{calc} is easier to extend to higher-order polynomial models, or to more than one x-variable, than the Cochran–Armitage score statistic (6.6). This flexibility motivates its greater use in standard logistic regression programs. As the sample size grows to infinity, however, both methods produce the same inference on β_1. We say in this case that the two procedures are **asymptotically equivalent**.

Example 7.4 Micronuclei in rodent cells – Logistic regression analysis
Return to the study of micronucleus induction by *tert*-butylhydroquinone (*t*BHQ) in Chinese hamster cells, discussed in Example 6.5. We consider here application of the logistic regression model (7.11) to the quantal response data in Table 6.4. Figure 7.11 gives a sample SAS program to perform the fit, using SAS PROC LOGISTIC. The resulting output (edited for presentation) appears in Fig. 7.12.

```
* SAS code to fit tBHQ data from Example 6.5;
 data tBHQ;
 input dose y @@;     n = 2000;
 cards;
 0      8     120   9     240   9
 480   18     600  39     720  44

 proc logistic;
 model y/n = dose / link=logit  covb ;
```

Fig. 7.11 SAS PROC LOGISTIC code for logistic regression fit to data in Example 6.5.

```
                          The SAS System
                      The LOGISTIC Procedure

Response Variable (Events): Y
Response Variable (Trials): N
Number of Observations: 6
Link Function: Logit

                         Response Profile
                  Ordered  Binary
                    Value  Outcome       Count
                        1  EVENT           127
                        2  NO EVENT      11873

                 Criteria for Assessing Model Fit
                               Intercept
              Intercept           and
Criterion       Only          Covariates    Chi-Square for Covariates
AIC           1409.964         1355.380              .
SC            1417.356         1370.165              .
-2 LOG L      1407.964         1351.380     56.584 with 1 DF (p=0.0001)
Score             .                .        53.427 with 1 DF (p=0.0001)

          Analysis of Maximum Likelihood Estimates
                   Parameter   Standard       Wald          Pr >
Variable   DF    Estimate       Error     Chi-Square    Chi-Square
INTERCPT    1     -5.8067       0.2345      613.0089        0.0001
DOSE        1     0.00283     0.000412       47.2266        0.0001

                Standardized
 Variable         Estimate       Odds Ratio
 INTERCPT             .             0.003
 DOSE             0.405001          1.003

                  Estimated Covariance Matrix
       Variable              INTERCPT                DOSE
       INTERCPT          0.0550033111         -0.000089399
       DOSE              -0.000089399         1.7001932E-7
```

Fig. 7.12 Output (edited) from SAS PROC LOGISTIC logistic regression fit to data in Example 6.5.

As Fig. 7.12 indicates, a significant dose response due to *t*BHQ is evidenced in these data. The LR/deviance statistic is found in the row labeled −2 LOG L: $G^2_{\text{calc}} = 56.584$, on 1 *df*. This tests the null hypothesis of no dose response, $H_0: \beta_1 = 0$, vs. the alternative of any difference, $H_a: \beta_1 \neq 0$; we reject H_0 when $G^2_{\text{calc}} > \chi^2_\alpha(1)$. At, say, $\alpha = 0.01$, we have $\chi^2_{0.01}(1) = 6.63$, and rejection is clear. (This is supported by the parenthetical output after the test statistic in the output: the P-value is listed at the program's default for very small P-values: any $P \leq 0.0001$ is reported as `p = 0.0001`.)

The score statistic for testing H_0 is provided on the computer output in the row labeled `Score`: $X^2_{calc} = 53.427$, on 1 *df*. As expected, this corroborates the LR inference and shows that the dose response is very significant. We also should expect the (signed) square root of this value to equal the Cochran–Armitage statistic from (6.6). Indeed, $\sqrt{53.427} = 7.309$, which matches our previous computation of $z_{calc} = 7.309$ from Example 6.5. (In effect, the score test in SAS PROC LOGISTIC provides another means of constructing the Cochran–Armitage trend statistic.) Again, a strongly significant result is obtained.

Note use of the `covb` option in the `model` statement in Fig. 7.11. This calls for display of the large-sample variance-covariance matrix of the ML estimators (at the bottom of Fig. 7.12). In particular, the estimated covariance of $\hat{\beta}_0$ and $\hat{\beta}_1$ is given as the off-diagonal term in this matrix display: $c_{01} = -8.94 \times 10^{-5}$. This value is useful in calculating **potency** estimators for quantal response data, as we will see in Section 7.3. □

One additional test we caution *against* using in this setting is the Wald test from Section 2.4.2. Although we recommend the Wald approach in many settings, it suffers from certain instabilities when employed to test $H_0: \beta_1 = 0$ for the simple linear logistic regression model (7.11). Recall that the statistic takes the form $W_{calc} = \hat{\beta}_1 / se[\hat{\beta}_1]$, and that rejection occurs when W_{calc} is greater than the upper-α critical point from a standard normal distribution. Unfortunately, as described by Hauck and Donner (1977), in this case W_{calc} can be shown to decrease to zero as the true value of β_1 increases sufficiently far from zero (that is, far away from H_0). As a result, the power of the Wald test actually *decreases* with extreme departures from H_0. Hauck and Donner note that the LR/deviance statistic G^2_{calc} does not suffer from this effect, and they, as we, recommend its use over Wald statistics in simple linear logistic regression. Væth (1985) gives general conditions for the Wald test to exhibit this poor performance; see also Fears *et al.* (1996).

7.2.2 Counts: Log-linear regression

For discrete data in the form of counts, we considered in Section 6.4 a Poisson parent distribution: $Y_{ij} \sim$ (indep.) $\mathcal{Poisson}(\mu_i)$, $i = 1, \ldots, n$; $j = 1, \ldots, r_i$. Recall that the Poisson rate parameters, μ_i, are assumed positive, so once again some restriction must be imposed on models used to describe μ_i

as a function of dose x. A common choice is to employ some form of exponential function, such as $\mu_i = \exp\{\hbar(x_i;\boldsymbol{\beta})\}$. If we choose for $\hbar(x_i;\boldsymbol{\beta})$ a simple linear predictor function, we have

$$\mu_i = \exp\{\beta_0 + \beta_1 x_i\} \tag{7.14}$$

or, in terms of the linear predictor,

$$\log(\mu_i) = \beta_0 + \beta_1 x_i,$$

which is known as a **log-linear model** for μ_i.

To use the log-linear model in (7.14) for dose-response testing, notice that the null hypothesis of homogeneity in the μ_is corresponds to $H_0{:}\beta_1 = 0$, while the alternative of a functional dose response corresponds to $H_a{:}\beta_1 \neq 0$. The LR test of H_0 vs. H_a is based on the Poisson log-likelihood

$$\ell(\mu_1, \ldots, \mu_n) = \sum_{i=1}^{n} r_i\left\{\bar{y}_{i+}\log(\mu_i) - \mu_i\right\} - \sum_{i=1}^{n}\sum_{j=1}^{r_i} \log(y_{ij}!)\ ,$$

where $\bar{y}_{i+} = \sum_{j=1}^{r_i} y_{ij}/r_i$ is the sample mean at the ith dose level. Substituting the log-linear specification (7.14) into this log-likelihood achieves

$$\ell(\beta_0, \beta_1) = \sum_{i=1}^{n} r_i\left\{\bar{y}_{i+}(\beta_0 + \beta_1 x_i) - \exp\{\beta_0 + \beta_1 x_i\}\right\} - \sum_{i=1}^{n}\sum_{j=1}^{r_i} \log(y_{ij}!). \tag{7.15}$$

The LR/deviance statistic is $G^2_{\text{calc}} = -2\{\ell(\tilde{\beta}_0, 0) - \ell(\hat{\beta}_0, \hat{\beta}_1)\}$, where $\hat{\beta}_0$ and $\hat{\beta}_1$ are unrestricted ML estimates based on maximizing (7.15), and $\tilde{\beta}_0$ is the ML estimate of β_0 under $H_0{:}\beta_1 = 0$. In this simple case, $\tilde{\beta}_0 = \log\{\bar{y}_{++}\} = \log\left\{\sum_{i=1}^{n}\sum_{j=1}^{r_i} y_{ij}/\sum_{i=1}^{n} r_i\right\}$. Unfortunately, $\hat{\beta}_0$ and $\hat{\beta}_1$ do not possess closed-forms, and computer calculation is required. Again, useful programs for this purpose include SAS PROC GENMOD, S-PLUS `glm`, and the GLIM statistical system.

Once calculated, $\hat{\beta}_0$ and $\hat{\beta}_1$ are substituted into (7.15), and the LR/deviance statistic G^2_{calc} is computed. In large samples, $G^2_{\text{calc}} \mathrel{\dot\sim} \chi^2(1)$ under H_0, so one rejects when $G^2_{\text{calc}} > \chi^2_\alpha(1)$. Notice that this is, again, a two-sided rejection region: the LR/deviance statistic ignores the sign of $\hat{\beta}_1$ and rejects H_0 in favor of any departure, $\beta_1 \neq 0$.

Other methods useful for testing $H_0{:}\beta_1 = 0$ under (7.14) include the Wald test and the score test. Both approaches yield inferences that are asymptotically equivalent to the LR test using G^2_{calc}. The Wald test is particularly simple to employ here: find the unrestricted ML estimator $\hat{\beta}_1$

and also its large-sample standard error, $se[\hat{\beta}_1]$. (These quantities are standard components of most log-linear model computer outputs.) Take $W_{calc} = \hat{\beta}_1/se[\hat{\beta}_1]$, and reject H_0 in favor of H_a when $W_{calc} > z_\alpha$. The corresponding P-value is approximately $1 - \Phi(W_{calc})$. Under a Poisson parent distribution on the observed counts, the Wald test is not known to be any less stable than the LR test (in contrast to the logistic regression case in Section 7.2.1); it may be recommended for use here as well. The score test is somewhat more difficult to compute, but would follow using the general approach from Section 2.4.3.

Example 7.5 Aquatic toxicity in C. dubia – *Log-linear model analysis*
Return to the study of reproductive toxicity in *C. dubia* after exposure to the herbicide nitrofen, discussed in Example 6.2. We consider here a log-linear regression analysis for the data given in Table 6.2. Specifically, take x_i = nitrofen dose (μg/l), and apply the log-linear model (7.14) to the observed offspring counts Y_{ij}. Figure 7.13 gives a sample SAS program to perform the fit, using SAS PROC GENMOD.

```
* SAS code to fit nitrofen data from Example 6.2;
data nitrofen;
input dose y @@;
cards;
  0 27     0 32     0 34     0 33     0 36
  0 34     0 33     0 30     0 24     0 31
 80 33    80 33    80 35    80 33    80 36
 80 26    80 27    80 31    80 32    80 29
160 29   160 29   160 23   160 27   160 30
160 31   160 30   160 26   160 29   160 29
235 23   235 21   235  7   235 12   235 27
235 16   235 13   235 15   235 21   235 17
310  6   310  6   310  7   310  0   310 15
310  5   310  6   310  4   310  6   310  5
proc genmod;
model y = dose / dist=poisson  link=log  type1;
```

Fig. 7.13 SAS PROC GENMOD program for log-linear regression fit for data in Example 6.2.

The corresponding output (edited for presentation) appears in Fig. 7.14. In particular, the LR/deviance statistic is generated from the `type1` option in the `model` statement in Fig. 7.13. This produces the output at the bottom of Fig. 7.14, where G^2_{calc} is given under `ChiSquare` in the row labeled DOSE: G^2_{calc} = 184.895. As noted above, this is the difference in deviances between the reduced model fit with just an Intercept term, and the full model including Dose: $D(\tilde{\beta}_0,0)$ = 312.484 and $D(\hat{\beta}_0,\hat{\beta}_1)$ = 127.589, respectively. (The comment that The scale parameter was held fixed in the output refers to an additional scale parameter used in generalized linear models; cf. Section 8.2.1. Under the Poisson parent model the scale parameter is a constant, set equal to 1.) The output also gives the associated (approximate) P-value, which is so small that the program lists it as 0.0000. (For presentation purposes, we write $P <$ 0.0001.) We find a significant dose response evidenced in these data.

```
                       The SAS System
                    The GENMOD Procedure
                     Model Information
       Distribution                          POISSON
       Link Function                         LOG
       Dependent Variable                    Y
       Observations Used                     50

          Criteria For Assessing Goodness Of Fit
Criterion                DF          Value        Value/DF
Deviance                 48       127.5887          2.6581
Pearson Chi-Square       48       111.9372          2.3320
Log Likelihood            .      2529.4687               .

              Analysis Of Parameter Estimates
Parameter  DF   Estimate   Std Err    ChiSquare  Pr>Chi
INTERCEPT   1     3.6352    0.0435    6973.0707  0.0000
DOSE        1    -0.0037    0.0003     176.1813  0.0000
NOTE:  The scale parameter was held fixed.

             LR Statistics For Type 1 Analysis
  Source        Deviance      DF   ChiSquare   Pr>Chi
  INTERCEPT     312.4840       0           .        .
  DOSE          127.5887       1    184.8953   0.0000
```

Fig. 7.14 Output (edited) from SAS PROC GENMOD simple log-linear regression fit for data in Example 6.2.

(A technical caveat: the SAS PROC GENMOD output lists a Log Likelihood as part of its Criteria For Assessing Goodness Of Fit. This is the maximized log-likelihood of the fitted model, and since it is a logarithm of a product of (estimated) probabilities, it should be less than zero. In Fig. 7.14, however, the quantity is given as the positive value 2529.4687, which is at first perplexing. For this Poisson model, however, SAS does not include quantities involving factorials in the log-likelihood. That is, the term $\sum_{i=0}^{k}\sum_{j=1}^{r_i}\log(y_{ij}!)$ is not subtracted out of the log-likelihood; cf. (7.15). This can lead the reported `Log Likelihood` in PROC GENMOD to display as positive.)

The square of the Wald statistic is given under `ChiSquare` in the row labeled `DOSE` in the `Analysis Of Parameter Estimates` section: $W^2_{calc} = (-0.0037/0.0003)^2 = 176.1813$. This statistic has 1 *df*, and its approximate *P*-value is also less than 0.0001 (see under `Pr>Chi` in the same section and row). In this example, the LR/deviance statistic and the Wald statistic lead to similar qualitative conclusions.

We can also examine whether additional curvature is evidenced in the dose response past the simple log-linear model for these data. To do so, we reapply the SAS code in Fig. 7.13, centering the dose variable,

```
dosebar  =  157;
doseadj  =  dose-dosebar;
```

adding a line in the DATA step to include the quadratic term,

```
dose2adj  =  doseadj*doseadj;
```

and refining the model statement, via

```
proc genmod;
model y = doseadj dose2adj/dist=poisson link=log type1;
```

The results from this fit (edited for presentation) appear in Fig. 7.15. There, the LR test for the additional importance of the quadratic term is found again under `ChiSquare`, now in the row labeled `DOSE2ADJ`: $G^2_{calc} =$ 71.7175. This is the difference in deviances between the reduced model fit with only `INTERCEPT` and `DOSEADJ` terms and that of the full model including `INTERCEPT`, `DOSEADJ`, and `DOSE2ADJ` terms. (In our $D(\cdot)$ notation, these are $D(\tilde{\boldsymbol{\beta}}) = 127.589$ and $D(\hat{\boldsymbol{\beta}}) = 55.8712$, respectively, where $\tilde{\boldsymbol{\beta}} = [\tilde{\beta}_0\ \tilde{\beta}_1\ 0]$ and $\hat{\boldsymbol{\beta}} = [\hat{\beta}_0\ \hat{\beta}_1\ \hat{\beta}_2]$; cf. Figs 7.15 and 7.16.) To test $H_0: \beta_2 = 0$ (assuming β_0 and β_1 are nonzero) vs. $H_a: \beta_2 \neq 0$, compare the LR/deviance statistic to a $\chi^2(1)$ reference distribution: the *P*-value is $P \approx P[\chi^2(1) > 71.7175] < 0.0001$. The quadratic term appears to add significantly in modeling the dose response, in addition to the linear and intercept terms.

Notice that the LS estimate of β_2 is given as −0.0000 with associated standard error 0.0000. Obviously, these terms are not equal to zero, they are simply less than 5×10^{-5} in absolute value. (The actual ML estimate is $\hat{\beta}_2 = 2.75\times10^{-5}$, with standard error 3.35×10^{-6}.) Transforming the dose term, say by dividing all terms by the maximum dose, would fix this display quirk (Exercise 7.26).

```
                    The SAS System
                 The GENMOD Procedure
                  Model Information
       Distribution                          POISSON
       Link Function                         LOG
       Dependent Variable                    Y
       Observations Used                     50

       Criteria For Assessing Goodness Of Fit
Criterion               DF           Value        Value/DF
Deviance                47         55.8712          1.1887
Pearson Chi-Square      47         51.3850          1.0933
Log Likelihood           .       2565.3275               .

           Analysis Of Parameter Estimates
Parameter   DF    Estimate  Std Err    ChiSquare  Pr>Chi
INTERCEPT    1      3.3173   0.0432    5899.3934  0.0000
DOSEADJ      1     -0.0049   0.0004     188.0746  0.0000
DOSE2ADJ     1     -0.0000   0.0000      67.1322  0.0000
NOTE:  The scale parameter was held fixed.

          LR Statistics For Type 1 Analysis
  Source        Deviance     DF   ChiSquare   Pr>Chi
  INTERCEPT     312.4840      0           .        .
  DOSEADJ       127.5887      1    184.8953   0.0000
  DOSE2ADJ       55.8712      1     71.7175   0.0000
```

Fig. 7.15 Output (edited) from SAS PROC GENMOD log-linear regression fit including quadratic term (`model y = doseadj dose2adj`) for data in Example 6.2.

The regression output in Fig. 7.15 from the full log-linear/quadratic model can be used to perform an LR test for *joint* significance of both linear and quadratic terms in the dose response. To test $H_0: \beta_1 = \beta_2 = 0$ against an alternative that assumes one or both β-parameters are nonzero, we construct the LR statistic as a difference of two deviances from the SAS output. This compares the deviance under a reduced model containing only an intercept term, $D(\tilde{\boldsymbol{\beta}})$, with that from the full model $D(\hat{\boldsymbol{\beta}})$, where now $\tilde{\boldsymbol{\beta}} = [\tilde{\beta}_0\ 0\ 0]$ and $\hat{\boldsymbol{\beta}} = [\hat{\beta}_0\ \hat{\beta}_1\ \hat{\beta}_2]$. At the bottom of Fig. 7.15, we find $D(\tilde{\boldsymbol{\beta}}) = 312.484$, while $D(\hat{\boldsymbol{\beta}}) = 55.8712$. Their difference is $G^2_{calc} = 256.6128$. As in Section 7.2.1, G^2_{calc} is distributed under H_0 as approximately χ^2, with df equal to the difference between the number of parameters under the full model and the number of nonzero parameters under the reduced model. Here $df = 3 - 1 = 2$, and the corresponding P-value is $P = P[\chi^2(2) > 256.6128] < 0.0001$.

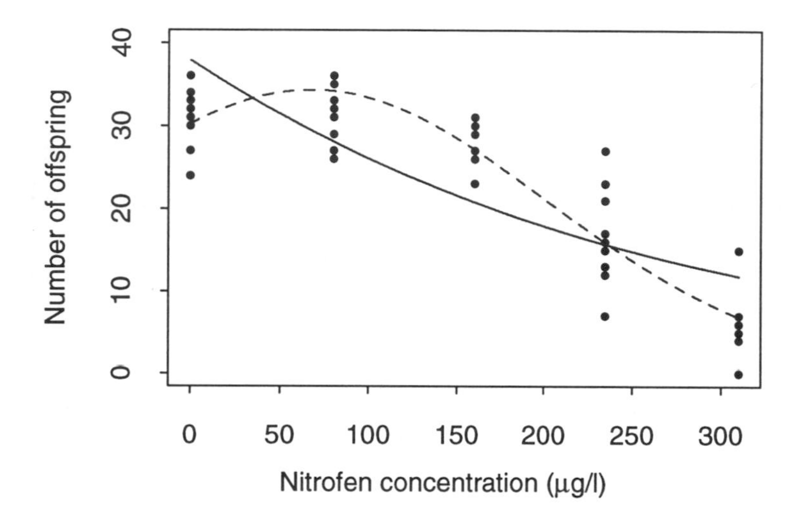

Fig. 7.16 Scatterplot of number of offspring in *C. dubia* exposed to differing concentrations of nitrofen; data from Examples 6.2 and 6.3. Superimposed are fitted log-linear models with only a linear nitrofen concentration term (solid curve) and with both linear and quadratic concentration terms (dashed curve).

It is interesting to examine how the estimate of β_1 in the model with only a linear (and intercept) term changes when an additional quadratic term is fitted. For the former fit, $\tilde{\beta}_1 = -0.0037$ in Fig. 7.14, while Fig. 7.15 gives $\hat{\beta}_1 = 0.0037$. This change in sign is not coincidental; it reflects the full log-linear/quadratic model's ability to accommodate an initial increase in the response at lower concentrations of nitrofen that was, in effect, ignored with the simple linear predictor; see the display in Fig. 7.16. As the figure helps illustrate, the quadratic term is a significant component in the model. □

7.2.3 Relation to generalized linear models

An important feature of the logistic and log-linear regression models described in the previous sections is that they can be modified to represent other nonlinear forms. Indeed, both models are special cases of a wider class of **generalized linear models** (GLiMs). Although we treat this class in greater detail in the next chapter, we note here some specific extensions for dose-response modeling with discrete data.

The basic precept of any GLiM is to take a simple linear model such as (6.1) and extend it in two ways: (a) consider generalizations to nonnormal parent distributions such as binomial or Poisson; and/or (b) consider generalizations to nonlinear functions that link the unknown parameters with some linear function of the x-variable(s). This latter quantity is known as the **link function**. For instance, with data in the form of proportions, $p_i = Y_i/N_i$, recall that the general dose-response model in (7.10) gave the ith response probability as $\pi_i = F(\beta_0 + \beta_1 x_i)$, where $F(\cdot)$ is some monotone function bounded between 0 and 1, such as a c.d.f. The special case considered in Section 7.2.1 is the logistic model (7.11). The link function in this case is the inverse of $F(\cdot)$, the **logit function** from Section 1.3.6: $g(\pi) = \log\{\pi/(1-\pi)\}$. That is, if $\pi = 1/(1+e^{-\eta})$, then $\eta = \log\{\pi/(1-\pi)\}$. A useful feature of this representation is that the **linear predictor**, $\eta = \beta_0 + \beta_1 x$, is equal to the natural logarithm of the odds ratio, $\log(OR) = \log\{\pi/(1-\pi)\}$; cf. Section 4.2.2. This has a natural interpretation in selected applications, primarily in (environmental) epidemiology (Breslow and Day, 1980, Section 6.1; Wallenstein and Bodian, 1987).

As we noted in Section 7.2.1, however, other forms for $F(\cdot)$ are possible. The probit regression model (Finney, 1971) arising from a normal c.d.f. provides an alternative to the logistic model (7.11), although the two models involve very similar functions and often will lead to similar results; see Example 7.4 (continued below), and also Exercise 7.21. Table 7.3 gives some additional, selected examples of GLiMs for quantal response data. In the table, the logistic and probit models are symmetric, so that $g(\pi) = -g(1-\pi)$, while the complementary log-log and complementary log models are asymmetric. McCullagh and Nelder (1989, Fig. 4.1) provide a graphic comparison of these and other link functions for proportion data.

Table 7.3 Selected GLiMs for dose response with data in the form of proportions

Regression model	$\pi = F(\eta)$	*Link function*	*Tolerance distribution*
Logistic	$\pi = \dfrac{1}{1+e^{-\eta}}$	$\eta = \log\left\{\dfrac{\pi}{1-\pi}\right\}$	Logistic
Probit	$\pi = \Phi(\eta)$	$\eta = \Phi^{-1}(\pi)$	Normal
Complementary log-log	$\pi = 1-\exp\{-e^{\eta}\}$	$\eta = \log\{-\log(1-\pi)\}$	Extreme-value
Complementary log	$\pi = \begin{cases} 0 & \text{if } \eta < 0 \\ 1-e^{-\eta} & \text{if } \eta \geq 0 \end{cases}$	$\eta = -\log(1-\pi)$	Exponential

Example 7.4 (continued) Micronuclei in rodent cells – Probit regression analysis

Recall the dose-response data on micronucleus induction after hamster cells' exposure to *tert*-butylhydroquinone (*t*BHQ), given in Table 6.4. A probit fit to these data is accomplished via SAS PROC LOGISTIC by replacing in Fig. 7.11 the `Model` option `/link=logit` with `/link=normit`. (One could also use SAS PROC PROBIT under its default link.) The results (edited for presentation) appear in Fig. 7.17, from which we achieve inferences almost identical to those under the logistic fit (cf. Fig. 7.12).

For these data, the estimated dose response is essentially identical under either the logit or the probit link. The estimated dose-response functions for the two GLiMs are presented in Fig. 7.18, where the strong similarity between the two prediction equations is evidenced. ❑

```
                          The SAS System
                       The LOGISTIC Procedure

Response Variable (Events): Y
Response Variable (Trials): N
Number of Observations: 6
Link Function: Normit
                          Response Profile
                    Ordered  Binary
                      Value  Outcome       Count
                          1  EVENT           127
                          2  NO EVENT      11873

                Criteria for Assessing Model Fit
              Intercept  Intercept and
Criterion       Only       Covariates     Chi-Square for Covariates
AIC           1409.964      1355.803           .
SC            1417.356      1370.588           .
-2 LOG L      1407.964      1351.803      56.161 with 1 DF (p=0.0001)
Score             .             .         53.427 with 1 DF (p=0.0001)

           Analysis of Maximum Likelihood Estimates
                 Parameter   Standard        Wald           Pr >
Variable   DF    Estimate       Error    Chi-Square     Chi-Square
INTERCPT    1    -2.7594       0.0808     1167.0510         0.0001
DOSE        1    0.00104     0.000149       48.8828         0.0001

                 Estimated Covariance Matrix
       Variable              INTERCPT                  DOSE
       INTERCPT          0.0065245609          -0.000010867
       DOSE              -0.000010867          2.2068588E-8
```

Fig. 7.17 Output (edited) from SAS PROC LOGISTIC probit regression fit to data in Example 6.5.

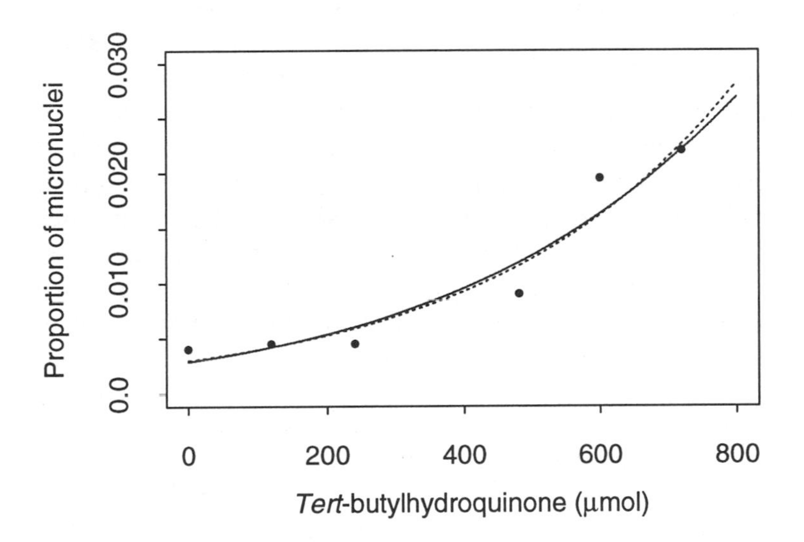

Fig. 7.18 Scatterplot of micronucleus induction versus *t*BHQ dose with the fitted probit (solid line) and logit (dashed line) regression curves superimposed.

For discrete data in the form of counts, GLiMs involve positive-valued dose-response functions that link the mean response, μ_i, with the linear predictor $\eta_i = \beta_0 + \beta_1 x_i$. The exponential function in (7.14) is a natural candidate for this construction. Alternative functions that may have subject-matter relevance in select applications include the simple square

$$\mu_i = (\beta_0 + \beta_1 x_i)^2,$$

and the absolute value

$$\mu_i = |\beta_0 + \beta_1 x_i|.$$

The latter form may be viewed as a special kind of segmented regression model for count data. An associated form is the truncated linear segment function

$$\mu_i = \begin{cases} 0 & \text{if } \beta_0 + \beta_1 x_i \leq 0 \\ \beta_0 + \beta_1 x_i & \text{if } \beta_0 + \beta_1 x_i > 0 \end{cases},$$

which is a special case of (7.3). Of course, for any of these forms to be useful, the underlying application should provide some motivation for their use, perhaps based on prior data, pilot studies, or other sources from the experimental literature.

7.3 POTENCY ESTIMATION FOR DOSE-RESPONSE DATA

Dose-response models are not limited to the simple issue of testing whether or not the response rises (or falls) with increasing doses of some environmental stimulus. Indeed, estimating other attributes of the dose response often is an important feature of the environmental study. For example, it is common with toxicological data to require some form of summary measure about the dose response for comparing different stimuli within test species, experimental protocols, etc. Many summary measures are possible, all taken under the general rubric of dose-response **potency estimators**. With a single potency measure, an investigator can make comparative statements about, or rankings among, a series of environmental agents as applied to a specific assay system. In the simplest case, where no significant dose response is evidenced after exposure to a particular agent, the potency is set equal to some boundary value that indicates no potent response (zero is common, although other boundary values are possible depending on the measure used to describe potency). For significant dose responses, some quantity departing from this boundary value is calculated from the experimental data.

Statistical issues dealing with estimators of potency are many and varied, with a history ranging back to **bioassay** of pesticides, herbicides, and other poisons (Bliss, 1935; Finney, 1952). The most obvious question, perhaps, is how to measure or define the potency in an unambiguous manner? A number of quantitative measures exists to answer this question, each based on a slightly different feature of the observed dose response. We will consider a few of the more useful ones in this section.

7.3.1 Median effective dose: ED_{50}, LD_{50}, TD_{50}

One of the earliest measures used to summarize dose response is the **median effective dose**, denoted as ED_{50}. (We studied this measure briefly in Exercise 3.16.) If the environmental agent is given as a concentration, we use the term **median effective concentration**, or EC_{50}. In either case, this is the quantity of the agent that is necessary to produce a response at the median (50%) level (Trevan, 1927). This can be 50% of the maximum possible response in a quantal response experiment, or 50% of the control response in a decreasing-response experiment, etc. For example, suppose we measure the proportion of times organisms respond to a toxic exposure, as in Section 7.2.1. Then, since the proportions are bounded between 0 and 1, the ED_{50} is the dose at which the response equals 0.50.

Obviously, the ED_{50} depends upon the outcome variable under study, and terminology has developed to indicate this. For instance, when simple lethality is the common outcome, the ED_{50} is more specifically the **median lethal dose**, or LD_{50}. Or, in chronic-exposure tumorigenicity experiments with laboratory animals, the ED_{50} is a **median tumorigenic dose**, or TD_{50}, the dose required to halve the probability of remaining tumor-free through the end of the experiment (Sawyer *et al.*, 1984). Notice that ED_{50} is an inverse measure of potency: a higher ED_{50} suggests a weaker agent, since a greater dose is required to produce the same median result.

An extensive literature has developed on use of ED_{50}, EC_{50}, LD_{50}, TD_{50}, etc.; see the early works by Irwin (1937) and Fieller (1940), and more modern discussions by Tamhane (1986), Finkelstein and Ryan (1987), Portier and Hoel (1987), Hamilton (1991), and Meier *et al.* (1993). Our goal here is to distill from this material useful features of the ED_{50} that allow for potency comparisons among environmental agents, particularly as they pertain to dose-response analysis. For example, when the functional form of the dose response is fully specified, the ED_{50} may be found as some combination of the dose-response parameters. From this, statistical estimators and confidence intervals may be derived.

Suppose we observed quantal response data, as in Section 7.2.1, where the underlying form of the dose response is the linear logistic form (7.11). Then, a parametric representation for the ED_{50} is found by setting the linear logistic equation equal to 0.5, and solving for x; that is, take ED_{50} equal to the value of x that solves $1/(1 + \exp\{-\beta_0 - \beta_1 x\}) = 0.5$. Assuming $\beta_1 \neq 0$, one finds $ED_{50} = -\beta_0/\beta_1$ (Exercise 7.30). The ML estimate of this quantity is $\hat{ED}_{50} = -\hat{\beta}_0/\hat{\beta}_1$. Large-sample $1 - \alpha$ confidence limits on the ratio $-\beta_0/\beta_1$ take the form

$$\hat{ED}_{50} + \frac{\gamma}{1-\gamma}\left\{\hat{ED}_{50} + \frac{c_{01}}{se^2[\hat{\beta}_1]}\right\} \pm \tag{7.16}$$

$$\frac{\mathfrak{z}_{\alpha/2}}{(1-\gamma)\,|\hat{\beta}_1|}\left(se^2[\hat{\beta}_0] + 2c_{01}\hat{ED}_{50} + se^2[\hat{\beta}_1]\hat{ED}_{50}^2 - \gamma\left\{se^2[\hat{\beta}_0] - \frac{c_{01}^2}{se^2[\hat{\beta}_1]}\right\}\right)^{1/2}$$

where $\gamma = \mathfrak{z}_{\alpha/2}^2 se^2(\hat{\beta}_1)/\hat{\beta}_1^2$ measures departure from symmetry in the distribution of $\hat{ED}_{50}$ ($\gamma \rightarrow 0$ indicates greater symmetry), and c_{01} is the estimated covariance between the ML estimators of β_0 and β_1. As we saw in Fig. 7.12, c_{01} is available from most logistic regression outputs. (The interval in (7.16) is based on a general approach for confidence intervals on ratios of parameters, known as **Fieller's theorem** (Fieller, 1940; Finney, 1952, Section 2.5). It is derived as part of Exercise 7.31.)

Example 7.6 Mortality in minnows exposed to fluoranthene – Logistic regression analysis

We illustrate LC_{50} estimation with an acute toxicity study in which fathead minnows were exposed to fluoranthene, a polycyclic aromatic hydrocarbon. (The data were kindly provided by Dr James Oris of Miami University.) In this study, 60 fathead minnows were exposed to the chemical at each of nine concentrations. Recorded were proportions of fish dying. (The data appear as part of the SAS code in Fig. 7.19.)

To analyze these data, Fig. 7.19 illustrates a linear logistic regression fit using SAS PROC PROBIT with the `d=logistic` option. Included is the `inversecl` option to output $LC_{100\rho}$ estimates associated with a range of values for ρ. The (edited) SAS output appears in Fig. 7.20.

Under the simple linear logistic regression model (7.11), Fig. 7.20 gives the ML parameter estimates for the minnow data, $\hat{\beta}_0$, $\hat{\beta}_1$, their standard errors, and the estimated covariance, c_{01}, between the two quantities. These are: $\hat{\beta}_0 = -6.2812$, $se[\hat{\beta}_0] = 0.61229$, $\hat{\beta}_1 = 0.6585$, $se[\hat{\beta}_1] = 0.05887$, and $c_{01} = -0.0352$. From these values, the $\widehat{LC}_{50}$ is

$$-\frac{-6.2812}{0.6585} = 9.5386 \ \mu\text{g/l}.$$

From (7.16), 95% confidence limits for the LC_{50} are calculated as $9.0132 < -\beta_0/\beta_1 < 10.420$. Notice that both the LC_{50} and its Fieller confidence limits are produced as part of the SAS PROBIT output in Fig. 7.20, under `Probability` at `0.50`.

Estimation of LC_{50} along with the asymptotic confidence limits (7.16) also can be performed directly in a programming environment such as S-PLUS. To illustrate, we display entry of the minnow data and the fit of the simple linear logistic regression model using the S-PLUS `glm` function in Fig. 7.21 (edited for presentation). Recall that the S-PLUS analysis is interactive, so program code and results occur on the same output. In our displays, we separate programming from output results with gray lines.

Comparing Fig. 7.21 with Fig. 7.20, we see that the coefficients from the two outputs are the same; however, different information is presented for both the model discrepancy measures and the estimated variance-covariance matrix of the parameter estimates. SAS provides the log-likelihood for a binary model while S-PLUS provides deviance summaries. Also, the SAS `covb` option gives the variance-covariance matrix of the parameter estimates while the S-PLUS `summary` defaults to generating the standard errors and

```
* SAS code to fit minnow mortality data;

data minnow;
   input conc mort n @@;
   cards;
31.0  60  60    14.5  60  60    11.8  46  60
11.2  47  60     7.5  10  60     3.9   1  60
 3.3   0  60     1.4   2  60     0.8   1  60

proc probit;
   model mort/n=conc / d=logistic covb inversecl;
```

Fig. 7.19 SAS PROC PROBIT code for logistic regression fit to minnow mortality data.

```
                      The SAS System
                     PROBIT Procedure

  Dependent Variable = MORT
  Dependent Variable = N
  Number of Observations = 9
  Number of Events   =    227
  Number of Trials   =    540
  Log Likelihood for LOGISTIC -117.2798078

 Variable  DF   Estimate  Std Err ChiSquare  Pr>Chi
 INTERCPT   1 -6.2811975 0.612913  105.0236  0.0001
 CONC       1 0.65850552 0.059665   121.807  0.0001

              Estimated Covariance Matrix
                      INTERCPT                CONC
  INTERCPT            0.375663           -0.035155
      CONC           -0.035155            0.003560

  Probability       CONC          95 % Fiducial Limits
                                    Lower         Upper
      0.01          2.5605          0.8870        3.7575
      0.05          5.0672          3.8917        5.9294
      0.10          6.2019          5.2370        6.9275
      0.20          7.4334          6.6747        8.0330
      0.50          9.5386          9.0132       10.0420
      0.70         10.8253         10.3166       11.3958
      0.90         12.8752         12.2223       13.7236
      0.95         14.0100         13.2305       15.0588
      0.99         16.5167         15.4123       18.0536
```

Fig. 7.20 Output (edited) from SAS PROC PROBIT logistic regression fit to minnow mortality data.

```
fconc<-c(31.0,14.5,11.8,11.2,7.5,3.9,3.3,1.4,.8)
ndead<-c(60,60,46,47,10,1,0,2,1)

nrisk<-rep(60,length(ndead))
fluor.df<-data.frame(conc=fconc,ndead=ndead,
                     nrisk=nrisk)

fluor.fit<-glm(cbind(ndead,nrisk-ndead)~conc,
               family=binomial(link=logit),
               data=fluor.df)
summary(fluor.fit)
```

```
Coefficients:
                 Value Std. Error   t value
(Intercept) -6.2811969 0.61266037 -10.25233
       conc  0.6585055 0.05964345  11.04070

    Null Deviance: 514.9841 on 8 degrees of freedom
Residual Deviance: 14.69952 on 7 degrees of freedom

Correlation of Coefficients:
     (Intercept)
conc -0.9612874
```

Fig. 7.21 S-PLUS `glm` program for logistic regression fit to minnow mortality data.

correlation. In the latter case, the covariance is found as the product of the correlation and the two standard errors: $c_{01} = (-0.9613)(0.6127)(0.0596) = -0.0351$. From these, the LC_{50} and Fieller confidence limits (7.16) can be calculated in S-PLUS (Fig. 7.22). This function also can be modified to estimate other levels of effect. For example, the function `dose.p` of Venables and Ripley (1997) finds the ED_{50} and other ED estimates, and provides associated confidence limits.

A scatterplot of fathead minnow mortality versus fluoranthene concentration is displayed in Fig. 7.23 along with a curve representing the fit of a simple linear logistic GLiM. The plot also illustrates the basic nature of the LC_{50} calculation: an inversion of the concentration–response relationship to estimate the specific concentration associated with a 50% response.

For an assay system where 50% lethality is unusual or uncommon, or when more subtle levels of population effects are of interest, it may be more appropriate to make potency comparisons across environmental agents using a lower effective dose, such as LC_{10} or LC_{01}. We explore this possibility further in Section 7.3.2. □

```
EC50.calc<-function(coef,vcov,conf.level=.95) {

# calculates conf. interval based on Fieller's theorem
# assumes link is linear in dose
       call <- match.call()
        b0<-coef[1]
        b1<-coef[2]
        var.b0<-vcov[1,1]
        var.b1<-vcov[2,2]
        cov.b0.b1<-vcov[1,2]
        alpha<-1-conf.level
        zalpha.2 <- -qnorm(alpha/2)
        gamma <- zalpha.2^2 * var.b1 / (b1^2)
        EC50 <- -b0/b1

const1 <- (gamma/(1-gamma))*(EC50 + cov.b0.b1/var.b1)
const2a <- var.b0 + 2*cov.b0.b1*EC50 + var.b1*EC50^2 -
                   gamma*(var.b0 - cov.b0.b1^2/var.b1)
const2 <- zalpha.2/((1-gamma)*abs(b1))*sqrt(const2a)

        LCL <- EC50 + const1 - const2
        UCL <- EC50 + const1 + const2
        conf.pts <- c(LCL,EC50,UCL)
        names(conf.pts) <- c("Lower","EC50","Upper")
        return(conf.pts,conf.level,call=call)
}

   EC50.calc(coefficients(fluor.fit),
         summary.glm(fluor.fit)$cov.unscaled)
```

```
$conf.pts:
    Lower      EC50    Upper
 9.013366 9.538564 10.04191

$conf.level:
[1] 0.95
```

```
$call:
EC50.calc(coef = coefficients(fluor.fit),
          vcov = summary.glm(fluor.fit)$cov.unscaled)
```

Fig. 7.22 S-PLUS function for estimating an EC_{50} along with confidence limits based on Fieller's theorem.

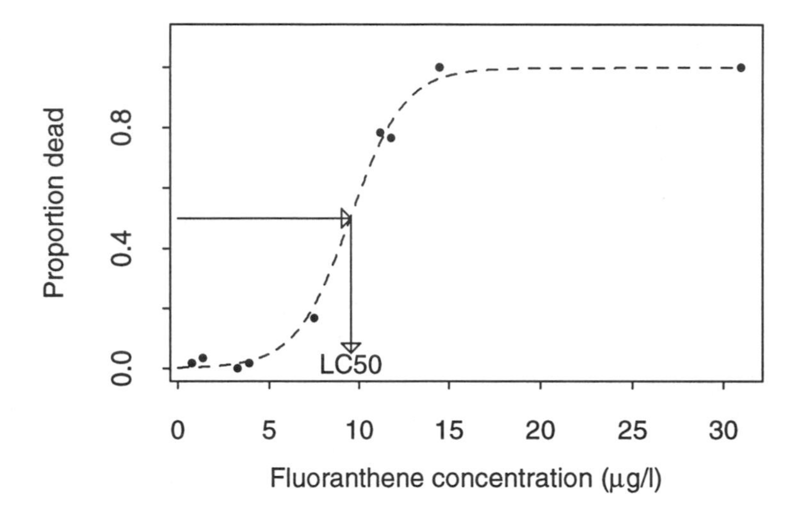

Fig. 7.23 Scatterplot of fathead minnow mortality versus fluoranthene concentration with the fitted linear logistic regression curve superimposed. The LC_{50} is found by estimating 50% mortality on the fitted logistic regression curve and then projecting this point down to the concentration axis.

For data in the form of counts, the log-linear model in (7.14) is employed for dose-response analysis; cf. Section 7.2.2. If the dose response to the environmental agent is expected to produce a decrease in the mean rate of response – as in Example 7.5 – then we define the ED_{50} as that dose producing a 50% drop from the mean control response. Suppose the control level is taken as $x = 0$. Under (7.14), we take $\mu = \exp\{\beta_0 + \beta_1 x\}$, which at $x = 0$ yields $\mu = \exp\{\beta_0\}$. A 50% reduction in this value is $\exp\{\beta_0\}/2$, so we solve for x in $\exp\{\beta_0 + \beta_1 x\} = \exp\{\beta_0\}/2$. This yields $ED_{50} = -\log(2)/\beta_1$. From dose-response data under a Poisson parent distribution, we find the ML estimate of β_1, and then substitute this value into the ED_{50} equation for the ML estimator $\widehat{ED}_{50} = -\log(2)/\hat{\beta}_1$. In large samples, approximate $1 - \alpha$ confidence limits on ED_{50} follow immediately from those on β_1: given the approximate limits $\hat{\beta}_1 \pm z_{\alpha/2} se[\hat{\beta}_1]$, we operate on these quantities to translate them into a set of limits on $ED_{50} = -\log(2)/\beta_1$:

$$\frac{-\log(2)}{\hat{\beta}_1 - z_{\alpha/2} se[\hat{\beta}_1]} < ED_{50} < \frac{-\log(2)}{\hat{\beta}_1 + z_{\alpha/2} se[\hat{\beta}_1]} .$$

(If $z_{\alpha/2}se[\hat{\beta}_1]$ is so large as to cause $\hat{\beta}_1 + z_{\alpha/2}se[\hat{\beta}_1] > 0$, then the calculation leads to a nonsensical result. This is appropriate, since in this case an approximate $1 - \alpha$ confidence interval for β_1 contains zero. That is, there is no significant evidence for a dose effect, and hence no reason to calculate a potency estimate.)

It is assumed implicitly that $\hat{\beta}_1 < 0$ in the construction of the ED_{50} estimate given above. With certain environmental data, however, it is possible to observe small increases in the response of the organism at low levels of a toxic stimulus, prior to observing a decline at more toxic doses. This phenomenon is known as **hormesis** (Stebbing, 1982; Calabrese and Baldwin, 1993), and it is a special case of an umbrella dose-response function. Bailer and Oris (1994) describe a simple generalization to the log-linear model for counts for use in potency estimation when hormetic effects were possible, by modeling the log-mean rate as a polynomial in dose: $\mu_i = \exp\{\beta_0 + \beta_1 x_i + \beta_2 x_i^2 + \cdots + \beta_k x_i^k\}$. For determining ED_{50}, this expression is set equal to a 50% decrease relative to the control response ($x = 0$), and is then solved for the concentration that induces that level of change. For example, if $k = 2$, then, as above, a 50% reduction in control response involves solving for x in the equation $\exp\{\beta_0 + \beta_1 x + \beta_2 x^2\} = \exp\{\beta_0\}/2$. This reduces to solving for x in the quadratic equation $\beta_2 x^2 + \beta_1 x + \log(2) = 0$. Substituting ML estimates for the β-parameters in this equation leads to a point estimate of the ED_{50}.

Although they are common for regression modeling with discrete data, log-linear and logistic models also are useful for estimating ED_{50}s with continuous responses. For instance, suppose an environmental toxin leads to decreases in body weight or biomass in an organism. A **three-parameter logistic model** for the mean biomass is

$$\mu_i = \frac{\mu_0}{1 + \left\{\frac{x_i}{ED_{50}}\right\}^{\beta}} \tag{7.17}$$

(Van Ewijk and Hoekstra, 1993). (Other logistic forms are also possible; see Chanter (1984).) Notice the explicit inclusion of ED_{50} as one of the parameters in (7.17); the other parameters are the spontaneous population mean rate, μ_0, and a dose-response parameter, β. When $\beta = 0$, no dose response is indicated in (7.17). If the data are normally distributed, then the LS estimates correspond to the ML estimates, and a nonlinear regression program such as SAS PROC NLIN or S-PLUS `nls` may be used to find point estimates, confidence intervals, significance tests for $\beta = 0$, etc.

In (7.17), there is a tacit assumption that if significant, the dose response is strictly decreasing, that is, $\beta \geq 0$. Van Ewijk and Hoekstra (1993) describe an extension of (7.17) that can incorporate a hormetic effect in the logistic dose response:

$$\mu_i = \frac{\mu_0 (1 + \gamma x_i)}{1 + \left[\frac{x_i}{\text{ED}_{50}}\right]^{\beta} \{2\gamma\text{ED}_{50} + 1\}}. \tag{7.18}$$

If the additional parameter γ is positive, the dose-response curve increases at low values of x_i before decreasing at higher values. At $\gamma = 0$, no hormesis occurs, and the model collapses to (7.17). Other extensions of the logistic curve exist that can accommodate a hormetic response; for example, Jolicoeur and Ponteir (1989) suggest a reciprocal biexponential form:

$$\mu_i = \frac{1}{\beta_0\exp\{\beta_1 x_i\} + \beta_2\exp\{-\beta_3 x_i\}}.$$

In all these cases, estimation proceeds via nonlinear LS methods, and requires computer calculation.

7.3.2 Other levels of effective dose

Traditionally, the ED_{50} has been used as a summary measure of the dose effect due to its central location on the dose-response curve. Many other response levels are possible, however, such as the ED_{01}, ED_{10}, ED_{90}, etc. Depending on the application, any of these may be appropriate for summary use. For example, estimation of low-dose effects for purposes of risk extrapolation may lead to consideration of lower levels of effective dose, such as the ED_{10} or even the ED_{01}. Indeed, Meier *et al.* (1993) show that in chronic exposure animal tumorigenicity studies, a form of TD_{01} possesses many desirable features, including low correlation with the maximum tolerated dose.

In general, the **effective dose ρ** is the dose that yields a $100\rho\%$ effect over the dose-response curve, for $0 < \rho < 1$; we denote this by $\text{ED}_{100\rho}$. For example, with quantal response under the linear logistic model (7.11), the $\text{ED}_{100\rho}$ is the value of x that solves $\rho = [1 + \exp\{-\beta_0 - \beta_1 x\}]^{-1}$, where $0 < \rho < 1$. This is

$$\text{ED}_{100\rho} = -\frac{1}{\beta_1}\left[\log\left\{\frac{\rho}{1-\rho}\right\} - \beta_0\right]. \tag{7.19}$$

Notice that if $\rho = 0.50$, (7.19) reduces to the ED_{50} found in Section 7.3.1 for the logistic model. A similar construction is possible for any GLiM with quantal response data; for example, under the probit model in Table 7.3, we solve for x in $\Phi(\beta_0 + \beta_1 x) = \rho$, resulting in

$$ED_{100\rho} = \frac{\mathfrak{z}_{(1-\rho)} - \beta_0}{\beta_1}. \tag{7.20}$$

(Recall that $\Phi^{-1}(\rho) = -\mathfrak{z}_\rho = \mathfrak{z}_{(1-\rho)}$.) For either model, the ML estimate $\widehat{ED}_{100\rho}$ is found from the ML estimates of the regression parameters, $\hat{\beta}_0$ and $\hat{\beta}_1$, by substituting these values into (7.19) or (7.20), as appropriate. Tamhane (1986) reviews of these and other estimation issues for measuring $ED_{100\rho}$.

To construct large-sample $1 - \alpha$ confidence limits on $ED_{100\rho}$ we return to Fieller's theorem (Fieller, 1940), as in (7.16). Indeed, as long as the $ED_{100\rho}$ is of the form $(\delta - \beta_0)/\beta_1$, where δ is some known constant, the form of the confidence limits from (7.16) remains unchanged (Exercise 7.47). For example, under the simple linear logistic model (7.11), the $ED_{100\rho}$ in (7.19) is of the form $(\delta - \beta_0)/\beta_1$, with $\delta = \log\{\rho/(1 - \rho)\}$, so a set of $1 - \alpha$ confidence limits on

$$\frac{1}{\beta_1}\left[\log\left\{\frac{\rho}{1 - \rho}\right\} - \beta_0\right]$$

is

$$\widehat{ED}_{100\rho} + \frac{\gamma}{1 - \gamma}\left\{\widehat{ED}_{100\rho} + \frac{c_{01}}{se^2[\hat{\beta}_1]}\right\} \pm \tag{7.21}$$

$$\frac{\mathfrak{z}_{\alpha/2}}{(1-\gamma)\,|\hat{\beta}_1|}\left(se^2[\hat{\beta}_0] + 2c_{01}\widehat{ED}_{100\rho} + se^2[\hat{\beta}_1]\widehat{ED}^2_{100\rho} - \gamma\left\{se^2[\hat{\beta}_0] - \frac{c^2_{01}}{se^2[\hat{\beta}_1]}\right\}\right)^{1/2}$$

where $\gamma = \mathfrak{z}^2_{\alpha/2} se^2(\hat{\beta}_1)/\hat{\beta}_1^2$ is again a positive measure of symmetry in the distribution of $\widehat{ED}_{100\rho}$ ($\gamma \to 0$ indicates greater symmetry).

Example 7.4 (continued) Micronuclei in rodent cells – Logistic regression analysis

For the data on micronucleus induction after hamster cells' exposure to *tert*-butylhydroquinone (*t*BHQ), we saw that the logistic model (7.11) yields ML estimates $\hat{\beta}_0 = -5.8067$ and $\hat{\beta}_1 = 0.0028$, with $se[\hat{\beta}_0] = 0.2345$, $se[\hat{\beta}_1] = 4.12\times10^{-4}$, and $c_{01} = -8.94\times10^{-5}$. The very low response rates in this assay make estimation of central EDs secondary to estimation of a very low ED. Here, quantities such as ED_{01} may be of greater interest for comparison purposes with other environmental mutagens.

From the β-parameter estimates, the ED_{01} is estimated as

$$-\frac{\log\left\{\frac{0.01}{0.99}\right\} - (-5.8067)}{0.0028} = 432.707 \ \mu\text{mol}.$$

Hence, using (7.21), 95% confidence limits for the ED_{01} are $352.778 < ED_{01} < 495.587$. □

A few brief warnings about the use of $ED_{100\rho}$ are warranted. First, confidence intervals using (7.21) generally grow wider for choices of ρ with $\rho \rightarrow 0$ or $\rho \rightarrow 1$, and larger samples sizes will be required to achieve the same level of accuracy as when $\rho \approx 0.5$. Second, negative $ED_{100\rho}$ values can be generated, although this is typically nonsensical when the concentrations are all positive. The effect occurs when the observed response at the control condition is less than $100\rho\%$. That is, the observed background response is already *above* the desired level of impact, $100\rho\%$, so the model is forced to drop to negative 'concentrations' to report an $ED_{100\rho}$. One must always keep the subject-matter conditions in mind when selecting the value of ρ. Third, much of our previous discussion focuses on models containing only one predictor variable, $\eta = \beta_0 + \beta_1 x$. The adequacy of this specification should be assessed as a precursor to potency estimation. Fourth, the choice of the link function is important if ρ is much less (or much greater) than 0.5. Different link functions can lead to $ED_{100\rho}$ values that differ by orders of magnitude for very small values of ρ. Fifth, if potency measures are used to rank different environmental hazards, the rankings will not necessarily be invariant to the choice of ρ in $ED_{100\rho}$. For example, if the simple linear logistic model is fitted to a series of hazards, then rankings based upon $ED_{100\rho}$ will be the same for all ρ only if the estimate of the slope parameter β_1 is the same for all hazards. It is often unwise to summarize a concentration-response pattern with a single number such as a potency index. As all these caveats indicate, with any nonlinear model the choice and use of the functional form must be made carefully, and with prudence.

When there is little or no motivation for choice of the logistic or probit (or any other) link function, nonparametric modeling and analysis of the $ED_{100\rho}$ are also possible. A popular nonparametric approach for estimating ED_{50} is the Spearman–Kärber method (Spearman, 1908; Kärber, 1931); described in detail by Finney (1952, Section 20.6). In addition, extensions of the Spearman–Kärber method exist that attempt to improve the estimation process, by correcting for outliers, allowing for sequential data collection, etc. These include the trimmed Spearman–Kärber method (Hamilton *et al.*, 1977), the **up-and-down method** for ED_{50} (Dixon and Mood, 1948) or for other ED_{100p}s (Durham and Flournoy, 1995), etc. A large literature exists in this area, and a complete review of these methods is beyond the scope of this chapter. We refer interested readers to books by Finney (1952, Chapter 20; 1971, Chapter 3), Govindarajulu (1988, Chapters 4, 7, 8), and Salsburg (1990, Chapter 3), or to the review by Hamilton (1991).

7.3.3* Other potency measures

Effective doses such as ED_{50} or ED_{01} are useful, easy-to-comprehend potency measures, and they are ubiquitous in toxicological dose-response analyses. Other measures exist, however, that provide useful interpretations and summary statements about the dose-response curve, especially when the curve is fully parameterized. For example, a common measurement of an environmental agent's potency is based on the rate of increase in the dose-response curve. Mathematically, this is the slope or tangent line of the curve at each dose level. If the dose response is assumed strictly linear, as in (6.1), then the slope is constant for all x; in (6.1) it is equal to β_1. Otherwise, the slope is changing with dose and either some specific dose must be chosen at which to measure the curve's slope, or some other parameter must be identified that represents a rate of change. Two possibilities for this measurement exist. If the dose response is modeled as a function of a simple linear predictor, $\beta_0 + \beta_1 x_i$, then β_1 represents a constant slope in the metric modeled by the dose-response function. For example, under the logistic model (7.11), β_1 is the slope in the logit metric: $\beta_0 + \beta_1 x_i = \text{logit}(\pi_i)$, where $\text{logit}(\pi) = \log\{\pi/(1-\pi)\}$. (The slope in the associated probability metric, $\beta_1\pi_i(1-\pi_i)$, is maximized at $\pi_i = 1/2$, that is, when x_i equals the median effective dose.) Alternatively, if the model is not linear in some metric, use the incremental rate of change in the dose response just past $x = 0$. That is, given a dose-response function $\mu(x)$ or $\pi(x)$, define the potency of an environmental agent as the dose-response slope at $x = 0$: this is typically the first derivative at zero, $\mu'(0)$ or $\pi'(0)$. This 'slope-at-zero' measure is motivated from low-dose extrapolation arguments: dose-response behavior at low doses of an environmental agent may approximate human responses to the agent, hence incremental change in dose response is of greatest interest near $x = 0$. For any differentiable dose-response function, the first derivative at $x = 0$ can be determined, and from this, estimates of the dose-response parameters, and then of $\mu'(0)$, can be calculated using some form of nonlinear LS regression.

The use of a slope-at-zero measure is not without controversy, however. For example, if $\mu(x)$ does not contain a linear term in x, for example, $\mu(x) = \beta_0 + \beta_1 x^{\beta_2}$, then for $\beta_2 > 1$, $\mu'(0)$ is identically 0. Thus, dose-response data well described by this $\mu(x)$ are always associated with small potency estimators. Again, we see that the choice of the dose-response function is highly influential on the ultimate value of any functionally derived potency estimator.

Example 7.3 (continued) Nonlinear potency estimation for Salmonella *mutagenesis*

For the *Salmonella* mutagenesis assay discussed in Examples 3.7 and 6.13, we noted that biomathematical considerations lead to highly nonlinear functions for this assay's dose response, as given in (7.6) or (7.7). For either of those forms, there is a linear predictor in dose, $\beta_0 + \beta_1 x_i$, at the kernel of the dose-response function representing the mutagenic effect. Thus, a natural potency estimator is the rate of change in this linear predictor, β_1. (The metric in which β_1 represents a linear slope is the complementary log from Table 7.3.)

Recall also that for small values of a, a first-order Taylor series approximation for $1 - e^{-a}$ gives $1 - e^{-a} \approx a$, so that both (7.6) and (7.7) may be approximated via less extreme forms. For example, (7.6) becomes

$$\mu_i \approx N_0 \exp\{-\beta_2 x_i\}(\beta_0 + \beta_1 x_i) \tag{7.22}$$

(Exercise 7.53). Here, the linear predictor $\beta_0 + \beta_1 x_i$ is approximately equal to the term representing the mutagenic effect, so that its slope, β_1, again may be viewed as a measure of potency, now on an approximately linear scale.

For the *Salmonella* assay, a natural extension of (7.22) that gives a slightly richer approximation of the nonmonotone response replaces the exponential term $\exp\{-\beta_2 x_i\}$ with the attenuation factor $\exp\{-\beta_2 x_i^\theta\}$, where we assume $\theta > 1$ (Leroux and Krewski, 1993). This yields

$$\mu_i \approx N_0 \exp\left\{-\beta_2 x_i^\theta\right\}(\beta_0 + \beta_1 x_i) . \tag{7.23}$$

Now, for any $\theta > 1$, the first derivative of (7.23) is

$$\frac{\partial \mu_i}{\partial x_i} \approx N_0 \exp\left\{-\beta_2 x_i^\theta\right\}\left(\beta_1 - \beta_0\beta_2\theta x_i^{\theta-1} - \beta_1\beta_2\theta x_i^\theta\right) . \tag{7.24}$$

Evaluating (7.24) at $x_0 = 0$, gives a slope-at-zero potency measure for the dose-response function in (7.23); this is $N_0\beta_1$. Thus, up to a constant, we see that β_1 approximates a measure of mutagenic potency from a number of different perspectives under the nonlinear form in (7.6). □

Notice that by measuring potency as a rate of change, we expect more potent agents to generate greater rates of change, slopes, etc. Thus the measure is directly comparable, in contrast to the inversely comparable features of the ED_{50}. To make it directly comparable, functions of the ED_{50} exist where more potent agents have larger values. Two examples include $1/ED_{50}$ and $\log\{(1/ED_{50}) + 1\}$ (Exercise 7.36). Portier and Hoel (1987) studied the latter potency estimator, and found that $\log\{(1/\hat{ED}_{50}) + 1\}$ is better approximated by a normal distribution than $\hat{ED}_{50}$.

We should note in passing that any parametric potency estimate based on rate of change, effective dose, etc., possesses an inherent shortfall: for some parametric dose-response functions, and under some distributional models, the dose-response parameters and/or the ED_{50} may not be calculable. Indeed, since computer iteration is almost always required to achieve the parameter estimates, there is no guarantee that the iteration will converge to stable estimates. For any dose-response model, warnings about highly nonlinear parametric forms remain valid: always use nonlinear fitting routines carefully and prudently.

It is also possible to define a potency measure in terms of inducing a specified absolute measure of change from the control response. In quantal response studies, **added risk** and **extra risk** of response associated with a specified level of some environmental exposure are often of interest. The added risk is the excess proportion of response relative to the control response: $AR(x_i) = \pi_i - \pi_0$. The extra risk is the added risk among those organisms that would not have responded under control conditions: $ER(x_i) = (\pi_i - \pi_0)/(1 - \pi_0)$. When used for regulatory purposes, permissible exposure limits may be defined as the concentration of a hazard that induces at most a specified added risk in the population, for example, a concentration associated with no more than a 10^{-3} added risk. Potency measures of this form are encountered frequently in quantitative risk assessment (West and Kodell, 1993; Bailer and Smith, 1994).

We close this section by noting that it is possible to define potency by some form of observed dose level, such as the no-observed-effect level (NOEL), the lowest-observed-effect level (LOEL), etc. As noted in Example 3.9, however, these various measures have fallen into disfavor, due to their typical reliance on multiple-comparison adjustments for proper determination, their inherent restrictions to the recorded dose levels of the experiment, and their generally limited information. (The former concern about multiplicity adjustment is not inherently bad, but since many forms of adjustment are possible, it is difficult to establish an unambiguous criterion for potency measurement with these quantities.) Many other concerns have been raised with use of NOELs (Suter, 1996), and we do not recommend observed-effect levels for use as potency measures.

7.4 COMPARING DOSE-RESPONSE CURVES

In environmental toxicity studies where k *separate* dose-response curves are under consideration, the investigator may be tempted to fix ρ and compare the dose-response curves using simply the $ED_{100\rho}$ associated with each curve. As we suggest above, however, this can lead to spurious inferences, since the conclusions regarding differences in dose-response curves may depend upon the selection of ρ. In fact, for most environmental applications, comparisons of potency measures such as $ED_{100\rho}$ are valid only under the assumption of *parallel* dose-response curves, an assumption which may not be tenable for the data under study.

We recommend comparing dose-response curves by constructing a single unifying model, and studying how the separate curves deviate under this model. For example, to compare dose-response curves among k populations, designate first a reference population and define a series of **indicator variables** for each of the other $k-1$ populations (Kleinbaum *et al.*, 1988). Using a model with a linear dose term, x, and letting the population at $k = 1$ serve as the reference population, the linear predictor with k populations is

$$\eta_{pi} = b_0 + b_1 x_{pi} + \sum_{j=2}^{k} b_j I_{\{p\}}(j) + \sum_{j=2}^{k} b_{k+j-1} x_{pi} I_{\{p\}}(j), \qquad (7.25)$$

where x_{pi} is the dose for the ith experimental unit in the pth population, and $I_{\{p\}}(j)$ is an indicator function that is 1 if the pth population corresponds to the jth population index, and 0 otherwise. In terms of each population's linear predictor, the model in (7.25) implies $\eta_{1i} = \beta_0 + \beta_1 x_{1i}$, $\eta_{2i} = (\beta_0 + \beta_2) + (\beta_1 + \beta_{k+1})x_{2i}, \ldots, \eta_{ki} = (\beta_0 + \beta_k) + (\beta_1 + \beta_{2k-1})x_{ki}$ $(i = 1, \ldots, n)$. The linear predictor in (7.25) can be embedded in any form of dose-response function – logistic, probit, log-linear, etc. – depending on the underlying data and on any other subject-matter motivation.

Equation (7.25) represents a form of **analysis of covariance** (ANCOVA) (Neter *et al.*, 1996, Section 25.2). The parameters $\beta_2, \ldots, \beta_k$ represent how the intercepts differ between the reference population and the remaining populations, while $\beta_{k+1}, \ldots, \beta_{2k-1}$ represent how the slopes differ between the reference population and the remaining populations.

Under (7.25), we assess differences between the dose-response curves by testing hypotheses about the βs. For example, the $2(k-1)$ *df* hypothesis

H_0: $\beta_2 = \cdots = \beta_k = \beta_{k+1} = \cdots = \beta_{2k-1} = 0$ represents no differences between the k dose-response curves, while the $k - 1$ *df* null hypothesis H_0: $\beta_{k+1} = \cdots = \beta_{2k-1} = 0$ represents parallel dose-response curves.

In general, as long as the null hypothesis of interest, H_0, implies a nesting of a simpler model within a more general model, a likelihood ratio test can be used to test H_0. We illustrate this construction in Example 7.7, with the comparison of logistic dose-response curves for $k = 2$ populations.

Example 7.7 Comparing dose-response curves for minnows exposed to fluoranthene

We illustrate the comparison of $k = 2$ dose-response curves by examining the survival experience of two populations of fathead minnows, *P. promelas*, exposed to fluoranthene, the polycyclic aromatic hydrocarbon studied in Example 7.6. The reference population is a control group (identified as `pop C` in Fig. 7.24, below) which corresponds to the data from Example 7.6. The second population is the first filial (F_1) generation from survivors of previous fluoranthene exposure (identified as `pop E` in Fig. 7.24, below). (The data were kindly provided by Dr James Oris of Miami University; they appear as part of the `data` step in Fig. 7.24.) We assume that the parents of fish in the exposed population are not represented in the control group, so that the two groups are statistically independent.

For both populations, 60 fathead minnows were exposed at each of nine fluoranthene concentrations. The proportion of dead fish out of the 60 exposed was the response of interest in this study. We fit a linear logistic model separately to each population, and also a single model to both populations allowing for the possibility of differing intercepts and/or slopes, as in (7.25). SAS code for these calculations is given in Fig. 7.24 with the (edited) output presented in Figs 7.25 and 7.26.

As Fig. 7.25 shows, the two populations exhibit different intercepts, interpreted as different background mortality rates. Further, the estimate of the slope in the exposed population, $\hat{\beta}_3 = 0.4099$, is lower than the estimate of the slope in the control population, $\hat{\beta}_4 = 0.6585$. (This difference is significant: $P = 0.0001$ from the `Type 1` LR test for `ICCONC` in Fig. 7.26.) Hence mortality increases less rapidly as a function of fluoranthene exposure in the population with parental exposure, relative to the control population. (To help visualize this effect, the data are plotted as part of Exercise 7.57.)

The separate linear logistic results in Fig. 7.25 also provide point estimates of the $LC_{100\rho}$s at select values of ρ. Notice, however, that the estimate of LC_{01} in the exposed population is nonsensical, since it is negative. This is not a surprise, nor is it a flaw in the model. The estimated

```
* SAS code to fit two-group minnow mortality data;
 data minnow2;
    input pop $ conc mort n @@;
    ic = pop="C";     icconc = ic*conc;
    cards;
 C  31.0  60  60   C  14.5  60  60   C  11.8  46  60
   ... other data omitted for space considerations ...
 E   3.3  10  60   E   1.4  11  60   E   0.8   8  60
 proc sort; by pop;
 proc probit; by pop;          * See Fig. 7.25;
    model mort/n=conc / d=logistic inversecl;
 proc genmod;                  * See Fig. 7.26;
 model mort/n = conc ic icconc / dist=bin
                                  link=logit type1;
```

Fig. 7.24 SAS program for logistic regression fit to the two-group minnow data.

```
                   The SAS System
                  PROBIT Procedure

   Dependent Variable = MORT
   Dependent Variable = N
   Number of Observations = 9
   Log Likelihood for LOGISTIC -117.2798078

 --------------------- POP=C -------------------------
   Variable  DF   Estimate  Std Err ChiSquare  Pr>Chi
   INTERCPT   1 -6.2811975 0.612913  105.0236  0.0001
   CONC       1 0.65850552 0.059665   121.807  0.0001

   Probability     CONC         95 % Fiducial Limits
                                  Lower        Upper
       0.01        2.5605        0.8870        3.7575
       0.10        6.2019        5.2370        6.9275
       0.50        9.5386        9.0132       10.0420
       0.90       12.8752       12.2223       13.7236
       0.99       16.5167       15.4123       18.0536

 --------------------- POP=E -------------------------
   Variable  DF   Estimate  Std Err ChiSquare  Pr>Chi
   INTERCPT   1 -2.6904391 0.237376  128.4617  0.0001
   CONC       1 0.40986424 0.032927  154.9398  0.0001

   Probability     CONC         95 % Fiducial Limits
                                  Lower        Upper
       0.01       -4.6471       -6.7571       -3.0817
       0.10        1.2034        0.0786        2.0810
       0.50        6.5642        5.9499        7.2041
       0.90       11.9251       10.9416       13.2066
       0.99       17.7755       16.0825       20.0641
```

Fig. 7.25 Output (edited) from SAS PROC PROBIT logistic regression for the two-group minnow data. POP=C indicates control group, POP=E indicates exposed group.

spontaneous response in the exposed population is $\exp\{\hat{\beta}_0\}/(1 + \exp\{\hat{\beta}_0\}) = e^{-2.69}/(1 + e^{-2.69}) = 0.0636$, and this is the smallest possible mortality rate the model allows for $x \geq 0$. Calling for an estimated $LD_{100\rho}$ at ρ less than 0.0636 forces the LC estimate to drop below $x = 0$.

Fluoranthene appears more toxic in the exposed population than in the control population, but perhaps only at lower doses: the $\widehat{LC}_{10}$ and $\widehat{LC}_{50}$ are smaller for the exposed population than the corresponding quantities for the control population, and these differences can be shown to be significant. Contrastingly, comparisons based upon the $\widehat{LC}_{90}$ and $\widehat{LC}_{99}$ would lead to the conclusion that fluoranthene toxicity was similar in the two populations. This incongruent inference illustrates the danger in comparing dose-response curves on the basis of only one single potency measure.

In Fig. 7.26, SAS continues to explore how these dose-response curves differ by fitting the logistic formulation of (7.25) with $k = 2$:

$$\text{logit}(\pi_i) = \beta_0 + \beta_1 x_i + \beta_2 I_{[\text{control}]}(i) + \beta_3 x_i I_{[\text{control}]}(i),$$

where $I_{[\text{control}]}(i)$ is 1 for the ith fish if it is from the control population, and 0 if it is from the exposed population. The model implies that the dose–response relationship in the exposed population is $\text{logit}(\pi_i) = \beta_0 + \beta_1 x_i$, while the dose–response relationship in the control population is $\text{logit}(\pi_i) = (\beta_0 + \beta_2) + (\beta_1 + \beta_3)x_i$. Thus, β_2 (β_3) indicates how the intercept (slope) in the control population differs from that in the control population. Estimates of these differences in intercepts and slopes between the two populations are given under `Analysis Of Parameter Estimates`, in the rows labeled `IC` and `ICCONC`. Notice that the parameter estimates associated with the `INTERCEPT` and `CONC` rows in Fig. 7.26 are exactly the intercept and slope terms associated with the analysis of the exposed population in Fig. 7.25. The intercept and slope terms associated with the control population, however, are available by manipulating results from both Fig. 7.25 and Fig. 7.26. Recall that the estimated slope of the linear logistic regression model in the control population is 0.6585, from Fig. 7.25. In Fig. 7.26, this is the sum of the exposed population slope estimate (the parameter estimate associated with `CONC`) and the slope difference estimate (`ICCONC`): $0.4099 + 0.2486 = 0.6585$.

Lastly, to study how the model was built across the two populations, a series of likelihood ratio tests of a hierarchy of models is given in Fig. 7.26 under `LR Statistics for Type 1 Analysis`. The sequence of models compared is: first,

$$\text{logit}(\pi_i) = \beta_0,$$

associated with the `INTERCEPT` term; then

$$\text{logit}(\pi_i) = \beta_0 + \beta_1 x_i,$$

associated with the CONC term; then

$$\text{logit}(\pi_i) = \beta_0 + \beta_1 x_i + \beta_2 I_{[\text{control}]}(i),$$

associated with the IC term; and finally,

$$\text{logit}(\pi_i) = \beta_0 + \beta_1 x_i + \beta_2 I_{[\text{control}]}(i) + \beta_3 x_i I_{[\text{control}]}(i)$$

associated with the ICCONC term. For example, the LR statistic to test the null hypothesis of parallel dose response in the two populations, $H_0{:}\ \beta_3 = 0$, against $H_a{:}\ \beta_3 \neq 0$, compares the full model

$$\text{logit}(\pi_i) = \beta_0 + \beta_1 x_i + \beta_2 I_{[\text{control}]}(i) + \beta_3 x_i I_{[\text{control}]}(i)$$

with a reduced model of the form

$$\text{logit}(\pi_i) = \beta_0 + \beta_1 x_i + \beta_2 I_{[\text{control}]}(i),$$

```
                     The SAS System
                  The GENMOD Procedure
                    Model Information
      Distribution                     BINOMIAL
      Link Function                    LOGIT
      Dependent Variable               MORT
      Dependent Variable               N
      Observations Used                18
      Number Of Events                 525
      Number Of Trials                 1080
          Criteria For Assessing Goodness Of Fit
 Criterion              DF          Value        Value/DF
 Deviance               14        38.3533          2.7395
 Log Likelihood          .      -312.8548               .
              Analysis Of Parameter Estimates
 INTERCEPT   1    -2.6904    0.2374   128.4617    0.0000
 CONC        1     0.4099    0.0329   154.9398    0.0000
 IC          1    -3.5908    0.6573    29.8455    0.0000
 ICCONC      1     0.2486    0.0681    13.3118    0.0003
 SCALE       0     1.0000    0.0000          .         .
 NOTE:  The scale parameter was held fixed.
             LR Statistics For Type 1 Analysis
   Source        Deviance    DF    ChiSquare   Pr>Chi
   INTERCEPT     909.0082     0            .        .
   CONC          105.2758     1     803.7323   0.0000
   IC             53.9397     1      51.3362   0.0000
   ICCONC         38.3533     1      15.5864   0.0001
```

Fig. 7.26 Output (edited) from SAS PROC GENMOD logistic regression fit to the two-group minnow mortality data with different slopes and intercepts.

leading to the LR statistic $G^2_{calc} = 15.5864$ ($P \approx 0.0001$). In fact, we see in Fig. 7.26 that all the parameters in this model differ significantly from zero. Thus, each level of complexity adds significantly to the model hierarchy, supporting the conclusion that the two concentration-response curves differ with respect to both slope and intercept. ❑

7.5 SUMMARY

In this chapter, the concept of parametric modeling for environmental response data is introduced. Examples of common parametric models are given for specific data settings, including polynomial and nonlinear regression models for continuous data, logistic regression models for proportion data, and log-linear regression models for count data. Within these contexts, measures of potency are introduced, including those based on median effective dose (ED_{50}), rate of change, and initial slope. In addition, a framework for comparing multiple dose-response curves is presented.

There are many forms of parametric functions available to model dose response; here, only selected, simpler forms for use with environmental data are introduced. An important area left unaddressed concerns dose-response modeling with serial correlated data, that is, data that display a correlation over increasing dose levels, or over time. Such a dose response is called a **growth curve**. Historically, very specific, nonlinear, parametric models have been applied to growth curve data, two examples include the **Gompertz curve** (Gompertz, 1825; Winsor, 1932; Pasternak and Shalev, 1992), or the **Michaelis–Menten model** (Michaelis and Menten, 1913). The Michaelis–Menten form has been shown in particular to be useful in modeling toxicokinetic activity when observing blood or tissue concentrations; see, for example, Becka *et al*. (1993). These toxicological applications are motivated by more than just environmental problems, however, and for greater detail readers are referred to the statistical literature on growth curves, for example, the text by Kshirsagar and Smith (1995).

EXERCISES

(Exercises with asterisks, e.g. 7.9*, typically are more advanced, or require greater knowledge and understanding of probability distributions and/or calculus.)

7.1. Describe some applications (from your own field of study, as appropriate) where an independent variable, x_i, is ordered and quantitative, leading to dose-response consideration. Explain. Are there any special forms of dose-response model used in your field that are similar to those illustrated herein?

7.2. Example 6.1 gives data on lymphocyte response in dolphins after exposure to PCBs. Consider a quadratic regression model for these data, and perform the fit. Is the quadratic term significant at $\alpha = 0.10$? Is the dose response significant at $\alpha = 0.10$? Would you see any differences in your results if you had not centered the predictor variables before fitting the quadratic model?

7.3. Exercise 6.8 gives data on hormone concentrations as a function of age in female Florida panthers. Repeat the quadratic analysis from Example 7.1 on these data. (Remember to use log-concentrations, and to adjust the age variable by its mean.) Is the additional quadratic term significant at $\alpha = 0.05$?

7.4. Return to the quadratic regression of log-hormone concentrations on age in male Florida panthers from Example 7.1. In the spirit of an exploratory analysis, consider the following (separate) modifications and extensions of the analysis in the example; in all cases, compare your results with those presented in Example 7.1.

a. Consider first the multicollinearity between $(x_i - \overline{x})$ and $(x_i - \overline{x})^2$. A simple yet useful quantity to assess multicollinearity is the observed correlation coefficient between $(x_i - \overline{x})$ and $(x_i - \overline{x})^2$; if this is far from zero, there may still be substantial multicollinearity present. If so, a more complex adjustment for multicollinearity is known as the **orthogonal polynomial** for the predictor variable (Snedecor and Cochran, 1980, Section 19.6). For a quadratic regression, the orthogonal polynomial construction retains at the first order the same centered x_is, but employs at the second order a new predictor quantity based on $(x_i - \overline{x})$. The full model is

$$\mu_i = \gamma_0 + \gamma_1(x_i - \overline{x}) + \gamma_2\left[(x_i - \overline{x})^2 - \frac{\tilde{m}_3}{\tilde{m}_2}(x_i - \overline{x}) - \tilde{m}_2\right],$$

where $\tilde{m}_k = \sum_{i=1}^{n}(x_i - \overline{x})^k/n$ is the kth observed central moment, $k = 2,3$. The new predictor quantities are designed to reduce multicollinearity to the fullest extent possible.

(i) Verify that the orthogonal polynomials reduce multicollinearity by comparing the observed correlation coefficients between $(x_i - \overline{x})$ and $(x_i - \overline{x})^2$, and between $(x_i - \overline{x})$ and $(x_i - \overline{x})^2 - (\tilde{m}_3/\tilde{m}_2)(x_i - \overline{x}) - \tilde{m}_2$.

(ii) Recalculate the regression fit using the orthogonal polynomials. Is the quadratic term still significant? Is the overall F-test now significant? (Operate at $\alpha = 0.05$.)

b. Since the original x_is are so widely spread, a simple change of the original x-scale may be appropriate to bring the predictor scale closer to an equi-spaced form, and thus reduce the possibly large influence of the observations at $x = 9$ and $x = 10$. The common choice here is the natural logarithm: let $u_i = \log\{x_i\}$, and find $\bar{u} = \sum_{i=1}^{n}\log\{x_i\}/n$. Fit the full model

$$\mu_i = \delta_0 + \delta_1(u_i - \bar{u}) + \delta_2(u_i - \bar{u})^2$$

and assess if the quadratic term is significant. Is the overall F-test now significant? (Operate at $\alpha = 0.05$.)

c. The residual for thc lower response ($y_2 = 0.39$) at age $x = 2$ is much larger in absolute value than any other residual. To assess the extent of this large magnitude on a standard scale we adjust this raw residual by its standard error, producing a **Studentized residual** (Neter *et al.*, 1996, Section 9.2), similar to the generalized Pearson residual from Section 2.7.1. Begin with the raw residuals $R_i = Y_i - \hat{Y}_i$. The standard error of R_i is $se[R_i] = \{\text{MSE}(1 - h_{ii})\}^{1/2}$, where h_{ii} is the ith diagonal element of the **hat matrix** of the regression: $\mathbf{H} = \mathbf{X}(\mathbf{X}'\mathbf{X})^{-1}\mathbf{X}'$. (The matrix $\mathbf{X}$ is the $n\times(P+1)$ **design matrix** of the regression; $\mathbf{X}'$ is its transpose.) These quantities are available from standard computer regression programs; SAS provides them via the INFLUENCE option in the MODEL statement under PROC REG, or via the H= option in the OUTPUT statement. (The INFLUENCE option also produces a 'Studentized residual' called RSTUDENT, but this is in fact the more complex **deleted Studentized residual**. Although this quantity is quite useful, we will not operate with it here. See Neter *et al.* (1996, Section 9.2) for more detail.) If the Studentized residual $R_i/\{\text{MSE}(1 - h_{ii})\}^{1/2}$ is larger in absolute value than about 2, the observation might be viewed as outlying in nature.

(i) Use a regression program to find the h_{ii} values from the hat matrix, under the quadratic model $\mu_i = \beta_0 + \beta_1(x_i - \bar{x}) + \beta_2(x_i - \bar{x})^2$.

(ii) Calculate the Studentized residual at $i = 2$, and assess if y_2 is an outlying value. If so, remove it from the data and recalculate the regression fit. Is the quadratic term still significant? Is the overall F-test now significant? (Operate at $\alpha = 0.05$.)

7.5. Whittaker *et al.* (1989) report data on frequency of chromosome malsegregation in the yeast *Saccharomyces cerevisiae* after exposure to the chemical solvent acetone. The data are:

Dose x_i (mg/ml)	0.0	37.4	44.5	51.4	58.2
Malsegregant freq. ($\times 10^6$), Y_i	0.2	8.3	19.0	38.3	20.0

It is common in this assay system to observe a nonmonotone dose response, where, as seen here, frequencies at higher doses often drop below a maximum response at a lower dose.

a. To model this effect, consider the quadratic polynomial, $P = 2$, in (7.1). (Since the data are frequencies, a stabilizing transformation may be appropriate. Here, use the natural logarithm: $V_i = \log\{Y_i\}$.) Fit this model to the transformed data. Is there a significant dose response at $\alpha = 0.05$?

b. Find the residuals, $v_i - \hat{v}_i$, from your quadratic fit in part (a). Plot the residuals as a function of dose. Is there any pattern that suggests a poor model fit?

c. Apply the recursive nonparametric procedure from Section 6.7 to these data, to test for increasing dose response. Set $\alpha = 0.05$. Does your result agree with what you found in part (a)?

d. Repeat the analysis in parts (a)–(c) using instead the square root transform $V_i = \sqrt{(Y_i)}$ Do the results differ?

7.6. Zimmermann and Mohr (1992) report data similar to those in Exercise 7.5, on the frequency of chromosome malsegregation in *S. cerevisiae* after exposure to the chemical 2-hydroxypropionitrile, in combination with a single sub-acute dose of a sister nitrile chemical, propionitrile. The data are:

Dose x_i (mg/ml)	0.0	0.96	1.21	1.45	1.69	1.93
Malseg. freq. ($\times 10^6$), Y_i	4.35	5.97	5.17	5.71	8.32	13.02
Dose x_i (mg/ml)	2.17	2.41	2.89	3.37	3.85	
Malseg. freq. ($\times 10^6$), Y_i	8.02	8.37	14.69	11.58	38.76	

These data illustrate an unusual but consistent effect with chromosome malsegregation frequencies: after certain chemical exposures (especially dual exposures), the dose response *inflects*: first increasing, then decreasing, then increasing. This suggests that a cubic model, $P = 3$, in (7.1) may be useful.

a. Fit the cubic model to these data (after transformation to log-frequencies, as in Exercise 7.5), and assess the features of the model fit. At $\alpha = 0.01$, is there a significant dose response?

b. Plot the data along with the fitted model. Does the fit appear reasonable? Are there any patterns in the residuals that suggest problems with the fit?

7.7*. In Example 7.2, the post-threshold response rises sharply. Fit an additional quadratic term to this segment, and assess its significance. Set $\alpha = 0.05$.

7.8. Show that under the continuity constraint $\beta_0 = \gamma_0 - \beta_1\tau$, the bilinear threshold model in (7.3) can be written as

$$\mu_i = \begin{cases} \gamma_0 & \text{if } x \le \tau \\ \gamma_0 + \beta_1(x_i - \tau) & \text{if } x > \tau . \end{cases}$$

7.9*. Extend the threshold linear model in (7.3) to account for *two* change points. Write out the function for this **trilinear threshold model**, expressed in terms of indicator functions when

a. jumps are expected at the join points of the segments;

b. a continuity constraint is imposed at the two join points.

7.10*. Describe how you would fit the trilinear threshold model in Exercise 7.9 when

a. jumps are expected at the join points of the segments;

b. a continuity constraint is imposed at the two join points.

7.11. Favor *et al.* (1990) report data on mutation frequencies in murine spermatogonia, after male mice were exposed to the mutagen ethylnitrosourea. Consider associated frequency data from two different laboratories:

Lab A

Dose x_i (mg/kg)	0.0	40	80	160	250
Mutant freq. ($\times 10^5$), Y_i	1.2	3.8	21.5	57.8	93.6

Lab B

Dose x_i (mg/kg)	0.0	25	50	750
Mutant freq. ($\times 10^5$), Y_i	0.8	0.1	9.4	33.2
Dose x_i (mg/kg)	100	150	200	250
Mutant freq. ($\times 10^5$), Y_i	43.1	59.3	68.7	69.8

Since the data are frequencies, a stabilizing transformation may be appropriate. Here, use the natural logarithm: $V_i = \log\{Y_i\}$.

a. Plot the data, overlaying the scatterplot for the two labs.

b. Fit a bilinear threshold model (7.3) to the transformed data separately for each laboratory. As in Example 7.2, constrain the model to be continuous at the join point τ. For Lab A, at what dose do you estimate the response passed its threshold? Is the post-threshold response significant at $\alpha = 0.05$? Are the results similar for Lab B?

c. Using your results in part (b), calculate residuals, $v_i - \hat{v}_i$, from the two separate model fits, and plot them against x_i. Do the patterns seems reasonable? If not, what would you do to correct the problem?

7.12*. In Example 7.2 or Exercise 7.11, how would you test for the occurrence of no threshold in the data? Apply your test to the data in Example 7.2 and Exercise 7.11. Set $\alpha = 0.05$.

7.13. Calculate the residuals, $y_i - \hat{y}_i$, from the fit of model (7.3) in Fig. 7.6, and plot them against x_i from Table 7.1.

a. Do you agree with the suggestion in Example 7.2 that there is an unaccounted-for trend in the values below the threshold point?

b. Calculate the residuals, $y_i - \hat{y}_i$, from the fit of model (7.4) in Fig. 7.8, and plot them against x_i from Table 7.1. Did use of the more advanced model improve the residual pattern?

7.14*. Obtain the derivatives of the trilinear threshold model in Exercise 7.9, with respect to all parameters in the model. (These might be required to fit a nonlinear regression to this model using SAS, S-PLUS, or other nonlinear regression software.)

7.15. In Example 7.2, how would you modify the analysis if the model is no longer continuous at $x = \tau$ (that is, the two segments do not necessarily meet at τ)? Perform such an analysis of the mercury toxicity data from Table 7.1. At $\alpha = 0.05$, does the new analysis differ qualitatively from that seen in Example 7.2?

7.16. Exercise 6.40 gives mutant counts in *S. typhimurium* after exposure to nitrofurantoin, similar to the data seen in Example 7.3. Apply the power transformation (7.8) to these data and fit the transformed mean model (7.9) to the transformed data. Note that the original doses were spaced geometrically, so apply a natural log transformation to the doses before fitting the model. Use consecutive-dose average spacing from (6.9) for the control dose.

a. For the fit, try the starting values: $\beta_0 = 5\times10^{-7}$, $\beta_1 = 1\times10^{-7}$, and $\beta_2 = 0.075$. Viewing β_1 as a measure of dose response for this model, does the dose response increase significantly at $\alpha = 0.05$? How does this compare with the results achieved in Exercise 6.40c?

b. Calculate the residuals, $v_i - \hat{v}_i$, from the fit, and plot them against the transformed doses. Is there a pattern evident in the residuals that suggests a poor model fit? If so, what might you do to correct it?

7.17. The variance-covariance matrix of a set of estimators can be used to find the standard errors of those estimators in a straightforward manner. The diagonal elements of the matrix are the estimated variances of the parameter estimator corresponding to that column, so the standard error is the square root of the estimated variance. Verify this feature in Fig. 7.12, by calculating the square roots of the diagonal elements under `INTERCEPT` and `SLOPE`, and checking that these are the same as the standard errors given under `Analysis of Maximum Likelihood Estimates`.

7.18. For the micronuclei data in Examples 6.5 and 7.4, find the six observed proportions, $p_i = y_i/N_i$, and from the parameter estimates from Fig. 7.12 find the fitted values, $\hat{\pi}_i = 1/[1 + \exp\{-\hat{\beta}_0 - \hat{\beta}_1 x_i\}]$. Plot the residuals $R_i = y_i - \hat{y}_i = N_i(p_i - \hat{\pi}_i)$ versus *t*BHQ dose. Do the residuals appear randomly scattered? What problems, if any, does the plot reveal?

7.19. Exercise 6.14 gives quantal response data on induction of chromosome aberrations in hamster cells after exposure to phenolphthalein. (The endpoint is similar to that studied in Example 7.4.) Fit a simple linear logistic model (7.11) to these data, and perform an LR test for increasing trend at $\alpha = 0.01$. How do your results compare with those from Exercise 6.14?

7.20. Exercise 6.16 gives quantal response data on liver damage after rats were exposed to an environmental toxin. Take as the predictor variable, x_i, the natural logarithm of dose. (Use consecutive-dose average spacing from (6.9) to define the control value.) Fit a simple linear-logistic model (7.11) to these data, and perform an LR test for increasing trend at $\alpha = 0.05$. How do your results compare with those from Exercise 6.16?

7.21. Compare the logit and probit models for proportion data by graphing the ratio of $probit\{\pi\} = \Phi^{-1}(\pi)$ to $\text{logit}\{\pi\} = \log\{\pi/(1 - \pi)\}$ over the range $0.025 < \pi < 0.975$. What pattern do you see, and what can you conclude about the similarities and/or differences between the two models?

7.22. Exercise 6.17 gives proportion data on cellular transformations as a function of exposure to two potentially toxic agents. Fit a simple linear logistic model (7.11) to each set of data, and perform an LR test for increasing trend at $\alpha = 0.10$. How do your results compare with those from Exercise 6.17?

7.23. Exercise 6.18 gives count data on numbers of avian hunting forays as a function of temperature. Set the predictor variable as the equi-spaced values 10, 15, 20, and fit a simple log-linear model (7.14) to these data. Test for increasing trend at $\alpha = 0.01$. How do your results compare with those from Exercise 6.18?

7.24. Exercise 6.20 gives count data on skin cancer response over time.

a. Set the predictor variable to the mid-points of the time intervals, and fit a simple log-linear model (7.14) to these data. Test for increasing trend at $\alpha = 0.10$. How do your results compare with those from Exercise 6.20?

b. Use of the actual years in part (a) is cumbersome, since the numbers are very large, but the difference between them is not. Set the predictor variable as the interval mid-point *minus* 1960, and reperform the fit. Do your results change?

7.25. Exercise 6.21 gives count data on tumorigenicity in rodents after exposure to 1,2-epoxybutane. Fit a simple log-linear model (7.14) to these data. Test for increasing trend at $\alpha = 0.01$. How do your results compare with those from Exercise 6.21?

7.26. Verify the claim in Example 7.5 that transforming the dose term by dividing all doses by the maximum dose, 310 μg/l, produces output with standard errors that do not read 0.0000 (as occurred in Fig. 7.15). How should you back-transform the parameter estimates to arrive at values on the original scale?

7.27. Repeat the analysis in Exercise 7.19, but now instead of the logit, use the following link functions from Table 7.3:

a. Probit.

b. Complementary log-log.

c. Complementary log.

7.28. Repeat the analysis in Exercise 7.20, but now instead of the logit, use the following link functions from Table 7.3:

a. Probit.

b. Complementary log-log.

c. Complementary log.

7.29. Repeat the analysis in Exercise 7.22, but now instead of the logit, use the following link functions from Table 7.3:

a. Probit.

b. Complementary log-log.

c. Complementary log.

7.30. Verify, as indicated in Section 7.3.1, that under a simple linear logistic model (7.11) with $\beta_1 \neq 0$, the ED_{50} is $-\beta_0/\beta_1$.

7.31*. Establish Fieller's theorem from (7.16) via the following steps:

a. Begin with two parameters, β_0 and β_1, for which we have estimators b_0 and b_1 that are unbiased; that is, $\text{E}[b_j] = \beta_j$ $(j = 0,1)$. Of interest is a set of confidence limits for $\text{ED}_{50} = -\beta_0/\beta_1$. For simplicity, write this as an unknown parameter, say $\theta = -\beta_0/\beta_1$. Let $D = b_0 + \theta b_1$. Show that D has expected value equal to zero.

b. Suppose that the standard error of each estimator in part (a) is given as $se[b_j]$. The squares of these quantities estimate the variance of each b_j. Show that the standard error of D, $se[D]$, is the square root of $se^2[b_0] + \theta^2 se^2[b_1] + 2\theta c_{01}$, where c_{01} is an estimate of the covariance of b_0 and b_1, $\text{Cov}[b_0, b_1]$. (*Hint*: what is $\text{Var}[D]$?)

c. With proper choice of estimators b_0 and b_1, $(D - \text{E}[D])/se[D] \mathrel{\dot\sim} \text{N}(0,1)$, in large samples. If so, a set of confidence limits could be derived from the relationship

$$|D - \text{E}[D]|/se[D] < z_{\alpha/2}.$$

Show that this is equivalent to $\{D - \text{E}[D]\}^2/se^2[D] < z^2_{\alpha/2}$.

d. Express the relationship in part (c) as $\{D - \text{E}[D]\}^2 < z^2_{\alpha/2} se^2[D]$. (Why does the inequality's direction remain the same?) Substitute the appropriate quantities from parts (a) and (b) into this expression for each quantity involving D. Show that this leads to a quadratic inequality in θ.

e. Take the inequality in part (d) and make it an equality. Solve for the two roots of this quadratic equation. It will be useful to define the quantity $\gamma = z^2_{\alpha/2} se^2(\hat{\beta}_1)/\hat{\beta}_1^2$.

f. Show that the two roots you found in part (e) may be written in the form of (7.16).

7.32*. Consider the quadratic logistic model $\pi(x) = 1/(1 + \exp\{-\beta_0 - \beta_1 x - \beta_2 x^2\})$. Derive an expression for ED_{50} in terms of the β-parameters. Do you experience any problem(s) in completing your derivation? If so, are there any constraints you could impose on the model that would overcome the problem(s)?

7.33. Estimate the ED_{50} for the quantal response in Exercises 6.14 and 7.19. Also give 99% confidence limits for ED_{50}, using (7.16).

7.34. Estimate the log-ED_{50} for the quantal response in Exercises 6.16 and 7.20. Also give 99% confidence limits for log-ED_{50}, using (7.16). How would you use your results to report ED information on the original dose scale?

7.35. Estimate the ED_{50} for both data sets in Exercises 6.17 and 7.22. Also give 99% confidence limits for ED_{50}, using (7.16).

7.36*. Suggest modification(s) of the ED_{50} to make it a directly comparable measure of potency; examples include $1/ED_{50}$ and $\log\{(1/ED_{50}) + 1\}$.

7.37*. The **delta method** is a form of Taylor series approximation for finding approximate variances of parameter estimators (Bishop *et al.*, 1975, Section 14.6). Use it to find the approximate variance associated with the two (and your) modifications in Exercise 7.36.

7.38. Under the simple log-linear model (7.14), $ED_{50} = -\log(2)/\beta_1$ is the dose that produces a 50% decrease from $x = 0$. Is this an inversely comparable or a directly comparable measure of potency?

7.39. In Example 7.5, it was seen that reproductive response in *C. dubia* after nitrofen exposure was significantly decreased. Use the results from that analysis to estimate the ED_{50} for reproductive decrease of nitrofen with this species. Can you include in your calculations a set of 95% confidence limits for ED_{50}?

7.40*. Under the log-linear model (7.14), define the $ED_{100\rho}$ as the dose that produces a 100ρ% decrease from the control mean response.

a. Find an expression for this value in terms of the β-parameters and describe how you would estimate it.

b. Can you construct a set of $1 - \alpha$ confidence limits on $ED_{100\rho}$ based on your estimator in part (a)?

7.41*. Under the three-parameter logistic model (7.17) for continuous dose responses, identify a condition on the parameter β that forces

(7.17) to represent a decreasing dose response. For this case, verify that the parameter ED_{50} in the model is in fact the median effective dose. View the effect here as a decrease from the control level at $x = 0$.

7.42. Van Ewijk and Hoekstra (1993) give data on biomass reductions in lettuce shoots after exposure to isobutylalcohol. Measured was the three-week biomass of two replicate plant shoots at various exposure concentrations:

Conc. x_i *(%)*	*0.0*	*0.32*	*1.0*	*3.2*	*10.0*	*32.0*	*100.0*
Biomass, Y_{ij}	1.126	1.096	1.163	0.985	0.716	0.560	0.375
	0.833	1.106	1.336	0.754	0.683	0.488	0.344

Assume the biomass values are independent and normally distributed with constant variance, and fit the three-parameter logistic model (7.17) to these data. Use log-concentration as your predictor variable, and employ consecutive-dose average spacing for the control value, from (6.9). Find point estimates and 95% confidence intervals for the ED_{50} and the effect parameter β. Is there a significantly decreasing dose response?

7.43*. Repeat the manipulations in Exercise 7.41 for the logistic hormesis model in (7.18).

7.44. Notice that the data in Exercise 7.42 increase slightly before decreasing over dose. This may be a hormetic effect. Repeat the analysis for Exercise 7.42 using the logistic hormesis model in (7.18).

7.45. Verify the expression in (7.20); that is, show that under a simple linear probit model, the $ED_{100\rho}$ is $\{\mathfrak{z}_{(1-\rho)} - \beta_0\}/\beta_1$. For this probit case, what is the ED_{50}?

7.46. Suppose $F(\eta)$ is an increasing, continuous function over $0 < \eta < 1$. Assume $F(\eta)$ also satisfies the constraint that $F(0) = 0.5$. Use $F(\eta)$ to model a quantal response via $\pi(x) = F(\beta_0 + \beta_1 x)$, as in (7.10). With this:

a*. Show that the ED_{50} for this model is always $ED_{50} = -\beta_0/\beta_1$.

b. Does the simple linear logistic model (7.11) satisfy these conditions on $F(\eta)$?

c. Does the simple linear probit model satisfy these conditions?

7.47*. Show that changing the ED of interest in Exercise 7.31 from ED_{50} to $ED_{100\rho}$ does not change the form of the confidence limits. That is, define $\theta = (\delta - \beta_0)/\beta_1$ for some known constant δ, and show that the steps in Exercise 7.31 lead to confidence limits of the same form as seen therein.

7.48*. Find the ED_{50} under the complementary log-log model $\pi(x) = 1 - \exp\{-\exp[\beta_0 + \beta_1 x]\}$ from Table 7.3.

7.49. Find point and 90% confidence interval estimates for the ED_{01} or the log-ED_{01}, as appropriate, with the quantal response data from:

a. Exercises 6.14 and 7.19.

b. Exercises 6.16 and 7.20.

7.50*. What is the initial rate of slope of the simple linear logistic model (7.11)? How does this compare, say, with the ED_{01}? What is a fundamental difference between the two as comparative measures of potency?

7.51*. Can you repeat the comparison in Exercise 7.50 for the simple linear probit model?

7.52*. Verify the suggestion in Example 7.3 that a first-order Taylor series approximation for $f(a) = 1 - e^{-a}$ (about the point $a = 0$) produces $f(a) \approx a$. What is the second-order Taylor approximation for $1 - e^{-a}$ about $a = 0$? How would you use this to approximate the relationships in (7.6) and/or (7.7)?

7.53. Repeat the power-transformed nonlinear model fit to the 1,3-butadiene data in Example 7.3, but now apply the approximation $1 - e^{-a} \approx a$ leading to (7.22). Specifically:

a. Replace $1 - \exp\{-\beta_0 - \beta_1 x_i\}$ in (7.6) with the simple linear predictor $\beta_0 + \beta_1 x_i$. With this, how does the power-transformed function (7.9) change?

b. Fit the power-transformed function you derived in part (a) to the power-transformed data in Table 7.2 via a nonlinear least squares regression routine. How does the new estimator of β_1 compare with the result from Example 7.3?

c. Assume the data in Table 7.2 are normally distributed with constant variances, and find a 95% confidence interval for β_1 based on the fit in part (b). How do the new limits compare with the results from Example 7.3?

7.54. Repeat Exercise 7.53 for the data in Exercises 6.40 and 7.16.

7.55. The analysis in Exercise 7.53 assumes, in effect, that $\theta = 1$ when using (7.23) to approximate (7.6). Extend this to the case of $\theta > 1$, by setting $\theta = 2$ in (7.23); cf. Myers *et al.* (1981). Then, using this approximation, repeat the steps in Exercise 7.53. How do the results compare?

7.56. Repeat Exercise 7.55 for the data in Exercises 6.40 and 7.16.

7.57. Return to the minnow mortality data from Example 7.7.

a. Graph mortality (Y) versus concentration (x) for each group (the control group graph appears in Fig. 7.23), overlaying the scatterplots. Does the graph support the logistic regression assumption?

b. Using the model fit illustrated in Fig. 7.26, calculate and graph the residuals $R_{pi} = Y_{pi} - \hat{Y}_{pi}$ vs. the predicted values $\hat{Y}_{pi}$ for these data. Do the residuals appear randomly scattered? What problems, if any, do you see with your plot?

7.58. Fit a common model to both the female panther data from Exercise 6.8 and the male panther data from Example 7.1, using the methods in Section 7.4. (Remember to use log-concentrations, and to adjust the age variable by its mean.) Test to see if the response relationship differs between female and males; set $\alpha = 0.01$. How, if at all, do female and male panthers differ?

8

Introduction to generalized linear models

As illustrated in previous chapters, environmental data occur in a multitude of forms: continuous measurements, ratios, counts, proportions, etc. When a normal parent assumption is inappropriate for any of these forms, the statistical analysis becomes necessarily more complex than that under normal distribution sampling. Examples include:

- Do increasing concentrations of an environmental toxin released in a watershed lead to exposure-related mortality among the watershed organisms? If mortality is measured as a count or proportion, can this be modeled in a functional manner?
- How do *in vivo* lymphocyte mutation frequencies compare in populations or groups of individuals exposed to different levels of ionizing radiation, and can any differences among the groups be modeled as a function of some concomitant variable, such as sex or age?

In some cases it is reasonable to assume that the measurements recovered from an environmental study are normally distributed (with homogeneous variances), and then to apply standard regression and/or analysis of variance (ANOVA) methods. Since both in multiple regression and ANOVA the mean response is written as a model that is linear in its unknown parameters, both methods fall under the general heading of **linear models**. In other settings, however, the data are clearly not normally distributed (as occurs with counts or proportions) or, even if normally distributed, often exhibit heterogeneous variances. Although transformations may help stabilize the data and make them appear at least approximately normal, they do not always guarantee good convergence to normality. Indeed, in some cases data transformation can actually reduce interpretability, when the transformed scale possesses no sensible context.

An alternative to normalizing transformations involves analyzing the data on the original scale of measurement. To do so, one modifies the normal-based linear model to account for nonnormal data structures. The new models require generalizations of the usual normal-theory models, and are known as **generalized linear models** (GLiMs). We have encountered some specific GLiMs in the preceding chapter: the logistic regression model (7.11) for proportion data, and the log-linear regression model (7.14) for counts are both special cases of GLiMs.

In this chapter, we provide a brief introduction to GLiM methodology, and illustrate some environmetric applications. A more complete presentation requires familiarity with advanced statistical principles, however, as in the introductory text by Dobson (1990) and in the definitive work by McCullagh and Nelder (1989). Indeed, the material we present below takes a step above the intermediate levels in previous chapters, and many of the details noted in Sections 8.2 and 8.3 are intended only for readers interested in studying the introductory technical aspects of GLiMs.

8.1 REVIEW OF CLASSICAL LINEAR MODELS

We begin with a short review of the concepts and terminology of the general linear model. We assume readers are familiar with the technical details underlying this material, as found in texts such as Snedecor and Cochran (1980), Moore and McCabe (1993), or Neter *et al.* (1996); our goal is only to highlight important features and establish notation.

We refer to the usual linear model based on normal theory as a **classical linear model**. In its basic form, this model expresses the mean response as a linear combination of a set of unknown parameters. That is, suppose we observe independent responses Y_{ij} from $j = 1, \ldots, n_i$ experimental units across each of $i = 1, \ldots, k$ different groups. Note that $n_i = 1$ is not uncommon, especially in observational studies where many different predictor variables are associated with only a single response variable. The model can accommodate this unreplicated setting as long as the number of observations, k, is larger than the number of unknown parameters.

A linear model relates the population mean of Y_{ij}, $\mu_i = \mathrm{E}[Y_{ij}]$, to some linear function of the unknown parameters. (Notice that μ_i does not vary with the index j; we assume that all replicate observations within a treatment level are identically distributed.) We have already seen examples of this in Chapter 7: the Pth-order regression model in (7.1), and the response surface equation noted at the end of Section 7.1.1. Although both these forms contain second-order or higher-order terms in the predictor variable(s), they are both valid linear models, since their β-parameters enter into the equation for the mean in a linear fashion. As such, we refer to the terms making up the model for μ_i as a **linear predictor**, or sometimes as a **linear prediction equation** when expressed as a relationship with μ.

Another example is the simple **one-factor ANOVA** model, where the linear predictor contains a grand mean across all populations, υ, and a fixed **treatment effect** term, τ_i, describing how the ith population differs from this grand mean: $\tau_i = \mu_i - \upsilon$ $(i = 1,\ldots, k)$. It is possible to write $\upsilon + \tau_i$ so the unknown β-parameters enter into the model in a linear fashion: use a series of **indicator variables** to classify each Y_{ij} into one of the k treatment groups. That is, over $m = 1, \ldots, k$, let $x_{ijm} = 1$ if $i = m$ (zero otherwise) for every $j = 1,\ldots, n_i$. Also, take a constant predictor variable to recover the grand mean: $x_{ij0} = 1$ for all i and j. With these, the one-factor model $\mu_i = \upsilon + \tau_i$ is equivalent to $\mu_i = \beta_0 x_{ij0} + \beta_1 x_{ij1} + \beta_2 x_{ij2} + \cdots + \beta_k x_{ijk}$, where β_0 corresponds directly to υ and $\beta_1, \ldots, \beta_k$ represent the effect of the τ_is $(i = 1,\ldots, k)$. Many other linear forms for modeling μ_i in this fashion are possible.

Experienced readers will note that when using x-indicator variables in the one-factor model, the last predictor variable, x_{ijk}, is superfluous. When combined with the term β_0 for the grand mean, the first $k-1$ predictor variables will consume all the information available for μ_i in this model. This is a direct consequence of the model parameterization, and is in effect an extreme form of multicollinearity. We refer to the effect as **aliasing** or as an **aliased predictor variable**. Computer outputs typically indicate this by leaving the parameter estimate or other summary information for β_k blank. (Actually, order is not important here: any set of $k-1$ predictor variables will consume all the information for μ_i when combined with β_0. It is common to label as aliased the last variable which is fitted, although different computer packages will have different criteria for such labeling.) In some cases the outputs also print a message stating that the model is **overparameterized**, or that the estimating equations are not of **full rank**. Although these technical terms are perhaps a bit unsettling, the information given for the other parameters is still valid. The easiest way to avoid such disturbances is to remove any aliased predictor variables from the prediction equation. For instance, in our one-factor ANOVA example this is accomplished by forcing $x_{ijk} = 0$ in the model for μ_i. This gives $\mu_i = \beta_0 x_{ij0} + \beta_1 x_{ij1} + \cdots + \beta_{k-1} x_{i,j,k-1}$. Depending on the nature of the x-indicators, however, this can also modify the interpretation of the β-parameters; see Neter *et al*. (1996, Section 16.11).

8.1.1 Estimation

Estimation of the various unknown parameters in the linear model proceeds via the method of least squares (LS) from Section 2.3.2: in this setting, the

LS criterion selects values to estimate the mean response by minimizing the sum of squared deviations

$$Q = \sum_{i=1}^{k} \sum_{j=1}^{n_i} (Y_{ij} - \hat{Y}_{ij})^2.$$

Here, $\hat{Y}_{ij}$ is the **predicted value** of Y_{ij}, which depends upon the form of the linear model. For example, the predicted value under the one-factor model $\mu_i = \upsilon + \tau_i$ is $\hat{Y}_{ij} = \hat{\upsilon} + \hat{\tau}_i$, where $\hat{\upsilon}$ and $\hat{\tau}_i$ are the LS estimators of the individual model parameters. For this simple model, $\hat{\upsilon} + \hat{\tau}_i$ collapses to the sample mean of the n_i replicate observations at the ith condition, $\overline{Y}_i$ ($i = 1, \ldots, k$). (Technically, if any aliased predictor variables are not removed from the model, then the model parameters, υ and τ_i, cannot be estimated; however, their sums, $\upsilon + \tau_i$, may still be estimable.)

For all but the simplest models, LS estimators and predicted values are found most efficiently via computer calculation. Packages such as SAS via PROC GENMOD, PROC GLM, and PROC REG, or S-PLUS via `lm`, `aov`, and `glm` facilitate the computations; for useful presentations on the S-PLUS functions, see Chambers and Hastie (1992b) or Spector (1994).

If, in addition, one is willing to impose distributional specifications on the observations, ML estimation of the unknown parameters is possible. For example, consider the common normal distribution specification, $Y_{ij} \sim$ (indep.) $N(\mu_i, \sigma^2)$; note the assumption of an equal variance, σ^2, for all observations. Under the normal parent model, ML estimators of the unknown linear parameters correspond to LS estimators. Indeed, for any classical linear model with prediction equation $\mu_i = \beta_0 x_{ij0} + \beta_1 x_{ij1} + \beta_2 x_{ij2} + \cdots + \beta_k x_{ijk}$, the $\hat{\beta}_m$-values can be shown to be both unbiased for the unknown parameters they are estimating – that is, $E[\hat{\beta}_m] = \beta_m$ – and normally distributed. In addition, they possess minimum variance among all linear unbiased estimators of the β_ms (Neter *et al.*, 1996, Section 1.6). Thus, inferences on the model parameters are conducted using the normal sampling distribution of the ML estimators. We have seen examples of this with the simple linear and polynomial regression models from Sections 6.1 and 7.1, respectively.

8.1.2 Diagnostics for model adequacy

Once a model fit has been performed, a number of model features require inspection. These include the quality of the model specification and of its underlying assumptions, and the accuracy of any model predictions. One way to do this is to examine the **residual values** from the fit. The **raw residuals** are defined as the deviations of the observations from the predicted values: $R_{ij} = Y_{ij} - \hat{Y}_{ij}$. Notice that the sum of squared residuals, $\sum_{i=1}^{k}\sum_{j=1}^{n_i}R_{ij}^2$, is the quantity Q that is minimized under the LS criterion. In this form, Q is referred to as the **residual sum of squares**. Since the residuals may be viewed as estimated errors of the LS fit, we often call the residual sum of squares the **error sum of squares** (SSE).

The SSE contains important information: under a normal parent distribution, SSE divided by the error *df* for the model yields an unbiased estimator of σ^2: the **mean square error**, MSE = SSE/df_E. For example, under the one-factor model, the $n_+ = \sum_{i=1}^{k}n_i$ observations provide total *df* of $df_{TOT} = n_+ - 1$, and the model fit over k treatment levels takes up $df_{MOD} = k - 1$ model *df*. What remains is represented by the residuals: $df_E = df_{TOT} - df_{MOD} = n_+ - k$. Thus the mean square error under the one-factor model is $\text{MSE} = \sum_{i=1}^{k}\sum_{j=1}^{n_i}R_{ij}^2/(n_+ - k)$.

Patterns in the residuals when plotted in various forms can help identify features of the model fit; for example, random scatter in a plot of R_{ij} against $\hat{Y}_{ij}$ indicates a reasonable model fit with respect to both the specification of E[Y] and the constant-variance assumption. By contrast, obvious curvilinearity in the plot suggests that higher-order polynomial terms or other predictor variables should be considered in the model. Another pattern, variable banding about $R = 0$, indicates departure from homogeneous variance. Many other model violations may be identified via residual graphics; for example, lack of independence among the observations is evidenced when a plot of R_{ij} vs. the order of observation displays a trend, while validity of the normal parent assumptions is assessed by graphing a histogram, boxplot, or normal probability plot of the R_{ij}s (Neter *et al.*, 1996, Section 3.3). Also, raw residuals may be standardized to study **outliers**: large standardized values for an R_{ij} can suggest that the observation departs too strongly from the data stream; cf. Exercise 7.4c and Section 8.3.2.

8.1.3 Inference and model assessment

Beyond characterizing the fit of the linear model, predicted values and residuals are employed to make inferences on the unknown parameters. As noted above, the MSE estimates the unknown variance, σ^2. It is also a crucial component of statistics that assess the model's features: both $1 - \alpha$ confidence intervals for the unknown linear parameters and tests of significance for various model terms are constructed using the MSE.

For example, a common question of interest is whether the prediction equation can be simplified. We first explored this question in Section 7.1.1 when we examined whether higher-order terms in a polynomial regression could be removed without appreciable loss in the quality of the model fit. We now provide a more formal introduction to this procedure. Suppose an investigator can postulate a simpler model that is a subset or special case of the more general linear model. We refer to the model with the larger number of terms as the **full model** while the model with the smaller number of terms is called the **reduced model**. The reduced model is obtained by eliminating a certain number of variables from the linear predictor associated with the full model. Suppose we denote SSE(FM) and $df_E(FM)$ as the residual sum of squares and the associated degrees of freedom, respectively, for the full model, and SSE(RM) and $df_E(RM)$ as the analogous quantities for the reduced model. To test the null hypothesis that the reduced model provides an adequate fit to the data, we form the **conditional F-statistic**

$$F_{\text{calc}} = \frac{\left\{\dfrac{\text{SSE}(RM) - \text{SSE}(FM)}{df_E(RM) - df_E(FM)}\right\}}{\left\{\dfrac{\text{SSE}(FM)}{df_E(FM)}\right\}}.$$

If the null hypothesis is true, $F_{\text{calc}} \sim F[df_E(RM) - df_E(FM), df_E(FM)]$. Reject this hypothesis when F_{calc} exceeds an upper-α critical point from this reference distribution.

Example 8.1 Interaction in a two-factor model

As a simple example of a reduced model comparison, consider the general **two-factor ANOVA** setting where two treatment factors are applied to the experimental units. Call these factor A and factor B. Assume the data are of the form Y_{ijm}, taken over $i = 1, \ldots, k_A$ levels of factor A and $j = 1, \ldots, k_B$ levels of factor B. Assume also that the design is balanced, with $m = 1, \ldots, n$ replicate experimental units observed at each A×B treatment combination. Thus there are $n k_A k_B$ total observations.

For the linear predictor, we write μ_{ij} in the standard two-factor, fixed-effects form

$$\mu_{ij} = \upsilon + \alpha_i + \beta_j + \gamma_{ij} \tag{8.1}$$

where υ represents the grand mean, the α_is parameterize the effect due to factor A, the β_js parameterize the effect due to factor B, and the γ_{ij}s are **interaction terms** that parameterize A×B crossed effects. Notice that model (8.1) includes $(k_A + 1)(k_B + 1)$ parameters: one for υ, k_A for the α-terms, k_B for the β-terms, and $k_A k_B$ for the γ-terms. Hence, since there are only $k_A k_B$ available cell means, we have overparameterized the data. In order to proceed with valid statistical analyses, some form of **identifiability constraint** or **side condition** must be placed on the linear parameters in (8.1). The common constraint is 'zero sums':

$$\sum_{i=1}^{k_A} \alpha_i = \sum_{j=1}^{k_B} \beta_j = \sum_{i=1}^{k_A} \gamma_{ij} = \sum_{j=1}^{k_B} \gamma_{ij} = 0 ,$$

although other forms are possible. For example, the GLIM system (Francis *et al.*, 1993) employs the 'first-cell' constraint $\alpha_1 = \beta_1 = \gamma_{i1} = \gamma_{1j} = 0$. Then, if $i = 1$ corresponds to a control or standard level, the other parameters are interpreted as deviations from that control level.

Interaction is an important concern in environmental problems, occurring when the response to one factor differs in a manner depending on the level of the other factor; see, for example, Ager and Haynes (1990), Butterworth and Quiring (1994), or Kroer *et al.* (1994). Viewed in terms of (8.1), interaction reflects the joint effects of two or more factors beyond what is predicted from the separate additive effects of each factor. **Synergistic** interaction reflects a response greater than that predicted from the additive effects of each factor, while **antagonistic** interaction reflects a response less than additive. For example, interaction between different toxins may represent an important component of the analysis of chemical mixtures as might be encountered at a toxic waste site. (We illustrate some additional features of synergy and antagonism in the discussion on cross-classified tables of proportions in Section 9.5.)

To test whether an interaction term contributes significantly to the linear model, the full model corresponds to (8.1), while the reduced model corresponds to $\mu_{ij} = \upsilon + \alpha_i + \beta_j$ with all the γ_{ij}s set equal to zero. That is, the reduced model hypothesis is $H_0: \gamma_{ij} = 0$ (for all i,j). Given normally distributed data with constant variances, we use a conditional F-statistic to test H_0. The test has $df_E(FM) = (n-1)k_A k_B$ and $df_E(RM) = n k_A k_B - k_A - k_B + 1$, so that the numerator df for the conditional F-test are $(k_A - 1)(k_B - 1)$; that is, the number of parameters specified under H_0. The denominator df are $df_E(FM) = (n-1)k_A k_B$. □

Statistical comparisons between full and reduced models also may be conducted by comparing the likelihoods under both models, via a likelihood ratio test. As we saw in Section 2.4.1, the log of the likelihood ratio (multiplied by –2) is then referred to a χ^2 distribution with df equal to the number of parameters reduced from the full model. For a classical linear model, it can be shown that this ratio takes on a form similar to the conditional F-statistic, above.

8.2 GENERALIZING THE CLASSICAL LINEAR MODEL

The classical linear model in Section 8.1 is an important statistical tool for environmental applications, but it is limited to those settings where the normal parent distribution is valid. Additionally, the model requires a strict identity between the mean parameter, μ_i, and the terms making up the linear predictor. If, for example, the response is a discrete count, its mean rate of response must be a positive function of the predictor variable, and the classical linear model may not apply. Or, even with normally distributed data, the mean may vary in some nonlinear fashion with an underlying predictor variable such as time or temperature. In its basic form, the classical linear model is unable to account for such effects.

It is possible, however, to generalize the linear model, and allow it to overcome these limitations. The generalizations accept forms of nonnormal data, and also link the mean response to the linear predictor in a possibly nonlinear fashion. We call the corresponding family **generalized linear models** (GLiMs). The basic precept of any GLiM is to extend the linear model in two ways: (a) generalize to nonnormal parent distributions such as binomial, Poisson, or gamma; and/or (b) generalize to nonlinear functions that link the unknown means of the parent distribution with the predictor variables. We begin with generalizations to nonnormal parent distributions.

8.2.1 Nonnormal responses

For many measured variables – such as weight changes or blood flow – a normal parent distribution might seem reasonable; for other measured quantities, however, the response may be far from normal. Indeed, many types of data encountered in the environmental sciences arise from counting processes, leading to nonnormal discrete observations (Ott, 1995). In order to account for a variety of different data structures, GLiMs allow the parent distribution of the data to belong to a wide class of probability functions. We have such a family in the **exponential class** from Section 1.4: for a random variable Y, with unknown parameter θ, the exponential class requires that the p.m.f. or p.d.f. of Y take the form given in (1.15):

$$\left.\begin{array}{c} p_Y(y) \\ \text{or} \\ f_Y(y) \end{array}\right\} = \exp\{\, T(y)\,\omega(\theta) - b(\theta) + c(y)\}.$$

Recall that when $T(y) = y$, the distribution is said to be in **canonical form**.

Traditionally, GLiMs have been based on exponential class members that are already in canonical form, and for which θ enters into the probability function via its **natural parameterization**, $\omega(\theta) = \theta$, so the natural parameter *is* θ. A common extension allows, however, for a possibly unknown **scale parameter**, $\varphi > 0$, achieving a specialized exponential class of p.m.f.s or p.d.f.s:

$$\left.\begin{array}{c} p_Y(y) \\ \text{or} \\ f_Y(y) \end{array}\right\} = \exp\left\{\frac{y\theta - b(\theta)}{a(\varphi)} + c(y,\varphi)\right\}, \tag{8.2}$$

where $a(\varphi)$, $b(\theta)$, and $c(y,\varphi)$ are known functions. In (8.2), the mean of Y, $\mu = \mathrm{E}[Y]$, is related to the parameter θ via $\mu = \partial b(\theta)/\partial\theta$, while the variance is $\mathrm{Var}[Y] = a(\varphi)\partial^2 b(\theta)/\partial\theta^2$. Also, operations with log-likelihoods are simplified: for the ith observation among a set of independent observations Y_i, the contribution to the log-likelihood is $\log\{p_Y(y_i)\}$ or $\log\{f_Y(y_i)\}$. Under (8.2), this becomes simply $[y_i\theta - b(\theta)]/a(\varphi) + c(y_i,\varphi)$.

When the scale parameter φ is known, we say $\partial^2 b(\theta)/\partial\theta^2$ is the **variance function** of Y, since it incorporates all the unknown parameter information in the variance term. Because μ and θ are related via $\mu = \partial b(\theta)/\partial\theta$, we often write the variance function as a function of μ, using the notation $\partial^2 b(\theta)/\partial\theta^2 = V(\mu)$.

In many cases, it is found that the scaling function $a(\varphi)$ simplifies to the form φ/w, where $w > 0$ is some known constant; this can simplify operations with p.m.f.s or p.d.f.s in the form of (8.2). Many other simplifying features exist for (8.2); McCullagh and Nelder (1989, Section 2.2.2) give additional details.

Example 8.2 The Poisson distribution as a member of the exponential class

We saw in Example 1.2 that the Poisson distribution could be written in the one-parameter exponential class form (1.15). To see that it also satisfies (8.2), recall that if $Y \sim \mathcal{Poisson}(\mu)$, then

$$p_Y(y) = \frac{\mu^y e^{-\mu}}{y!} I_{\{0,1,\ldots\}}(y) \,.$$

From Example 1.2, the natural parameter for the Poisson is $\theta = \log\{\mu\}$, so let $\mu = e^\theta$. Then, the p.m.f. may be re-expressed in the exponential form

$$\exp\left\{y\theta - e^\theta + \log\left[\frac{I_{\{0,1,\ldots\}}(y)}{y!}\right]\right\}.$$

There is no additional scale parameter entering into this function, so we set $\varphi = 1$, and take, trivially, $a(\varphi) = \varphi = 1$. For $b(\theta) = e^{\theta}$ and $c(y,\varphi) = c(y,1) = \log\{I_{\{0,1,\ldots\}}(y)/y!\}$, and it is clear that the Poisson p.m.f. is of the exponential form (8.2). (A technical caveat: when y is a nonnegative integer the indicator function $I_{\{0,1,\ldots\}}(y)$ equals 1 and the function $c(y,1)$ is well defined. When y is some other value, however, the indicator function is 0, and thus the function $c(y,1)$ here attempts to evaluate the natural logarithm of zero. Though this is technically impossible, we can appeal to a limiting argument for the evaluation: recognize that as its argument approaches 0, the natural logarithm approaches $-\infty$. Evaluated in the exponent of the p.m.f., this drives the probability to an infinitesimal value, the limiting value of which is itself 0. This is precisely what the probability mass should be when y is not a nonnegative integer.) Notice that $\partial b(\theta)/\partial\theta = \partial e^{\theta}/\partial\theta = e^{\theta}$ is indeed equal to μ, and that $\mathrm{Var}[Y] = a(\varphi)\partial^2 b(\theta)/\partial\theta^2 = (1)\partial^2 e^{\theta}/\partial\theta^2 = e^{\theta}$ simplifies also to $\mathrm{Var}[Y] = \mu$, as expected. □

Example 8.3 The gamma distribution as member of the exponential class

Suppose we take $Y \sim \mathcal{G}amma(\alpha, \beta)$, where α and β are positive parameters, as in Section 1.3.2. The associated p.d.f. is

$$f_Y(y) = \frac{y^{\alpha} e^{-y/\beta}}{y\,\Gamma(\alpha)\,\beta^{\alpha}} I_{(0,\infty)}(y).$$

The mean parameter here is $\mathrm{E}[Y] = \mu = \alpha\beta$, and the variance is $\alpha\beta^2 = (\alpha\beta)^2/\alpha = \mu^2\varphi$, where $\varphi = 1/\alpha$. Written in terms of μ and φ, the p.d.f. is

$$f_Y(y) = \frac{y^{1/\varphi}\exp\{-(\mu\varphi)^{-1}y\}}{y\,\Gamma(1/\varphi)\,(\mu\varphi)^{1/\varphi}} I_{(0,\infty)}(y)\,,$$

which is equivalent to

$$f_Y(y) = \exp\left\{\frac{-(y/\mu) - \log(\mu)}{\varphi} + \frac{\log(y/\varphi)}{\varphi} - \log[y\,\Gamma(1/\varphi)\,I_{(0,\infty)}(y)]\right\}.$$

Writing the natural parameter as $\theta = -1/\mu$ shows that this density is in the form (8.2), for $a(\varphi) = \varphi$, $b(\theta) = -\log\{-\theta\}$, and $c(y,\varphi) = \varphi^{-1}\log(y/\varphi) - \log\{y\,\Gamma(1/\varphi)I_{(0,\infty)}(y)\}$. (A similar caveat applies as in Example 8.2 regarding the natural logarithm of the indicator function, $\log\{I_{(0,\infty)}(y)\}$.) □

Exercise 8.6 explores other members of the natural exponential class, including the binomial, geometric, and normal distributions.

8.2.2 Linking the mean response to the predictor variables

A second generalization of the classical linear model from Section 8.1 involves extending the relationship between the mean parameter, μ_i, and the linear predictor. For notational convenience, we denote any general linear predictor as η_i, for example, $\eta_i = \upsilon + \tau_i$ for the one-factor ANOVA model or $\eta_i = \beta_0 + \beta_1(x_i - \bar{x}) + \beta_2(x_i - \bar{x})^2 + \cdots + \beta_P(x_i - \bar{x})^P$ for the polynomial regression model (7.1). For simplicity, the subscripts on μ and η will be omitted unless the index is required for clarity.

For a classical linear model, the mean, μ, and the linear predictor, η, are identically related, $\mu = \eta$. GLiMs extend this relationship by allowing some monotone function of μ_i to equal the linear predictor. This is written as $g(\mu) = \eta$, where $g(\cdot)$ is the **link function**. In some cases, the link is trivial; for example, $\mu = \eta$ is just the **identity link**, $g(\mu) = \mu$. No matter what its form, however, the link function defines the scale over which the systematic effects of the model, represented by η, are modeled as additive.

The link function represents an important generalization of the linear model, especially when some constraints on the nature of μ suggest certain forms for $g(\mu)$. For instance, in Example 8.2 the Poisson mean μ must be positive. Thus we require a link function that takes a strictly positive quantity to a linear predictor that can take on any real value. A common choice is the natural logarithm, $g(\mu) = \log(\mu)$, although we explored a few other possibilities at the end of Section 7.2.3. This motivates use of the term **log-linear model** for this specific form of GLiM (as in Section 7.2.2; see also Sections 8.4.4 and 9.4). Or, with binomial data the parameter of interest is the response proportion, π, for which the link function must take a value between 0 and 1 to the real line. Table 7.3 lists a series of link functions that satisfy this requirement.

Since the link function is assumed monotone, it is possible to derive the **inverse link function**, say $g^{-1}(\eta)$, which characterizes the mean as a function of the linear predictor. For simplicity, we write $h(\eta) = g^{-1}(\eta)$, so that $\mu_i = h(\eta_i)$. For example, with the Poisson distribution, the inverse link to $g(\mu) = \log(\mu) = \eta$ is $\mu = g^{-1}(\eta) = e^{\eta}$. Or, for the binomial distribution, the inverse links in Table 7.3 are the c.d.f.s being used to describe the tolerance distributions for the GLiM.

We close this section by noting a special link function that can be applied with any GLiM, which, because of its inherent elegance, is often the first choice when no other motivation is available. Recall that the natural parameter, θ, in (8.2) is viewed as the most compact parameterization of the unknown information in the data. Commonly, however, θ is not equal to the mean, μ, but is instead some function of it. In Example 8.2, for instance, the natural parameter is $\theta = \log\{\mu\}$. In general, if the relationship between θ and μ represents a monotone function of μ, it can be employed as a link function. (Typically, we ignore constants such as 1/2 or –1.) In such a case, the link is called a **canonical link**. Table 8.1 gives some common GLiMs with their canonical links; see also McCullagh and Nelder (1989, Section 2.2).

Table 8.1 Selected GLiMs with canonical link functions

Distribution	*Mean*	$b(\theta)$	*Natural parameter*	*Canonical link*
Normal:				
$Y \sim N(\mu,\sigma^2)$	μ	$\theta^2/2$	$\theta = \mu$	Identity: $g(\mu) = \mu$
Gamma:				
$Y \sim \mathcal{G}amma(\alpha,\beta)$	$\mu = \alpha\beta$	$-\log(-\theta)$	$\theta = -1/\mu$	Reciprocal: $g(\mu) = 1/\mu$
Binomial:				
$Y \sim \mathcal{B}in(N,\pi)$	$E[Y/N] = \pi$	$\log(1+e^{\theta})$	$\theta = \log\left\{\frac{\pi}{1-\pi}\right\}$	Logit: $g(\pi) = \text{logit}\{\pi\}$
Poisson:				
$Y \sim \mathcal{P}oisson(\mu)$	μ	e^{θ}	$\theta = \log\{\mu\}$	Log: $g(\mu) = \log\{\mu\}$

In the next section, we introduce details for defining, fitting, and deriving inferences from GLiMs. We use as a guide our review of classical linear models from Section 8.1. It will be seen that the machinery employed in classical linear models extends quite naturally to GLiMs. We note, however, that the next section is intended as a technical introduction for readers interested in the procedural aspects of generalized linear modeling, and few pedagogic illustrations are provided. Readers more concerned with specific examples and applications of GLiMs in environmetric settings may wish to move forward to the formal examples in Section 8.4, and use Section 8.3 only for background reference.

8.3 GENERALIZED LINEAR MODELS

As in Section 8.1, suppose we observe mutually independent responses Y_{ij} from $j = 1, \ldots, n_i$ replicate experimental units across each of $i = 1, \ldots, k$ different groups. (The replicate index j is not required; we include it only for generality.) The distribution of Y_{ij} is assumed to follow one of the exponential class forms in (8.2). That is, the p.m.f. or p.d.f. for any Y_{ij} is

$$\exp\left\{\frac{y_{ij}\theta_i - b(\theta_i)}{a(\varphi_i)} + c(y_{ij}, \varphi_i)\right\}. \tag{8.3}$$

In (8.3), the natural and scale parameters may vary (in a deterministic fashion) with the treatment index i, but not with the replicate index j.

For inferential and predictive purposes, a linear model is postulated that relates the population mean, $\mu_i = \mathrm{E}[Y_{ij}]$, to a **linear predictor**, η_i, of unknown regression parameters via the monotone link function $g(\mu_i) = \eta_i$. Notice that the mean parameter is a function of θ_i via $\mu_i = \partial^2 b(\theta_i)/\partial\theta_i^2$, so it does not vary with the index j; that is, we assume all replicate observations within a treatment level are identically distributed. We write the linear predictor as a linear combination of $P + 1 > 1$ predictor variables, x_{im}, and $P + 1$ unknown β-parameters:

$$\eta_i = \beta_0 x_{i0} + \beta_1 x_{i1} + \cdots + \beta_P x_{iP} . \tag{8.4}$$

(Generally, $x_{i0} = 1$ if an intercept term is incorporated in the model.)

As noted above, an implication of this structure is that the mean response, μ_i, may be written as

$$\mu_i = g^{-1}(\eta_i) = h(\eta_i) = h(\beta_0 x_{i0} + \beta_1 x_{i1} + \cdots + \beta_P x_{iP}),$$

where $h(\eta) = g^{-1}(\eta)$ is the inverse link function. Obviously, (8.4) can include polynomial forms such as (7.1). With proper selection of the x-variables, however, it can also include one-factor, two-factor, or other forms of ANOVA-type models, as in Section 8.1 or Example 8.1. In general, the unknown β-parameters enter into (8.4) in a linear fashion. It is the vector of β-parameters, $\boldsymbol{\beta} = [\beta_0 \ldots \beta_P]$, to which statistical inferences are directed. For example, if a particular parameter, say β_3, is seen to be significantly different from zero, then the individual predictor variable x_{i3} contributes significantly to the GLiM.

8.3.1 Estimation

Estimation of the β-parameters is accomplished via maximum likelihood. Since the Y_{ij}s are assumed independent, the likelihood function $\mathcal{L}(\theta_1, \ldots, \theta_k \mid \mathbf{Y})$ is the product of individual terms of the form in (8.3). The corresponding log-likelihood is a sum over i and j of the exponent terms from (8.3):

$$\ell(\theta_1, \ldots, \theta_k) = \sum_{i=1}^{k}\sum_{j=1}^{n_i}\left\{\frac{y_{ij}\theta_i - b(\theta_i)}{a(\varphi_i)} + c(y_{ij}, \varphi_i)\right\}.$$

Assuming the scale parameters φ_i are known, the ML estimates of θ_i are found by maximizing $\ell(\theta_1, \ldots, \theta_k)$ with respect to $\theta_1, \ldots, \theta_k$. Under the GLiM, however, the parameters of interest are the β_ms, not the θ_is. These are linked to the latter through the linear predictor and the link function:

$$\begin{aligned} &\text{linear predictor:} && \eta_i = \beta_0 x_{i0} + \beta_1 x_{i1} + \cdots + \beta_P x_{iP} \\ &\text{link function:} && g(\mu_i) = \eta_i \\ &\text{mean parameter:} && \partial b(\theta_i)/\partial\theta_i = \mu_i . \end{aligned}$$

That is, each θ_i may be written as an implicit function of the β_ms (and of the predictor variables). Thus it is appropriate to denote the likelihood and log-likelihood as functions of the $\boldsymbol{\beta}$-vector: $\mathcal{L}(\boldsymbol{\beta} \mid \mathbf{Y})$ and $\ell(\boldsymbol{\beta})$, respectively.

To find the ML estimate of β_m, we maximize $\ell(\boldsymbol{\beta})$ with respect to it. This is accomplished by setting the first derivative of $\ell(\boldsymbol{\beta})$ with respect to each β_m equal to zero, $m = 0, \ldots, P$. Using differential calculus (Exercise 8.8), this produces the P estimating equations

$$\sum_{i=1}^{k} \sum_{j=1}^{n_i} \frac{x_{im}\, h'(\eta_i)}{a(\varphi_i)\, V(\mu_i)} (y_{ij} - \mu_i) = 0 . \tag{8.5}$$

In (8.5), $h'(\eta_i) = \partial h(\eta_i)/\partial\eta_i$ is the first derivative of $h(\eta_i)$, η_i is the linear predictor from (8.4), and $V(\mu_i)$ is the variance function from Section 8.2.1: $\mathrm{Var}[Y_{ij}] = a(\varphi_i)\, V(\mu_i)$.

Solution of (8.5) for all $m = 0, \ldots, P$ yields the joint ML estimate of $\boldsymbol{\beta}$. Unfortunately, this generally requires computer iteration. The ML estimating equations (8.5) correspond to a form of weighted least squares where the weights are chosen iteratively, based on the previous iteration's β-estimates. The method is known as **iteratively reweighted least squares** (McCullagh and Nelder, 1989, Section 2.5). The computations are accomplished, for example, using the package GLIM (Francis *et al.*, 1993) which stands for Generalized Linear Interactive Modeling and is designed specifically for GLiMs, the SAS procedure PROC GENMOD, or the S-PLUS routine `glm`. (We warn the reader that the similarly named SAS procedure PROC GLM is useful for the classical linear model as described in Section 8.1, but it is not intended for use with more general GLiMs.)

As a technical aside, note that SAS PROC GENMOD fits GLiMs via a Newton–Raphson algorithm to maximize the log-likelihood directly (SAS Institute Inc., 1993). In contrast, the S-PLUS `glm` routine uses iteratively reweighted least squares. Also, potential differences in final estimates may occur among different packages when the scale parameter φ is estimated differently. By default, SAS uses maximum likelihood to estimate φ, but S-PLUS uses a scaled deviance function to achieve a method of moments estimate. The latter approach is only a user-specified option (`dscale`) in SAS. As such, subtle differences in parameter estimates may result when using different GLiM computer packages.

Note that for small values of n_i, the ML estimators of $\boldsymbol{\beta}$ are biased, for which computational corrections can be applied. These will vary, depending on the nature of the linear predictor, the link function, any identifiability constraints on the model (such as the sum of coefficients being equal to zero, etc.; see Section 8.1.3), and the underlying parent distribution (Cordeiro and McCullagh, 1991).

8.3.2 Diagnostics for model adequacy

As with the classical linear model, it is important to inspect the results of a GLiM fit for quality and accuracy. Given the vector of ML estimators $\hat{\boldsymbol{\beta}} = [\hat{\beta}_0 \ldots \hat{\beta}_P]$, the ML estimator of the linear predictor is $\hat{\eta}_i = \hat{\beta}_0 x_{i0} + \cdots + \hat{\beta}_P x_{iP}$. Applying the inverse link function to $\hat{\eta}_i$ yields ML estimates of the mean response as simply $\hbar(\hat{\beta}_0 x_{i0} + \cdots + \hat{\beta}_P x_{iP})$. Since $\mu_i = \mathrm{E}[Y_{ij}]$, these quantities are taken as the **predicted values** $\hat{Y}_{ij} = \hbar(\hat{\beta}_0 x_{i0} + \cdots + \hat{\beta}_P x_{iP})$, from which **raw residuals**

$$R_{ij} = Y_{ij} - \hat{Y}_{ij} = Y_{ij} - \hbar(\hat{\beta}_0 x_{i0} + \cdots + \hat{\beta}_P x_{iP})$$

are calculated. (A note of caution: with binomial data, the parameter of interest is $\mathrm{E}[Y_{ij}/N_{ij}] = \pi_i$, not $\mathrm{E}[Y_{ij}] = \mu_i = N_{ij}\pi_i$. Thus the link function relates π_i, not μ_i, to η_i. In terms of the original observations, the predicted values in this case are written as $\hat{Y}_{ij} = N_{ij}\hbar[\hat{\beta}_0 x_{i0} + \cdots + \hat{\beta}_P x_{iP}]$.)

Diagnostic plots and other summary operations on the residuals (of whatever form) help indicate potential model violations, as discussed in Section 8.1.2. Models that incorporate additional dispersion may be appropriate in this case. To assess potential excess variability, a raw residual may be scaled by dividing it by the estimated standard deviation of Y_{ij}; we termed the result a **Pearson residual** in Section 2.7.1. For GLiMs, one employs

$$R_{ij}^P = \frac{Y_{ij} - \hat{Y}_{ij}}{\sqrt{V(\hat{\mu}_i)}},$$

where $V(\hat{\mu}_i)$ is the estimated variance function of the GLiM. Summing the squared Pearson residuals over $i = 1, \ldots, k$ and $j = 1, \ldots, n_i$ produces a summary measure of residual variation, often referred to as the **Pearson**

goodness-of-fit statistic: $X^2 = \sum_{i=1}^{k}\sum_{j=1}^{n_i}(R_{ij}^P)^2$. Some authors use X^2 to quantify goodness-of-fit, since in large samples $X^2 \,\dot{\sim}\, \chi^2(n_+ - P)$ when the model is specified correctly. In general, however, the quality of the large-sample χ^2 approximation varies greatly, depending on the parent distribution of the Y_{ij}s.

Other residual definitions are possible, including Studentized and deviance-based versions. The deviance-based form is $R_{ij}^D = |Y_{ij} - \hat{Y}_{ij}|(Y_{ij} - \hat{Y}_{ij})^{-1}\sqrt{d_{ij}}$, where d_{ij} is the (i,j)th contribution to the deviance function. It can be more stable than R_{ij}^P, and is entering greater use; see McCullagh and Nelder (1989, Section 2.4), Duffy (1990), and the references therein.

8.3.3* Goodness-of-link

One new concern that arises when assessing GLiM adequacy is the quality of the link assumption. This is an issue of **goodness-of-link**: how well does the assumed link function represent the true relationship between μ and η? To test goodness-of-link, Pregibon (1985) gave a general methodology that embeds the hypothesized link into an extended family of link functions. That is, suppose we extend the link function to, say, $\eta_i = g(\mu_i;\gamma)$, where the parameter γ indexes the extended family. If at some special value of γ, such as $\gamma = 0$ or $\gamma = 1$, the extended family reduces to the particular link function of interest, then a test for the goodness-of-link involves testing if γ is significantly different from that specific value. (Notice that significance here indicates a poor link specification.) Examples of such families are common. For instance, under a binomial parent model numerous forms are available, each of which contains the canonical logit link from Table 8.1; Morgan (1993, Chapter 4) provides a useful review. For example, the extended link function

$$g(\pi;\gamma) = \log\left\{\frac{(1-\pi)^{-\gamma} - 1}{\gamma}\right\} \tag{8.6}$$

contains the logit link at $\gamma = 1$ (Aranda-Ordaz, 1981). Or

$$g(\pi;\gamma) = \log\left\{\frac{\pi + \gamma(1-\pi)}{1-\pi}\right\}, \tag{8.7}$$

contains the logit link at $\gamma = 0$ (Whittemore, 1983).

For Poisson parent models, the canonical log-linear model represented by (7.14) may be embedded in a power family $g(\mu;\gamma) = (\mu^\gamma - 1)/\gamma$, for $\gamma > 0$. The log-linear link obtains in the limit as $\gamma \to 0$. Under this power-function link, the inverse link is $h(\eta) = (1 + \gamma\eta)^{1/\gamma} I_{[-1/\gamma,\infty)}(\eta)$. (Notice the truncation below $\eta = -1/\gamma$, in order to satisfy the constraint that $\mu > 0$.)

Once an extended link function $g(\mu;\gamma)$ is adopted, some null ('good') link is set via $H_0{:}\,\gamma = \gamma_0$ Then, testing for goodness-of-link within the extended family involves the new linear predictor

$$\eta_i = \beta_0 x_{i0} + \beta_1 x_{i1} + \cdots + \beta_P x_{iP} + \delta z_i, \tag{8.8}$$

where $\delta = \gamma_0 - \gamma$ and $z_i = g'(\mu_i;\gamma_0)$. (The latter term is $\partial g(\mu_i;\gamma)/\partial\gamma$ evaluated at $\gamma = \gamma_0$.) Testing goodness-of-link at $\gamma = \gamma_0$ is then tantamount to testing $\delta = 0$ (Pregibon, 1985). Because the z_i terms usually involve the unknown β-parameters, however, a preliminary iteration is required for the fit: use the desired link at $\gamma = \gamma_0$ to form an estimate for z_i, and denote this by $\hat{z}_i$. This can be, for example, $g'(\tilde{\mu}_i;\gamma_0)$, where $\tilde{\mu}_i$ is the restricted ML estimate of μ_i under $H_0{:}\,\gamma = \gamma_0$. Then, run the model fit using the extended linear predictor (8.8), with $\hat{z}_i$ in place of z_i. Iterate to improve the estimate of z if interest exists in calculating ML estimates of $\boldsymbol{\beta}$ or δ (Pregibon, 1980).

To test for goodness-of-link, Pregibon (1985) recommends use of the score test of $\delta = 0$, since it is identical to the score test of $\gamma = \gamma_0$. The goodness-of-link hypothesis corresponds to $H_0{:}\,\delta = 0$, under which the score statistic is approximately $\chi^2(1)$. Reject the goodness-of-link when the statistic exceeds the upper-α critical point $\chi^2_\alpha(1)$. Here again, computer calculation is required for the computations; useful programming code for constructing score tests in GLIM is given in Pregibon (1982).

Some caveats regarding this methodology are in order. First, the nature of the link family, or of the design matrix associated with the predictor variables, may hinder the analysis by aliasing $\hat{z}_i$ with some linear combination of the predictor variables. This can occur if the null model leads to fitted values for $\hat{\mu}$ that are essentially constant (Pregibon, 1980), aliasing $\hat{z}_i$ with the grand mean, or if the linear predictor contains only one significant qualitative predictor variable, for example, a dichotomous indicator. Also, incorrect specification of the parent distribution can lead to unstable or incorrect inferences, even when the link function is identified correctly (Lachenbruch, 1990). For settings where the distributional specification is uncertain or unknown, more complex testing methodology is necessary; for example, Cheng and Wu (1994) propose a form of semi-parametric goodness-of-link test, based on quasi-likelihood methods that do not require a fully specified likelihood.

8.3.4* Inference and model assessment

Once calculated, the ML estimates $\hat{\boldsymbol{\beta}}$ are employed to perform standard statistical inferences on $\boldsymbol{\beta}$. Approximate standard errors and estimated covariances among the $\hat{\beta}_m$s are constructed using large-sample likelihood theory, as described, for example, in Cox (1988). We calculate standard errors based on estimates of the Fisher information elements for each $\hat{\beta}_m$ ($m = 0, \ldots, P$). The information elements are collected together into the Fisher information matrix for $\boldsymbol{\beta}$, $\mathbf{F}_{\boldsymbol{\beta}}$. From Section 2.3.4, the mth diagonal element of $\mathbf{F}_{\boldsymbol{\beta}}$ here has the form

$$\mathfrak{F}_m = \mathrm{E}\left[\frac{-\partial^2\ell(\beta_0, \ldots, \beta_P)}{\partial\beta_m^2}\right] = \sum_{i=1}^{k} \frac{n_i\, x_{im}^2}{a(\varphi_i)V(\mu_i)}[\hbar'(\eta_i)]^2 \tag{8.9}$$

and the (t,u)th off-diagonal element is

$$\mathfrak{F}_{tu} = \mathrm{E}\left[\frac{-\partial^2\ell(\beta_0, \ldots, \beta_P)}{\partial\beta_t\partial\beta_u}\right] = \sum_{i=1}^{k} \frac{n_i\, x_{it}\, x_{iu}}{a(\varphi_i)V(\mu_i)}[\hbar'(\eta_i)]^2 \tag{8.10}$$

(Exercises 8.13 and 8.14). The Fisher information matrix is assembled from (8.9) and (8.10), and the inverse of this matrix, $\mathbf{F}_{\boldsymbol{\beta}}^{-1}$, is used to approximate the variance-covariance matrix of $\hat{\boldsymbol{\beta}}$. An approximate standard error for the ML estimator of β_m, $se[\hat{\beta}_m]$, is the square root of the mth diagonal element of this inverse matrix.

Typically, $\mathbf{F}_{\boldsymbol{\beta}}^{-1}$ will itself contain unknown β-parameters. If so, we substitute the ML estimates for these values, and use the resulting, estimated inverse matrix $\hat{\mathbf{F}}_{\boldsymbol{\beta}}^{-1}$. Also, the large-sample covariance of $\hat{\beta}_t$ and $\hat{\beta}_u$ is estimated by the (t,u)th element of $\hat{\mathbf{F}}_{\boldsymbol{\beta}}^{-1}$. All the standard computer packages noted above display these $\hat{\mathbf{F}}_{\boldsymbol{\beta}}^{-1}$-based quantities as part of their standard or optional outputs.

Using the large-sample standard errors, approximate $1-\alpha$ Wald intervals on any β_m may be formed: $\hat{\beta}_m \pm \mathfrak{z}_{\alpha/2} se[\hat{\beta}_m]$. (If simultaneous intervals on $J > 1$ parameters are desired, one can adjust the critical point via a Bonferroni correction: $\mathfrak{z}_{\alpha/(2J)}$.) Or, to test $\mathrm{H_0}$: $\beta_m = 0$ vs. $\mathrm{H_a}$: $\beta_m \neq 0$, a Wald test, a likelihood ratio (LR) test, or a score test may be constructed, using the appropriate likelihood-based values. For instance, the (two-sided) Wald statistic $W_{calc} = |\hat{\beta}_m|/se[\hat{\beta}_m]$ is referred in large samples to a standard normal distribution. Reject $\mathrm{H_0}$ in favor of $\mathrm{H_a}$ when W_{calc} exceeds $\mathfrak{z}_{\alpha/2}$. Or, the LR statistic is the difference $G^2_{calc} = -2\ell(\tilde{\boldsymbol{\beta}}) + 2\ell(\hat{\boldsymbol{\beta}})$, where $\tilde{\boldsymbol{\beta}}$ is the restricted ML estimator of $\boldsymbol{\beta}$ under $\mathrm{H_0}$: $\beta_m = 0$. In large samples, G^2_{calc} is referred to a $\chi^2(1)$ distribution; if the test statistic exceeds $\chi^2_\alpha(1)$, we reject $\mathrm{H_0}$ in favor of $\mathrm{H_a}$.

As noted in Sections 2.4.1 and 7.2.1, the LR statistic is essentially a difference in deviances, $G^2_{calc} = D(\tilde{\boldsymbol{\beta}}) - D(\hat{\boldsymbol{\beta}})$, where $\tilde{\boldsymbol{\beta}}$ is the restricted ML estimator of $\boldsymbol{\beta}$ under the null hypothesis of interest, H_0. When viewed in this form, the LR/deviance statistic summarizes the importance of various terms in the model. If the specification of model parameters under H_0 nests that model within the fuller model containing all parameters, this is similar to the comparison between reduced and full models discussed in Section 8.1.3. Then, since the nested model involves fewer parameters, its deviance should be larger than that of the fuller model. The difference in these deviances is G^2_{calc}, and we can test the significance of the increase by comparing G^2_{calc} to an upper-α critical point from its large-sample reference distribution, here χ^2 with *df* equal to the number of parameters specified in H_0. When G^2_{calc} is significant, some combination of predictor variables associated with the fuller model contributes significantly to it, and should be retained. Across a series of nested sub-models, this relationship can be summarized in an **analysis of deviance table**, mimicking the effect seen in an ANOVA table for a classical linear model. We provide some examples of the use of analysis of deviance tables in Section 8.4.

In small samples, the χ^2 approximation for the LR/deviance statistic may suffer, and some form of adjustment to the test statistic is appropriate. The most common adjustment divides G^2_{calc} by a correction factor that makes its null distribution approximate $\chi^2(1)$ more closely. This is known as a **Bartlett correction**, after Bartlett's pioneering work in the area (Bartlett, 1937). For LR tests with GLiMs, Cordeiro (1983) provided Bartlett correction factors, and Cordeiro *et al*. (1994b) gave some extensions and a review. The methodology may be extended to score statistics and Wald statistics for selected GLiMs (Cordeiro *et al.*, 1993; 1994a).

8.4 EXAMPLES AND ILLUSTRATIONS

8.4.1 Constant-variance continuous data (normal random variables)

GLiMs for continuous data whose variance is constant often are based on a normal parent distribution. Under the normal model, the GLiM scale parameter, φ, is simply the variance σ^2. Also, since the normal mean, μ, may take on any real value, no restrictions are required on the link function. From Table 8.1, the canonical link is the simple identity $\eta = \mu$, but the choice of link may be driven by other subject-matter concerns.

In general, a GLiM using the normal parent model may be viewed as a special case of a nonlinear model, and fitted using methods similar to those discussed in Chapter 7. For example, one can use SAS PROC GENMOD instead of PROC GLM or PROC REG for a linear or polynomial predictor, or, if the link is properly defined, instead of PROC NLIN for a more complex nonlinear model.

Example 8.4 Quadratic regression – Panther hormone concentrations
In Example 7.1 and Exercises 6.7 and 7.4, we studied hormone concentrations in male Florida panthers (*Felis concolor coryi*) as a function of the animals' ages, x. The response variable was the logarithm of the ratio of blood-level estriadol to blood-level testosterone (each in pg/ml), and it was taken as approximately normally distributed. $n = 8$ observations gave a sample mean age of $\bar{x} = 4.875$ years. (Refer to Exercise 6.7 for the data or see the SAS code in Fig. 8.1.)

In Example 7.1, we used SAS PROC REG to fit a quadratic model to these data. Here, Fig. 8.1 gives SAS code using PROC GENMOD to perform the same fit. The resulting output (edited for presentation) appears in Fig. 8.2. There, we find a number of features corresponding to the familiar ANOVA-type output from Fig. 7.2. First and foremost, the deviance is identically equal to the Pearson X^2 statistic (see under `Criteria For Assessing Goodness Of Fit`) and they both equal the residual sum of squares (SSE) for the full model (`MODEL1`) in Fig. 7.2. In fact, these equalities are guaranteed under a normal parent model using the identity link. Second, under `Analysis Of Parameter Estimates` we see the ML estimates (and their standard errors) coincide with the results in Fig. 7.2: $\hat{\beta}_0 = -0.2346$ (0.1526), $\hat{\beta}_1 = 0.0555$ (0.0241), and $\hat{\beta}_2 = -0.0325$ (0.0115). The associated two-sided test statistics under `ChiSquare` in `Analysis Of Parameter Estimates` are exactly the squares of the t-statistics under `T for H0:Parameter=0` in Fig. 7.2, but the corresponding P-values in Fig. 8.2 are slightly off! This occurs because PROC GENMOD uses a large-sample χ^2-approximation for the P-value, while PROC REG and PROC GLM use the exact t distribution. The effect for these data is important: we saw in Fig. 7.2 that the overall F-test was only marginally significant ($P = 0.0905$), and that only the quadratic term (`X2`) showed any significant effect ($P = 0.0367$). In Fig. 8.2, however, the large-sample χ^2 approximation overstates the significance of the results: the P-values are *lower* than their (correct) counterparts in Fig. 7.2. (Recall that some adjustments to the model may be appropriate here, as illustrated in Exercise 7.4.)

We also include in Fig. 8.1 an option that displays the estimated variance-covariance matrix of the ML estimators. (The option is part of the `model` statement: `covb`.) The resulting matrix appears at the bottom of the output.

```
* SAS code to fit Panther data from Exercise 6.7;
data mpanther;
input age ratio @@;
y = log(ratio);
   agebar = 4.875 ;
         x = age - agebar;
         x2 = x*x;
cards;
 1  0.42    2  0.39    2  0.54    3  0.65
 3  0.75    9  0.56    9  0.53   10  0.48
proc genmod;
   model y = x x2 / dist=normal  link=identity
                    dscale       covb;
```

Fig. 8.1 SAS PROC GENMOD code for quadratic regression fit to data in Exercise 6.7.

```
                    The SAS System
                 The GENMOD Procedure
                   Model Information
      Distribution                        NORMAL
      Link Function                       IDENTITY
      Dependent Variable                  Y
      Observations Used                   8

       Criteria For Assessing Goodness Of Fit
Criterion                DF          Value        Value/DF
Deviance                  5         0.1238          0.0248
Pearson Chi-Square        5         0.1238          0.0248
Log Likelihood            .         4.9440               .

           Analysis Of Parameter Estimates
Parameter  DF   Estimate   Std Err   ChiSquare  Pr>Chi
INTERCEPT   1    -0.2346    0.1526      2.3653  0.1241
X           1     0.0555    0.0241      5.3207  0.0211
X2          1    -0.0325    0.0115      8.0026  0.0047
SCALE       0     0.1573    0.0000           .       .
NOTE:   The scale parameter was estimated by
        the square root of DEVIANCE/DOF.

            Estimated Covariance Matrix
 Parameter     INTERCEPT             X            X2
 INTERCEPT    0.02327822    0.00257820   -0.00163312
         X    0.00257820    0.00057965   -0.00020860
        X2   -0.00163312   -0.00020860    0.00013214
```

Fig. 8.2 Output (edited) from SAS PROC GENMOD quadratic regression fit to data in Exercise 6.7.

The matrix's diagonal elements are just the squares of the standard errors, while the off-diagonal elements are the estimated covariances. For example, $\mathrm{Cov}[\hat{\beta}_1, \hat{\beta}_2]$ is estimated as –0.0002086. This value can be useful in operations involving joint inferences on β_1 and β_2.

We close by noting an important, additional option included in Fig. 8.1's PROC GENMOD `model` statement: `dscale`. This calls for estimation of the scale parameter via the deviance, which we know from above identically equals the SSE. (See the `NOTE` following the `Analysis Of Parameter Estimates` output.) For the normal parent model, this would be appropriate, since we know that MSE = SSE/df_E is an unbiased estimator of the unknown variance σ^2. Indeed, we see in Fig. 8.2 that the `Parameter Estimate` for `Scale` is 0.1573. This should be identically equal to $\sqrt{\mathrm{MSE}}$ (PROC GENMOD views the `scale` parameter for the normal parent model as $\varphi^{1/2} = \sigma$), and this is indeed the case in Fig. 7.2; see `Root MSE` under `MODEL1`. If we had *not* specified the `dscale` option, PROC GENMOD would estimate $\varphi^{1/2}$ via maximum likelihood, and the values would not coincide exactly. (The $\hat{\beta}_m$s would still be unbiased estimates, but their standard errors would be underestimated, leading to potentially spurious inferences.) In general, we advise caution when applying PROC GENMOD with normally distributed data. ❑

8.4.2 Quadratic-variance continuous data (gamma random variables)

A common GLiM for positive-valued continuous data where the variance is not constant involves a gamma parent distribution. The standard formulation for the gamma distribution is given in Section 1.3.2, with parameters $\alpha > 0$ and $\beta > 0$. As seen in Example 8.3, however, the gamma can be reparameterized for a GLiM by taking its mean parameter as $\mu = \alpha\beta$, and its exponential class scale parameter as $\varphi = 1/\alpha$. In this form, the population variance is quadratic in μ, that is, $\alpha\beta^2 = \mu^2\varphi$.

Since its variance is quadratic in μ, the gamma possesses an important feature: the ratio of its population standard deviation to its mean is constant in μ, $(\varphi\mu^2)^{1/2}/\mu = \varphi^{1/2}$. This ratio is referred to as the population **coefficient of variation** (CV), which is used to compare the relative variability between two populations. For those situations where the data are thought to have a constant CV, the gamma model often is considered (McCullagh and Nelder, 1989, Chapter 8; Maul, 1991).

From Table 8.1, the canonical link for a gamma GLiM is the reciprocal, $g(\mu) = 1/\mu$. The corresponding inverse link is also in a reciprocal form, but in order to satisfy the model constraint that $\mu > 0$, we truncate the linear predictor if it reaches or drops below zero: $\mu = \eta^{-1} I_{(0,\infty)}(\eta)$. An alternative link function that does not suffer this problem is the natural logarithm $g(\mu) = \log(\mu)$, with inverse link $\mu = e^{\eta}$. In the latter case, we could call this a form of **log-linear model** for μ, as in Section 7.2.2, except that the data are now continuous measurements with constant CV.

Example 8.5 Gamma GLiM for mutant frequencies
Direct *in vivo* study of human toxic response to environmental agents is performed via a variety of protocols. The scope and nature of these methods evolves rapidly with increasing biomedical technology, and certain background features and aspects of a new protocol or system must be understood in order for the system to be used efficiently. For example, Finette *et al.* (1994) report data on possible gender and age differences in human teenagers and children studied for mutational damage at the human hypoxanthine phosphoribosyl transferase (*hprt*) locus. The goal is to gain from the data analysis an understanding of the factors influencing spontaneous *hprt* mutant frequencies.

The data given in Table 8.2 represent results from such a **background incidence study**. Fifty infants, children, and teenagers were assayed via an *in vivo* T-lymphocyte mutation assay to determine their individual *hprt* mutation frequencies, as a function of their ages and genders. Specific interest was directed at identifying whether gender plays a role in *hprt* mutation. (Previous studies *in adults* have determined that *in vivo hprt* mutant frequencies rise as an individual ages. It is thought that the effect should occur also in children, although determination of this issue is a secondary goal of the study.) A plot of the data from Table 8.2 exhibits increases in *hprt* mutant frequency with increasing age, but little evidence is seen of any difference between males and females. More critically, response variability appears to increase with increasing age (Exercise 8.17), a candidate for a gamma GLiM.

Within this context, our goal is to compare the F gender classification to the M gender classification in Table 8.2. Although this may appear to be a simple two-sample comparison, possible age-related effects complicate the analysis. Hence, it is appropriate here to consider an **analysis of covariance** (ANCOVA) that adjusts the F vs. M comparisons for the age variable. It is common in ANCOVA to center the predictor variable: $x = \text{age} - \overline{\text{age}}$, where $\overline{\text{age}}$ is the sample mean of the ages pooled across both genders. (In Table 8.2, $\overline{\text{age}} = 5.8344$.) This simplifies between-level comparisons of the gender variable (Neter *et al.*, 1996, Section 25.2).

Table 8.2 Mutant *hprt* frequencies ($\times 10^6$) for normal human infants, children, and teenagers

Gender	*Age*	*MF*	*Gender*	*Age*	*MF*	*Gender*	*Age*	*MF*
F	0.75	0.3	F	13.2	4.0	M	4.2	2.5
F	0.83	0.4	F	13.9	1.2	M	4.4	2.6
F	1.0	0.4	F	13.9	1.2	M	4.7	0.7
F	1.2	0.77	F	15.2	4.3	M	5.16	0.7
F	2.0	2.0				M	5.2	2.2
F	2.0	2.9	M	0.08	1.3	M	6.4	5.4
F	3.1	0.7	M	0.2	0.3	M	8.1	4.1
F	3.2	1.0	M	0.75	0.5	M	8.2	0.81
F	3.2	1.8	M	0.75	1.9	M	8.7	0.5
F	4.0	0.45	M	0.8	1.1	M	8.9	2.2
F	5.0	0.4	M	1.0	0.32	M	9.7	2.1
F	6.5	8.0	M	1.0	1.3	M	10.0	2.3
F	7.5	1.4	M	1.0	4.4	M	10.7	2.0
F	7.7	8.7	M	1.3	3.1	M	12.0	5.5
F	9.0	9.0	M	1.4	2.2	M	12.5	4.0
F	10.9	0.8	M	1.5	0.85	M	13.1	0.2
F	11.0	1.9	M	1.7	1.7	M	13.2	7.0

Data from Finette *et al*. (1994).

```
* SAS code to fit Mut. Freq. data ;

data MutFrq;
input gender $ age y @@;
   agebar =  5.8344;
         x = age - agebar;
cards;
f  0.75  0.3   f  0.83  0.4  f  1.00  0.4
   ... other data omitted for space considerations ...
m  13.1  0.2   m  13.2  7.0

proc genmod;
     class gender;
     model y = x gender / dist=gamma   link=log
                          type3;
contrast 'f vs. m' gender 1 -1;
```

Fig. 8.3 SAS PROC GENMOD program for gamma GLiM fit to data in Table 8.2.

In formulating this model, we have made the implicit assumption that a linear relationship exists between mutant frequency and age, and this relationship is the same (parallel, on whatever link scale is chosen) for males and females. This is a common assumption in an ANCOVA; nonetheless, it should be evaluated in practice, by at least a visual inspection (Exercise 8.17), but also more formally by the methods in Section 7.4.

For the distribution of the mutant frequencies, we consider a gamma parent model. To illustrate use of a noncanonical link, we will employ a log-linear model, where the link is based on the natural logarithm. Figure 8.3 presents SAS PROC GENMOD code to perform the log-linear fit under a gamma GLiM. (The data presentation is suppressed to conserve space; see Table 8.2 for the full data.) Note that we have centered the predictor variable, age – $\overline{\text{age}}$, and we include a `class` statement to identify the classification variable `gender` to index the gender factor.

An important feature in Fig. 8.3 is the order given in the `model` statement. We specify that the quantitative covariate (here, age – $\overline{\text{age}}$) is fitted first in the sequential order, before the classification variable used to define the factor of interest (here, gender). By making this order specification, the model fit adjusts the factor-level comparison for the effect of the covariate. As part of the model statement, we must include a number of options. The first is the distribution specification, `dist=gamma`. Next, we specify the noncanonical link `link=log`. We also request a `type3` analysis, which outputs the LR/deviance statistic for each predictor variable when fitted *last* in the model. This is analogous the `Type III SS` or **partial sums of squares** produced in a classical linear model fit, using SAS PROC GLM. The `type3` analysis is appropriate for the ANCOVA in this example, since it assures that the test for `gender` will be conducted after the `age` term has been fitted.

Under PROC GENMOD, the `type3` output appears as a series of G^2_{calc}-statistics, each listing the scaled difference in deviances between the full model with all terms fitted, and a reduced model fit without the particular term of interest. Since these are true LR/deviance statistics, they are referred to large-sample χ^2 distributions with *df* equal to the number of parameters carried by each listed term. Note that the `type3` analysis should not be confused with a `type1` output, which produces a form of analysis of deviance table, as mentioned in Section 8.3.4.

We also include in the PROC GENMOD code a `contrast` statement, after the `model` statement. The particular contrast of interest here is the 1 *df* comparison between the F and M gender levels. The results for this test should correspond to the 1 *df* test associated with the `gender` parameter estimate under `Analysis of Parameter Estimates`. The use of the `contrast` statement in this example illustrates how a GLiM can incorporate contrast comparisons between levels of a classification variable (here, gender). By fitting the centered age predictor first in the model, any contrasts among levels of gender are adjusted for age, as desired in an ANCOVA. The output (edited for presentation) appears in Fig. 8.4.

```
                    The SAS System
                 The GENMOD Procedure

                   Model Information
       Distribution                        GAMMA
       Link Function                       LOG
       Dependent Variable                  Y
       Observations Used                   50

       Criteria For Assessing Goodness Of Fit
Criterion               DF          Value        Value/DF
Deviance                46        37.2654          0.7929
Pearson Chi-Square      46        37.7115          0.8024
Log Likelihood           .       -86.4991               .

           Analysis Of Parameter Estimates
Parameter  DF   Estimate   Std Err   ChiSquare  Pr>Chi
INTERCEPT   1     0.7780    0.1536     25.6607  0.0000
X           1     0.0830    0.0279      8.8600  0.0029
GENDER   F  1     0.0127    0.2397      0.0028  0.9577
GENDER   M  0     0.0000    0.0000           .       .
SCALE       1     1.4864    0.2705           .       .
NOTE:  The scale parameter was estimated by
       maximum likelihood.

          LR Statistics For Type 3 Analysis
          Source          DF   ChiSquare  Pr>Chi
          X                1      8.6079  0.0033
          GENDER           1      0.0028  0.9577

              CONTRAST Statement Results
     Contrast     DF   ChiSquare  Pr>Chi  Type
     F vs. M       1      0.0028  0.9577  LR
```

Fig. 8.4 Output (edited) from SAS PROC GENMOD gamma GLiM for data in Table 8.2.

As an aside, note that this analysis can be replicated in S-PLUS using the `glm` function and the commands

```
glm(mut.freq~I(age-mean(age))+sex,
    family=Gamma(link=log),data=mutant.df)
```

where `age`, `gender` and `mut.freq` are variables from Table 8.2 stored in the data frame `mutant.df`. Recall from Section 8.3.1, however, that

SAS and S-PLUS differ in their default estimates of the unknown scale parameter: SAS uses maximum likelihood, but S-PLUS uses the deviance function divided by the appropriate degrees of freedom. As such, small differences will result between the two default outputs in the final estimates and standard errors. (To estimate the scale parameter via the deviance in SAS, use the `dscale` option in the `model` statement.)

In the output from Fig. 8.4, the test for a gender effect appears in the material under `LR Statistics For Type 3 Analysis`. There, since `gender` was listed last in the `model` statement, the test for a gender effect (adjusted for age) comprises the last row of information: we see that the scaled change in deviance to a reduced model not including `GENDER` is $G^2_{calc} = 0.0028$, on 1 *df*. This is not significant ($P \approx 0.9577$), suggesting a lack of any gender effect in these data, after adjustment for the age effect.

(A technical note: SAS actually uses the difference in scaled deviances, $\varphi D(\tilde{\boldsymbol{\beta}}) - \varphi D(\hat{\boldsymbol{\beta}}) = \varphi D(\texttt{INTERCEPT,X}) - \varphi D(\texttt{INTERCEPT,X,GENDER})$ in our notation, to calculate G^2_{calc}. This assumes that $\mathrm{Var}[Y_{ij}]$ has the form φ/w_i, for some known weights $w_i > 0$ (McCullagh and Nelder, 1989, Section 2.3). Thus if $\varphi \neq 1$, quantities listed as `Deviance` in the PROC GENMOD output must be multiplied by the appropriate value of φ or $\hat{\varphi}$ before taking the difference as G^2_{calc}. If we had attempted to calculate the LR/deviance statistic for the gender effect directly from the deviance values, we would have needed to include $\hat{\varphi}$ in the calculations.)

The insignificant difference between genders is corroborated by the single *df* contrast results under `CONTRAST Statement Results:` the LR statistic and P-value from the `type3` analysis are identical to the contrast results. This is expected in this analysis: since the gender classification factor had only two levels, there is only 1 *df* of information available in the analysis, so the `type3` fit of GENDER summarizes completely the comparison between the two levels. Another consequence of this effect is seen under `Analysis Of Parameter Estimates`, where the parameter estimate and standard error for GENDER = M are given as identically zero. SAS has allocated the single *df* of information between genders to the F level (it comes before M alphabetically). Here is a case where the predictor information for M is **aliased** with the previous terms in the model.

For this analysis, we conclude that there does not appear to be a significant difference between female and male mutational response, when adjusted for possible difference in age. □

The gamma GLiM is gaining recognition for use with continuous environmental data where the variance changes with the square of the mean;

for example, Styer (1994) gives an illustration with atmospheric deposition data. McCullagh and Nelder (1989, Section 8.4) furnish some general examples, and for guidance on designing studies under a gamma GLiM see Burridge and Sebastiani (1994). In general, the gamma GLiM is a useful model and we encourage its use when appropriate in environmental studies.

8.4.3 Binary/'quantal' data (binomial random variables)

We have illustrated GLiM analyses for quantal response data previously in Section 7.2.1 with the logistic regression model, and more generally in Section 7.2.3. In both sections, emphasis was on modeling a dose response with a binomial parent distribution for the data, where only quantitative predictor variables were involved in the model fit. Clearly, however, the fuller techniques of general linear modeling may be applied to binary data, with the linear model fit on a logit scale, or on a probit scale, or on a complementary log-log scale, etc. Here are some examples.

Example 8.6 Logistic analysis of covariance for aquatic toxicity to chlorpyrifos
Similar to the experiment in Example 5.6, consider an aquatic survival study employing the infaunal copepod *Amphiascus tenuiremis*, where animal mortality was recorded after acute exposure to the agricultural pesticide chlorpyrifos (Green *et al.*, 1996). The pesticide is an organophosphate that can prove toxic to aquatic organisms after run-off into aquatic ecosystems.

Three different life stages of *A. tenuiremis* were studied: full adult (A), copepodite (C – an intermediate stage between full adult and juvenile), and nauplius (N – the larval or juvenile form). Of interest was whether the three different life stages responded in a different manner to the acute pesticide exposure, over a series of increasing doses. The data are proportions of surviving copepods, Y, out of a total number, N, exposed at each dose–stage combination. They appear in Table 8.3 (see also Fig. 8.7).

We model the survival proportions in Table 8.3 via a binomial parent distribution. With no strong subject-matter motivation for choice of link function, we choose the canonical logit link in Table 8.1. Since it is expected that the copepods will exhibit decreasing survival after pesticide exposure, the dose variable in this experiment serves as a quantitative covariate. Adjustment for the dose covariate is necessary, however, before testing for differences in survival among the three life stages. This is, again, an ANCOVA, similar to that used in Example 8.5.

Table 8.3 Aquatic survival data (Y = survivors/N = total animals) in *Amphiascus tenuiremis* after exposure to chlorpyrifos

Stage = adult (A)			*Stage = copepodite(C)*			*Stage = nauplius(N)*		
Dose	*Y*	*N*	*Dose*	*Y*	*N*	*Dose*	*Y*	*N*
0	23	25	0	25	25	0	19	20
0	23	25	0	24	25	0	17	20
0	25	25	0	22	25	0	18	20
0	22	25	0	23	25	0	16	20
0	23	25	0	25	25	0	18	20
0	25	25	0	23	25	0	19	20
6	22	25	19	22	25	22	14	20
6	20	25	19	18	25	22	15	20
6	21	25	19	21	25	22	14	20
26	24	25	37	17	25	23	18	20
26	25	25	37	22	25	23	16	20
26	23	25	37	20	25	23	15	20
51	20	25	62	12	25	39	13	20
51	22	25	62	20	25	39	12	20
51	22	25	62	19	25	39	14	20
69	13	25	65	13	25	45	14	20
69	10	25	65	17	25	45	10	20
69	15	25	65	16	25	45	8	20
74	8	25	84	13	25	59	3	20
74	10	25	84	14	25	59	1	20
74	14	25	84	8	25	59	8	20
88	3	25	98	5	25	59	5	20
88	2	25	98	7	25	59	3	20
88	2	25	98	8	25	59	3	20
126	3	25	122	1	25	71	0	20
126	0	25	122	2	25	71	1	20
126	3	25	122	4	25	71	1	20
129	0	25				84	0	20
129	2	25				84	0	20
129	0	25				84	1	20

Data from Green *et al.* (1996).

We use SAS PROC GENMOD to fit the logit link. We begin by fitting a simple logit model that includes a linear dose term and two indicator variables for the copepod stages (the indicators are implicit in the use of the SAS CLASS statement). This can be written as

$$\text{logit}(\pi_{ij}) = \beta_0 + \beta_1 u_i + \beta_2 I_{\{A\}}(j) + \beta_3 I_{\{C\}}(j)$$

where π_{ij} is the probability of death for copepods in stage j exposed to (mean-centered) dose $u_i = x_i - \bar{x}$, and $I_{\{.\}}(j)$ is the **indicator variable** that equals 1 when the stage index equals the value in the subscript, and 0 otherwise. (For example, $I_{\{C\}}(j) = 1$ when j = C, and 0 otherwise.) The parameters of this model can be interpreted as multiplicative effects on the odds of death; for example, the odds of death increase by a multiplicative factor of $\exp\{\beta_1\}$ for each unit increase in dose within any fixed copepod stage. As in Example 8.5, this ANCOVA model makes the strong assumption that the effect of dose is the same for all stages; we will explore this assumption later in the example.

The SAS code appears in Fig. 8.5, and it exhibits features similar to the gamma GLiM ANCOVA code in Fig. 8.3. In particular, the sequence of terms entered in the `model` statement reflects the ANCOVA by fitting first the quantitative covariate, `dose`. Fit second in sequence, the (three-level) single factor `stage` is adjusted for `dose`. There are three new issues to note, however: (a) we now have three contrasts of interest, corresponding to all three **pairwise comparisons** among the three life stages, A vs. C, A vs. N, and C vs. N; (b) we invoke the `type1` test option in the `model` statement, rather than `type3` as in Fig. 8.3, in order to illustrate an **analysis of deviance table** (the reader can verify that the `type3` option produces the same LR/deviance statistic for testing the 2 *df* `stage` effect as that seen below); and (c) for simplicity, we employed SAS PROC STANDARD to center the dose covariate u_i about its pooled mean. As used in Fig. 8.5, PROC STANDARD changes u_i to $u_i - \bar{u}$ internally within SAS, so that all future calls to the variable u involve the mean-centered quantities. The output (edited for presentation) appears in Fig. 8.6.

```
* SAS code for aquatic survival data from Table 8.3;
 data asg;
 input stage $  u  y  n @@;
 p=y/n;
 cards;
 a   0     23    25     a    0     23    25
   ... other data omitted for space considerations ...
 n   84     0    20     n    84     1    20

 proc standard data=asg mean=0 out=ag2;
     var u;

 proc genmod data=ag2;
  class stage;
  model y/n = u stage / dist=b link=logit type1;
  contrast 'A vs. C' stage 1 -1  0 ;
  contrast 'A vs. N' stage 1  0 -1 ;
  contrast 'C vs. N' stage 0  1 -1 ;
```

Fig. 8.5 SAS PROC GENMOD program for logistic fit to data in Table 8.3.

```
                    The SAS System
                 The GENMOD Procedure

                  Model Information
      Distribution                         BINOMIAL
      Link Function                        LOGIT
      Dependent Variable                   Y
      Dependent Variable                   N
      Observations Used                    87
      Number Of Events                     1142
      Number Of Trials                     2025

             Class Level Information
             Class     Levels  Values
             STAGE          3  a c n

       Criteria For Assessing Goodness Of Fit
Criterion                DF         Value       Value/DF
Deviance                 83      183.9291         2.2160
Pearson Chi-Square       83      176.9357         2.1318
Log Likelihood            .     -900.5968              .

           Analysis Of Parameter Estimates
Parameter  DF   Estimate   Std Err   ChiSquare   Pr>Chi
INTERCEPT   1    -0.5050    0.0984     26.3686   0.0000
U           1    -0.0478    0.0021    515.0734   0.0000
STAGE   a   1     1.2800    0.1496     73.1778   0.0000
STAGE   c   1     1.5742    0.1533    105.5005   0.0000
STAGE   n   0     0.0000    0.0000           .        .
NOTE:  The scale parameter was held fixed.

          LR Statistics For Type 1 Analysis
  Source        Deviance     DF   ChiSquare   Pr>Chi
  INTERCEPT    1156.7642      0           .        .
  U             315.8087      1    840.9555   0.0000
  STAGE         183.9291      2    131.8796   0.0000

            CONTRAST Statement Results
     Contrast      DF    ChiSquare   Pr>Chi   Type
      A vs. C       1       3.9902   0.0458   LR
      A vs. N       1      78.4465   0.0000   LR
      C vs. N       1     116.0137   0.0000   LR
```

Fig. 8.6 Output (edited) from SAS PROC GENMOD logistic fit to data in Table 8.3.

In the output from Fig. 8.6, the analysis of deviance table appears as the material under `LR Statistics For Type 1 Analysis`. There, working our way up from the full model with dose (`U`) and `STAGE`, we see that the change in deviance to a reduced model not including `STAGE` is

$G^2_{\text{calc}} = 131.88$, on 2 *df*. This is very significant ($P < 0.0001$), suggesting a strong stage effect in these data, after adjustment for the dose effect.

The contrast analysis under `CONTRAST Statement Results` also illustrates strongly significant differences among the three life stages. Of course, the all-pairwise construction requires some correction for multiplicity, such as a Bonferroni adjustment. Here, this translates to multiplying the observed P-values by the number of comparisons (Exercise 6.32b). The result shows that the adult (A) and intermediate (C) stages do not differ significantly ($P \approx 3 \times 0.0458 = 0.1374$), but both the A vs. N and C vs. N comparisons are very significant, with, minimally, $P < 0.0003$ (that is, 3×0.0001) for each. (Recall that all these inferences are adjusted for the very strong dose effect, via the ANCOVA construction.)

Figure 8.7 illustrates the fit of this model to the copepod data. We see clear evidence of decreasing survival with increasing chlorpyrifos exposure: the nauplii appear most sensitive in that they exhibit poorest survival in response to the chemical insult. The figure also brings the quality of this first model into question, since it appears that survival is systematically overpredicted in the juvenile stage at higher exposures, but underpredicted for the copepodites at higher exposures. As a next step in the analysis of these data, we modify our previous model to allow for higher-order dose terms and for dose effects to differ across copepod stages. Three additional models are fitted using the S-PLUS `glm` function, as displayed in Fig. 8.8. The first model, `copepod.fit`, corresponds to the simple model discussed above. The second model, `copepod.fit2`, allows the slope of the dose term to differ with copepod stage. This can be written as

$$\text{logit}(\pi_{ij}) = \beta_0 + \beta_1 u_i + \beta_2 I_{\{A\}}(j) + \beta_3 I_{\{C\}}(j) + \beta_4 u_i I_{\{A\}}(j) + \beta_5 u_i I_{\{C\}}(j),$$

where $u_i = x_i - \bar{x}$ is the mean-centered dose. The third model, `copepod.fit3`, allows u_i to enter with both linear and quadratic terms, and the associated parameters can differ across the various copepod stages. This expands upon the second model by adding three additional parameters to accommodate the quadratic terms. Figure 8.8 presents a formal comparison of these three models using the S-PLUS `anova` function. This function performs a series of model comparisons in which the change in deviance can be evaluated as a function of additional parameters.

From the analysis in Fig. 8.8, we see that the addition of separate dose parameters for the various stages (`copepod.fit` versus `copepod.fit2`) leads to a significant reduction ($P < 0.0001$) in the deviance. A further significant reduction ($P < 0.0001$) occurs with addition of quadratic terms.

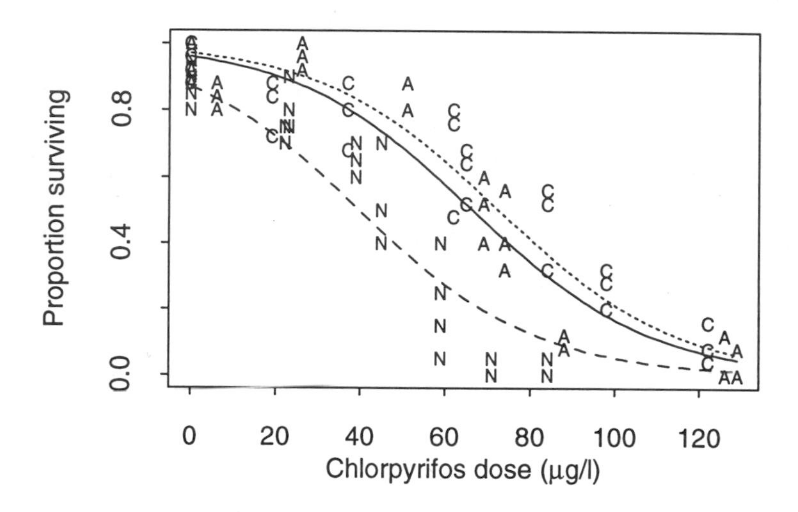

Fig. 8.7 Plot of the observed proportion and ANCOVA-predicted proportion of copepods at various stages ('A'/dotted line = adult; 'C'/solid line = intermediate; 'N'/dashed line = juvenile or nauplius) surviving exposure to various concentrations of chlorpyrifos. Data from Table 8.3.

A summary of the parameter estimates obtained from fitting the third model is also given in Fig. 8.8. This yields the following predicted logits for the various copepod stages:

```
juvenile:      predicted logit(p_iN) = -0.5071 -0.0938u_i - 0.00088u_i^2
intermediate:  predicted logit(p_iC) =  1.0213 -0.0350u_i - 0.00012u_i^2
adult:         predicted logit(p_iA) =  0.9843 -0.0457u_i - 0.00022u_i^2
```

The associated prediction curves are displayed in Fig. 8.9. Compared with Fig. 8.7, we see the addition of higher-order dose terms along with separate dose parameters for each stage yields a qualitatively better fit to the data.

A possible simplification to our model is suggested from Fig. 8.9 and from the coefficients of the predicted log-odds from above. While it appears that both intermediate and adult stages differ from the juvenile stage, it does not appear that the intermediate and adult stages differ from each other. This could be assessed by comparing the current full model with a reduced model containing common parameters for the intercept, linear and quadratic terms in both the intermediate and adult stages. Such a reduced model essentially combines the two stages into a single stage for the analysis (Exercise 8.22).

```
copepod.lst<-scan(what=list(stage="",dose=0,y=0,nsurv=0))
A    0    23    25    C    0    25    25    N    0    19    20
         ... other data omitted for space considerations ...
A  129     0    25                          N   84     1    20

copepod.df<-data.frame(copepod.lst)
copepod.df$iA<-as.numeric(copepod.df$stage=="A")
copepod.df$iC<-as.numeric(copepod.df$stage=="C")
copepod.fit1<-glm(cbind(y,nsurv-y)~ I(dose-mean(dose)) + iA + iC,
              family=binomial(link=logit),data=copepod.df)

copepod.fit2<-glm(cbind(y,nsurv-y)~ I(dose-mean(dose))*(iA + iC),
              family=binomial(link=logit),data=copepod.df)

copepod.fit3<-glm(cbind(y,nsurv-y)~ (I(dose-mean(dose)) +
              I((dose-mean(dose))^2)) *(iA + iC),
              family=binomial(link=logit),data=copepod.df)

anova(copepod.fit1,copepod.fit2,copepod.fit3,test="Chisq")
```

```
Analysis of Deviance Table
Response: cbind(y, nsurv - y)
Model  Resid. Df Resid. Dev Df Deviance       Pr(Chi)
    1        83   183.9291
    2        81   154.1944  2 29.73473 3.492900e-07
    3        78   129.3039  3 24.89052 1.627607e-05
```

```
summary(copepod.fit3)
```

```
Coefficients:
                                    Value   Std. Error   t value
               (Intercept) -0.5071041791 0.1356760682 -3.737610
        I(dose - mean(dose)) -0.0938132223 0.0096979248 -9.673536
    I((dose - mean(dose))^2) -0.0008770721 0.0002366171 -3.706715
                        iA  1.4914504520 0.1937024537  7.699698
                        iC  1.5284498745 0.1904798914  8.024206
     I(dose - mean(dose)):iA  0.0481300815 0.0102795837  4.682104
     I(dose - mean(dose)):iC  0.0587773040 0.0101854209  5.770729
 I((dose - mean(dose))^2):iA  0.0006605230 0.0002521107  2.619972
 I((dose - mean(dose))^2):iC  0.0007538493 0.0002485815  3.032604

(Dispersion Parameter for Binomial family taken to be 1 )

    Null Deviance: 1156.764 on 86 degrees of freedom
Residual Deviance: 129.3039 on 78 degrees of freedom
```

Fig. 8.8 Output (edited) from S-PLUS `glm` logistic fit to data in Table 8.3.

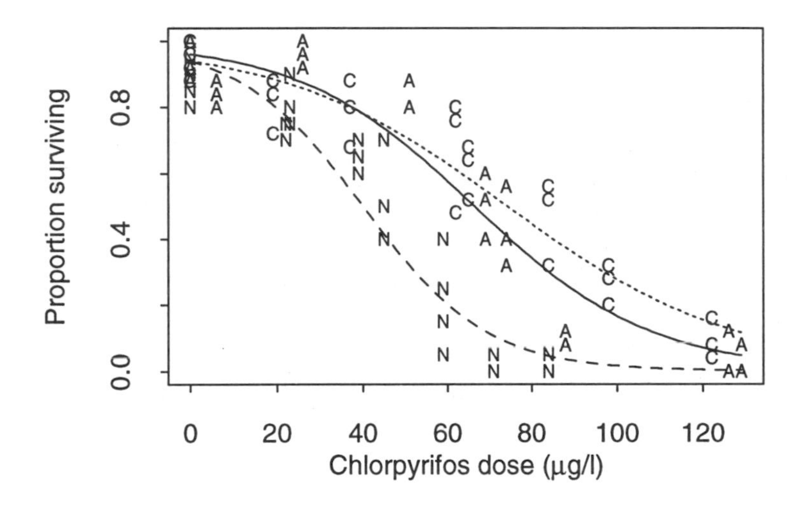

Fig. 8.9 Plot of the proportion of copepods at various stages ('A'/dotted line = adult; 'C'/solid line = intermediate; 'N'/dashed line = juvenile or nauplius) surviving exposure to various concentrations of chlorpyrifos, with improved prediction equations overlaid.

In this example, we have hinted at some of the intricacies that can be incorporated into GLiMs. Indeed, questions of model complexity are as much philosophical as statistical in nature, where improved fit due to large numbers of parameters is balanced against additional difficulty of interpretation introduced in such models. For the aquatic toxicity data in Table 8.3, the additional complexity appears to have been justified, since the effort resulted in a better understanding of the factors affecting *A. tenuiremis* survival after exposure to this pesticide. □

Example 8.7 Logistic regression incidental tumor test

An interesting and important application of logistic regression methodology occurs in environmental carcinogenicity testing, as described in Section 6.3.2. For example, in laboratory studies with rodents such as undertaken by the US National Toxicology Program (NTP), the animals are exposed in a chronic fashion to some environmental agent over an extended period, usually 104–105 weeks. A typical experimental design for such a study exposes 50 animals to one of three or four doses (including a concomitant control). Each animal's response is coded as 1 if it exhibits the tumor, or as 0 if it does not. Pooled within each dose group, these independent binary observations sum to a binomial random variable, with parameters N = {number of animals/exposure level}, and π = {probability of carcinogenic response at that exposure}. The study's goal is identification of increasing carcinogenic dose response in π after exposure to the agent.

The statistical analysis of such data may seem straightforward: simply employ the Cochran–Armitage trend test for proportions from Section 6.3.1. An important complication is introduced, however, by the extended exposure and testing period: tumor incidence rates may vary with animal age, allowing animals dying early in the study to portray different incidence and susceptibility patterns than those dying later. Hence, differences in mortality among the animals represent complicating factors in the analysis of tumor incidence. If the study involves a long chronic exposure period – such as 15–24 months – **intercurrent mortality** patterns may occur, where the different exposure groups exhibit differential survival.

If the mortality patterns in a long-term carcinogenicity study are roughly equal across all exposure groups, the Cochran–Armitage test from Section 6.3.1 remains valid, although its sensitivity may be compromised (Ryan, 1985). If the intercurrent mortality is severe, however, it can act to diminish greatly the sensitivity and power of the trend test (Bailer and Portier, 1988).

To account for effects of intercurrent mortality, an adjustment is required for different survival patterns seen among differently treated groups of animals. (Recall from Section 6.3.2 the discussion on survival-adjusted trend tests. We also discuss methods for testing if multi-level data exhibit heterogeneous survival patterns in Chapter 11.) Indeed, these **survival-adjusted analyses** are the norm for long-term carcinogenicity studies; Dinse (1994) provides a good review. To assess increasing tumor incidence, the simplest such adjustment records the time to tumor and also the age at which the animal dies or is sacrificed, and incorporates these variables into a regression analysis. Unfortunately, in most carcinogenicity experiments, the tumors are **occult**; that is, they develop and grow inside the animal and are identified histopathologically only after the animal dies or is sacrificed at the end of the study. Data analysis of the actual time-course of the cancer's progression is not possible in this case, unless additional assumptions are made about the **context of observation** in which the tumor was recorded, for example, whether an observed tumor in a particular animal contributed to that animal's death (Peto *et al.*, 1980; McKnight and Wahrendorf, 1992), and/or additional **interim sacrifices** are made that provide greater information on tumor development over time (McKnight and

Crowley, 1984; Dinse, 1988). Indeed, we discuss in Example 11.4 a survival-adjusted test based on the assumption that the tumor is highly lethal. This is, of course, a very extreme instance of known tumor lethality.

An important special case where a logistic GLiM may be employed for survival-adjusted testing occurs when it is reasonable to assume that tumors observed in animals dying before the end of the study are strictly nonlethal, that is, occurrence of the tumor in no way affects the animals' survival patterns. We call these **incidental tumors**. (Note that 'incidental' here refers to the tumor's lethality, not to the carcinogenic response with which it may be associated.) For such tumors, Dinse and Lagakos (1983) describe an adjustment for intercurrent mortality that employs the individual death time for each animal as a covariate in the linear predictor, along with a predictor variable for dose. Once recorded, these values are fitted to the individual binary observations in a logistic regression GLiM. That is, take two predictor variables, x_1 = time on study, and x_2 = dose (or log-dose, etc.), and form the linear predictor $\eta_{ij} = \beta_0 + \beta_1 x_{ij1} + \beta_2 x_{ij2}$ for each of the $j = 1, \ldots, n_i$ binary observations Y_{ij} in the $i = 1, \ldots, k$ dose groups. Regress Y_{ij} on a logit scale against the linear predictor η_{ij}. This models the **tumor onset probability** as logistic in dose, x_1, and survival time, x_2: $\pi_{ij} = [1 + \exp\{-\beta_0 - \beta_1 x_{ij1} - \beta_2 x_{ij2}\}]^{-1}$.

Under this model, the null hypothesis $H_0{:}\beta_2 = 0$ represents the lack of a tumor effect over dose, while $H_a{:}\beta_2 \neq 0$ represents a dose-related tumorigenic response. (One could also test for a strictly increasing response with increasing dose, $H_a{:}\beta_2 > 0$.) Dinse and Lagakos (1983) recommend the score statistic from this logistic regression fit for testing H_0. Asymptotically equivalent to this test is the LR statistic of H_0, which is occasionally simpler to identify.

To illustrate, the data in Table 8.4 are results of an NTP (1987b) study of the chemical repellent 1,4-dichlorobenzene. The study followed-up on the sub-chronic experiment discussed in Example 6.4, and it considered murine carcinogenesis after chronic gavage exposure to the chemical over a 104-week period. We will consider the specific issue of assessing an exposure-related increase in hepatocellular adenomas for male mice. These tumors are some of the least lethal studied by the NTP (Portier *et al.*, 1986). Hence, the incidental tumor analysis may be appropriate for survival-adjusted analyses of benign hepatocellular tumorigenicity. Data for time on study, x_1 (in weeks), and tumor incidence ($Y = 1$ if animal exhibited tumor, 0 otherwise) as a function of dose, x_2 (in mg/kg body weight), appear in Table 8.4. The summary proportions at each exposure are 3/50 (6%) at 0 mg/kg, 13/50 (26%) at 300 mg/kg, and 16/50 (32%) at 600 mg/kg.

Table 8.4 Murine tumorigenicity after exposure to 1,4-dichlorobenzene. Variables are time on study (x_1, in weeks), and tumor incidence ($Y = 1$ if animal exhibited tumor, 0 otherwise), listed by dose, x_2

$x_2 = 0$ mg/kg				$x_2 = 300$ mg/kg				$x_2 = 600$ mg/kg			
x_1	Y	x_1	Y	x_1	Y	x_1	Y	x_1	Y	x_1	Y
11	0	104	0	8	0	104	1	11	0	104	1
11	0	104	0	10	0	104	0	11	0	104	1
11	0	104	0	11	0	104	0	11	0	104	0
38	0	104	0	11	0	104	0	11	0	104	1
59	0	104	0	11	0	104	1	11	0	104	0
67	0	104	0	11	0	104	0	11	0	104	1
74	0	104	0	11	0	104	0	11	0	104	1
85	1	104	0	11	0	104	1	12	0	104	0
85	0	104	0	26	0	104	1	84	0	104	1
88	0	104	0	34	0	104	1	86	0	104	0
89	0	104	0	80	0	104	0	88	1	104	1
90	1	104	0	90	0	104	0	89	0	104	0
90	0	104	0	90	0	104	1	90	1	104	1
93	0	104	0	92	0	104	0	90	0	104	0
94	0	104	1	93	0	104	0	91	0	104	0
97	0	104	0	98	0	104	0	91	1	104	1
97	0	104	0	101	0	104	0	93	0	104	1
98	0	104	1	103	1	104	1	95	0	104	0
98	0	104	0	104	0	104	0	97	0	104	1
99	0	104	0	104	0	104	0	104	1	104	0
99	0	104	0	104	1	104	0	104	0	104	0
102	0	104	0	104	1	104	0	104	0	104	0
104	0	104	0	104	1	104	1	104	0	104	0
104	0	104	0	104	0	104	1	104	0	104	0
104	1	104	0	104	0	104	0	104	1	104	0

Data from NTP (1987b, Table C2).

To analyze these data using the logistic regression incidental tumor test, we take the binary observations from Table 8.4 and fit to them a logistic regression model with linear predictor $\beta_0 + \beta_1 x_1 + \beta_2 x_2$, where x_1 is time on study (in weeks) and x_2 is dose.

The incidental tumor analysis may be performed in S-PLUS, using the `glm` function. Figure 8.10 presents the code and S-PLUS output. The analysis is conducted via a sequence of S-PLUS function invocations: first `glm`, then `summary` and `anova`. In the figure, the analysis of deviance table appears in response to the call to `anova`. There, working our way up from the full model with terms for time on study (`X1`) and dose (`X2`), we see that the change in deviance to a reduced model not including `X2` is $G^2_{\text{calc}} = 8.0316$, on 1 *df*. This is significant ($P \approx 0.0046$), suggesting an important dose effect in these data after adjustment for survival differences. We conclude that there is a strong, dose-dependent relationship between chemical exposure and hepatocellular carcinogenicity in these animals.

```
tumor.lst<-scan(what=list(dose=0,time=0,y=0))
11 0 0 104 0 0  8 0 300 104 1 300  11 0 600 104 1 600
          ... other data omitted for space considerations ...

tumor.df<-data.frame(tumor.lst)

tumor.fit<-glm(y~dose+time,family=binomial(link=logit),
               data=tumor.df)
summary(tumor.fit)
```

```
Deviance Residuals:
       Min         1Q     Median          3Q      Max
 -1.097395 -0.8206817 -0.5886815 -0.06352017 2.348401

Coefficients:
                  Value   Std. Error   t value
(Intercept) -7.400518981 2.8509479454 -2.595810
       dose  0.002413828 0.0008839569  2.730708
       time  0.055395038 0.0277976033  1.992799

Residual Deviance: 137.8663 on 147 degrees of freedom
```

```
anova(tumor.fit,test="Chisq")
```

```
Analysis of Deviance Table
Binomial model
Response: y
Terms added sequentially (first to last)
     Df Deviance Resid. Df Resid. Dev     Pr(Chi)
NULL                   149   160.5651
dose  1  7.05758       148   153.5076 0.007893063
time  1 15.64132       147   137.8663 0.000076563
```

```
tumor.fit2<-glm(y~time+dose,family=binomial(link=logit),
                data=tumor.df)

anova(tumor.fit2,test="Chisq")
```

```
Analysis of Deviance Table
Binomial model
Response: y

Terms added sequentially (first to last)
     Df Deviance Resid. Df Resid. Dev     Pr(Chi)
NULL                   149   160.5651
time  1 14.66734       148   145.8978 0.000128249
dose  1  8.03156       147   137.8663 0.004596931
```

Fig. 8.10 Output (edited) from S-PLUS `glm` logistic GLiM fit to data in Table 8.4.

The predicted tumor onset probability apperas in Fig. 8.11 as a function of dose and survival time. The graph shows that older mice at higher doses are more prone to hepatocellular carcinogenicity: almost 50% of mice exposed to the high dose are predicted to exhibit tumors by 104 weeks on study.

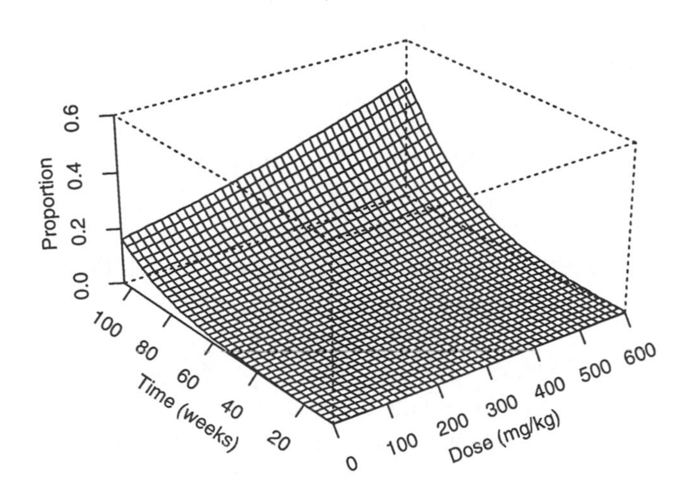

Fig. 8.11 Display of the predicted probability of tumor onset as a function of dose and time, using parameter estimates from Fig. 8.10.

Logistic regression analysis of incidental tumor data can be extended to include other explanatory factors as covariables (Dinse and Haseman, 1986), giving it great flexibility. For cases where the tumor is strictly incidental to the animal's mortality, it represents an important approach for assessing environmental carcinogenicity. It does suffer, however, when the incidental lethality assumption is invalid: if the death rate for tumor-free animals increases over increasing exposures, the incidental tumor test loses sensitivity to detect a true increase in tumor incidence (Dinse, 1994).

Research is ongoing into methods for assessing exposure-related carcinogenicity with tumors of intermediate lethality. For instance, we described in Section 6.3.2 a modification of the Cochran–Armitage trend test that employs polynomial weights based on the observed times on study (Bailer and Portier, 1988; Bieler and Williams, 1993). This method can increase power to detect carcinogenic effects in the presence of moderate to strong tumor lethality, although the power is diminished somewhat when the carcinogenic stimulus is weak (Tan and Zhang, 1996). Other promising avenues include incorporation of relevant **biomarkers** (McMillan and Silvers, 1996) and other aspects of tumor onset (McKnight, 1991; Kodell *et al.*, 1994) in the analysis, employing interim sacrifice data to estimate specific toxicological parameters (Williams and Portier, 1992), and trend tests that model constant mortality differences (Dinse, 1993), constant mortality ratios (Lindsey and Ryan, 1994), or proportional tumor prevalences (Kodell *et al.*, 1997) among the experimental animals. ❑

8.4.4 Count data (Poisson random variables)

We have illustrated GLiM analyses for count data previously in Section 7.2.2 with the log-linear model, and more generally in Section 7.2.3. In both sections, emphasis was on modeling a dose response under a Poisson parent distribution for the data, where only quantitative predictor variables were involved in the model fit. The more flexible techniques of generalized linear modeling may be applied also to count data, with the linear model fit on some positive scale such as log-linear or absolute value. For an example, see the quadratic analysis of the *C. dubia* aquatic toxicity data in Fig. 7.15. Or consider the following:

Example 8.8 Interaction in a two-factor model for count data
Consider again the problem of detecting a significant interaction between two qualitative factors as in Example 8.1, but now where the data are Poisson counts. For instance, Van Beneden (1994) describes an experiment where DNA is taken from softshell and hardshell clams and transfected into murine cells in culture in order to study the ability of the murine host cells to indicate selected damage to the clam DNA. (Mouse cells are much easier to culture than clam cells. If the transgenic technology can integrate successfully the clam DNA into the murine host cells, then laboratory study of *in vivo* aquatic toxicity in clams can be made more efficient.)

As part of an initial experiment to study the transfection technology, clam DNA from the two different species was transfected into separate plates containing murine NIH3T3 cell cultures. The response is the number of focal lesions seen on the plates after a fixed period of incubation. Ostensibly, the goal is to assess whether there are differences in response between the two species.

An additional consideration here is whether certain compounds modify the yield if included on the plate environment, and whether any species×modifier interaction is present. (That is, do the modifying compounds exhibit differential effects on the lesion yield, depending on which species of clam is studied?) Thus, in terms of the two-factor ANOVA from Example 8.1, we have factor A distinguishing the two different species

($i = 0$ for softshell, $i = 1$ for hardshell), and factor B indicating whether or not the culture medium contained the modifier ($j = 0,1$). From an experiment with the inhibiting modifier dexamethasone, observed counts were given as $Y_{00} = 3$, $Y_{01} = 0$, $Y_{10} = 20$, $Y_{11} = 1$.

Notice that there is no replication reported in this experiment. Under the classical linear model, such a design could not be accommodated unless the interaction term were dropped from the model, otherwise there would be more parameters to estimate (including the unknown variance, σ^2) than data points with information. Under a Poisson GLiM, however, four data points provide sufficient information to fit a simple log-linear model with the two-factor linear predictor (8.1). (Recall that the Poisson variance equals the mean, so there is no need to estimate variance separately from the other parameters in the linear predictor.) We will see, however, that the model is **saturated**, leading to certain unusual features in the computer output.

Applying a Poisson log-linear model with a two-factor/interaction linear predictor, $\eta_{ij} = \upsilon + \alpha_i + \beta_j + \gamma_{ij}$, to these data yields the SAS PROC GENMOD code displayed in Fig. 8.12. The resulting output (edited for presentation) appears in Fig. 8.13. There, the analysis of deviance table is located under `LR Statistics For Type 1 Analysis`. Working our way up from the full model with terms for species, dexamethasone, and and their interaction, we see that the change in deviance to a reduced model not including the interaction is $G^2_{calc} = 0.2732$, on 1 *df*. This is not significant ($P \approx 0.6012$), suggesting a lack of any species×dexamethasone interaction in these data.

Since the interaction is not significant, we can continue up the table, and assess the significance of the next term in the sequence. Based on the order given in the `model` statement in Fig. 8.12, this is the dexamethasone effect: we see the difference in deviance between a full model with both species and dexamethasone (and a grand mean) and a reduced model without the dexamethasone effect is $G^2_{calc} = 24.96$, on 1 *df*. Since this is significant ($P < 0.0001$), we stop at this point and conclude that a dexamethasone effect appears in these data. (Reversing the order to test for the species effect achieves similar results; see Exercise 8.31.)

Certain features in this model fit are worth highlighting. First, as we noted above, the model is saturated by inclusion of the interaction term in the linear predictor. In effect, we are estimating four features – the grand mean, species effect, dexamethasone effect, and the species×dexamethasone interaction – using just four data points. This represents one way to attain a **saturated model**. The PROC GENMOD output indicates this by reporting the deviance as 0.00. (This is precisely correct: the deviance is the difference

```
 * Log-linear model with interaction for clam
data;

 data clam;
 input species $ modif y @@;
 cards;
 s 0  3  s 1  0
 h 0 20  h 1  1

 proc genmod;
 class species modif;
 model y = species modif species*modif /
             dist=poisson   link=log   type1;
```

Fig. 8.12 SAS PROC GENMOD program for two-factor log-linear GLiM with interaction.

```
                  The SAS System
                The GENMOD Procedure

                 Model Information
       Distribution                        POISSON
       Link Function                       LOG
       Dependent Variable                  Y
       Observations Used                   4

               Class Level Information
               Class     Levels  Values
               SPECIES        2  h s
               MODIF          2  0 1

       Criteria For Assessing Goodness Of Fit
Criterion                 DF        Value      Value/DF
Deviance                   0       0.0000             .
Pearson Chi-Square         0       0.0000             .
Log Likelihood             .      39.2105             .
NOTE:  The scale parameter was held fixed.

            LR Statistics For Type 1 Analysis
 Source            Deviance    DF   ChiSquare  Pr>Chi
 INTERCEPT          40.4165     0           .       .
 SPECIES            25.2304     1     15.1861  0.0001
 MODIF               0.2732     1     24.9572  0.0000
 SPECIES*MODIF       0.0000     1      0.2732  0.6012
```

Fig. 8.13 Output (edited) from SAS PROC GENMOD fit to two-factor log-linear GLiM.

between twice the log-likelihood of the current model and twice the log-likelihood of the fullest possible model. Since we have fitted the fullest possible model, the deviance difference must be zero.)

Second, and more importantly, the lack of a significant interaction in these data may be due to a lack of sensitivity in the test statistic. As with any of the GLiMs studied in this chapter (except the classical linear models in Section 8.1), the inferences here are based on large-sample statistical arguments. With so very few observed counts from this experiment, especially in the dexamethasone groups, the LR/deviance tests may have very little power to detect a true interaction. Further experimentation that generates larger data sets would be needed to corroborate the finding of a noninteractive effect between species and dexamethasone. □

8.5 SUMMARY

In this chapter, a brief introduction is provided to the construction and use of generalized linear models (GLiMs) for environmetric analyses. Beginning with a review of concepts from the classical, normal-based linear model, GLiMs are seen to provide extensions on two fronts. First, they extend the parent distribution to an exponential class of probability functions that includes such diverse forms as the binomial, Poisson, gamma, and normal. Second, they permit incorporation of nonlinear relationships between the population mean and the predictor variables. Through these generalizations, the GLiM formulation allows the statistical machinery developed for the classical linear model to be applied to a much wider class of models. Examples include analysis of variance and covariance on logit or log-linear scales, regression analyses under a gamma parent model, and polynomial regressions similar in nature to those discussed for dose-response modeling in Chapter 7.

EXERCISES

(Exercises with asterisks, e.g. 8.4*, typically are more advanced, or require greater knowledge and understanding of probability distributions and/or calculus.)

8.1. Example 6.1 gives data on lymphocyte response in dolphins after exposure to PCBs. For the quadratic polynomial regression model fit to these data in Exercise 7.2, calculate the residuals $Y_{ij} - \hat{Y}_{ij}$ and verify that the sum of their squares equals the residual sum of squares from your original output. Plot the residuals against $\hat{Y}_{ij}$. Is any additional pattern evident?

8.2. Repeat Exercise 8.1 for the quadratic polynomial fit performed on the female Florida panther data in Exercises 6.8 and 7.3.

8.3. Verify that the numerator *df* for the conditional F-statistic in Example 8.1 are $(k_A - 1)(k_B - 1)$.

8.4*. How could you write the one-factor ANOVA x-variable $x_{ijm} = 1$ if $i = m$ (zero otherwise) from Section 8.1, in terms of the indicator function notation from Section 1.3.1, $I_{\mathscr{A}}(\cdot)$?

8.5. Identify an environmental application (from your own field of study, as appropriate) where assessing interaction between two factors (Example 8.1) is an important consideration.

8.6*. Show that the following distributions are all members of the exponential class, as characterized by (8.2):

a. Geometric with mean parameter μ.

b. Binomial with sample size parameter N and probability parameter π.

c. Negative binomial with mean parameter μ and dispersion parameter $\delta = 1$.

d. Negative binomial with mean parameter μ and dispersion parameter $\delta = 2$.

e. Exponential with mean parameter μ.

f. Normal with mean parameter μ and variance parameter σ^2.

8.7*. Find the canonical link function for those distributions in Exercise 8.6 that are not already identified in Table 8.1.

8.8*. Use differential calculus to show that for the log-likelihood based on (8.3), the derivative with respect to β_m, $\partial\ell(\boldsymbol{\beta})/\partial\beta_m$, is as given in the left side of (8.5), by using the following arguments:

a. Recognize that the derivative of $\ell(\boldsymbol{\beta})$ is the derivative of a finite double sum. Thus $\partial\ell(\boldsymbol{\beta})/\partial\beta_m$ is the double sum of a derivative of the individual log-likelihood terms. Show that the individual derivatives are of the form

$$\frac{\partial}{\partial\beta_m}\left(\frac{y_{ij}\theta_i - b(\theta_i)}{a(\varphi_i)} + c(y_{ij}, \varphi_i)\right) = \frac{1}{a(\varphi_i)}\left(y_{ij}\frac{\partial\theta_i}{\partial\beta_m} - \frac{\partial b(\theta_i)}{\partial\beta_m}\right).$$

b. Use the chain rule to evaluate the initial derivative term in part (a), $\partial\theta_i/\partial\beta_m$, as

$$\frac{\partial\theta_i}{\partial\beta_m} = \frac{\partial\theta_i}{\partial\mu_i}\frac{d\mu_i}{d\eta_i}\frac{\partial\eta_i}{\partial\beta_m}.$$

c. For the first term in part (b), $\partial\theta_i/\partial\mu_i$, recall that under (8.2), $\mu_i = b'(\theta_i) = \partial b(\theta_i)/\partial\theta_i$. Show that this implies $\partial\mu_i/\partial\theta_i = b''(\theta_i)$. Under the relationship $\text{Var}[Y_{ij}] = a(\varphi_i)V(\mu_i) = a(\varphi_i)b''(\theta_i)$, recognize that $V(\mu_i) = b''(\theta_i)$, so $\partial\mu_i/\partial\theta_i$ is $V(\mu_i)$. From this, argue that $\partial\theta_i/\partial\mu_i = 1/V(\mu_i)$.

d. For the second term in part (b), $d\mu_i/d\eta_i$, recall that the inverse link function relates μ_i and η_i via $\mu_i = h(\eta_i)$. Assuming $h(\cdot)$ is differentiable, what is $d\mu_i/d\eta_i$ in terms of $h(\eta_i)$?

e. For the third term in part (b), $\partial\eta_i/\partial\beta_m$, use the relationship in (8.4) to show that $\partial\eta_i/\partial\beta_m = x_{im}$.

f. Collect together the results from parts (c)–(e) to write $\partial\mu_i/\partial\beta_m$ in part (b) as $h'(\eta_i)x_{im}/V(\mu_i)$.

g. Use the chain rule to evaluate the second term in part (a), $\partial b(\theta_i)/\partial\beta_m$, as

$$\frac{\partial b(\theta_i)}{\partial\beta_m} = \frac{\partial b(\theta_i)}{\partial\theta_i}\frac{\partial\theta_i}{\partial\beta_m}.$$

Use your results in parts (c) and (f) to write this as $\mu_i h'(\eta_i)x_{im}/V(\mu_i)$.

h. Collect together your results form parts (f) and (g) to simplify the right side of the equation in part (a).

i. Sum the quantities in part (h) over $i = 1, \ldots, k$ and $j = 1, \ldots, n_i$. This should be the double sum in (8.5).

8.9. Into what specific form does the Pearson residual R_{ij}^P simplify for the case where $Y_{ij} \sim \mathcal{Poisson}(\mu_i)$? What is the form of the associated X^2-statistic?

8.10. Into what specific form does the Pearson residual R_{ij}^P simplify for the case where $Y_{ij} \sim N(\mu_i, \sigma^2)$? What is the form of the associated X^2-statistic?

8.11. Consider the Aranda-Ordaz (1981) extended link in (8.6).

a. Verify that the logit link from Table 8.1 obtains for $\gamma = 1$.

b*. Find the link function that obtains in the limit as $\gamma \to 0$.

8.12. Consider the Whittemore (1983) extended link in (8.7).

a. Verify that the logit link from Table 8.1 obtains for $\gamma = 0$.

b. Find the link function that obtains for $\gamma = 1$.

8.13*. Use differential calculus to show that for the log-likelihood based on (8.3), the second derivative with respect to β_m, $\partial^2\ell(\boldsymbol{\beta})/\partial\beta_m^2$, is as given in (8.9). In particular:

a. Show that, using the chain rule, $\partial \hbar'(\eta_i)/\partial\beta_m = x_{im}\hbar''(\eta_i)$.

b. Assume that $V(\mu)$ is a differentiable function of μ. Use the chain rule to show that the *third* derivative of $b(\theta_i)$ is

$$b^{(3)}(\theta_i) = \frac{\partial b''(\theta_i)}{\partial\theta_i} = \frac{\partial V(\mu_i)}{\partial\theta_i} = \frac{\partial V(\mu_i)}{\partial\mu_i}\frac{\partial\mu_i}{\partial\theta_i} = V(\mu_i)\frac{\partial V(\mu_i)}{\partial\mu_i}.$$

[*Hint*: see Exercise 8.8c. For simplicity, write $\partial V(\mu_i)/\partial\mu_i$ as $V'(\mu_i)$.]

c. Use the results in parts (a) and (b) to show that

$$\frac{\partial V(\mu_i)}{\partial\beta_m} = \frac{\partial b''(\theta_i)}{\partial\beta_m} = \frac{x_{im}\, b^{(3)}(\theta_i)\, \hbar'(\eta_i)}{V(\mu_i)} = x_{im}\, \hbar'(\eta_i)\, V'(\mu_i).$$

d. Use the above results to show that

$$\frac{\partial}{\partial\beta_m}\frac{\hbar'(\eta_i)}{a(\varphi_i)V(\mu_i)} = \frac{x_{im}}{a(\varphi_i)V(\mu_i)}\left(\hbar''(\eta_i) - [\hbar'(\eta_i)]^2\frac{\partial\log\{V(\mu_i)\}}{\partial\mu_i}\right)$$

and that

$$\frac{\partial}{\partial\beta_m}\frac{\mu_i\hbar'(\eta_i)}{a(\varphi_i)V(\mu_i)} = \frac{x_{im}}{a(\varphi_i)V(\mu_i)}\left(\mu_i\hbar''(\eta_i) - \mu_i[\hbar'(\eta_i)]^2\frac{\partial\log\{V(\mu_i)\}}{\partial\mu_i} + [\hbar'(\eta_i)]^2\right).$$

[*Hint*: see Exercise 8.8f.]

e. From part (d), show

$$\frac{\partial}{\partial\beta_m}\left\{\frac{x_{im}\,\hbar'(\eta_i)}{a(\varphi_i)V(\mu_i)}(Y_{ij}-\mu_i)\right\} = \frac{(x_{im})^2}{a(\varphi_i)V(\mu_i)}\left\{-[\hbar'(\eta_i)]^2 + (Y_{ij}-\mu_i)\left[\hbar''(\eta_i) - [\hbar'(\eta_i)]^2\frac{\partial\log\{V(\mu_i)\}}{\partial\mu_i}\right]\right\}$$

Thus, show that $-\partial^2\ell/\partial\beta_m^2$ is the sum over $i = 1, \ldots, k$ and $j = 1, \ldots, n_i$ of

$$\frac{x_{im}^2}{a(\varphi_i)V(\mu_i)}\left\{[\hbar'(\eta_i)]^2 - (Y_{ij}-\mu_i)\left[\hbar''(\eta_i) - [\hbar'(\eta_i)]^2\frac{\partial\log\{V(\mu_i)\}}{\partial\mu_i}\right]\right\}.$$

f. Recall that $\mathrm{E}[Y_{ij} - \mu_i] = \mathrm{E}[Y_{ij}] - \mu_i = 0$, so using part (e), $\mathrm{E}[-\partial^2\ell/\partial\beta_m^2]$ reduces to (8.9).

8.14*. As necessary, repeat or mimic the arguments in Exercise 8.13 to establish the relationship in (8.10).

8.15*. Recall from Example 8.4 that the large-sample χ^2-approximation was used in Fig. 8.2 for the *P*-values associated with the pointwise hypotheses $H_0: \beta_m = 0$, but that an exact *t* distribution-based *P*-value is used in the corresponding analysis in Fig. 7.2. Under what 'condition' would the two analyses coincide?

8.16*. In Example 8.4, had we failed to use the `dscale` option in SAS PROC GENMOD, we would have incorrectly estimated the scale parameter, which is the constant population standard deviation σ, via maximum likelihood. The results would have given the estimated scale parameter under the full quadratic model as 0.1244 (instead of the correct value of $\sqrt{\text{MSE}} = 0.1573$), and the standard errors would all have been smaller than the values given in Fig. 8.2 by a factor of 0.1244/0.1573 = 0.791. Notice that this is equal to $(8/5)^{-1/2}$. What is going on here?

8.17. Return to the mutant frequency data in Example 8.5.

a. Plot the *hprt* mutant frequencies as a function of age. Do you agree with the contention in the text that response variability appears to be increasing with age? Also, does your plot suggest that the log-mutant frequencies between genders are parallel, as required for the ANCOVA in the example?

b. Calculate the residuals $R_{ij} = Y_{ij} - \hat{Y}_{ij}$ for these data, and graph them vs. the predicted values $\hat{Y}_{ij}$. Do the residuals appear randomly scattered? What problems, if any, do you see with your plot?

8.18. Return to the mutant frequency ANCOVA in Example 8.5 and re-analyze the data using the canonical link for the gamma GLiM. Do the results change substantively? Compare the interpretation of the parameters from these two different link functions.

8.19*. Suppose data are sampled randomly from a parent gamma distribution for which a GLiM is considered. Derive an extended link function that includes both the canonical link from Table 8.1 and the logarithmic link employed in Example 8.5.

a. Provide the details for a goodness-of-link test under this extended link function, as discussed in Section 8.3.3.

b. Apply your test in part (a) to the data in Example 8.5. Does the logarithmic link exhibit significant departure from goodness-of-link? (Set $\alpha = 0.05$.)

8.20. Bond *et al.* (1994) report data on the frequency of mutations in mice after exposure to the two environmental toxins 1,3-butadiene and ethylnitrosourea (ENU). Mutant frequencies for the murine *lacZ*$^-$ gene were given in three different tissues: lung, liver, and bone marrow. Consider the following data:

Exposure	*Tissue: Lung*	*Liver*	*Bone marrow*
Vehicle control	4.4	2.4	3.0
1,3-butadiene	9.1	3.1	1.8
ENU	16.2	13.0	73.1

a. Employ a gamma GLiM to assess whether a significant tissue×exposure interaction exists. Set $\alpha = 0.05$.

b. The data on ENU exposure are actually included as a form of **positive control** group. Remove it from the analysis and center your attention on the mutagenic effects of 1,3-butadiene. Determine whether a significant tissue×exposure interaction exists with the reduced data. Set $\alpha = 0.05$.

8.21. Return to the aquatic toxicity data in Example 8.6. Recall that there was no strong motivation for use of the logit link.

a. Reanalyze these data using a probit link. Do the results change substantively? What would be the interpretation of the parameter estimates from this fit?

b*. Reanalyze these data using the Aranda-Ordaz link from (8.6). Do the results change substantively? What would be the interpretation of the parameter estimates from this fit?

c*. Under the extended link in (8.6), test goodness-of-link for the logistic model with these data. Set $\alpha = 0.10$.

8.22*. In Example 8.6, we suggested that a simplification from a model in which unique intercept, linear, and quadratic parameters were fitted to the three copepod stages might be possible. As described therein, test to see if a reduced model with common intercept, linear, and quadratic terms for both intermediate and adult stages provides an adequate description of these data.

8.23. Au *et al.* (1994) report data on proportions of chromosome translocations (a form of genetic damage) after joint exposure to heavy metals and γ-radiation. Prior to two fixed exposures of 75 cGy γ-rays, human lymphocytes in culture were exposed to one of $k = 5$ doses of nickel acetate (in μmol). After incubation, the cells were examined for translocations, and the total proportion of cells with translocations was reported at each nickel acetate exposure level for four replicate experiments. The data are:

Dose (μ*mol*)	*Experiment:* *1*	2	3	4
0.0 (Control)	$\frac{7}{50}$	$\frac{7}{50}$	$\frac{8}{50}$	$\frac{11}{50}$
0.1	$\frac{6}{50}$	$\frac{9}{70}$	$\frac{3}{30}$	$\frac{10}{50}$
1.0	$\frac{10}{50}$	$\frac{6}{34}$	$\frac{15}{66}$	$\frac{13}{50}$
10.0	$\frac{16}{64}$	$\frac{7}{36}$	$\frac{13}{40}$	$\frac{12}{60}$
100.0	$\frac{14}{59}$	$\frac{5}{41}$	$\frac{5}{21}$	$\frac{15}{79}$

a. Consider a logistic regression GLiM for these data, and assess whether there is significant interaction between dose and experiment. Set $\alpha = 0.05$.

b. Based on your results in part (a), how would you proceed to assess whether the data exhibit an increasing trend with dose? Perform such a test. Set $\alpha = 0.05$.

8.24. Return to Exercise 8.23 and reanalyze the data using the complementary log link from Table 7.3. Do the results change?

8.25*. In Exercise 8.23, the choice of a logistic link function was arbitrary. Indeed, Exercise 8.24 employed a complementary log link. Notice that the extended link function from (8.7) contains both these links as special cases.

a. Appeal to (8.7) and perform a goodness-of-link test for the logistic link with these data. Set $\alpha = 0.10$.

b. Appeal to (8.7) and perform a goodness-of-link test for the complementary log link with these data. Set $\alpha = 0.10$.

c. What do you conclude from the results in parts (a) and (b)?

8.26*. Repeat Exercise 8.25 on:

a. The cellular transformation data in Exercises 6.17a and 7.22.

b. The cellular transformation data in Exercises 6.17b and 7.22.

8.27. Return to the environmental carcinogenesis study in Example 8.7. Ignore the time-on-study information in x_1, and test for trend without survival adjustment using the Cochran–Armitage test from Section 6.3.1. Set $\alpha = 0.05$. Do the results change?

8.28. In a carcinogenesis study similar to that in Example 8.7, the US National Toxicology Program examined the response of female mice to 1,3-butadiene (NTP, 1984). (We studied the mutagenicity of this chemical in Example 6.13.) Of interest in this study is the chemical's carcinogenicity after inhalation exposure. Mice were exposed to the chemical via inhalation over 61 weeks. Three exposures were used: a zero-dose control and two test groups at 625 ppm and 1250 ppm. Data on observed mortality, x_1 (in weeks), and lung tumor incidence, Y (1 if animal exhibited tumor, 0 otherwise), as a function of dose, x_2, are:

$x_2 = 0$ *ppm*				$x_2 = 625$ *ppm*				$x_2 = 1250$ *ppm*			
x_1	Y	x_1	Y	x_1	Y	x_1	Y	x_1	Y	x_1	Y
2	0	61	0	2	0	56	0	20	0	61	0
6	0	61	0	2	0	56	0	22	0	61	1
30	0	61	0	29	0	57	1	40	0	61	1
61	0	61	0	29	0	58	0	40	0	61	0
61	0	61	0	32	0	58	1	43	0	61	1
61	0	61	0	33	0	59	0	47	0	61	1
61	0	61	0	37	0	60	0	50	1	61	1
61	0	61	0	38	0	60	1	53	0	61	1
61	0	61	0	39	0	61	0	53	0	61	0
61	0	61	0	40	0	61	0	55	0	61	1
61	0	61	1	41	0	61	1	55	0	61	0
61	0	61	0	46	0	61	1	56	0	61	1
61	0	61	0	48	0	61	0	56	0	61	0
61	0	61	0	48	0	61	0	58	1	61	1
61	1	61	0	52	0	61	1	58	1	61	1
61	0	61	1	52	0	61	1	58	1	61	0
61	0	61	0	53	0	61	0	59	0	61	1
61	0	61	0	53	0	61	0	59	1	61	1
61	0	61	0	54	0	61	1	60	1	61	1
61	0	61	0	55	0	61	1	61	0	61	1
61	0	61	0	56	0	61	0	61	0	61	0
61	0	61	0	56	0	61	1	61	0	61	1
61	0	61	0	56	1	61	1	61	0	61	0
61	0	61	0	56	0			61	1	61	1
61	0			56	0			61	0		

Analyze these tumorigenicity data via the logistic regression incidental tumor test, as described in Example 8.7. Set $\alpha = 0.01$.

8.29. Return to the environmental carcinogenesis data in Exercise 8.28. Ignore the time-on-study information in x_1, and test for trend without a survival adjustment using the Cochran–Armitage test from Section 6.3.1. Set $\alpha = 0.01$. Do the results change substantively?

8.30*. In the following analyses, a log-linear link was chosen purely for convenience. Appeal to the extended power link function $g(\mu;\gamma) = (\mu^{\gamma} - 1)/\gamma$, for $\gamma > 0$, in Section 8.3.3 and test goodness-of-link to assess the quality of the log-linear ($\gamma = 0$) assumption. Set $\alpha =$ 0.10:

a. The avian foray data in Exercise 7.23.
b. The skin cancer data in Exercise 7.24.
c. The tumorigenicity data in Exercise 7.25.
d. The aquatic toxicity data in Example 7.5. (Operate with the quadratic predictor used in Fig. 7.15.)
e. The DNA transfection data in Example 8.8.

8.31. Recall that interest in Example 8.8 also concerned the nature of any difference in species' response. To assess this, reverse the order of the model terms in the two-factor/interaction fit. That is, specify the model sequence as dexamethasone first, then species, then their interaction. Do the conclusions change from those in the example?

9

Analysis of cross-classified tabular/categorical data

One of the simplest ways to report environmental data is to count the number of times a specific event occurs and categorize those counts based on some environmental or biological classification. For such data, an important question of interest is whether an observed classification for response is consistent with the theoretical classification that attempts to explain the underlying effect. For example:

- Do observed genotypic frequencies in a population under study conform to theoretical predictions based upon a standard model, such as Hardy–Weinberg equilibrium?

In other instances, there may be two classification schemes or qualitative factorizations examined for their mutual effect on the environmental response. Or, the differential response of two or more populations with respect to a single classification variable may be of interest. For instance:

- Two different toxins released into an ecosystem can each induce toxicity-related mortality to organisms inhabiting the system. The two toxins represent the different factors. When toxic exposures occur in tandem, the number of deaths may be much greater than if either exposure occurred separately. How do we quantify this increase, and assess its significance?
- Mutations in subjects exposed to an environmental toxin may differ from those that occur spontaneously in an unexposed population. The mutant counts over different genetic categories represent a *spectrum* of mutations for each population. Is an observed difference between the two populations' spectra significant?

Classification data in the form of counts often are tabulated for easy display, from which obvious patterns in the response may become apparent. Statistical methods can formalize this analysis, in order to test for departures from simple effects or from theoretical predictions of a particular pattern of importance. In this chapter we review briefly the statistical assumptions underlying the analysis of tabular count data. We also discuss some special tabular forms and effects that occur with environmental toxicity data, and describe some associated methods for their analysis. To do so, we draw upon more extensive discussions on categorical data analysis, as found in a number of advanced texts (Bishop *et al.*, 1975; Read and Cressie, 1988; Santner and Duffy, 1989; Agresti, 1990). Interested readers are encouraged to examine these sources for details that exceed the scope of this chapter.

9.1 *R*×*C* CONTINGENCY TABLES

To analyze and interpret environmental count data, we arrange the counts by category or population in a table. We have already encountered such a construction in Section 4.1.7, where we compared two binomial proportions. There, we arranged the observations in a 2×2 contingency table, where the two populations being compared represented the columns of the table, and the distinction between success and failure – the two possible outcomes under binomial sampling – represented the rows.

In many studies with environmental data, the 2×2 construction requires extension to either $C > 2$ columns for more than two populations under comparison, or to $R > 2$ rows for more than two outcome categories. For example, we encountered a form of 2×C contingency table in Section 6.3 (although it was not presented in this fashion), where trends were studied in a dichotomous response (an $R = 2$ level factor) across $C = k + 1$ dose levels. More generally, given some specific outcome variable representing occurrences of some event, we say that there are $R \geq 2$ row categories, and $C \geq 2$ column categories. This leads to an ***R*×*C* contingency table**, which takes the form:

	Col. category 1	*Col. category 2*	···	*Col. category C*	*TOTAL*
Row category 1	Y_{11}	Y_{12}	···	Y_{1C}	Y_{1+}
Row category 2	Y_{21}	Y_{22}	···	Y_{2C}	Y_{2+}
⋮	⋮	⋮		⋮	⋮
Row category R	Y_{R1}	Y_{R2}	···	Y_{RC}	Y_{R+}
TOTAL	Y_{+1}	Y_{+2}	···	Y_{+C}	Y_{++}

The number of counts in row category i ($i = 1, \ldots, R$), and in column category j ($j = 1, \ldots, C$) is denoted by Y_{ij}. Thus the table represents a **cross-classification** of the data among the various row and column categories. (Notice the calculation of the row sums $Y_{i+} = \sum_{j=1}^{C} Y_{ij}$ and of the column sums $Y_{+j} = \sum_{i=1}^{R} Y_{ij}$ on the margins of the table, in similar form to the 2×2 table from Section 4.1.7. From these, the total count is defined as $Y_{++} = \sum_{i=1}^{R} Y_{i+} = \sum_{j=1}^{C} Y_{+j} = \sum_{i=1}^{R}\sum_{j=1}^{C} Y_{ij}$. These summary values are important for the statistical procedures we describe below.)

Perhaps the most common statistical question for data arranged in a cross-classified form is whether the row and column categories affect the outcome variable in an independent fashion. That is, if an event occurs in row category i, does that affect whether it is cross-classified in column category j, or column category $j - 1$, or column category $j + 3$, etc.? We often describe the null hypothesis for such a condition as one of **independence**, that is, the row and column classifications are independent with respect to the count variable being recorded. If one of the classification factors – say, columns – has its marginal totals fixed by design, then it is more appropriate to assess whether the pattern of response across rows varies from column to column. The null condition in this case translates to one of **homogeneity** in row response across columns. In either case, the null effect represents a lack of **interaction** between the categories, as we will develop in Section 9.4.

9.2 STATISTICAL DISTRIBUTIONS FOR CATEGORICAL DATA

9.2.1 Poisson parent distributions

The standard statistical distribution used to represent count data is the Poisson distribution, as described in Sections 1.2.5 and 4.1.8. Thus for an $R\times C$ table of counts Y_{ij}, we assume $Y_{ij} \sim$ (indep.) $\mathcal{Poisson}(\mu_{ij})$, $i = 1, \ldots, R$; $j = 1, \ldots, C$. For this model, no constraints are imposed on the marginal counts; each of the $R\times C$ table entries is independently determined. In general, we will employ the Poisson as the parent distribution for any Y_{ij} in an $R\times C$ contingency table.

9.2.2 Full multinomial parent distributions

Recall that when sampling under a Poisson distribution for Y_{ij}, if we condition the analysis on the observed total count Y_{++}, the resulting conditional distribution for the observations becomes multinomial: for an $R\times C$ contingency table, this is $\{Y_{ij}\} \mid Y_{++} \sim \mathcal{Multinom}(Y_{++},\{\pi_{ij}\})$, where π_{ij} is the conditional probability of observing a response in the (i,j)th cell of the $R\times C$ table.

The multinomial parent distribution has a natural motivation for environmental analyses of a single population in which more than one attribute is used to classify a response. For example, an investigator evaluating relative species composition in a lake may capture a large predetermined number of fish (Y_{++}). These fish might then be cross-classified by two categories – say, species and age – to determine if any interrelationship existed between the two classification categories.

Conditioning to yield a multinomial parent distribution can produce useful interpretations under certain null and alternative models. For instance, for cases where the ith level of the row factor and the jth level of the column factor can be quantified by some predictor variable x_{ij}, McCullagh and Nelder (1989, Section 6.4) model the mean count as $E[Y_{ij}] = \exp\{\beta_0 + \beta_1 x_{ij}\}$. They consider the null hypothesis of equal means, $H_0\colon \beta_1 = 0$ vs., say, the one-sided alternative $H_0\colon \beta_1 > 0$. By conditioning on Y_{++}, they note that the full multinomial distribution for $\{Y_{ij}\} \mid Y_{++}$ is independent of the nuisance parameter β_0, under H_0. This simplifies construction of conditional test statistics and P-values.

Conditional on Y_{++}, full multinomial sampling requires $\sum\sum \pi_{ij} = 1$, where here and in subsequent equations the double sums are taken over all possible values of i and j. In terms of the μ_{ij}s, this implies $\mu_{++} = Y_{++}$. From this, one can show also that $\pi_{ij} = \mu_{ij}/\mu_{++}$ (Exercise 9.2). When this condition holds, conditional inferences on the π_{ij} and unconditional inferences on the μ_{ij} are identical (Read and Cressie, 1988, Section 3.4); that is, any statistical conclusion(s) reached about row×column independence based upon the μ_{ij}s will be the same as those based conditionally upon the π_{ij}. Technically, we say that the log-likelihoods under either distribution have the same core or **kernel** function as regards the unknown parameters. In essence, the log-likelihoods are identical, with the exception of constants that do not involve the model parameters (Bishop *et al.*, 1975, Section 3.2). The corresponding likelihood-based inferences produce identical results under either model. Thus for practical purposes, whether one conditions on Y_{++} is inconsequential: the statistical conclusions are the same.

9.2.3 Product multinomial parent distributions

There is a third distributional assumption that can be made for the Y_{ij}s that leads also to inferences identical to those from the Poisson or full multinomial distributions. It is known as the **product multinomial distribution**, and it is best understood from the perspective of comparing an R-level categorical response variable among C different populations.

That is, in an $R \times C$ contingency table, suppose that the columns distinguish C different populations, to be compared across R different outcome categories among the rows of the table. The simple 2×2 table in Section 4.1.7 took this form, where $C = 2$ populations or groups were compared across $R = 2$ outcomes (success or failure). A critical feature of that table was that the column sums – Y_{+j} in the notation of this chapter – were viewed as fixed, since they represented the total number of experimental units examined from each population. For example, suppose an acute toxicity study is conducted at C different toxin concentrations, including a control. The number of organisms placed on test at each concentration, Y_{+j}, is fixed by the experimenter. At each concentration level, the number of dead organisms and the number of survivors is recorded (so $R = 2$), and the pattern of response among the C concentrations is studied to determine if it differs between dead organisms and survivors.

Extended to the case of $R > 2$ outcomes, if we continue to view the column sums as fixed, each column delineates R different categories from a single population or comparison group. In this case, we view any column's distribution as a (simple) R-category multinomial, since it represents R independent outcomes from the same population. That is, at any fixed level j, $[Y_{1j}\ Y_{2j} \dots\ Y_{Rj}] \mid Y_{+j} \sim \mathcal{Multinom}(Y_{+j}, [\tau_{1j}\ \tau_{2j} \dots\ \tau_{Rj}])$, where $\tau_{ij} = \mu_{ij}/Y_{+j}$. (Notice that by assuming each of the column totals is fixed, the grand total, Y_{++}, must be fixed also.) Extended across $C > 2$ independent populations, the likelihood function for the entire set of R columns of counts is the product of the C simple multinomial likelihoods; hence the name 'product multinomial' for this distribution.

A necessary aspect of product multinomial sampling is that the column probabilities must sum to 1, that is, $\tau_{1j} + \tau_{2j} + \cdots + \tau_{Rj} = 1$, at each $j = 1, \dots, C$. Since we write $\tau_{ij} = \mu_{ij}/Y_{+j}$, this implies that $\mu_{+j} = Y_{+j}$ at each $j = 1, \dots, C$. Summing over j then gives $\mu_{++} = Y_{++}$. Under this condition, the product multinomial log-likelihood kernel is equivalent to the Poisson and to the full multinomial, and so likelihood-based inferences for $R \times C$ tables are all identical under any of the three parent distributions.

This identical-inference feature is important for analysis of $R \times C$ contingency tables. It provides a luxury of choice from among an integrated set of statistical methods for $R \times C$ tables, unencumbered by subtle specifics such as which or whether certain marginal totals are fixed or not. Although it is therefore relatively unimportant whether we choose a Poisson or multinomial distribution for the Y_{ij}s, we will for technical purposes and for clarity's sake continue to employ a Poisson parent distribution for Y_{ij} unless specific sampling aspects of the problem demand otherwise.

9.2.4 Null hypothesis of independence/homogeneity

Under the Poison assumption that $Y_{ij} \sim$ (indep.) $\mathcal{Poisson}(\mu_{ij})$, $i = 1, \ldots, R$; $j = 1, \ldots, C$, the null hypothesis of independence (that is, of no interaction between rows and columns) can be written in a straightforward manner. As suggested in Section 9.1, if it makes no difference whether an event occurring in row category i is cross-classified in column category j, we expect the frequency of occurrences in the ith row to be unaffected if we rearrange the columns. In other words, knowledge that an observation falls in a particular column does not affect our ability to predict its row location. In this case, the probability of a randomly selected observation having been classified into the ith row category is simply the marginal row probability μ_{i+}/μ_{++} $(i = 1, \ldots, R)$. The same effect occurs if viewed across rows for a fixed column: the probability of a response being classified into the jth column category can be summarized as the marginal column probability μ_{+j}/μ_{++} $(j = 1, \ldots, C)$. Thus the probability of occurrence in any (i,j)th cell of the table is then the product of the two marginal probabilities: $\mu_{ij}/\mu_{++} = (\mu_{i+}/\mu_{++})(\mu_{+j}/\mu_{++})$. (That is, apply the multiplication rule of probability from Section 1.1: multiply two independent probabilities to find the probability of their joint occurrence.) Collected together for all RC cells of the table, this defines a null hypothesis of row×column independence:

$$H_0: \mu_{ij} = \mu_{++}\left(\frac{\mu_{i+}}{\mu_{++}}\right)\left(\frac{\mu_{+j}}{\mu_{++}}\right), \text{ for all } i,j,$$

or, more simply,

$$H_0: \mu_{ij} = \frac{\mu_{i+}\mu_{+j}}{\mu_{++}}, \text{ for all } i,j. \tag{9.1}$$

The corresponding alternative of no independence ('some' interaction) is $H_a: \mu_{ij} \neq \mu_{i+}\mu_{+j}/\mu_{++}$, for some i,j. Notice that if we condition the sampling scenario on the observed grand total, Y_{++}, then under the constraint that $\mu_{++} = Y_{++}$, the null hypothesis in (9.1) becomes $H_0: \pi_{ij} = \pi_{i+}\pi_{+j}$ (for all i,j), where π_{i+} and π_{+j} are the marginal probabilities of response, for example, $\pi_{i+} = \sum_{j=1}^{C} \pi_{ij}$. As expected, the Poisson and multinomial formulations produce coincidental results.

Under product multinomial sampling, the null hypothesis in (9.1) is replaced by a form of **homogeneity** among the C populations or groups being compared. That is, if the populations (columns) are viewed as independent of outcome categories (rows), then the probability of response must be equal – that is, homogeneous – across columns within any row. We write this as

$$\mathrm{H_0}: \tau_{i1} = \tau_{i2} = \cdots = \tau_{iC} \quad (\text{for all } i = 1, \ldots, R), \tag{9.2}$$

where $\tau_{ij} = \mu_{ij}/Y_{+j}$. This is equivalent to the null hypothesis in (9.1) (Exercise 9.6).

Although fairly complex in nature, the independence/homogeneity hypotheses (9.1) or (9.2) may be assessed via a variety of statistical tests, as we describe in the next section.

9.3 STATISTICAL TESTS OF INDEPENDENCE IN $R \times C$ TABLES

9.3.1 Pearson's chi-square test and goodness-of-fit testing

The basic paradigm used to test departures from independence in categorical data tables is based on an historic inferential methodology: **goodness-of-fit testing**. The approach was developed by Karl Pearson (1900), who identified a value for each observation that corresponded ideally to what that observation should be under some specific biological classification, such as independence/homogeneity. Then, these 'expected' observations are compared to the actual observations, to see how much they differ quantitatively. Too large a difference between the two, on average, would suggest a failure in the goodness-of-fit.

Viewed in terms of some null hypothesis $\mathrm{H_0}$, denote the expected values under $\mathrm{H_0}$ as E_{ij}. Then, Pearson's measure of departure for each observation, Y_{ij}, is $(Y_{ij} - E_{ij})^2/E_{ij}$. This is the original form of a (squared) **Pearson residual** for Y_{ij}, as seen earlier in Sections 2.7.1 and 8.3.2. Summing the squared residuals over all the observations yields the familiar **Pearson chi-square statistic** for goodness-of-fit, $X^2_{\text{calc}} = \sum\sum(Y_{ij} - E_{ij})^2/E_{ij}$. The null hypothesis is rejected when X^2_{calc} is too large, representing too great a departure in the Y_{ij}s from the expected values in E_{ij}.

The expected counts, E_{ij}, in the goodness-of-fit construction may arise from some specified theoretical consideration, or be implicit from a null hypothesis such as (9.1) above. Before detailing the latter case, we illustrate a general form of goodness-of-fit test where the biological theory suggests values for the expected counts.

Example 9.1 Goodness-of-fit for theoretically specified expected values – Hardy–Weinberg equilibrium

In an ecological study of population genetics in fathead minnows, the animals' genotypic frequencies were analyzed in order to evaluate whether **Hardy–Weinberg (HW) equilibrium** existed in the population

(Schlueter *et al.*, 1995). The Hardy–Weinberg law states that heredity, acting in the absence of the influence of other factors, will not change allelic or genotypic frequencies in a randomly mating population (Hardy, 1908; Weinberg, 1908). (For additional background on the Hardy–Weinberg theory, see Weir (1990), Lovell (1995), and the references therein.)

Viewed in the context of cross-classified data, the allelic contribution of male minnows to their offspring is crossed with the allelic contribution of female minnows. We can display the resulting theoretical genotypic population probabilities for the offspring in tabular form, as in Table 9.1.

Table 9.1 Offspring genotypes resulting from random mating of males and females in the population for two alleles (A, B) at a single genetic locus. Row (column) entries correspond to the allelic contribution of females (males)

	*Male allele**	
*Female allele**	A (p)	B (q)
A (p)	AA (p^2)	AB (pq)
B (q)	AB (pq)	BB (q^2)

*Table entries are genotypes and (in parentheses) genotypic probabilities.

In the fathead minnow study, the genetic structure of the animals' isocitrate dehydrogenase (IDH) system was studied in detail. The corresponding genetic locus was *IDH-1**. In its simplest formulation, the *IDH-1** locus exhibits a simple two-allele structure, as in Table 9.1. Denote the two alleles by A and B. The study produced the following genetic frequencies among the three possible outcome genotypes: 51 (AA), 415 (AB), and 888 (BB). Let p be the probability of observing allele A in the population, and q be the probability of observing allele B in the population. From Table 9.1, we see that HW equilibrium implies the following probabilities of occurrence for the three possible genotypes: $\text{P}(AA) = p^2$, $\text{P}(AB) = 2pq$, and $\text{P}(BB) = q^2$. Notice that $p + q = 1$.

To evaluate whether the HW equilibrium predictions are consistent with the observed genotypes, we apply the X^2 goodness-of-fit statistic. In effect, we have a 3×1 table of counts, with $R = 3$ genotypes, and (trivially) $C = 1$ locus. Let $i = 1, 2, 3$ index the three genotypes AA, AB, and BB, respectively. To find the expected counts under HW equilibrium, recognize that the number of A alleles observed is $N_A = (2)(51) + 415 = 517$. Similarly, the number of B alleles is $N_B = (2)(888) + 415 = 2191$. Since the total number of alleles observed is $N_{TOT} = (2)(51 + 415 + 888) = 2708$, an estimate of the A allele population probability is $\hat{p} = N_A / N_{TOT}$; for the B allele, it is $\hat{q} =$

$N_B/N_{TOT} = 1 - \hat{p}$. With the fathead minnow *IDH-1** data, these are $\hat{p}$ = 517/2708 = 0.191 and $\hat{q}$ = 2191/2708 = 0.809, respectively.

Under HW equilibrium, we expect the probability of observing an *AA* genotype to be p^2, which here is estimated as $(0.191)^2 = 0.036$. Similarly, we expect the HW probability of observing a *BB* genotype to be $(0.809)^2 = 0.655$, and of observing *AB* to be (2)(0.191)(0.089) = 0.309. The HW expected genotypic counts are these estimated probabilities multiplied by the total number of fish sampled: $E_{11} = (0.191^2)(1354) = 49.395$, $E_{21} = (0.309)(1354) = 418.438$, and $E_{31} = (0.809^2)(1354) = 886.167$. (Notice that the E_{ij}s sum to 1354.) Applying the X^2 goodness-of-fit statistic to the data $Y_{11} = 51$, $Y_{21} = 415$, and $Y_{31} = 88$, produces

$$X^2_{\text{calc}} = \frac{(51 - 49.395)^2}{49.395} + \frac{(415 - 418.438)^2}{418.438} + \frac{(888 - 886.167)^2}{886.167}$$

or simply $X^2_{\text{calc}} = 0.052 + 0.028 + 0.004 = 0.084$. We reference this goodness-of-fit statistic against a χ^2 distribution; the null model *df* are calculated by taking the number of observed categories, here $R = 3$, and subtracting from it the number of constraints imposed by the probability model and the number of nonredundant parameters estimated separately from the data. Since the multinomial model constrains $\pi_{11} + \pi_{21} + \pi_{31} = 1$, and we also estimated p (but formed q via $\hat{q} = 1 - \hat{p}$), we find $df = 3 - 1 - 1 = 1$. (This form of *df* differencing is particular to the case where E_{ij} is determined from a theoretical model prediction (Agresti, 1990, Section 3.2.4). With cross-classified $R \times C$ tables used to study the independence effect in (9.1), the *df* calculation is somewhat different.) In any case, for the fathead minnow *IDH-1** data, the *P*-value is approximately $P[\chi^2(1) \geq 0.084] = 0.772$. There appears to be no significant departure from HW predictions at this locus. ❑

The form of goodness-of-fit testing seen in Example 9.1 is quite common for settings where counts are compared to some theoretical standard, such as HW equilibrium. It is also the case, however, that questions regarding independence relationships between two qualitative variables as in (9.1) can be assessed using the same form of Pearson X^2 statistic. For the null hypothesis of independence in an $R \times C$ table, the expected values are constructed as the sample analogs of the null effect: replace each μ_{ij} in (9.1) with its sample estimate based on the Y_{ij}s. This produces

$$E_{ij} = \frac{Y_{i+}Y_{+j}}{Y_{++}} = Y_{++}\left(\frac{Y_{i+}}{Y_{++}}\right)\left(\frac{Y_{+j}}{Y_{++}}\right). \tag{9.3}$$

Under the homogeneity null hypothesis in (9.2), an analogous argument leads to the same expected values in (9.3).

In large samples the corresponding X^2_{calc}-statistic is distributed approximately as χ^2 with $(R-1)(C-1)$ *df*, so we write $X^2_{calc} \dot{\sim} \chi^2\{(R-1)(C-1)\}$. The *df* here correspond to the number of Y_{ij}s in the $R \times C$ table that may be modified without affecting the row and column totals; see Agresti (1990, Section 3.3). Alternatively, one can find the *df* as RC – {number of constraints in the model} – {number of parameters being estimated}. Under independence, there is 1 constraint and $(R-1)+(C-1)$ parameters are estimated; under homogeneity, there are $(C-1)+1$ constraints and $(R-1)$ parameters are estimated. In either case, we find $df = (R-1)(C-1)$.

The fit to the independence null model (9.1) is poor when X^2_{calc} is too large, relative to the referent χ^2 distribution. Thus the corresponding P-value is approximately $P[\chi^2\{(R-1)(C-1)\} \geq X^2_{calc}]$; reject H_0 when this P drops below a specified significance level, α, or, equivalently, when X^2_{calc} exceeds an upper-α critical point from $\chi^2\{(R-1)(C-1)\}$.

Note that in the 2×2 case, the Pearson X^2_{calc} statistic corresponds to the test statistic from (4.8); see Exercise 9.8.

Example 9.2 Aquatic toxicity to azinphosmethyl

In Example 5.6, we studied aquatic toxicity in *Amphiascus tenuiremis* to the pesticide azinphosmethyl. The marker of aquatic toxicity was the capacity of *A. tenuiremis* to produce young that grow to the adult stage; Table 5.3 gives reproductive results for numbers of female offspring that survive to adulthood (after a 26-day maturation). In Exercise 5.13 we considered similar results for male offspring. Since the number of offspring surviving to adulthood is determined independently for males and females at the different azinphosmethyl doses, we can combine the male and female data, and question whether the pattern of offspring survival rates is similar between sexes over dose. That is, is the sex of the offspring independent of dose when considering toxicity-moderated maturation?

The combined data, pooled over separate replicates, are given in Table 9.2 as a 3×2 contingency table. Therein, we see an apparent difference in total response rates (more females survive). More importantly, however, is whether the pattern of survival over dose is homogeneous. To assess this, we assume a Poisson sampling model is appropriate for the combined data, and compute X^2_{calc}. The calculations require the expected values E_{ij} given by (9.3); these are displayed in Table 9.3.

Table 9.2 Offspring counts of *Amphiascus tenuiremis* after exposure to azinphosmethyl

	Sex of offspring	
Dose	*Female*	*Male*
Control	363	103
10 µg/l azinphosmethyl	157	99
40 µg/l azinphosmethyl	219	83

Based on the E_{ij}s in Table 9.3, we find $X^2_{\text{calc}} = \sum\sum(Y_{ij} - E_{ij})^2/E_{ij} = 22.609$ for testing homogeneity/independence with the data in Table 9.2. There are $(3-1)(2-1) = 2$ *df*, so the approximate P-value is $P \approx \text{P}[\chi^2(2) > 22.609] < 0.001$. Thus, at $\alpha = 0.01$ we find significant evidence that sex of offspring appears to exhibit a differential response to dose of the pesticide.

Table 9.3 Expected numbers of *A. tenuiremis* offspring after exposure to azinphosmethyl, under dose×sex independence

	Sex of offspring	
Dose	*Female*	*Male*
Control	336.30	129.7
10 µg/l azinphosmethyl	184.75	71.25
40 µg/l azinphosmethyl	217.95	84.05

These computations may be performed via standard statistical computing packages. In particular, X^2_{calc} can be determined using SAS PROC FREQ, as given in Fig. 9.1. The data from Table 9.2 are entered in the SAS DATA step such that the row and column variables are defined along with the count in each cross-classified cell. Calculation of X^2_{calc} is in response to invocation of PROC FREQ: the `table` command defines the classification variables `dose` and `sex`, while the `weight` statement identifies a variable in the data set containing the counts in each cross-classified cell. We control the output with certain options in the `table` command (the terms following the option separator '`/`'): to suppress construction of certain summary percentages we add the `nocol` and `nopct` options, to request construction of expected counts as in Table 9.3 we include the `expected` option, and to request X^2_{calc} we include the `chisq` option. The resulting output (edited for presentation) is displayed in Fig. 9.2. Notice in the figure that the entries from both Tables 9.2 and 9.3 are displayed in the frequency table, including the corresponding percentages of observed females and males at each dose.

```
* SAS code for A. tenuiremis data from Table 9.2;

data azinph;
input sex $ dose y @@;
cards;
f 0 363   f 10 157   f 40 219
m 0 103   m 10  99   m 40  83

proc freq;
     weight y;
     table dose*sex / nocol nopct expected chisq;
```

Fig. 9.1 SAS PROC FREQ program for testing dose×sex independence in *A. tenuiremis*.

```
                     The SAS System
                   The FREQ Procedure

                  TABLE OF DOSE BY SEX
        DOSE       SEX
        Frequency|
        Expected |
        Row Pct  |f       |m       |  Total
        ---------+--------+--------+
               0 |    363 |    103 |    466
                 |  336.3 |  120.7 |
                 |  77.90 |  22.10 |
        ---------+--------+--------+
              10 |    157 |     99 |    256
                 | 184.75 |  71.25 |
                 |  61.33 |  38.67 |
        ---------+--------+--------+
              40 |    219 |     83 |    302
                 | 217.95 | 84.053 |
                 |  72.52 |  27.48 |
        ---------+--------+--------+
        Total         739      285     1024

        STATISTICS FOR TABLE OF DOSE BY SEX
Statistic                     DF     Value          Prob
--------------------------------------------------------
Chi-Square                     2     22.609        0.001
Likelihood Ratio Chi-Square    2     22.022        0.001

Sample Size = 1024
```

Fig. 9.2 Output (edited) from SAS PROC FREQ program for testing dose×sex independence in *A. tenuiremis*.

```
azin.df<-data.frame(sex=c("f","f","f","m","m","m"),
                    dose=as.factor(c(0,10,40,0,10,40)),
                    count=c(363,157,219,103,99,83))

crosstabs(count~sex+dose,data=azin.df)
```

```
+----------+
|N         |
|N/RowTotal|
|N/ColTotal|
|N/Total   |
+----------+
sex     |dose
        |0      |10     |40     |RowTotl|
-------+-------+-------+-------+-------+
f       |363    |157    |219    |739    |
        |0.491  |0.212  |0.296  |0.72   |
        |0.779  |0.613  |0.725  |       |
        |0.354  |0.153  |0.214  |       |
-------+-------+-------+-------+-------+
m       |103    | 99    | 83    |285    |
        |0.361  |0.347  |0.291  |0.28   |
        |0.221  |0.387  |0.275  |       |
        |0.101  |0.097  |0.081  |       |
-------+-------+-------+-------+-------+
ColTotl|466    |256    |302    |997    |
        |0.47   |0.25   |0.29   |       |
-------+-------+-------+-------+-------+
Test for independence of all factors
     Chi^2 = 22.60909 d.f.= 2 (p=1.23168e-05)
     Yates' correction not used
```

Fig. 9.3 Output (edited) from S-PLUS `crosstabs` function for testing dose×sex independence in *A. tenuiremis*.

In Fig. 9.2, X^2_{calc} and its associated P-value are presented in the row labeled `Chi-Square` under `STATISTICS FOR TABLE OF DOSE BY SEX`. This result agrees with our previous calculation.

The X^2 analysis can be replicated in S-PLUS using the `crosstabs` function, as illustrated in Fig. 9.3. The tabular data are entered into a data frame with both of the classification variables, `sex` and `dose`, defined as factors. (This must be stated explicitly in the definition of the data frame for the `dose` variable, since variables with numeric levels are not viewed by default as qualitative factors.) Invocation of the `crosstabs` function is similar to formula expressions encountered in previous S-PLUS modeling illustrations: the response is to the left of the tilde ('~') and the predictor/classification variables are to the right. Alternatively, if the data were entered as 363 individual case records – for example, 363 data lines with sex = `f` and dose = `0`, or sex = `m` and dose = `10`, etc. – then the `crosstabs` call would not contain any variable to the left of the tilde; use the command `crosstabs(~sex+dose,data=....)`.

Inferences from Fig. 9.3 are similar to those seen above: at all doses, there are significantly more female offspring than expected, and significantly fewer male offspring than expected. □

9.3.2 Extensions of the chi-square test: Power-divergence statistics

The appeal to a large-sample χ^2 approximation for X^2_{calc} requires that a large enough number of observations be recorded, not only for the total count Y_{++}, but also for the marginal counts Y_{i+} and Y_{+j}. Under row×column independence/homogeneity, operations on the marginal totals impact the expected counts E_{ij}. Thus, a standard rule of thumb for applying the χ^2 approximation to X^2_{calc} is that $E_{ij} > 1$ for all cells and, to help ensure general large-sample effects, that the average expected cell count is greater than 5. For samples where this is not achieved, the χ^2 approximation to X^2_{calc} generally underestimates the effect of any departure from H_0. (That is, the test loses sensitivity to detect true interactions between rows and columns.)

It can be shown, however, that Pearson's X^2 is only one member of a larger family of goodness-of-fit statistics, known as **power-divergence statistics** (Read and Cressie, 1988, Section 1.2). These employ generalizations of the Pearson residuals to compare the observed and expected values in an $R{\times}C$ table. Specifically, for any real number λ, the raw **power-divergence residual** is $Y_{ij}\{(Y_{ij}/E_{ij})^{\lambda} - 1\}$, where under (9.1) or (9.2), E_{ij} is given by (9.3). (Alternative definitions of the E_{ij}s are possible when employing the power-divergence residuals, although they are much more complex to compute and implement. For details see Reed and Cressie (1988, Section 3.4) or Lee and Shen (1994).) Aggregating the raw power-divergence residuals over all RC cells in the table, and standardizing for the exponent λ, we obtain the general power-divergence statistic

$$C^2_{\lambda} = \frac{2}{\lambda(\lambda+1)} \sum_{i=1}^{R} \sum_{j=1}^{C} Y_{ij}\left\{\left(\frac{Y_{ij}}{E_{ij}}\right)^{\lambda} - 1\right\}. \tag{9.4}$$

In large samples, $C^2_{\lambda} \mathrel{\dot\sim} \chi^2\{(R-1)(C-1)\}$ for any value of λ. The large-sample approximation is most accurate in the range $0.5 \le \lambda \le 1$; in particular, at $\lambda = 1$, $C^2_1 = X^2_{\text{calc}}$, so (9.4) recovers Pearson's X^2. Outside this range, a correction factor is recommended when appealing to a χ^2 approximation for C^2_{λ} (Read and Cressie, 1988, Section H.1).

For certain choices of λ, the χ^2 approximation for the power-divergence statistic can be superior to that for Pearson's X^2 (that is, at $\lambda = 1$). These include what is known as the Cressie–Read statistic at $\lambda = 2/3$:

$$C^2_{2/3} = \frac{9}{5} \sum_{i=1}^{R} \sum_{j=1}^{C} Y_{ij} \left\{ \left(\frac{Y_{ij}}{E_{ij}} \right)^{2/3} - 1 \right\}, \tag{9.5}$$

and the power-divergence statistic corresponding to $\lambda = 1/2$

$$C^2_{1/2} = \frac{8}{3} \sum_{i=1}^{R} \sum_{j=1}^{C} Y_{ij} \left\{ \left(\frac{Y_{ij}}{E_{ij}} \right)^{1/2} - 1 \right\}. \tag{9.6}$$

Both $C^2_{2/3}$ and $C^2_{1/2}$ possess relatively stable approximations to a $\chi^2\{(R-1)(C-1)\}$ reference distribution. The latter form provides a slightly better approximation in small samples (Read, 1993), and may be recommended for general use.

Example 9.2 (continued) Aquatic toxicity to azinphosmethyl
For the azinphosmethyl toxicity data in Table 9.2, we can also calculate power-divergence statistics to assess the independence of dose and sex. Note that in (9.5) and (9.6), the only additional quantities we need are the ratios Y_{ij}/E_{ij}, where the E_{ij}s are given in Table 9.3. These ratios are given in Table 9.4.

Table 9.4 Ratio of observed to expected numbers, Y_{ij}/E_{ij}, of offspring of *A. tenuiremis* after exposure to azinphosmethyl, from Table 9.2. Expected values are calculated under sex×dose independence (Table 9.3)

	Sex of offspring	
Dose	*Female*	*Male*
Control	1.079	0.794
10 µg/l azinphosmethyl	0.850	1.389
40 µg/l azinphosmethyl	1.005	0.988

Using the ratios in Table 9.4, we find $C^2_{2/3} = 22.383$ and $C^2_{1/2} = 22.282$, each with 2 *df*. Both statistics are similar to $X^2_{calc} = 22.609$ calculated earlier. Indeed, the corresponding P-values are essentially identical: $P < 0.001$. As above, significant departure from independence is suggested.

While power-divergence test statistics such as $C^2_{2/3}$ and $C^2_{1/2}$ are straightforward to calculate by hand, they are not readily available in common statistical computing packages. In particular, neither SAS nor S-PLUS calculates them directly. Instead, the expected counts E_{ij} can be generated by SAS PROC FREQ, S-PLUS `print.crosstabs`, or by the GLIM system. These are then extracted for use in constructing the power-divergence statistic by writing and implementing some sort of user-defined function. (One can, of course, read the E_{ij}s from the output for use in simple hand calculations.)

For example, we can calculate $C^2_{2/3}$ and $C^2_{1/2}$ in S-PLUS by employing a relationship between (9.1) and features of GLiMs from Chapter 8. It can be shown that predicted values from a GLiM with a Poisson parent distribution, a log link, and a linear predictor with only additive terms (no interactions) in the model are identical to E_{ij}s generated under the independence null hypothesis (9.1); we discuss this relationship in more detail in Section 9.4. In Fig. 9.4 we take advantage of this relationship to construct expected counts in S-PLUS which are then employed in a user-defined function, `power.diverg`, to calculate the power-divergence statistics.

The results in Fig. 9.4 are identical to the hand calculations given above. Similar outcomes are possible in SAS by using PROC GENMOD with the `/obstats` option to generate the E_{ij}s, and then calculating the statistic with direct coding in a new `data` step. □

One limiting form of power-divergence statistic we do not recommend corresponds to another, established form of test for (9.1). As $\lambda \rightarrow 0$, C^2_λ converges to the generalized likelihood ratio statistic, G^2_{calc}, based on subtracting twice the constrained, maximum log-likelihood under (9.1) from twice the unconstrained, maximized log-likelihood. Applied to $R \times C$ contingency tables, this statistic has been shown to overemphasize deviations in cells with very small observed counts. As such, it is known to be quite unstable in small samples; see, for example, Cressie and Read (1989), or our own work (Piegorsch and Bailer, 1994), and the references therein. We do not recommend the LR statistic for general use with $R \times C$ contingency tables. (Notice that the output in Fig. 9.2 includes a line for `Likelihood Ratio Chi-Square`, which is in fact G^2_{calc}. Although it corroborates the result using X^2_{calc} – due primarily to the large sample sizes with these data – the LR test can nonetheless operate in an unstable manner with other $R \times C$ tables.)

A general rule of thumb for employing the χ^2 approximation to power-divergence statistics is to require that $Y_{++}/\{(R-1)(C-1)\} \geq 10$. That is, the total sample size should be at least 10 times larger than the *df* in the $R \times C$ table. For the power-divergence forms in (9.5) and (9.6), this can be relaxed to $Y_{++}/\{(R-1)(C-1)\} \geq 5$. For the azinphosmethyl data in Example 9.2, the χ^2 approximation is reasonable, since $Y_{++}/\{(R-1)(C-1)\} = 997/\{(3-1)(2-1)\} = 997/2 = 498.5$ is well above 5 or 10.

```
az.fit<-glm(count~sex+dose,family=poisson(link=log),data=azin.df)
expected<-predict(az.fit,type="response")

power.diverg<-function(observed,expected,NR=2,NC=2,lambda=1) {
# Power-divergence Statistic
# NR (NC) = number of rows (columns) in the table
# lambda = 1 (default to Pearson X2)
  c2.lambda<- 2/(lambda*(lambda+1)) *
        sum(observed*((observed/expected)^lambda - 1))
  p.value<- 1 - pchisq(c2.lambda,df=(NR-1)*(NC-1))
  data.frame(C2.lambda=c2.lambda,lambda=lambda,p.value=p.value)
}

# Pearson X2 ==> lambda = 1
power.diverg(azin.df$count,expected,
             NR=length(unique(azin.df$sex)),
             NC=length(unique(azin.df$dose)))
```

```
     C2.lambda   lambda      p.value
  1  22.60906      1       1.231699e-05
```

```
# Cressie-Read ==> lambda = 2/3
power.diverg(azin.df$count,expected,lambda=2/3,
             NR=length(unique(azin.df$sex)),
             NC=length(unique(azin.df$dose)))
```

```
     C2.lambda    lambda     p.value
  1  22.38332   0.6666667  1.37887e-05
```

```
# Power-divergence ==> lambda = 1/2
power.diverg(azin.df$count,expected,lambda=1/2,
             NR=length(unique(azin.df$sex)),
             NC=length(unique(azin.df$dose)))
```

```
     C2.lambda    lambda     p.value
  1  22.28192      0.5     1.450583e-05
```

Fig. 9.4 User-defined S-PLUS power-divergence function and associated output (edited) applied to testing dose×sex independence in *A. tenuiremis*.

An exception to this rule of thumb occurs when a large number of the table's cells (more than about 20%) contain only zeros or ones. In this case, the $R{\times}C$ table is said to be highly **sparse**, and the asymptotic χ^2 approximation for even the Power-Divergence family can be adversely affected. More advanced test statistics are required in this case, as we note in Section 9.3.4.

9.3.3 Fisher's exact test for homogeneity in $R \times C$ tables

As we saw previously, it is possible to compute exact P-values in contingency tables via Fisher's exact test. In Section 4.1.7, we applied the exact test to assess homogeneity between proportions in a simple 2×2 table, and it is a straightforward extension to consider exact tests for the general $R \times C$ contingency table. Therein, an exact test computes all possible tabular configurations of the observed data that could be generated under the condition that the row and column marginal totals are held fixed. (As in the 2×2 case, this test is conditional on the observed row and column totals.) The corresponding conditional P-value is the probability of recovering a tabular configuration as extreme as or more extreme than that actually observed. If P is very small, reject the null hypothesis in (9.2).

To find the conditional P in the general $R \times C$ case, find the probability of each possible tabular configuration as extreme as or more extreme than that observed, conditional on the row and column sums Y_{i+} and Y_{+j}. The probabilities are based on an extension from the 2×2 case of the hypergeometric distribution:

$$P[Y_{11} = y_{11}, Y_{12} = y_{12}, \ldots, Y_{RC} = y_{RC} \mid Y_{i+} = y_{i+}, Y_{+j} = y_{+j}] = \frac{\prod_{i=1}^{R} (y_{i+})! \prod_{j=1}^{C} (y_{+j})!}{(y_{++})! \prod_{i=1}^{R} \prod_{j=1}^{C} y_{ij}!}; \quad (9.7)$$

cf. the description in Section 1.5.2 (in (9.7) the binomial coefficients have been simplified into factorial operators). For any observed table, we compute the X^2_{calc} measure of departure from independence for all possible $R \times C$ tables with the same row and column totals. The two-sided exact P-value is the sum of the hypergeometric probabilities (9.7) corresponding to all those tables whose X^2_{calc} measures are larger than or equal to the observed table's (Roff and Bentzen, 1989; Agresti, 1990, Section 3.5).

Unfortunately, when the values of Y_{i+} and Y_{+j} grow large, the calculations required for the exact P-value can become unwieldy. As in the 2×2 case, computational algorithms are then available to facilitate the calculations (Agresti *et al.*, 1979; Mehta and Patel, 1983; Pagano and Trichler, 1983). For example, the $R \times C$ exact test may be performed in SAS PROC FREQ with the `/exact` option in the `tables` sub-command. Alternatively, the

STATXACT system (Mehta and Patel, 1991; Russek-Cohen, 1994) performs a number of exact tests, including that for $R \times C$ tables.

9.3.4* Testing independence/homogeneity in sparse $R \times C$ tables

As we noted in Section 9.3.2, the reference to a χ^2 distribution for calculating P-values or other statistical inferences from power-divergence test statistics may not be appropriate in selected settings. Of greatest concern is the case when the $R \times C$ table exhibits many low or zero counts. This situation is known as **sparseness**; it is more properly characterized as a situation where many of the E_{ij} are small, especially if many are less than 1 (Lewontin and Felsenstein, 1965). In such settings, one can apply Fisher's exact test from Section 9.3.3, since the hypergeometric calculations for the P-value are unaffected by sparseness in the data table.

A slightly simpler alternative to Fisher's exact test for sparse tables involves determination of an alternative reference distribution for the X^2 test statistic, or for some functions thereof. Such a construction was suggested by Zelterman (1987): begin with

$$D_{\text{calc}} = X^2_{\text{calc}} - \sum_{i=1}^{R} \sum_{j=1}^{C} \frac{Y_{ij}}{E_{ij}},$$

where the E_{ij}s are the expected values from (9.3). Next, set

$$\mu_D = \frac{Y_{++}(R-1)(C-1)}{(Y_{++}-1)} - RC,$$

and calculate

$$\sigma_D^2 = \frac{2Y_{++}}{Y_{++}-3}\left\{\frac{(R-1)(Y_{++}-R)}{Y_{++}-1} - \zeta_R\right\}\left\{\frac{(C-1)(Y_{++}-C)}{Y_{++}-1} - \zeta_C\right\} + \frac{4\zeta_R\zeta_C}{Y_{++}-1},$$

where

$$\zeta_R = \frac{Y_{++}\sum_{i=1}^{R}\left(\sum_{j=1}^{C} Y_{ij}\right)^{-1} - R^2}{Y_{++} - 2} \quad \text{and} \quad \zeta_C = \frac{Y_{++}\sum_{j=1}^{C}\left(\sum_{i=1}^{R} Y_{ij}\right)^{-1} - C^2}{Y_{++} - 2}.$$

Use these in the statistic

$$Z_D = \frac{D_{\text{calc}} - \mu_D}{\sigma_D} \tag{9.8}$$

to test independence. Z_D is referenced in large samples to a standard normal distribution, so the corresponding approximate P-value is $2\{1 - \Phi(|Z_D|)\}$. Z_D possesses good power to detect departures from independence in $R{\times}C$ tables exhibiting extreme sparseness. We advocate its use for settings where $R{\times}C$ data are sparse and computing resources are too limited to apply Fisher's exact test.

Example 9.3 Assessing heterogeneity between mutational spectra
Sparse data tables occur often in environmental studies of mutation at the DNA level. The studies are performed to construct and analyze the spectrum of these mutations over, say, $R \geq 2$ genetic categories of interest, and the data are counts of how often a particular mutant is observed in each category. The counts are displayed in a categorical data table, with each row representing a different mutant site or category. If there are C treatments or groups whose spectra are under comparison, such data will constitute an $R{\times}C$ contingency table.

Unfortunately, sequencing large numbers of mutants for identification and classification is time- and resource-intensive. If the experimental subjects are large animals (including humans), very few mutants will be sampled and categorized. This can lead to sparse $R{\times}C$ data, since many of the table's cells will be zeros or ones.

The mutational spectra in Table 9.5 exhibit this phenomenon. The data, discussed by Cariello *et al.* (1994), represent single base substitution patterns in the human *p53* gene from three different forms of human cancer: small-cell lung cancer (SCLC), nonsmall-cell lung cancer (NSCLC), and colon cancer. Differences among the pattern of mutant spectral response for these cancers may suggest differences in the mutagenic processes involved in tumor development or progression, so comparison of the spectral patterns is of interest.

Table 9.5 Mutant spectra of *p53* gene mutations in small-cell lung carcinomas (SCLC), nonsmall-cell lung carcinomas (NSCLC), and colon cancers

Mutation	*SCLC*	*NSCLC*	*Colon*	*Row totals*
GC→TA	6	16	0	22
AT→GC	4	0	3	7
GC→AT	1	5	14	20
GC→CG	1	4	1	6
AT→TA	0	2	1	3
Col. totals	12	27	19	58

Data from Cariello *et al.* (1994).

Since the sampling scheme in Table 9.5 views the column totals as fixed, the product multinomial distribution from Section 9.2.3 is valid. Comparison of the spectral patterns across the $C = 3$ cancer types amounts to testing homogeneity of the pattern of response, as given in (9.2).

Unfortunately, seven of the possible 15 mutant–cancer combinations in Table 9.5 are so rare that no more than one subject was recorded at that combination. Indeed, under the independence assumption, five of the expected cell counts, E_{ij} from (9.3), are near or below 1.0 (Table 9.6). With about one-third of the table's cells representing low expected values, we would classify this table as sparse. Hence the Z_D-statistic in (9.8) may be more appropriate than any of the χ^2-based power-divergence statistics for identifying spectral heterogeneity.

Table 9.6 Expected numbers, E_{ij}, of *p53* gene mutations in Table 9.5, under independence between mutation and cancer type

Mutation	*SCLC*	*NSCLC*	*Colon*	*Row totals*
GC→TA	4.55	10.24	7.21	22
AT→GC	1.45	3.26	2.29	7
GC→AT	4.14	9.31	6.55	20
GC→CG	1.24	2.79	1.97	6
AT→TA	0.62	1.40	0.98	3
Col. totals	12	27	19	58

Applied to the data in Table 9.5, we find $X^2_{\text{calc}} = 33.645$ and $\sum\sum Y_{ij}/E_{ij} = 15.062$. Thus $D_{\text{calc}} = 18.583$. Also, $\mu_D = -6.860$ and $\sigma^2_D = 13.711$. From these values, $Z_D = 6.871$. Referred to a standard normal distribution, the approximate P-value here is very small: $P < 0.001$. Strongly significant departure from homogeneity is evidenced, and we conclude that the spectra from the three different tumor types are different. (Notice that the unadjusted value of $X^2_{\text{calc}} = 33.645$ for these data is also very significant: $P \approx \text{P}[\chi^2(4) > 33.645] < 0.001$, so that for this table of counts, the sparseness does not appear to degrade the χ^2-based statistic as much as might be expected. See Exercise 9.17.)

Unfortunately, as in Example 9.2, the calculations required for Z_D are not readily available in many statistical computing packages. In particular, neither S-PLUS nor SAS calculates Z_D directly. We can, however, calculate the statistic via user-defined S-PLUS functions. First, Fig. 9.5 gives the definition of a data frame, `mutant.p53.df`, that contains the frequency data. The `levels` of the classification variables are defined to override default behavior of alphabetical sorting of categorical variable levels. These construct the contingency table, `mutant.tab`. Notice the printed warning from S-PLUS concerning the interpretation of the Pearson statistic,

`Some expected values are less than 5, don't trust stated p-value,`

due to the extreme sparseness in this table.

We present the S-PLUS function for Z_D in Fig. 9.6. Applied to the *p53* mutant spectra data in Table 9.5, the S-PLUS results appear in Fig. 9.7. The expected counts and Pearson statistic are constructed using the same technique as described in Fig. 9.4 (predictions from a log-link GLiM under a Poisson parent distribution; see Section 9.4). The figure displays values of $D_{\text{calc}} = 18.580$ and $Z_D = 6.870$ that correspond (within round-off error) to those calculated above. ❑

```
mutant.p53.df<-data.frame(
 mutation=as.factor(rep(c("GC->TA","AT->GC","GC->AT",
                          "GC->CG","AT->TA"),rep(3,5))),
 cancer=as.factor(rep(c("SCLC","NSCLC","Colon"),5),
 count=c(6,16,0,4,0,3,1,5,14,1,4,1,0,2,1)))

levels(mutant.p53.df$mutation)<-
         c("GC->TA","AT->GC","GC->AT","GC->CG","AT->TA")
levels(mutant.p53.df$cancer)  <-c("SCLC","NSCLC","Colon")
mutant.tab<-crosstabs(count~mutation+cancer,data=mutant.p53.df)
crosstabs(count ~ mutation + cancer, data = mutant.p53.df)
```

```
      mutation|cancer
              |SCLC   |NSCLC  |Colon  |RowTotl|
      -------+-------+-------+-------+-------+
      GC->TA | 3     | 0     | 4     |7      |
      -------+-------+-------+-------+-------+
      AT->GC | 1     | 2     | 0     |3      |
      -------+-------+-------+-------+-------+
      GC->AT |14     | 5     | 1     |20     |
      -------+-------+-------+-------+-------+
      GC->CG | 1     | 4     | 1     |6      |
      -------+-------+-------+-------+-------+
      AT->TA | 0     |16     | 6     |22     |
      -------+-------+-------+-------+-------+
      ColTotl|19     |27     |12     |58     |
      -------+-------+-------+-------+-------+
   Test for independence of all factors
      Chi^2 = 33.64526 d.f.= 8 (p=4.710024e-05)
      Yates' correction not used
 Some expected values are less than 5, don't trust stated p-value
```

Fig. 9.5 Output (edited) from S-PLUS `crosstabs` program for testing the homogeneity of *p53* mutant spectra in Table 9.5.

```
zd<-function(X2,ctable,observed,expected) {
# X2 = value of Pearson's X2
# ctable = cross-classified table (from "crosstabs")
# observed = observed counts
# expected = expected counts (under independence)
  NR <- nrow(ctable)
  NC <- ncol(ctable)
  row.sums <- apply(ctable,1,sum)
  col.sums <- apply(ctable,2,sum)
  total.sum<- sum(row.sums)
 Dstat <- (X2 - sum(observed/expected))
 muD<-(total.sum*(NR-1)*(NC-1)/(total.sum-1)) - NR*NC
 part1 <-  2*total.sum/(total.sum - 3)
 part2 <- (NR-1)*(total.sum-NR)/(total.sum-1) -
    (total.sum*sum(1/row.sums) - NR^2)/(total.sum-2)
 part3 <- (NC-1)*(total.sum-NC)/(total.sum-1) -
    (total.sum*sum(1/col.sums) - NC^2)/(total.sum-2)
 part4 <- 4/(total.sum-2)*
    (total.sum*sum(1/row.sums) - NR^2)/(total.sum-2)*
    (total.sum*sum(1/col.sums) - NC^2)/(total.sum-2)
  sigma2D <- part1*part2*part3 + part4
  ZD <- (Dstat - muD)/sqrt(sigma2D)
 data.frame(D=Dstat,mu=muD,sigma=sqrt(sigma2D),Z=ZD,
                                      p=2*(1-pnorm(ZD)))
}
```

Fig. 9.6 S-PLUS Z_D-statistic function.

```
mutant.fit<-glm(count~mutation+cancer,family=poisson(link=log),
                data=mutant.p53.df)
mutant.pearson.x2<- sum(resid(mutant.fit,type="pearson")^2)
mutant.expected<-predict(mutant.fit,type="response")

zd(mutant.pearson.x2,mutant.tab,mutant.p53.df$count,
                                              mutant.expected)
```

```
         D        mu    sigma        Z            p
 1 18.57962 -6.859649 3.702812 6.870256 6.408651e-12
```

Fig. 9.7 Output (edited) from S-PLUS `zd` function for testing the homogeneity of *p53* mutant spectra in Table 9.5.

9.4 LOG-LINEAR MODELS AND RELATIONSHIPS TO GENERALIZED LINEAR MODELS

Since we have imposed a Poisson assumption on the parent distributions for the Y_{ij} in an $R \times C$ table, it is natural to consider Poisson-based GLiMs for the data. From Table 8.1, the canonical link for Poisson-distributed data is the logarithm, $g(\mu) = \log\{\mu\}$. Under this link function, the independence hypothesis on the cell means μ_{ij} corresponding to (9.1) becomes $\log\{\mu_{ij}\} = -\log\{\mu_{++}\} + \log\{\mu_{i+}\} + \log\{\mu_{+j}\}$ for all i and j. This has the form $\log\{\mu_{ij}\} = \upsilon + \alpha_i + \beta_j$, suggesting the more general, ANOVA-type GLiM

$$\log\{\mu_{ij}\} = \upsilon + \alpha_i + \beta_j + \delta_{ij} \tag{9.9}$$

($i = 1, \ldots, R; j = 1, \ldots, C$). The model relates the natural logarithm of the mean to a linear predictor, so it is a **log-linear model** as in Sections 7.2.2 and 8.4.4. Independence in (9.1) is recovered when

$$\mathrm{H}_0\colon \delta_{ij} = 0 \text{ for all } i \text{ and } j, \tag{9.10}$$

that is, when *no significant interaction* is exhibited in the log-linear model.

Since the number of parameters in (9.9) is greater than the number of observations available to fit the model, some sort of identifiability constraint on the linear parameters is required. (Otherwise, some estimated parameters will give redundant values; we encountered similar effects in Examples 8.1 and 8.8.) Examples of such constraints include zeroing out the initial parameters, $\alpha_1 = \beta_1 = \delta_{i1} = \delta_{1j} = 0$ for all i and j (known as a **reference cell coding**), and the more common constraint that the parameters sum to zero over each level of the two factors, $\sum_{i=1}^{R}\alpha_i = \sum_{j=1}^{C}\beta_j = \sum_{i=1}^{R}\delta_{i1} = \sum_{j=1}^{C}\delta_{1j} = 0$. The choice between these, or any other form of identifiability constraint, is usually arbitrary (Agresti, 1990, Section 6.1).

Notice that the identifiability constraints for (9.9) reduce a set of $1 + R + C + RC$ parameters down to a set of $1 + (R-1) + (C-1) + (R-1)(C-1) = RC$ parameters. Since there are RC independent observations in the $R \times C$ table, (9.9) will still **saturate** the table (cf. Example 8.8): in effect, each independent datum fits one parameter. To test (9.10), we fit the unrestricted, saturated model (9.9) and also a model restricted via (9.10), then calculate the twice the difference in their maximized log-likelihoods. The difference in df between both model fits is the difference in the number of fitted parameters: $df = (R-1)(C-1)$. In large samples this log-likelihood difference is referred to a χ^2 distribution with $(R-1)(C-1)$ df. This is a form of likelihood ratio (LR) statistic, and is in fact exactly the LR statistic G^2_{calc} from Section 9.3.2. Thus the GLiM formulation of no independence/no interaction can recover statistics introduced above for simple tabular classifications.

Recall, however, that we do not recommend the LR statistic for testing independence, either in its contingency table form or in this unified GLiM form. For either case, the statistic exhibits poor small-sample stability, and can incur false-positive errors far more often than the specified nominal significance rate. (Of course, the LR/deviance statistic is useful for many other statistical problems, as we have seen throughout this text.)

An alternative for $R \times C$ contingency tables is to calculate the **predicted values** for the cell means, $\hat{\mu}_{ij}$, based on the fitted GLiM. (Predicted values are referred to also as **fitted values**.) With these, calculation of the more stable Pearson X^2-statistic is possible. Take, for example, the simple no-interaction model under (9.10): $\log\{\mu_{ij}\} = \upsilon + \alpha_i + \beta_j$. Fitted via a GLiM, the predicted values for this model are $\hat{\mu}_{ij} = Y_{i+}Y_{+j}/Y_{++}$. These are identical to the E_{ij}s in (9.3); that is to say, under a no-interaction model this simple log-linear GLiM recovers the expected values for testing (9.1). Thus the Pearson X^2-statistic is $X^2_{\text{calc}} = \sum_{i=1}^{R}\sum_{j=1}^{C}(Y_{ij} - \hat{\mu}_{ij})^2/\hat{\mu}_{ij}$, where the cell mean estimate $\hat{\mu}_{ij}$ is based on the reduced model $\log\{\mu_{ij}\} = \upsilon + \alpha_i + \beta_j$. We illustrated similar calculations for $C^2_{1/2}$ and Z_D using S-PLUS in Figs 9.4 and 9.6, respectively.

Example 9.2 (continued) Aquatic toxicity to azinphosmethyl

Returning to the azinphosmethyl toxicity data in Table 9.2, we recalculate X^2_{calc} for testing dose×sex independence using a log-linear model. Figure 9.8 illustrates output from SAS PROC GENMOD for a Poisson log-linear GLiM using (9.9). The output (edited for presentation) includes X^2_{calc} at `Pearson Chi-Square` under `Criteria For Assessing Goodness Of Fit`. We see $X^2_{\text{calc}} = 22.609$, corresponding to our direct calculations above.

In Fig. 9.8 we include output from the `/obstats` option. This is a set of per-observation statistics (see `Observation Statistics`), which includes the predicted values $\hat{\mu}_{ij}$ (see the column under `Pred`, and note the similarities with the values in Table 9.3), the raw residuals $Y_{ij} - \hat{\mu}_{ij}$ (see the column under `Resraw`), and the Pearson residuals under the Poisson parent model, $(Y_{ij} - \hat{\mu}_{ij})/\sqrt{\hat{\mu}_{ij}}$ (see the column under `Reschi`). Summing the squares of these latter values once again recovers $X^2_{\text{calc}} = 22.609$. This is the same quantity as given under `Chi-Square` in Fig. 9.2. (The `/obstats` option produces more per-observation statistics than are presented in Fig. 9.8; users are encouraged to explore the full features of the `/obstats` option at their discretion.)

```
                     The SAS System
                  The GENMOD Procedure
                    Model Information
      Distribution                        POISSON
      Link Function                       LOG
      Dependent Variable                  Y
      Observations Used                   6
               Class Level Information
               Class     Levels  Values
               ROW            3  1 2 3
               COL            2  1 2

        Criteria For Assessing Goodness Of Fit
Criterion              DF          Value      Value/DF
Deviance                2        22.0218       11.0109
Pearson Chi-Square      2        22.6091       11.3045
Log Likelihood          .      4377.7523             .

           Analysis Of Parameter Estimates
Parameter  DF   Estimate   Std Err   ChiSquare  Pr>Chi
INTERCEPT   1     4.4314    0.0764   3360.6245  0.0001
ROW     1   1     0.4338    0.0739     34.4769  0.0001
ROW     2   1    -0.1652    0.0850      3.7835  0.0518
ROW     3   0     0.0000    0.0000           .       .
COL     1   1     0.9528    0.0697    186.7243  0.0001
COL     2   0     0.0000    0.0000                   .
NOTE:   The scale parameter was held fixed.

                Observation Statistics
  Y        Pred    ...     Resraw      Reschi      Resdev
 363   336.3027    ...    26.6973      1.4558      1.4371
 157   184.7500    ...   -27.7500     -2.0416     -2.0962
 219   217.9473    ...     1.0527      0.0713      0.0713
 103   129.6973    ...   -26.6973     -2.3442     -2.4325
  99    71.2500    ...    27.7500      3.2875      3.1028
  83    84.0527    ...    -1.0527     -0.1148     -0.1151
```

Fig. 9.8 Selected output (edited) from SAS PROC GENMOD log-linear fit for data in Table 9.2.

We conclude by noting that there are a number of possible residuals available from a GLiM fit; some of these are given in the SAS output in Fig. 9.8. For example, a **deviance residual** corresponds to the signed square root of each observation's contribution to the deviance statistic (McCullagh and Nelder, 1989, Section 2.4.3). (Recall from Section 7.2.1 that the difference in deviances is also the LR statistic, G^2_{calc}.) Summing the squared deviance residuals produces the deviance statistic. The per-observation deviance residuals are given as part of the output in Fig. 9.8, under

`Resdev`. The row labeled `Deviance` under `Criteria For Assessing Goodness Of Fit` gives the full deviance statistic $G^2_{\text{calc}} = 22.022$; this compares with the similar quantity given as `Likelihood Ratio Chi-Square` in Fig. 9.2. ◻

More complex log-linear GLiMs may not produce predicted values with simple closed forms such as (9.3), but they may still be fitted via computer. Software that performs these calculations includes SAS PROC GENMOD (via the `/obstats` option) or the GLIM system. For any estimable log-linear model, the generalized Pearson statistic is

$$X^2_{\text{calc}} = \sum_{i=1}^{R} \sum_{j=1}^{C} \frac{(Y_{ij} - \hat{\mu}_{ij})^2}{\hat{\mu}_{ij}},$$

where $\hat{\mu}_{ij}$ is the predicted value for the mean Poisson response in the (i,j)th cell. To test the goodness-of-fit of the fitted model, refer X^2_{calc} in large samples to a χ^2 distribution. The *df* here are calculated as the difference between the number of nonredundant parameters fitted and the number of nonredundant parameters from a fully saturated model.

9.5 TABLES OF PROPORTIONS

The constructions described in Section 9.1 for testing independence in two-way categorical data tables can be extended to three-way or even higher-dimensional tables when more than two categorical factors are studied. Extra dimensions induce many more forms of independence and interaction, however, leading to greater complexities in the test statistics. Even the simple case of interaction in a 2×2×2 table can lead to interpretational tangles, as was first discussed by Bartlett (1935). A complete discussion on higher-order contingency tables is beyond the scope of this chapter, however, and we refer the reader to more advanced texts, such as Agresti (1990, Section 5.2), for more details.

One special case of a three-way table we do consider is the $R \times C$ **table of proportions**, which is essentially a re-expression of a three-way $R \times C \times 2$ table with the third dimension (or 'margin') fixed by sampling. For an $R \times C$ table of proportions, the third dimension is dichotomous – say success vs. failure, or response vs. nonresponse – and the total number of responses and nonresponses is fixed. To illustrate, consider the simplest case of a 2×2 table of proportions: the row and column category variables each represent two levels of two different factors, and the datum at each combination of levels is a count of the number of responses, divided by the total number of responses and nonresponses (Piegorsch *et al.*, 1988). Table 9.7 illustrates the design schematic. (See also Example 9.4 below.)

Table 9.7 General form of a 2×2 table of proportions

	Factor B	
Factor A	*Level 0*	*Level 1*
Level 0	$\frac{Y_{00}}{N_{00}}$	$\frac{Y_{01}}{N_{01}}$
Level 1	$\frac{Y_{10}}{N_{10}}$	$\frac{Y_{11}}{N_{11}}$

In the general $R \times C$ case, each cell represents a proportion, Y_{ij}/N_{ij}. Under a parent binomial model, $Y_{ij} \sim$ (indep.) $\mathcal{B}in(N_{ij}, \pi_{ij})$, $i = 0,1$; $j = 0,1$, the observed proportions Y_{ij}/N_{ij} estimate the response probabilities π_{ij}.

9.5.1 Joint action

For a two-factor table of proportions, it is often of interest to determine if the two factors interact in some systematic fashion, in effect increasing (or decreasing) the response probability beyond that predicted by the single-factor responses. For example, in a 2×2 table of proportions, suppose $i = 0$ and $j = 0$ represent background levels of the row and column factors, respectively. If $i = 1$ and $j = 1$ represent some form of exposure or other positive condition above the background level, then the response probabilities π_{10} and π_{01} are the single-factor effects, relative to the background probability π_{00}. Of interest is how the **joint action** of the two factors, quantified by the joint response probability π_{11}, relates to the single-factor response probabilities π_{01} and π_{10}.

Different joint-action models for π_{11} are possible, depending on the interactive effects of the two factors. A general theory of joint action was developed by Hewlett and Plackett in a pair of seminal articles (Plackett and Hewlett, 1952; Hewlett and Plackett, 1959), in which they distinguished among forms of joint activity based upon the site-specific action of the factors. To illustrate, suppose each factor's levels represent a dose of some toxic agent, and the response is some detrimental effect in the experimental unit under study. If the two agents act upon the same site and produce a response via similar mechanisms, the joint action is *similar*. For example, two environmental carcinogens may affect the same target organ, inducing the same form of cancer in that organ. On the other hand, if the two agents act at different sites (or at the same site but with distinguishably different mechanisms), the joint action is *dissimilar*, or also called *independent*. For example, radiation exposure may damage DNA, while (joint) exposure to an environmental mutagen may interfere with DNA repair. (Note that to indicate dissimilarity in joint action, the term *independent* is being used differently here than in Section 9.1.)

Once the form of joint action – similar or independent – has been identified, a null hypothesis can be formulated in terms of the two agents' interaction. Hewlett and Plackett defined no interaction as a form of *simple* action, while any interaction was called *complex* action. From this characterization, we can construct statistical models to express the different forms of joint action, and to assess the action exhibited in a two-factor table of proportions.

9.5.2 Simple independent action

A useful null model for joint action corresponding to one of the Hewlett–Plackett forms is the combination of simple action and independent action, termed **simple independent action**, where the sites or mechanisms of action are thought to be independent, and the ('simple') null effect is no interaction. Following Finney (1971, Section 11.5), we can describe this condition in terms of the nonresponse probabilities $\theta_{ij} = 1 - \pi_{ij}$. Consider a simple background vs. exposure comparison in a 2×2 table of proportions (Table 9.7), where experimental units in the (1,1)th cell represent the joint exposure. Under simple independent action, the probability of joint response, π_{11}, can be written in terms of the single-exposure probabilities. Let the proportion of experimental units that are expected to respond to the row factor be π_{10}. Under the independence assumption that the joint actions are at different sites or mechanisms, the remaining proportion $(1 - \pi_{10})$ of jointly exposed units has a further proportion expected to respond to the column factor of π_{01}. Thus the probability of responding to the joint exposure is the sum of these quantities: $\pi_{11} = \pi_{10} + (1 - \pi_{10})\pi_{01} = \pi_{10} + \pi_{01} - \pi_{10}\pi_{01}$. The joint probability of nonresponse is $\theta_{11} = 1 - \pi_{11} = 1 - \pi_{10} - \pi_{01} + \pi_{10}\pi_{01} = (1 - \pi_{10})(1 - \pi_{01}) = \theta_{10}\theta_{01}$. Adjusting each of these quantities relative to the nonzero background probability, θ_{00}, gives a relationship among **relative risks** of nonresponse: $\theta_{11}/\theta_{00} = (\theta_{10}/\theta_{00})(\theta_{01}/\theta_{00})$, that is,

$$\frac{\theta_{10}\,\theta_{01}}{\theta_{11}\,\theta_{00}} = 1,$$

or, equivalently,

$$\theta_{11}\,\theta_{00} \;=\; \theta_{10}\,\theta_{01}\,. \tag{9.11}$$

This is a multiplicative model in the nonresponse probabilities, and it is the null model corresponding to simple independent action.

Applying a logarithmic transform and rearranging (9.11) produces an additive relationship among the log-nonresponse probabilities:

$$\mathrm{H_0{:}}\log\{\theta_{10}\} + \log\{\theta_{01}\} - \log\{\theta_{00}\} - \log\{\theta_{11}\} = 0. \tag{9.12}$$

Notice that the relationship in (9.12) is essentially a form of **interaction contrast** in the log-nonresponse. This appears similar to the generalized ANOVA forms seen in the logistic regression and log-linear models from Chapters 7 and 8, and it is possible to extend the simple independent action model into a special form of GLiM under a binomial likelihood. We explore this further in Section 9.5.3.

When the contrast in (9.12) is zero, simple independent action holds. When it is positive, the response probability π_{11} will be larger than expected, relative to the single-exposure response. This form of interactive departure is often referred to as **synergy** or **synergism** between the two factors; see Blot and Day (1979) or Darroch and Borkent (1994). (In some settings, one factor may have no effect on the experimental units, but when combined with another factor that does, the contrast in (9.12) may still be positive. In this case, some authors choose to characterize the departure as **potentiation** of the second factor by the first; see, for example, Piegorsch and Margolin (1989). Technically, of course, this is only a special form of synergism.) When the contrast in (9.12) is negative, the interactive departure is referred to as **antagonism**.

In terms of the probabilities of response, synergy occurs if $\pi_{11} > \pi_{10} + \pi_{01} - \pi_{10}\pi_{01}$, that is, the probability of joint-exposure response exceeds that predicted by a simple additive model under the independent action assumption. Synergy can be expressed alternatively in terms of the nonresponse probabilities: $\theta_{11} < \theta_{10}\,\theta_{01}$. Then, antagonism is expressed as $\theta_{11} > \theta_{10}\,\theta_{01}$.

Under a binomial assumption on the observed response counts, Y_{ij}, unrestricted maximum likelihood estimators for the θ_{ij}s are the nonresponse proportions $q_{ij} = (N_{ij} - Y_{ij})/N_{ij}$. These are used in the contrast estimator of (9.12):

$$\log\{q_{10}\} + \log\{q_{01}\} - \log\{q_{00}\} - \log\{q_{11}\}\,,$$

which can be written as a **log{*W*} statistic** (Wahrendorf *et al.*, 1981; Weinberg, 1986):

$$\log\{W\} = \log\left\{\frac{q_{10}q_{01}}{q_{00}q_{11}}\right\}. \tag{9.13}$$

As with the contrast in (9.12), when $\log\{W\} > 0$ the value of the estimated proportion response in the joint-action cell is greater than predicted under simple independent action in (9.12), and a positive departure (synergy) is suggested. When $\log\{W\} < 0$, the joint action is lower than predicted under (9.12), and negative departure (antagonism) is suggested.

When the cell sizes, N_{ij}, are large, the standard error for the estimated contrast in (9.13) is

$$se[\log\{W\}] = \left\{\frac{1-q_{10}}{N_{10}q_{10}} + \frac{1-q_{01}}{N_{01}q_{01}} + \frac{1-q_{00}}{N_{00}q_{00}} + \frac{1-q_{11}}{N_{11}q_{11}}\right\}^{1/2}, \tag{9.14}$$

from which a Wald statistic can be constructed to test the null hypothesis of simple independent action in (9.12). Divide $\log\{W\}$ by the standard error in (9.14), and refer the resulting statistic to a standard normal distribution. Thus if $Z_W = \log\{W\}/se[\log\{W\}]$, an approximate two-sided P-value for testing H_0 is $2\{1 - \Phi(|Z_W|)\}$. One-sided departures may be assessed by calculating the corresponding one-sided P-values.

Alternative forms for the $\log\{W\}$-statistic also have been proposed, based on rearrangements of the log-nonresponse contrast in (9.12). These are discussed, for example, by Hogan *et al.* (1978) and Piegorsch *et al.* (1986). The alterations provide no substantial improvement in stability or power over the $\log\{W\}$-statistic, however, and for testing departures from simple independent action in 2×2 tables of proportions we recommend use of $\log\{W\}$ via (9.13) and (9.14).

Example 9.4 Joint action in the E. coli *fluctuation assay*

The **fluctuation assay** involves laboratory study of mutagenesis in the bacterium *Escherichia coli*, where the bacteria present a growth pattern after mutations cause a change in their dependence on the amino acid histidine. (The bacteria are engineered to be histidine-dependent.) When they are cultured in histidine-free media, mutant reversions to histidine independence can accumulate, causing a cloudy reaction in wells containing the culture. (Notice the similarity to the *Salmonella* assay from Example 3.7.) This formulation allows for sensitive assays of *E. coli* mutagenicity in response to toxic environmental exposures (Collings *et al.*, 1981).

The data in this form of fluctuation assay are proportions, Y_{ij}/N_{ij}, of test wells indicating a mutagenic (growth) response in the bacteria. The probability of response in any well is taken as π_{ij}, where i and j index differing experimental conditions, such as chemical exposure levels. Denote the associated nonresponse probability as $\theta_{ij} = 1 - \pi_{ij}$. The total number of wells, N_{ij}, is fixed by the design. The data are modeled via a binomial parent distribution, and the methods above for testing joint action are applicable.

To illustrate, Table 9.8 presents *E. coli* data from a mutagenesis study of sodium azide, where the bacteria were exposed to none or 5.2 μm of the metal (Piegorsch and Margolin, 1989). The mutagenic effect of sodium azide was thought to be increased after joint exposure to the heavy metal chromium-VI, and so a joint-action study was incorporated into the experimental design. The data in Table 9.8 represent the corresponding 2×2 table of proportions. A plot of these proportions is given in Fig. 9.9.

Table 9.8 Proportions of *E. coli* test wells indicating mutation in fluctuation test after joint exposure to sodium azide and chromium-VI

	Chromium-VI	
Sodium azide	*Control*	*5.2 μm*
Control	$\frac{10}{50}$	$\frac{10}{50}$
0.04 μm	$\frac{19}{50}$	$\frac{37}{50}$

Data from Piegorsch and Margolin (1989).

It is expected that the two compounds in this study act in a substantially different manner when inducing genotoxicity in *E. coli*, so that a null model of simple independent action (9.12) is appropriate. Applying the $\log\{W\}$-statistic from (9.13) to the proportions in Table 9.8 yields

$$\log\{W\} = \log\left\{\frac{31}{50}\right\} + \log\left\{\frac{40}{50}\right\} - \log\left\{\frac{40}{50}\right\} - \log\left\{\frac{13}{50}\right\}$$

or $\log\{W\} = 0.869$. The associated standard error from (9.14) is $se[\log\{W\}] = \sqrt{0.079} = 0.281$. The test statistic is the ratio of these two quantities: $Z_W = 3.09$ ($P \approx 0.002$). The low P-value implies clear departure from simple independent action, indicating significant mutagenic potentiation after joint exposure to sodium azide and chromium-VI. □

9.5.3* Simple independent action when $R > 2$ or $C > 2$: Relationships to GLiMs

When the $R \times C$ table of proportions has more than two rows or more than two columns, we extend the simple independent action model from (9.12) by requiring it to hold in all cells of the table representing a joint exposure:

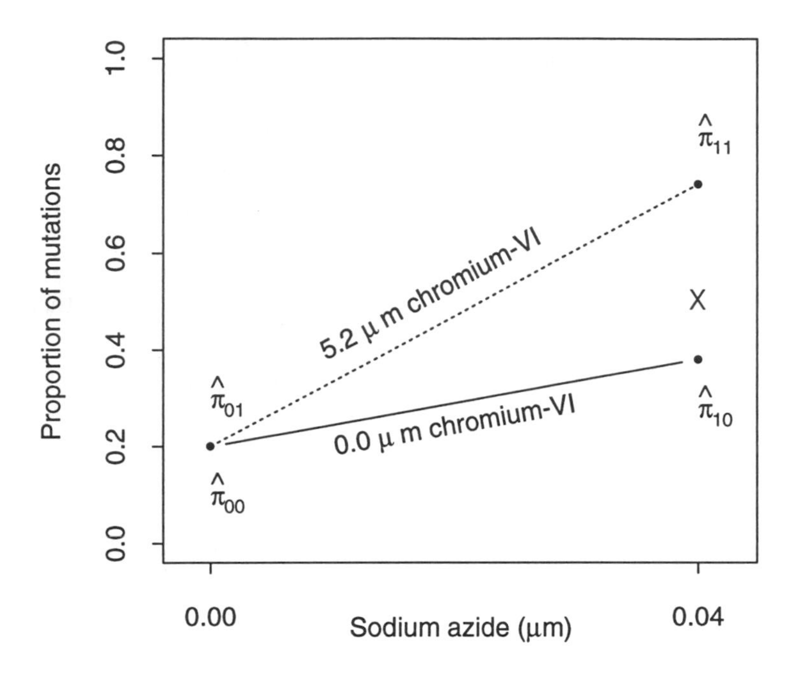

Fig. 9.9 Plot of sample proportions (•) from Table 9.8. The estimated joint action point under SIA is marked by an X.

$$H_0: \log\{\theta_{i0}\} + \log\{\theta_{0j}\} - \log\{\theta_{00}\} - \log\{\theta_{ij}\} = 0 \quad \text{(for all } i > 0 \text{ and } j > 0). \tag{9.15}$$

To test (9.15), we can extend the log{ W }-statistic by incorporating a GLiM. Since the Y_{ij}s are assumed binomial, however, we do not consider a log-linear model as in Section 9.4, but instead construct a complementary log GLiM from Table 7.3. As given by Wahrendorf *et al.* (1981), the simple independent action model may be written as a special case of the complementary log-linear form

$$-\log\{1 - \pi_{ij}\} = -\log\{\theta_{ij}\} = \upsilon + \alpha_i + \beta_j + \omega_{ij}, \tag{9.16}$$

where we impose the identifiability constraint $\alpha_0 = \beta_0 = \omega_{i0} = \omega_{0j} = 0$ for all $i > 0$ and $j > 0$. (That is, constrain all main-effect and interaction parameters associated with the single-response cells to equal zero, a form of reference cell coding, as in Section 9.4 Notice that the resulting model in (9.16) is fully saturated.) This is recognized as a special form of GLiM for binomial data, with a standard ANOVA-type linear predictor, but with a complementary log link function, $g(\pi) = -\log\{1 - \pi\}$.

Under the complementary log GLiM in (9.16), simple independent action holds when $\omega_{ij} = 0$ for all $i > 0$ and $j > 0$. Indeed, in the simple 2×2 case $\omega_{11} = \log\{\theta_{10}\} + \log\{\theta_{01}\} - \log\{\theta_{00}\} - \log\{\theta_{11}\}$ is again the simple independent action contrast in (9.12). Thus, to test the global null hypothesis of simple independent action at all joint exposures, we employ a likelihood ratio/deviance test of whether the interaction parameters, ω_{ij}, all equal zero. As with any LR/deviance statistic, one subtracts twice the constrained, maximized log-likelihood with $\omega_{ij} = 0$ for all i and j, from twice the unconstrained, maximum log-likelihood under (9.16). In large samples, this statistic is referenced to a χ^2 distribution with $(R-1)(C-1)$ *df*.

Unfortunately, the likelihood computations for the LR test, or for finding maximum likelihood estimates of the ω-parameters (or other parameters) require computer calculation. These may be performed in many software packages, including the GLIM system, S-PLUS, and SAS. Since the complementary log model in (9.16) does not represent a standard link function for binomial-based GLiMs, however, the computer user must employ self-constructed link functions to fit the model. For GLIM computing, a complementary log function is given by Wacholder (1986); we illustrate similar SAS code in the following example.

Example 9.4 (continued) Joint action in the E. coli *fluctuation assay*
We illustrate the calculation of the complementary log GLiM in SAS PROC GENMOD with the *E. coli* mutagenicity data from Table 9.8. To implement this GLiM in SAS, we construct two user-defined functions, the link function in `fwdlink` and the inverse link function in `invlink`. In Fig. 9.10, these two user-defined functions `fwdlink` and `invlink` are constructed as part of the PROC GENMOD operations, based on the following details.

Since the complementary log link function is

$$\eta_{ij} = g(\pi_{ij}) = -\log(1 - \pi_{ij})$$

where π_{ij} is the mean proportion response ($i = 0,1$; $j = 0,1$), we take

```
fwdlink link = -log(1 - _mean_);
```

in Fig. 9.10. For binomial data in PROC GENMOD, the program-level variable `_mean_` represents π_{ij}. Also, if η_{ij} is the linear predictor

$$\eta_{ij} = \upsilon + \alpha_i + \beta_j + \omega_{ij} ,$$

then equating η_{ij} with $g(\pi_{ij})$ and inverting yields the inverse link function

$$\pi_{ij} = h(\eta_{ij}) = 1 - \exp(-\eta_{ij}).$$

Note that since this quantity defines a probability under the GLiM, it must be contained between 0 and 1. Thus we truncate $\hbar(\eta_{ij})$ at 0 if $\eta_{ij} < 0$. To perform the calculations in practice, we replace 0 with 10^{-6} in the SAS code, and take

```
invlink ilink = prob;
```

where the intermediate statements

```
if _xbeta_ < 1.e-6  then  prob = 1.e-6 ;
      else prob = 1 - exp(-_xbeta_) ;
```

are required before defining `invlink` (Fig. 9.10). Note that `_xbeta_` is the SAS program-level variable for η_{ij}.

Figure 9.10 also illustrates the need to code directly a set of regression indicator variables (`a0`, `a1`, ..., `w11`) to imitate an analysis where the ('first-index') reference cell identifiability constraints $\alpha_0 = \beta_0 = \omega_{00} = \omega_{10} = \omega_{01} = 0$ are imposed. (Unless constructed otherwise, SAS imposes *last*-index reference cell coding: $\alpha_1 = \beta_1 = \omega_{01} = \omega_{10} = \omega_{11} = 0$.) To ensure the first-index reference cell constraint, we fit only those regression indicators in the `model` statement not zeroed out under the constraint: that is, fit only `a1`, `b1`, and `w11`. The output from this analysis is displayed in Fig. 9.11.

In Fig. 9.11, the test of $H_0: \omega_{11} = 0$ in (9.16) involves the interaction term `W11`. Both the LR/deviance statistic and the Wald statistic for evaluating H_0 appear as part of the SAS output. These provide an assessment of simple independent action for the *E. coli* data. The LR/deviance test statistic is $G^2_{calc} = 11.2004$; see `ChiSquare` for `W11` under `LR Statistics For Type 1 Analysis`. With an associated *P*-value of approximately 0.0008, we reject H_0 and conclude that departure from simple independent action is indicated for mutagenic response to joint sodium azide and chromium-VI exposure.

Alternatively, Wald statistics for H_0 are available under `Analysis Of Parameter Estimates`. The Wald statistic is $W^2_{calc} = 9.5380$, under `ChiSquare` for `W11`. The associated *P*-value of 0.0020 suggests significant departure from H_0, corroborating the LR result. Notice also that $\sqrt{9.5380} = 3.088$ corresponds to the value of Z_W calculated in the previous section. Indeed, the listed `Parameter Estimate` for `W11` corresponds to the ML estimator for ω_{11}, and we find $\hat{\omega}_{11} = 0.869$ is the value of $\log\{W\}$ calculated above for these data. The positive sign of this estimator implies that the departure from H_0 is a synergistic one: joint exposure to sodium azide and chromium-VI appears to *increase* significantly the probability of mutagenic response above that predicted by simple independent action. □

```
* SAS code to fit log-q GLiM to data in Table 9.8;
data ecoli;
input naaz chrom y n  a0 a1  b0 b1  w00 w01 w10 w11;
cards;
0     0    10  50  1 0  1 0  1 0 0 0
0     5.2  10  50  1 0  0 1  0 1 0 0
0.04  0    19  50  0 1  1 0  0 0 1 0
0.04  5.2  37  50  0 1  0 1  0 0 0 1
proc genmod;
      if _xbeta_ < 1.e-6  then  prob = 1.e-6;
      else prob = 1 - exp(-_xbeta_) ;
fwdlink link = -log(1 - _mean_) ;
invlink  ilink = prob ;
model y/n = a1 b1 w11/ d=binomial type1;
```

Fig. 9.10 SAS PROC GENMOD complementary log program for evaluating simple independent action between sodium azide and chromium-VI in Table 9.8.

```
                     The SAS System
                  The GENMOD Procedure

                   Model Information
      Distribution                          BINOMIAL
      Link Function                         USER
      Dependent Variable                    Y
      Dependent Variable                    N
      Observations Used                     4

           Analysis Of Parameter Estimates
Parameter  DF   Estimate  Std Err   ChiSquare  Pr>Chi
INTERCEPT   1    0.2231    0.0707      9.9586  0.0016
A1          1    0.2549    0.1314      3.7646  0.0523
B1          1   -0.0000    0.1000      0.0000  1.0000
W11         1    0.8690    0.2814      9.5380  0.0020
SCALE       0    1.0000    0.0000           .       .
NOTE:  The scale parameter was held fixed.

         LR Statistics For Type 1 Analysis
  Source         Deviance    DF   ChiSquare  Pr>Chi
  INTERCEPT       41.8331     0           .       .
  A1              13.4739     1     28.3592  0.0000
  B1              11.2004     1      2.2735  0.1316
  W11              0.0000     1     11.2004  0.0008
```

Fig. 9.11 Output (edited) from SAS PROC GENMOD complementary log fit of *E. coli* data in Table 9.8.

In (9.16) if either or both of the factors under study are quantitative, a response surface model for the joint action may be fitted. For example, suppose the two factors are recorded at dose levels r_i and c_j. Assuming $R > 3$ and $C > 3$, one can model ω_{ij} as a second-order model that includes quadratic and cross-product terms: $\omega_{ij} = \delta_0 + \delta_1 r_i + \delta_{11} r_i^2 + \delta_2 c_j + \delta_{22} c_j^2 + \delta_{12} r_i c_j$. This allows the joint action to vary in a second-order fashion. Specific forms of the response surface will indicate specific forms of departure from simple independent action; e.g. Piegorsch *et al.* (1988) illustrated an elliptic paraboloid for ω_{ij} with data on cell differentiation after exposure to two different immunoactivating stimuli (Exercise 9.27). This response surface extended upwards parabolically in all directions from a minimum point. Where the surface intersected the r–c plane, simple independent action was indicated. Above the plane, the response was positive, and synergistic departure from simple independent action was indicated, while below it the departure was antagonistic. Thus the simple second-order model allowed for a variety of local and global departures, depending on the nature of the estimated δ-parameters.

The GLiM formulation also allows for multi-factor simple independent action models, for use when the joint effects of three or more different stimuli are under study. In the three-factor case, the fully saturated model includes all second- and third-order interactions:

$$-\log\{\theta_{ijk}\} = \upsilon + \alpha_i + \beta_j + \gamma_k + \delta_{ij} + \varphi_{ik} + \lambda_{jk} + \omega_{ijk}$$

for $i = 0, \ldots, R-1$; $j = 0, \ldots, C-1$; and $k = 0, \ldots, L-1$. Under a reference cell identifiability constraint,

$$\alpha_0 = \beta_0 = \gamma_0 = \delta_{i0} = \delta_{0j} = \varphi_{i0} = \varphi_{0k} = \lambda_{j0} = \lambda_{0k} = \omega_{i0k} = \omega_{ij0} = \omega_{0jk} = 0,$$

this model can test for simple independent action among the three factors (Exercise 9.26). For example, full, three-way simple independent action holds when all second- and third-order interactions are zero. The model is necessarily more complex than the two-factor model presented in (9.16), however, and we will not discuss it further. We refer interested readers to the discussion on multi-factor models in Piegorsch *et al.* (1988, Section 5).

9.5.4* Simple similar action

Another form of simple joint action under the Hewlett and Plackett (1959) formulation is the combination of simple action and similar action, referred to as **simple similar action**. Here, the sites or mechanisms of action between two environmental stimuli are thought to be similar in function, and the parametric relationship describing the null effect of no interaction differs substantively from that for simple independent action in Section 9.5.2. Since the joint action of the two stimuli under study is similar, the model essentially equates the effectiveness or **potency** of one stimulus with some multiple of the potency of the second. To see this, suppose each stimulus is administered or recorded at quantified doses: r_i for the first and c_j for the second ($i = 0, \ldots, R-1$; $j = 0, \ldots, C-1$). (Typically, r_0 and c_0 are taken as zero-dose controls.) The assumption of similar action suggests that any measure of dose from the second stimulus can be converted into an equivalent measure of dose from the first stimulus via some scaling constant, $r = \rho c$. The scaling constant, ρ, is the **relative potency** of the first stimulus to the second. Under simple similar action, we model the probability of joint response at any dose combination using this ρ-parameter for relative potency.

Denote the probability of joint response as $\pi(r,c)$, for any dose combination (r,c). Since $r = \rho c$, exposure to the combination (r,c) is equivalent to exposure at the combination $(r + \rho c, 0)$. Thus we can relate the probability $\pi(r,c)$ to an additive combination of the doses, say $\pi(r,c) = F(r + \rho c)$, where $F(\cdot)$ is some nondecreasing function that satisfies $0 \le F(x) \le 1$ for all x. For example, $F(\cdot)$ can be a form of c.d.f., such as the standard normal distribution c.d.f, $F(x) = \Phi(x)$, or the standard logistic distribution c.d.f., $F(x) = 1/(1 + e^{-x})$.

Departure from simple similar action under this parametric formulation is determined by testing for goodness-of-fit to the assumed model when no interaction is fitted between the two doses. Specifically, suppose the logistic model is adopted for the probability function $F(\cdot)$. We observe proportions Y_{ij}/N_{ij} at dose levels r_i and c_j ($i = 0, \ldots, R-1$; $j = 0, \ldots, C-1$), and we assume a simple similar logistic response:

$$\pi(r_i, c_j) = 1/[1 + \exp\{-\upsilon - \beta(r_i + \rho c_j)\}].$$

This can be rewritten as

$$\log\left\{\frac{\pi(r_i, c_j)}{1 - \pi(r_i, c_j)}\right\} = \upsilon + \beta r_i + \beta\rho c_j, \qquad (9.17)$$

which is recognized as a form of GLiM: a logistic regression on r_i and c_j. Indeed, viewed as a GLiM, $F(\cdot)$ serves as an inverse link function. If the logistic specification for $F(\cdot)$ is correct, a significant lack of fit suggests significant departure from simple similar action.

To test lack of fit, one can fit the proportion data to the logit model in (9.17) using, SAS PROC LOGIST, PROC GENMOD, S-PLUS `glm`, etc. A simple test of fit finds the predicted values $\hat{\mu}_{ij} = N_{ij}(1 + \exp\{-\hat{\upsilon} - \hat{\beta}(r_i + \hat{\rho} c_j)\})$ and calculates a Pearson goodness-of-fit statistic

$$X^2_{\text{calc}} = \sum_{i=1}^{R} \sum_{j=1}^{C} \frac{(Y_{ij} - \hat{\mu}_{ij})^2}{\hat{\mu}_{ij}}.$$

Significant lack of fit is suggested when the associated P-value, $P \approx \text{P}[\chi^2(RC-3) > X^2_{\text{calc}}]$, is smaller than some pre-defined α level. Other tests for goodness-of-fit are available with the logistic regression GLiM, and these may be more accurate in small samples; see Tsiatis (1980) or Lemeshow and Hosmer (1982) for more detail.

A critical assumption in this logistic formulation for simple similar action is that the assumed inverse link function $F(\cdot)$ must be correct (at least approximately) for the goodness-of-fit approach to provide a valid test of simple similar action. If not, then large values of the goodness-of-fit statistic may be due to incorrect link specification, and not to departure from simple similar action. Unfortunately, there is no easy way to test for logistic model adequacy while testing simultaneously for departure from simple similar action. One can apply a separate **goodness-of-link** or **model adequacy** test to see if the logit link is valid (Pregibon, 1985; Cheng and Wu, 1994) as in Section 8.3.3, but this requires proper specification of the linear predictor. In effect, one would have to test the link using a full model linear predictor; for example, expanding (9.17) to include an interaction term: $\eta_{ij} = \upsilon + \beta(r_i + \rho c_j) + \gamma r_i c_j$. If the link test is not significant, one could then move on to test the goodness-of-fit of the simple independent action model under the logistic specification. If the link test is significant, one must consider another link instead, searching until an acceptable goodness-of-link result is achieved. Besides the extensive computational acrobatics this effort requires, there is the salient statistical concern that the familywise false-positive rate will grow as more and more link tests are performed, leading to the possibility that a valid link will be discarded due to Type I error inflation. Based on these concerns, we advise caution when testing for simple similar action with quantal response data. Whenever possible, consider a link specification that has been validated previously in the literature or by earlier studies, in order to avoid concerns about link inadequacy.

We close this section by noting that it is possible for an observed joint-action response to deviate from one form of simple action, and yet not deviate from the other. That is, the response may deviate from simple independent action and still display no deviation from simple similar action. As such, and similar to the concerns noted above on proper link specification, the declaration of the null model must be given careful consideration. Application of a particular joint-action model is contingent upon its relevance to the experimental structure and to any biological or ecological assumptions made about the response, the experimental units, and the sampling framework under which the data are garnered.

9.6 SUMMARY

In this chapter, basic methods are described for analyzing data that are cross-classified into tables of counts. The methods allow for testing independence between two categorical classification variables or for testing homogeneity of response for one categorical variable across two or more populations, via the well-established Pearson X^2-test. It is seen that X^2_{calc} may be recovered by fitting the categorical data via a special form of generalized linear model, specifically, a log-linear model in the response rates. Alternatives to the X^2 statistic are presented for special cases, such as small sample sizes or sparse data tables with many zeros and ones. These include a useful class of power-divergence statistics that include X^2 as a special case.

Methods are described also for cross-classified tables of proportions, where interest centers on assessing the degree of interaction between the two classification variables. When the interaction produces an increase in the response proportion above that predicted from the two individual exposure categories, the effect is viewed as a form of synergy between the two exposure variables. Here again, the statistical model is as a special case of generalized linear model, from which various point estimates and test statistics may be derived.

EXERCISES

(Exercises with asterisks, e.g. 9.2*, typically are more advanced, or require greater knowledge and understanding of probability distributions and/or calculus.)

9.1. Return to the simple 2×2 table of counts from Section 4.1.7. Rewrite that table symbolically in the form of an R×C table given in Section 9.1.

9.2*. Suppose a cross-classified set of counts, Y_{ij} ($i = 1, \ldots, R; j = 1, \ldots, C$), is given a parent distribution that is multinomial, conditional on the grand total Y_{++}, with response probabilities $\pi_{ij} = \mu_{ij}/Y_{++}$, for some values $\mu_{ij} > 0$.

a. Show that the sum of the π_{ij}s over $i = 1, \ldots, R$ and $j = 1, \ldots, C$ is 1. (*Hint*: go back to the definition of the multinomial p.m.f. from Section 1.5.1, and recall that the p.m.f. sums to 1. You will need to employ the multinomial theorem from Section 1.5.1.)

b. Using the result in part (a), show $\mu_{++} = Y_{++}$.

9.3*. Suppose a cross-classified set of counts, Y_{ij}, is given a Poisson parent distribution with means μ_{ij}, as in Section 9.2.1. Write out the likelihood function with respect to the μ_{ij}s. Show that this is essentially identical to a likelihood where the counts are assumed multinomial, conditional on the grand total Y_{++}, with response probabilities $\pi_{ij} = \mu_{ij}/Y_{++}$.

9.4*. Suppose a cross-classified set of counts, Y_{ij} ($i = 1, \ldots, R; j = 1, \ldots, C$), has a parent distribution that is a product of C independent multinomials, conditional on the column totals Y_{+j}, with response probabilities $\tau_{ij} = \mu_{ij}/Y_{+j}$ at each j (for some values $\mu_{ij} > 0$). Show that:

a. $\sum_{i=1}^{R} \tau_{ij} = 1$ for each $j = 1, \ldots, C$. (*Hint*: see Exercise 9.2.)

b. $\mu_{+j} = Y_{+j}$ for each $j = 1, \ldots, C$.

c. $\mu_{++} = Y_{++}$.

9.5*. Extend the results of Exercise 9.3 to the case where the Y_{ij}s are given a product multinomial distribution as in Exercise 9.4.

9.6. Show that for $\tau_{ij} = \mu_{ij}/Y_{+j}$, the null hypotheses in Equations (9.1) and (9.2) are equivalent. (*Hint*: use results from Exercise 9.4.)

9.7. Given the expected values E_{ij} from (9.3), show that:

a. $\sum_{i=1}^{R} E_{ij} = Y_{+j}$ for every $j = 1, \ldots, C$.

b. $\sum_{j=1}^{C} E_{ij} = Y_{i+}$ for every $i = 1, \ldots, R$.

c. $\sum_{i=1}^{R} \sum_{j=1}^{C} E_{ij} = Y_{++}$.

9.8. Show that when $R = C = 2$, the Pearson X^2-statistic collapses to the square of the test statistic in (4.8).

9.9. Show that when $\lambda = 1$, the power-divergence statistic C_1^2 is identical to X^2. (*Hint*: use the results from Exercise 9.7.)

9.10. Exercise 5.5 gives discrete data on severity of liver tissue damage after rats were exposed to a toxin at various doses (Yanagawa *et al.*, 1994). Those data were part of a larger experiment, in which the severity of damage was listed on a graded scale from none, through trace levels, up to severe. Taking severity as an additional explanatory factor (along with dose), we can view the data as a 4×4 contingency table (see the display that follows). If dose had no effect on severity, we would expect the pattern of severity response to be homogeneous across doses, and the null model in (9.1) would

Dose	*None*	*Severity* *Trace*	*Moderate*	*Severe*	*Row totals*
0	16	1	0	0	17
0.5	15	4	0	0	19
2	10	8	0	1	19
10	11	3	7	1	22
Col. totals	52	16	7	2	77

apply. To assess this:

a. Which probability model (Poisson, full multinomial, or product multinomial) best describes the data presented above?

b. Calculate the Pearson X^2-statistic for these data, and test the null hypothesis of homogeneity/independence. Set $\alpha = 0.05$.

c. Calculate the power-divergence statistic $C^2_{1/2}$ for these data, and use it to test the null hypothesis of homogeneity/independence. Set $\alpha = 0.05$. Compare the result to your inference from part (b).

d. Perform Fisher's exact test on these data using the extended hypergeometric distribution, as discussed in Section 9.3.3. Set $\alpha = 0.05$. Compare the result to that from part (b).

e. There are many zeros and ones in this contingency table. Based on the expected values E_{ij} required to calculate X^2_{calc} or $C^2_{1/2}$, or using the sparseness measure $Y_{++}/\{(R-1)(C-1)\}$, would you classify this table as sparse? If so, perform a reanalysis using the Z_D-statistic from (9.8).

9.11. (Montelone *et al*., 1992) An experiment was conducted to study the spectrum of base pair mutations in two strains of the yeast *Saccharomyces cerevisiae* at seven sites in the yeast genome. The data may be arranged in a 7×2 contingency table, with site as the row variable and strain as the column variable:

[Site]	*Strain* *RAD3*	*rad3-102*	*Row totals*
[18]	17	1	18
[27]	2	8	10
[29]	5	7	12
[64]	0	7	7
[83]	11	4	15
[88]	10	0	10
[90]	121	56	177
Col. totals	166	83	249

As in Example 9.3, if the pattern of mutations is similar across groups, we would conclude that the two strains exhibit essentially identical mutational spectra. This is a question of homogeneity under product multinomial sampling, as given by (9.2). Use the power-divergence statistic $C^2_{1/2}$ to assess this null effect. Set $\alpha = 0.01$.

9.12*. Show that the limiting form of the power-divergence statistic as $\lambda \to 0$ is the LR statistic for testing the null model in (9.1).

9.13*. (Read and Cressie, 1988) The power-divergence residuals can be redefined to give them interpretation as measures of each cell's departure from independence, but such that their sum is still equal to C^2_λ. Construct this as follows:

a. Begin by defining the modified residual as $h_\lambda(Y_{ij}, E_{ij}) = Y_{ij}\{(Y_{ij}/E_{ij})^\lambda - 1\} + \lambda(E_{ij} - Y_{ij})$, where E_{ij} is the expected value from (9.3). This is the raw power-divergence residual, plus the quantity $\lambda(E_{ij} - Y_{ij})$. Show that for any λ, the sum of these additional quantities is zero: $\sum_{i=1}^R \sum_{j=1}^C \lambda(E_{ij} - Y_{ij}) = 0$. (*Hint*: use the results from Exercise 9.7.)

b. Show that for any λ and at any i and j, the modified residual is always nonnegative: $h_\lambda(Y_{ij}, E_{ij}) \geq 0$. In particular, show that if $Y_{ij} = E_{ij}$, then $h_\lambda(Y_{ij}, E_{ij}) = 0$.

c. Now sum the modified residuals over all i and j, and let

$$A^2_\lambda = \frac{2}{\lambda(\lambda+1)} \sum_{i=1}^{R} \sum_{j=1}^{C} h_\lambda(Y_{ij}, E_{ij}).$$

Show that $A^2_\lambda = C^2_\lambda$. [*Hint*: use the result from part (a).]

9.14*. (Read and Cressie, 1988) The core value of the test statistic Z_D is related to a generalized power-divergence statistic. To see this, recall that $D_{\text{calc}} = X^2_{\text{calc}} - \sum_{i=1}^R \sum_{j=1}^C (Y_{ij}/E_{ij})$. Then:

a. Show that D_{calc} can be written as

$\sum_{i=1}^R \sum_{j=1}^C E_{ij}^{-1}\{Y_{ij} - E_{ij} - (1/2)\}^2 - RC - \sum_{i=1}^R \sum_{j=1}^C (4E_{ij})^{-1}$.

(*Hint*: $RC = \sum_{i=1}^R \sum_{j=1}^C E_{ij}/E_{ij}$.)

b. Recall that the modified power-divergence residuals from Exercise 9.13 have the form $h_\lambda(u,v) = u\{(u/v)^\lambda - 1\} + \lambda(v - u)$. Show that $B^2_1 = D_{\text{calc}} + RC + \sum_{i=1}^R \sum_{j=1}^C (4E_{ij})^{-1}$, where

$$B^2_\lambda = \frac{2}{\lambda(\lambda+1)} \sum_{i=1}^{R} \sum_{j=1}^{C} h_\lambda(Y_{ij} - \tfrac{1}{2}, E_{ij}).$$

c. Combine the results of parts (a) and (b) to show that $B_1^2 = \sum_{i=1}^{R}\sum_{j=1}^{C} E_{ij}^{-1}\{Y_{ij} - E_{ij} - (1/2)\}^2$, a 'generalized' X^2-statistic.

9.15. As in Example 9.3, Cariello *et al.* (1994) reported on a study of mutations in the human *p53* gene isolated from bladder cancers of smokers and of nonsmokers. Collected together over site and form of mutation, the spectra comprise a 26×2 contingency table:

Mutation	*Smokers*	*Nonsmk.*	*Mutation*	*Smokers*	*Nonsmk.*
396 G→C	1	0	422 G→A	0	1
438 G→A	0	1	481 G→A	1	0
503 A→G	1	0	542 G→A	1	0
574 C→T	1	0	586 C→T	1	0
637 C→T	1	0	638 G→C	1	0
701 A→G	0	1	707 G→A	1	0
722 C→G	0	1	742 C→T	0	1
743 G→A	1	0	744 G→A	1	0
772 G→A	0	1	773 A→G	0	1
818 G→A	1	0	839 G→C	3	3
844 C→T	1	0	853 G→A	1	1
853 G→C	1	0	856 G→A	0	2
859 G→A	2	0	860 A→C	1	0

If the pattern of mutations is similar between smokers and nonsmokers, we would conclude that smoking does not affect the mutational spectrum in bladder cancers, at least in the *p53* gene. This is a question of homogeneity under product multinomial sampling, as in (9.2). Since the data here are highly sparse, use the Z_D-statistic from (9.8) to test for homogeneity. Set $\alpha = 0.10$.

9.16. (Cherian, 1994) Mammalian metallothionein proteins bind strongly with heavy metals present in the environment, such as zinc and copper. To study whether metallothionein activity indicates heavy metal presence in human tissues, an experiment was performed to measure metallothionein levels in human testicular cancers by staining the tissues and categorizing the degree of metallothionein stain (none, trace, moderate, strong). Cross-classified by stage of tumor progression (I, II, or III, in increasing progression), the data comprise a 3×4 contingency table (see the display on page 456). If tumor stage/severity were unrelated to metallothionein activity, we would consider the two categorical variables independent and the null model in (9.1) would apply. To assess this answer the following:

	Metallothionein staining				
Tumor stage	*None*	*Trace*	*Moderate*	*Strong*	*Row totals*
I	2	5	0	2	9
II	8	0	6	5	19
III	1	0	0	3	4
Col. totals	11	5	6	10	32

a. Which probability model (Poisson, full multinomial, or product multinomial) would best describe these data?

b. Calculate the Pearson X^2-statistic for these data, and use it to test the null hypothesis of independence. Set $\alpha = 0.10$.

c. Calculate the power-divergence statistic $C^2_{1/2}$ for these data, and use it to test the null hypothesis of independence, at $\alpha = 0.10$. Compare the result to your inference from part (b).

d. This contingency table is made up of many small numbers, including quite a few zeros and ones. Based on the expected values E_{ij} required to calculate X^2_{calc} or $C^2_{1/2}$, or using the sparseness measure $Y_{++}/\{(R - 1)(C - 1)\}$, would you classify this table as sparse? If so, perform a reanalysis using the Z_D-statistic from (9.8).

9.17. Return to the data in Table 9.5. Calculate the sparseness measure $Y_{++}/\{(R - 1)(C - 1)\}$ for these data. Is it large enough (greater than 10) to motivate use of the χ^2 approximation for X^2? If not, is it large enough (greater than 5) to motivate use of the χ^2 approximation for $C^2_{1/2}$? If so, calculate $C^2_{1/2}$ for these data and compare the resulting inference to that achieved in Example 9.3.

9.18. In a study of the relationship between insect stage and type of tree attacked for tree-boring insects, data were collected in a cross-classified manner as follows:

		Tree type		
Insect stage	*Pine*	*Hemlock*	*Spruce*	*Row totals*
Juvenile	20	1	1	22
Sub-adult	11	16	2	29
Adult	9	0	2	11
Col. totals	40	17	5	62

Test the hypothesis that tree type and insect stage are independent. Set $\alpha = 0.01$. Which test statistic did you select, and why? Would any statistic(s) be ill advised for testing independence with these data? Why?

9.19. Using the log-linear model (9.9), show that the no-interaction condition $\delta_{ij} = 0$ (for all i,j) from (9.10) implies independence as in (9.1). Impose the reference cell identifiability constraints $\alpha_1 = \beta_1 = \delta_{i1} = \delta_{1j} = 0$ (for all $i > 1, j > 1$).

9.20*. Repeat Exercise 9.19 under the zero-sum identifiability constraints $\sum_{i=1}^{R}\alpha_i = \sum_{j=1}^{C}\beta_j = \sum_{i=1}^{R}\delta_{i1} = \sum_{j=1}^{C}\delta_{1j} = 0$.

9.21. How would you write the null effect of simple independent action in Section 9.5.2 if the background response, π_{00}, were identically equal to zero (that is, if no response is possible when neither stimulus is applied)? Is this still interpretable as some form of multiplicative relationship on the complementary log scale?

9.22. A study of joint action of the mutagen 9-aminoacridine (9-AA) and the metal potassium chromate (K-Cr) gave the following response proportions in the *E. coli* fluctuation test (as in Example 9.4):

	K-Cr (μm)	
9-AA (μm)	*Control*	*2.9*
Control	$\frac{0}{192}$	$\frac{3}{192}$
40	$\frac{49}{192}$	$\frac{119}{192}$

a. Assume that the 9-AA and K-Cr affect mutagenesis in the bacteria in a different manner, so that the simple independent action model is valid. Use the $\log\{W\}$-statistic in (9.13) to test whether any significant departure from simple independent action is evidenced with these data, at $\alpha = 0.01$.

b. Repeat the analysis on part (a) using an appropriate GLiM.

9.23. Verify the operations leading to (9.11). That is:

a. Show that if $\theta_{ij} = 1 - \pi_{ij}$ and if $\pi_{11} = \pi_{10} + (1 - \pi_{10})\pi_{01}$, then $\theta_{11} = \theta_{01}\theta_{10}$.

b. Form the relative risks of nonresponse $RR_{ij} = \theta_{ij}/\theta_{00}$, and show $RR_{11} = RR_{01}RR_{10}$ leads to (9.11).

9.24. The measure of departure from simple independent action in (9.12) may be viewed as an interaction parameter, and as such may be estimated directly (rather than tested for departure from zero). In the 2×2 case, this would be, say, $\omega_{11} = \log\{\theta_{10}\} + \log\{\theta_{01}\} - \log\{\theta_{00}\} - \log\{\theta_{11}\}$. An estimate of this measure is $\log\{W\}$ from (9.13).

a. What would you suggest as a large-sample $1 - \alpha$ confidence interval for the interaction parameter, based on the $\log\{W\}$-

statistic? (*Hint*: $\log\{W\}$ is the maximum likelihood estimator of the interaction parameter ω_{11}.)

b. Apply your result in part (a) to the *E. coli* mutagenicity data from Example 9.4. Do your calculations support the conclusions made in the example?

9.25*. The interaction parameter in Exercise 9.24 can be extended to larger tables of proportions, as in (9.15): $\omega_{ij} = \log\{\theta_{i0}\} + \log\{\theta_{0j}\} - \log\{\theta_{00}\} - \log\{\theta_{ij}\}$. Using the complementary log-linear model (9.16), show that the no-interaction condition $\omega_{ij} = 0$ (for all i,j) implies simple independent action as in (9.15). Impose reference cell identifiability constraints $\alpha_0 = \beta_0 = \omega_{i0} = \omega_{0j} = 0$ (for all $i > 0, j > 0$).

9.26. A study similar to that from Exercise 9.22 was conducted to assess the joint action of the potential mutagens hydroxyurea and caffeine, with and without additional X-ray bombardment. The data were proportions of human lymphocytes in culture that exhibited chromosome damage after joint exposures (Piegorsch and Margolin, 1989). The *pair* of 2×2 tables representing the proportion outcomes are:

At 0 Gy X-ray bombardment:

	Caffeine	
Hydroxyurea	*Control*	*2.5×10^{-3} mol*
Control	$\frac{3}{200}$	$\frac{12}{200}$
2.5×10^{-3} mol	$\frac{12}{200}$	$\frac{17}{200}$

and at 4.0 Gy X-ray bombardment:

	Caffeine	
Hydroxyurea	*Control*	*2.5×10^{-3} mol*
Control	$\frac{29}{200}$	$\frac{78}{200}$
2.5×10^{-3} mol	$\frac{41}{200}$	$\frac{135}{200}$

Assume that the mutagenic effects of all three stimuli are different in human lymphocytes, so that the simple independent action model is valid.

a. Compute the $\log\{W\}$-statistic in (9.13) for each 2×2 table, and test whether any significant departure from simple independent action is evidenced. Set your per-comparison significance level to $\alpha = 0.01$.

b. Since these data were garnered from a single experiment, one should adjust the inferences in part (a) by applying a Bonferroni correction. Perform such an adjustment, with overall familywise error rate set to $\alpha = 0.01$. Do your results change in any substantive way from part (a)?

c. Using $\log\{W\}$ as a measure of the departure from simple independent action, does it appear that the departure is comparable in each table, or are the departures (if any) different? (If the departures differ, then the synergistic departure in each table differs at differing levels of X-ray exposure. This is a form of *three-way interaction*; cf. Section 9.5.3.)

d*. Analyze these data via the multi-way complementary log GLiM given in Section 9.5.3. Test for a significant three-way interaction, at $\alpha = 0.01$.

9.27. (Piegorsch *et al.*, 1988) A study of cell differentiation in human leukocytes recorded the proportion of cells (out of 200 cells examined) that changed structure (or 'differentiated') after joint exposure to two chemical agents, tumor necrosis factor (TNF) and interferon-γ (IFγ). Four levels of each agent were administered (including a zero-dose control), producing the following 4×4 table of proportions:

	TNF (U/ml)			
IFγ (U/ml)	*0*	*1*	*10*	*100*
0	$\frac{11}{200}$	$\frac{22}{200}$	$\frac{31}{200}$	$\frac{102}{200}$
4	$\frac{18}{200}$	$\frac{38}{200}$	$\frac{68}{200}$	$\frac{171}{200}$
20	$\frac{20}{200}$	$\frac{52}{200}$	$\frac{69}{200}$	$\frac{180}{200}$
100	$\frac{39}{200}$	$\frac{69}{200}$	$\frac{128}{200}$	$\frac{193}{200}$

The two cell differentiating agents operate at different cellular levels, and so the simple independent action model is valid for describing joint action.

a. View each of the $(4 - 1)(4 - 1) = 9$ combinations of joint exposures as separate 2×2 tables of proportions (that is, for any joint exposure, construct a 2×2 table of proportions by relating that joint exposure outcome to the associated single- or no-exposure outcomes). Compute the $\log\{W_{ij}\}$-statistic in (9.13) for each 2×2 table (i = 1,2,3; j = 1,2,3), and test whether any significant departure from simple independent action is evidenced. Set your per-comparison significance level to $\alpha = 0.10$.

b. Do you see any pattern in the values of $\log\{W_{ij}\}$ computed in part (a)?

c. Since the nine separate $\log\{W_{ij}\}$-statistics all have at least one proportion in common (which one?), one should adjust the inferences in part (a) for multiplicity. Apply a Bonferroni adjustment, with overall familywise error rate set to $\alpha =$ 0.10. Do your results change in any substantive way from part (a)?

d. Piegorsch *et al.* (1988) modeled the joint action for these data via a second-order regression model in log-dose, and calculated maximum likelihood estimates under a complementary log link, as in (9.16). This produced an estimated joint-action equation of

$$\hat{\omega}(\mathrm{TNF},\mathrm{IF}\gamma) = 0.088 - 0.447\log\{\mathrm{TNF}\} + 0.519(\log\{\mathrm{TNF}\})^2 - 0.134\log\{\mathrm{IF}\gamma\} + 0.098(\log\{\mathrm{IF}\gamma\})^2 + 0.156\log\{\mathrm{TNF}\}\log\{\mathrm{IF}\gamma\}.$$

At each observed value of $\log\{\mathrm{TNF}\}$ plot this function on a $\log\{\mathrm{IF}\gamma\}$ scale. (Calculate the estimated joint action over only positive values of TNF and IFγ, to avoid taking a logarithm of zero.) What pattern do you see?

e. Repeat the plot performed in part (d), but now plot at each observed value of $\log\{\mathrm{IF}\gamma\}$ on a $\log\{\mathrm{TNF}\}$ scale. Are the patterns similar? Can you estimate where the predicted joint response reaches a minimum?

f*. Verify the computations in part (d) via a GLiM fit. (Use consecutive-dose average spacing from (6.9) to define the control values.) With this, test for global (that is, any) departure from simple independent action, at $\alpha = 0.10$.

9.28*. Consider a joint-action experiment where the two stimuli *inhibit* a toxic response. That is, if the 'success' outcomes for the binomial data are a toxic response such as death, the stimuli act to inhibit death, or increase the health of the organism, etc. In this case, the simple independent action model as developed in (9.11) is invalid (since the stimuli affect 'success' responses in an inverse manner). It is possible, however, to develop a simple independent action model under inhibitory stimuli in a simple manner. Explain how this can be done, and give the simple independent action condition in terms of the 'success' probabilities, π_{ij}. (*Hint*: recognize the symmetry in binomial response data.)

9.29*. In Section 4.2.2, we presented the odds ratio as a summary measure of association for use in two-group categorical data analysis. In a 2×2 contingency table, there is a single odds ratio, often written as $OR = (\pi_{11}\pi_{22})/(\pi_{12}\pi_{21})$.

a. Show that independence in (9.1) implies that $OR = 1$.

b. In an $R \times C$ contingency table, a set of $(R - 1)(C - 1)$ odds ratios provides a complete description of association/dependence in the table (Agresti, 1990, Section 2.2.6). Suggest a possible set of such odds ratios. What does independence in an $R \times C$ table imply about these odds ratios?

c. Suppose simple independent action holds in a 2×2 table of proportions. Find the odds ratio relating the row and column factors.

d. Suppose that simple similar action holds for a 2×2 table of proportions. Find the odds ratio relating the row and column factors. (*Hint*: use (9.17) to find the four cell probabilities, $\pi_{ij} = \pi(r_i, c_j)$, and let $r_0 = c_0 = 0$ and $r_1 = c_1 = 1$.)

e. Discuss how the different uses of the word *independent* are highlighted here, considering the implication of the result in part (a) (that is, independence implies $OR = 1$) on the result in part (c).

10

Incorporating historical control information

In select instances, past data on a particular endpoint can improve further analysis of the environmental phenomenon under study. The goal is to incorporate the historical data on control responses into the statistical analysis, improving the power of the test to detect a significant effect of the treatment, exposure, etc. Some classic examples are as follows:

- In environmental carcinogenesis testing, historical control data can add valuable information when assessing exposure-related tumor incidence with very rare cancers. If only a small number of tumors is observed in either the concurrent control or treatment group, a simple two-sample comparison may not possess sufficient power to detect a small but true tumorigenic effect. Use of appropriate historical control information in this case can increase the test's power to detect the small effect.
- When an environmental test protocol is short and/or simple enough to allow a large historical database to accumulate on control responses, one can employ the historical data in ongoing, current tests of environmental effects. For example, the specific locus mutagenesis assay employs a large database of historical control data along with the concurrent control sample when testing for exposure-related mutagenesis (Selby and Olson, 1981).

In this chapter, we discuss some basic methods for incorporating historical control data in environmental toxicity testing. We begin by reviewing some fundamental guidelines and protocol similarities that must be met in order for historical data to be compatible with the current experiment.

10.1 GUIDELINES FOR USING HISTORICAL CONTROL DATA

Historical information on environmental phenomena can be misused, and care must be taken when including historical data in any statistical analyses. For example, nomenclature conventions and diagnostic criteria must be calibrated carefully, so as to allow meaningful comparisons among similar effects. Also, outside sources of variability, such as laboratory differences or changes in susceptibility of a strain or species over time, must be isolated and controlled. Based on suggestions by Pocock (1976), we recommend that the following restrictions be in place when forming a database of historical outcomes, and when employing these data in a statistical analysis:

1. The historical data must be gathered by the same research organization (and preferably the same research team) that conducts or oversees the current study.
2. The study protocol must remain fixed throughout the period covering the historical and current studies; this includes the method of scoring outcomes of interest.
3. The historical and concurrent control groups must be comparable with regard to age, sex, and other explanatory factors. Where doubt exists as to such comparability, the burden of proof rests upon the research team performing the study.
4. There must be no detectable systematic differences in response between the various control groups.

In practice, selected aspects of these criteria may be difficult to achieve. For example, if a study has a multi-location/multi-center design, it is difficult for the same research team to conduct each experiment. In such situations, the coordinating institution must strive to satisfy criterion 1 to the best extent possible. Or, if a study protocol has been improved to incorporate new advances in biotechnology, differences between the old and the improved protocol may invalidate criterion 2.

In general, although these restrictions may seem stringent, they are intended to prevent improper or cavalier incorporation of historical data in environmental decision-making. For example, if a laboratory's histopathological review has changed over time such that slides read previously as noncancerous are now viewed as, say, benign tumors, the more recent control tumorigenicity rates may increase relative to the older control rates. The corresponding concurrent control rates would appear lower relative to the historical controls. This **temporal heterogeneity** is an artifact of the change in laboratory practice, however, and is not associated with any underlying biological or environmental phenomenon. Under this scenario the Type I error rates of any tests incorporating the historical control database may be inflated, since the historical control adjustment has artificially lowered the background tumorigenicity rate.

The investigator must acknowledge at all times the basic premise of any control-vs.-treatment comparison: a properly constructed, concurrent control represents the most appropriate group to which any treatment may be compared. Any historical data brought under consideration must match the concurrent control with respect to this comparability; cf. criterion 3. If comparability is compromised, inclusion of historical data is unwise. (In some cases, however, it may be possible to include some form of covariate adjustment in the statistical model to correct for poor comparability; for example, adjusting observed responses for age if the experimental units are studied at different ages or lifespans.) Improper use of historical data can invalidate the statistical analysis, by inducing increases in variability similar to the effect of unrecognized overdispersion, as discussed in Section 6.5. As a result, the sensitivity of the corresponding analysis is compromised (Krewski *et al.*, 1988; Haseman, 1992), defeating the very purpose for historical considerations.

Statistically, data comparability may be viewed as a form of **exchangeability**, where any permutation of any subset(s) of the historical data results in the same statistical effects (Margolin and Risko, 1984). We will employ some limited features of exchangeability in the analyses presented below, but the technical details of the topic reach beyond the scope of this chapter. For a greater treatment on exchangeability, see the review by Draper *et al.* (1993).

Perhaps the greatest use of historical control data in environmental biology is to expand limited control information and enhance statistical power when testing for an increase (or decrease) in the observed response compared to a concurrent control. We present in this chapter two forms of analysis that incorporate historical data in simple two-sample, control-vs.-treatment comparisons, the first for normal parent distributions, and the other for binomial parent distributions. We then move on to a short discussion on use of historical controls to test for trend in the response over increasing levels of a quantifiable dose. In all cases, the methodology will involve a form of **hierarchical model**, where some added statistical variability in the unknown parameters is incorporated hierarchically into the statistical model; cf. Example 1.1. In certain applications, and as we see below, this is also known as a **superpopulation model** (Gaver *et al.*, 1990).

10.2 TWO-SAMPLE HYPOTHESIS TESTING – NORMAL DISTRIBUTION SAMPLING

Consider the simple problem of comparing two populations, a control and a treatment group, as in Section 4.1. Index the control group as $i = 0$ and the treatment as $i = 1$. Assume the data include n_i replicates in each group. To incorporate historical control data in the analysis, we mimic a construction due to Pocock (1976) that is discussed in detail by Margolin and Risko (1984). Assume that the current two-sample data may be written in an ANOVA-style format,

$$Y_{ij} = \mu + \theta + i\beta + \varepsilon_{ij}, \tag{10.1}$$

where heterogeneous variances are accepted in the error terms, $\varepsilon_{ij} \sim$ (indep.) $N(0,\sigma_i^2)$ ($i = 0,1$; $j = 1, \ldots, n_i$). (The model can be simplified if variance homogeneity, $\text{Var}[\varepsilon_{ij}] = \sigma^2$, is felt to be a reasonable assumption.) In addition, a **random effect** assumption is made that incorporates variability into the common θ-term: $\theta \sim N(0,\sigma_\theta^2/2)$, independent of ε_{ij}. The parameter θ models current-experiment variability. We incorporate a similar parameter in the historical control assumption, below. (The division by 2 in $\text{Var}[\theta]$ is done for convenience; it will lead to simplifications in the statistical analysis.) The parameter β is the unknown, fixed-effect, treatment parameter corresponding to the difference in mean response between the two groups, that is, $\beta = E[Y_{1j}] - E[Y_{0j}] = \{\mu + E[\theta] + \beta\} - \{\mu + E[\theta]\} = (\mu + \beta) - \mu$. Recall from Section 9.4 that this is a form of **reference cell coding**. Of interest is testing $H_0{:}\beta = 0$ vs. $H_a{:}\beta > 0$. (Two-sided tests are also possible, after appropriate modifications.)

The historical control data are incorporated by assuming that they can be summarized as a single average value, $\overline{Y}_c$, based on n_c observations. Common terminology for this is a **superpopulation model**, here taken as

$$\overline{Y}_c = \mu + \theta_c + \varepsilon_c, \tag{10.2}$$

where $\varepsilon_c \sim N(0,\sigma_c^2/n_c)$, independent of $\theta_c \sim N(0,\sigma_\theta^2/2)$. The term 'superpopulation' signifies simply that the historical controls are drawn from some large, idealized control population, independent of and encompassing that for the concurrent control.

Notice that in the model assumptions for (10.1) and (10.2), the variance parameters are assumed known. For σ_c^2 and σ_θ^2 this is not unreasonable, since we expect a large database of control information to be available. If variances are estimated from this database, the *df* associated with the variance estimates are expected to be large enough to be essentially infinite, approximating known information. For the current experiment, the variances also are assumed known. This assumption is made primarily for mathematical simplicity. If inappropriate, sample variances can be substituted for the unknown σ_i^2-values, changing the associated reference distributions from standard normal or χ^2, to t or F, as necessary.

The superpopulation model yields a maximum likelihood (ML) estimate for β that is a linear combination of the sample means:

$$b = \overline{Y}_1 - \hat{\theta}\overline{Y}_0 - (1 - \hat{\theta})\overline{Y}_c ,$$

where $\overline{Y}_0$ and $\overline{Y}_1$ are the sample control and treatment means, respectively, from the current data, $\overline{Y}_c$ is the historical control mean, and $\hat{\theta}$ is an estimate of θ based on the known variance components $\sigma_0^2, \sigma_\theta^2, \sigma_c^2$:

$$\hat{\theta} = \frac{\dfrac{\sigma_c^2}{n_c} + \sigma_\theta^2}{\dfrac{\sigma_c^2}{n_c} + \sigma_\theta^2 + \dfrac{\sigma_0^2}{n_0}} .$$

Notice that b contrasts the treatment information in $\overline{Y}_1$ with a weighted average of the two control means, with $\hat{\theta}$ as the weighting factor. If there were no historical control information, $n_c = 0$ and $\hat{\theta} = 1$. Contrastingly (and far less desirable), if there were no concurrent control information, $n_0 = 0$ and $\hat{\theta} = 0$.

With these values, and setting $\Omega_{01} = (\sigma_1^2/n_1)/(\sigma_0^2/n_0)$, the standard error of b under $H_0{:}\,\beta = 0$ is

$$se_0[b] = \frac{1}{\dfrac{n_0}{(1+\Omega_{01})\sigma_1^2} + \left\{(1+\Omega_{01})\sigma_\theta^2 + \dfrac{\Omega_{01}\sigma_0^2}{n_0} + \dfrac{\Omega_{01}^2\sigma_1^2}{n_1} + \dfrac{\sigma_c^2}{n_c}(1+\Omega_{01})^2\right\}^{-1}}$$

(Margolin and Risko, 1984). In large samples, reject H_0 in favor of H_a when the Wald statistic $W_{calc} = b/se_0[b]$ exceeds z_α.

We should warn that in practice these calculations may produce only limited increases in statistical power to detect $\beta > 0$. For a similar hierarchical model, Hoel (1983) found that power increases were moderate, only achieving substantial improvements when the historical variances were drastically lower than those for the current data, that is, when σ_c^2 was much smaller than $\min\{\sigma_0^2, \sigma_1^2\}$.

10.3 TWO-SAMPLE HYPOTHESIS TESTING – BINOMIAL SAMPLING

For incorporating historical data in the two-sample binomial setting it is possible to construct a hierarchical model similar to that used in the preceding section, but with slightly better sensitivity. We assume that the current data satisfy $Y_i \sim$ (indep.) $\mathcal{B}in(N_i,\pi_i)$, $i = 0,1$. The hypotheses of interest are $H_0{:}\pi_0 = \pi_1$ vs. $H_a{:}\pi_0 < \pi_1$. (Again, two-sided alternatives may be considered, with appropriate modifications.) We assume further that the control response probability π_0 is itself distributed as a beta random variable, from Section 1.3.1: $\pi_0 \sim \mathcal{Beta}(a,b)$. This is similar to a model proposed by Tarone (1982b), where the hierarchical assumption on π_0 accounts for additional information from the historical data. As noted in Section 1.2.2, the beta distribution for π_0 induces a beta-binomial distribution on the marginal distribution of the historical counts, Y_c, taken over N_c historical animals (Hoel, 1983).

We use Y_c and N_c to estimate the hierarchical parameters, a and b. For simplicity, method of moments (MoM) estimation for a and b is common, although some additional control information is required for the method to produce valid estimators. This is usually in the form of a sample variance, s^2, for the control proportions. Then, under the hierarchical beta assumption the marginal mean and variance of the historical control count are $E[Y_c \,|\, N_c,a,b] = N_c a/(a+b)$ and $\mathrm{Var}[Y_c \,|\, N_c,a,b] = N_c ab(a+b+N_c)/\{(a+b)^2(a+b+1)\}$. It will be convenient to reparameterize a and b into $\mu = a/(a+b)$ and $\varphi = 1/(a+b)$, so that $E[Y_c \,|\, N_c, a, b] = N_c\mu$ and $\mathrm{Var}[Y_c \,|\, N_c, a, b] = N_c\mu(1-\mu)(1+N_c\varphi)/(1+\varphi)$. Then, equating the first moment of Y_c/N_c to the historical proportion gives, simply, $\mu = Y_c/N_c$. Equating the variance of Y_c/N_c to the sample variance requires $N_c^{-1}\mu(1-\mu)(1+N_c\varphi)/(1+\varphi) = s^2$. Solving for μ and φ yields MoM estimators $\mu = Y_c/N_c$ and

$$\varphi = \frac{s^2 - N_c^{-1}\mu(1-\mu)}{\mu(1-\mu) - s^2} .$$

Transforming back to a and b via the inverse relationship $a = \mu/\varphi$, $b = (1-\mu)/\varphi$ gives

$$a = \mu\frac{\mu(1-\mu) - s^2}{s^2 - N_c^{-1}\mu(1-\mu)}$$

and

$$b = (1-\mu)\frac{\mu(1-\mu) - s^2}{s^2 - N_c^{-1}\mu(1-\mu)} ,$$

where $\mu = Y_c/N_c$ from above. If the estimated values of a and/or b drop below their lower bounds of zero, simply truncate them to zero.

To test H_0 vs. H_a, a statistic incorporating a and b that measures departure from H_0 is

$$T_{calc} = \frac{(N_0 + a + b)Y_1 - N_1(Y_0 + a)}{a + b + N_+} ,$$

where $N_+ = N_0 + N_1$ (Hoel and Yanagawa, 1986; Krewski *et al.*, 1988). Unfortunately, the small-sample distribution of T_{calc} can be highly skewed and is difficult to approximate. To overcome this, Hoel and Yanagawa (1986) perform the analysis conditional on the concurrent control value, Y_0. They note that in this case, a conditional analysis is locally most powerful against alternative configurations near H_0. Conditional on Y_0, Hoel and Yanagawa show that the standard error of T_{calc} is

$$se[T_{calc} \mid Y_0, a, b] = \left\{ \frac{N_0 + a + b}{N_0 + a + b + 1} p_0(1 - p_0) \left[N_1 - \frac{N_1^2}{a + b + N_+} \right] \right\}^{1/2},$$

where $N_+ = N_0 + N_1$ and p_0 is an updated estimator of π_0 based on the historical control information: $p_0 = (Y_0 + a)/(N_0 + a + b)$. In large samples, the Wald statistic $W_{calc} = T_{calc}/se[T_{calc} \mid Y_0; a, b]$ is approximately standard normal, and one rejects H_0 in favor of H_a when $W_{calc} > z_\alpha$. The one-sided P-value is $P \approx 1 - \Phi(W_{calc})$.

In small samples, the exact conditional distribution of T_{calc} can be calculated directly. Note, however, that the one-sided rejection region based on T_{calc} rejects H_0 if T_{calc} grows too large. Since T_{calc} is a linear function of Y_1, one can also reject H_0 when Y_1 is too large. This allows for calculation of the exact conditional P-value using the conditional distribution of Y_1 (Hoel, 1983). Specifically,

$$\mathrm{P}[Y_1 = y \mid Y_0, a, b] = \frac{N_1!\, \Gamma(Y_0 + y + a)\, \Gamma(N_+ - Y_0 - y + b)\, \Gamma(N_0 + a + b)}{y!\,(N_1 - y)!\, \Gamma(Y_0 + a)\, \Gamma(N_0 - Y_0 + b)\, \Gamma(a + b + N_+)} \qquad (10.3)$$

(Hoel and Yanagawa, 1986), so the upper one-sided P-value is

$$P = \sum_{y=y_1}^{N_1} \mathrm{P}[Y_1 = y \mid Y_0, a, b]\,. \qquad (10.4)$$

In (10.4), the individual probabilities are calculated using (10.3). Reject H_0 in favor of H_a when $P < \alpha$.

Example 10.1 Environmental carcinogenesis of benzene

To illustrate the Hoel–Yanagawa method of incorporating historical data, we turn to an environmental carcinogenicity study of the simple aromatic hydrocarbon, benzene. (Benzene is a component of many industrial chemicals, including petroleum and crude oil. It is ubiquitous in the environment, and is therefore important to study for its potential toxicity.)

As reported by the US National Toxicology Program (1986), the study considered murine carcinogenesis after gavage exposure to the chemical over a 105-week chronic exposure period. Multiple-site carcinogenicity was evidenced in male and female B6C3F$_1$ mice (Huff, 1989), and the chemical was considered by the NTP to exhibit clear evidence of carcinogenicity.

We will consider hepatocellular adenomas in female mice. The control response in this study showed only 1 of 49 animals (2.04%) with a liver adenoma at the end of the 105-week study. In a group of 50 mice exposed to 50 mg/kg benzene, the hepatocellular adenoma rate increased to 5/50 (10.0%). Of interest is testing whether the treated group's response probability, π_1, significantly exceeds the control response probability, π_0, that is, testing $H_0:\pi_0 = \pi_1$ vs. $H_a:\pi_0 < \pi_1$.

Fisher's exact test applied to these data yields a one-sided P-value of 0.106. This suggests an insignificant comparison, even though the observed response in the exposed group is almost five times as great as that in the control group. Can incorporation of historical data improve the statistical sensitivity here? To assess this, consider the NTP's reported historical control rate for hepatocellular adenomas in female B6C3F$_1$ mice: 47 control animals out of a total 1176 exhibited hepatocellular adenomas in previous two-year NTP studies, with an historical standard deviation of s = 2.55% (NTP, 1986, Table F15). We incorporate these values into the Hoel–Yanagawa conditional two-sample test.

To determine the historical parameter values a and b, the MoM approach employs the historical information on murine hepatocellular adenomas: μ = 47/1176 and $s^2 = 0.00065$. The corresponding MoM calculations give a = 2.44 and b = 58.63, and the associated test statistic is

$$T_{\text{calc}} = \frac{(49 + 2.44 + 58.63)(5) - (50)(1+2.44)}{2.44 + 58.63 + 99} = \frac{378.30}{160.07}$$

or simply 2.363. The conditional standard error is

$$se[T_{\text{calc}} \,|\, Y_0, a, b] = \sqrt{\frac{110.07}{111.07}(0.0313)(0.9687)\left\{50 - \frac{2500}{160.07}\right\}}\ ,$$

which calculates to 1.016. From these, the Wald statistic is W_{calc} = 2.363/1.016 = 2.33. This yields an approximate one-sided P-value of 0.010.

We programmed the W_{calc}-statistic into an S-PLUS function (Fig. 10.1), and applied it to the hepatocellular adenoma tumorigenicity data from the NTP study. The results appear in Fig. 10.2. This analysis mirrors the calculations reported above.

```
Wc.stat<-function(Y0,N0,Y1,N1,YC,NC,S2) {

# Hoel-Yanagawa conditional two-sample test
# Y0 (Y1) = no. of successes in concurrent control (treatment)
#           groups
# N0 (N1) = no. of trials in the concurrent control (treatment)
#           groups
# YC (NC) = number of successes (trials) in histor. control data
#      S2 = variance in historical control proportions
#
  mu.C <- YC/NC
  phi.C <- (S2 - mu.C*(1-mu.C)/NC) / (mu.C*(1-mu.C) - S2)
  a.C <- mu.C/phi.C
  b.C <- (1-mu.C)/phi.C
  N.plus <- N0 + N1
   temp <- N0 + a.C + b.C
   T.C <- (temp*Y1 - N1*(Y0 + a.C)) / (N1+temp)
   p.0 <- (Y0+a.C)/temp
   se.TC <- sqrt(temp/(temp+1)*p.0*(1-p.0)*(N1 - N1^2/(N1+temp)))
  W.C <- T.C/se.TC
  data.frame(a=a.C,b=b.C,mu=mu.C,phi=phi.C,Tc=T.C,
             se.Tc=se.TC,Wc=W.C,pvalue=1-pnorm(W.C))
}
```

Fig. 10.1 S-PLUS function for W_{calc}-statistic.

```
Y0<-1; N0<-49; Y1<-5; N1<-50
YC<-47; NC<-1176; S2<-0.00065
Wc.stat(Y0,N0,Y1,N1,YC,NC,S2)
```

```
         a       b      mu    phi     Tc  se.Tc     Wc   pvalue
1 2.4417 58.6537 0.03997 0.0164 2.3635 1.0158 2.3267 0.009991
```

Fig. 10.2 Output (edited) from S-PLUS `Wc.stat` function for assessing carcinogenicity of benzene.

It is instructive here to contrast this approximate P-value with the exact conditional value available by employing (10.4). Applied to these data, (10.4) is

$$P = \sum_{y=5}^{50} \frac{50!\,\Gamma(3.44+y)\,\Gamma(156.63-y)\,\Gamma(110.07)}{y!\,(50-y)!\,\Gamma(3.44)\,\Gamma(106.63)\,\Gamma(160.07)}.$$

The computations required to determine this quantity require a computer, particularly to evaluate the gamma function. (The factorial operators are related to the gamma function; they may be calculated via $m! = \Gamma\{m-1\}$.)

The exact conditional value based on these data is found to be $P = 0.014$, showing fairly good agreement with the approximate value based on W_{calc}, above. From either calculation, we see a reasonably strong departure from the null hypothesis, adding to the evidence that benzene is indeed a mammalian carcinogen.

We end this example with some caveats. First, the use of simple proportion-based tests may not be appropriate with chronic exposure data, since the simple proportions fail to take into account any intercurrent mortality, that is, differences in survival between the exposed and unexposed animals. (We discuss selected methods that adjust for intercurrent mortality in Section 11.4; see also Example 8.7.) In this particular case, these was no significant difference in survival between the control group and the 50 mg/kg exposure group (NTP, 1986, Table 14), so survival adjustment was not critical.

Also, it is not clear that all of the four criteria for proper inclusion of historical data – listed in Section 10.1 – were met for the NTP historical database. The data were collected from a number of different laboratories that all conduct the NTP carcinogenesis assay. Although the data were all gathered under NTP's standard protocol, it is tacitly acknowledged that laboratory-specific factors often undermine the assumed constancy of protocol, pathology, and other experimental aspects of the NTP carcinogenesis assay (Haseman *et al.*, 1989). This concern violates criterion 2. Indeed, the NTP conclusions on murine carcinogenicity of benzene were *not* based on explicit use of the historical data (although they did mention the historical rates in their discussion), stemming in part from violations of the inclusion criteria. □

As we noted in Section 3.5.2, multi-disciplinary input on a chemical's carcinogenicity is required from numerous sources, and the ability of a single statistical test to identify a carcinogenic response should never be used as sole motivation for such a complex decision, even if the test is enhanced with historical data.

10.4 TREND TESTING WITH HISTORICAL CONTROLS

One can also use historical data to test for increasing trend when an environmental agent is studied over a range of quantifiable measures, such as exposure doses. Different methodologies are appropriate for different data structures, however. For instance, if the data are continuous measurements possessing normal distributions, the superpopulation model from

Section 10.2 can be used. The approach is essentially a trend test extension of the two-sample method described above, and is detailed by Pocock (1976).

Similarly, when the data are proportions based on a binomial parent distribution, the methods from Section 10.3 can be extended to test for trend. The literature for this sampling frame is quite extensive, however, and a complete review is beyond the scope of this section. Krewski *et al.* (1988) give an excellent review of the methodology through the late 1980s; additional sources include Prentice *et al.* (1992) on animal carcinogenicity experiments, and Ryan (1993) on developmental toxicity experiments. As these and many other authors note, development of appropriate models and methods for incorporating historical control data in trend testing is an ongoing research effort; further study is necessary to identify stable and sensitive statistical technologies in this area.

Interestingly, methods for incorporating historical data when testing for trend in count data have not been studied as extensively as those for proportion data. The approaches that have been proposed, however, provide good examples of how new statistical methodology develops in response to problems in environmental toxicology. We present the details as a formal example.

Example 10.2 Historical controls in trend tests with count data

Tarone (1982a) proposed an historical control model for count data based on a hierarchical gamma-Poisson assumption. Specifically, assume the current data are Poisson-distributed counts: $Y_{ij} \sim$ (indep.) $\mathcal{Poisson}(\mu_i)$, where μ_i is the mean rate of response at the ith exposure level ($i = 0, \ldots, k$), and $j = 1, \ldots, r_i$ indexes replicate observations at each exposure level. For the doses, the initial dose corresponds to a control level, x_0, and the other doses increase monotonically from it: $x_0 < x_1 < \cdots < x_k$. Of interest is testing $H_0{:}\, \mu_0 = \mu_1 = \cdots = \mu_k$ vs. the increasing trend hypothesis H_a given in (6.2). From the closure properties of the Poisson distribution under addition, the distribution of the per-exposure sums is also Poisson: $Y_{i+} \sim$ (indep.) $\mathcal{Poisson}(r_i\mu_i)$, $i = 0, \ldots, k$.

Assume also that a group of previous experiments has provided a large set of control data of equal quality and comparability to the current data. Denote these historical control counts as V_{mn}, where $m = 1, \ldots, M$ indexes the M historical control studies, and $n = 1, \ldots, N_m$ indexes the replicate counts from each historical study. Tarone notes that while it may be acceptable to assume independent Poisson parent distributions for each of the individual historical control observations (that is, for fixed m, $V_{m1}, \ldots, V_{mN_m}$ are sampled independently from the same Poisson distribution), it is likely that the control group rate μ_0 will vary from study to study. This can be modeled hierarchically, as in Section 10.3. Tarone suggests a gamma distribution for the hierarchical distribution of μ_0, which induces a negative binomial distribution on the group sums: $V_{m+} \sim$ i.i.d. $\mathcal{NB}(N_m\lambda\,,\,\delta/N_m)$, where $\lambda > 0$ is the mean hierarchical control rate, and $\delta > 0$ is a form of dispersion parameter accounting for study-to-study variability.

Tarone's (1982a) historical model for counts leads to a simple score test for trend in the current data; the trend statistic has the general form

$$T = \sum_{i=0}^{k} x_i\,(Y_{i+} - r_i\tilde{Y}_0)\ . \qquad (10.5)$$

In (10.5), r_i is the number of replicate observations per exposure level, and $\tilde{Y}_0$ is an estimator of the mean response under H_0 that incorporates the historical data: $\tilde{Y}_0 = \{\delta^{-1} + Y_{++}\}/\{(\lambda\delta)^{-1} + r_+\}$, with $Y_{++} = \sum_{i=0}^{k}\sum_{j=1}^{r_i} Y_{ij}$ and $r_+ = \sum_{i=0}^{k} r_i$. λ and δ are parameters of the negative binomial distribution for V_{m+} above. (In the case where no historical data are taken, set $\lambda = \delta = 0$, so that (10.5) collapses to the numerator of the Cochran–Armitage trend statistic (6.8); see Exercise 10.7.) Dividing T in (10.5) by its large-sample standard error, $se[T]$, leads to a trend statistic, $z_{\text{calc}} = T/se[T]$. If $\tilde{Y}_0$ is constructed correctly, the distribution of z_{calc} approaches a standard normal as the sample sizes grow large. Then, one rejects H_0 in favor of an increasing trend when $z_{\text{calc}} > \mathfrak{z}_\alpha$. The associated P-value is $P \approx 1 - \Phi(z_{\text{calc}})$.

Specification of $\tilde{Y}_0$ and of the corresponding standard error $se[T]$ are crucial to this construction. If the hierarchical parameters λ and δ are known, the standard error is

$$se[T] = \left\{ \frac{\tilde{Y}_0}{(\lambda\delta)^{-1} + r_+} \left[\sum_{i=0}^{k} r_i x_i^2 - \left(\sum_{i=0}^{k} r_i x_i \right)^2 \right] \right\}^{1/2} .$$

More typically, however, the hierarchical parameters are unknown. In this case Tarone (1982a) suggests replacing λ and δ by their ML estimates, found using the V_{m+}s. The ML estimate for λ is the pooled historical mean,

$$\hat{\lambda}_{\text{MLE}} = \frac{\sum_{m=1}^{M} V_{m+}}{\sum_{m=1}^{M} N_m},$$

while the ML estimate for δ is the solution to the nonlinear equation

$$\sum_{m=1}^{M} N_m \left\{ \log\left(1 + \hat{\lambda}_{\text{MLE}}\delta\right) + \psi\left(\frac{N_m}{\delta}\right) - \psi\left(V_{m+} + \frac{N_m}{\delta}\right) \right\} = 0 .$$

Notice the similarity to (6.16). If the historical control information is extensive, the substitution of the ML estimates in $\tilde{Y}_0$ and in $se[T]$ does not critically affect the large-sample normal reference distribution of z_{calc}: continue to reject H_0 in favor of an increasing trend when $z_{\text{calc}} > z_{\alpha}$. If the historical control information is limited, however, Tarone (1982a) notes that some instability may arise in the large-sample normal approximation, and caution is urged.

In an extension of Tarone's groundbreaking work, Leroux *et al.* (1994) confirm that in small samples, the ML estimators $\hat{\lambda}_{\text{MLE}}$ and $\hat{\delta}_{\text{MLE}}$ can generate instabilities in the score test, including inflated Type I error rates and power fluctuations. They propose an extended score statistic based on generalized estimating equations (GEEs; cf. Section 2.7.2) for λ and δ. Their measure of trend, T, has the same form as (10.5), as does their estimator of mean response under H_0, $\tilde{Y}_0$. To estimate λ and δ, however, they propose GEE-type estimators based on iterative solution of two equations for λ and δ. Leroux *et al.* show that their GEE statistic has dramatically improved small-sample stability, even when the original data values, Y_{ij}, depart from Poisson distribution sampling. The added stability *and* robustness of their GEE approach are strong motivating features in favor of the method's use. Leroux *et al.* note, however, that further study and development of this advanced technology for incorporating historical controls is needed before it can be recommended unequivocally. □

10.5 SUMMARY

In this chapter, a brief introduction is given on use of historical control data for selected environmental applications. Basic guidelines are provided for use of historical data, which, although somewhat strict, are necessary to avoid misinterpretation or misuse of the historical information. It is emphasized that the concurrent control group is the most appropriate source of information against which a treatment group may be compared. When historical data are unavailable, the analyses collapse to basic control-vs.-treatment comparisons, as in Chapter 4. By contrast, when experimental resources are so severe as to restrict collection of concurrent control data, the historical methods may be considered as surrogates for the concurrent control. In such instances, however, comparability between historical controls and some idealized, concurrent, 'baseline' group is still necessary, along with appropriately modified analogs of the other criteria from Section 10.1. These are often difficult to ensure in practice, and it is generally inadvisable to employ historical data as a replacement for concurrent information.

Details on the historical control models are provided for the two-sample (control-vs.-treatment) testing problem, with emphasis on normally distributed or binomially distributed data. Extensions for the use of historical data in trend testing also are noted, with an example of a model for count data used to illustrate the basic issues. Emphasis is placed on the fact that historical data must be incorporated into the analysis with prejudice, in order to avoid destabilizing the very sensitivity and accuracy that the additional information is intended to provide.

EXERCISES

(Exercises with asterisks, e.g. 10.7*, typically are more advanced, or require greater knowledge and understanding of probability distributions and/or calculus.)

10.1. Verify the method of moments (MoM) equations for the hierarchical beta-binomial model for historical information described in Section 10.3. In particular, set $\mu = a/(a + b)$ and $\varphi = 1/(a + b)$, and equate $\mu = Y_C/N_C$ and $N_C^{-1}\mu(1 - \mu)(1 + N_C\varphi)/(1 + \varphi) = s^2$. Show that this yields the stated forms for μ and φ, and then back-transform to find a and b.

10.2. For Exercise 4.10, additional information on the proportion response of micronucleated cells in Swiss Webster mice used in the study is available via historical controls. In particular, suppose previous data show a mean historical control count of 101 micronucleated cells out of 76 000 studied, with an historical variance of $s^2 = 2.132$. Assuming the conditions for proper incorporation of these historical data are met, reanalyze the data from Exercise 4.10, using the historical information. Do the historical data change the inference?

10.3. Verify Fisher's exact test P-values reported in Example 10.1, as described in Section 4.1.7.

10.4. In an environmental carcinogenesis study similar to that discussed in Example 10.1, the US National Toxicology Program reported on the carcinogenic potential of the chemical solvent stabilizer 1,2-epoxybutane. An exposure group of 50 male rats received 200 ppm of the chemical, via inhalation, for 105 weeks (NTP, 1988), while an independent control group of another 50 rats was untreated (Exercise 4.20). One particular tumor of interest that exhibited increased occurrences in rats was leukemia. The control animals gave a response proportion of 25/50 (50%) leukemias, while the exposed animals produced 31/50 (62%) leukemias.

a. Perform a simple two-group analysis to test whether these values suggest an increase in leukemia response due to chemical exposure. Set $\alpha = 0.05$.

b. Historical information on this tumor in rats indicates that the control response in this study may be somewhat high. For leukemias, the historical control data show an overall incidence of 583/1977 (29.94%) leukemias in male rats (NTP, 1988, Table A.4c), with an observed standard deviation of $s = 11.59\%$. Assuming the conditions for proper incorporation of these historical data are met, perform a reanalysis of these data using the historical information, mimicking the approach in Example 10.1. Does your inference from part (a) change?

10.5. In an environmental carcinogenesis study similar to that discussed in Example 10.1, the US NTP reported on the carcinogenic potential of the chemical 1,3-butadiene. An exposure group of 50 female mice received 200 ppm of the chemical, via inhalation, for 105 weeks (NTP, 1993a), while an independent control group of another 50 mice was untreated (Example 4.1). Interest in murine tumorigenicity here included ovarian granulosa cell tumors. The control animals gave a response proportion of 1/49 (2%) tumors (one mouse's ovarian tissue was found to be insufficient for tumor identification), while the exposed animals produced 6/50 (12%) tumors.

a. Perform a simple two-group analysis to test if there is an increased response due to chemical exposure. Set $\alpha = 0.05$.

b. Historical information on this tumor in mice shows that it is somewhat rare: the overall incidence is 1/548 (0.18%) granulosa cell tumors in female mice (NTP, 1993a, Table B4f), with an observed standard deviation of $s = 0.30\%$. Assuming the conditions for proper incorporation of these historical data are met, perform a reanalysis of these data using the historical information, mimicking the approach in Example 10.1. Does your inference from part (a) change?

10.6. Suppose you have data in the form of counts, as in Section 10.4. Develop a method for incorporating historical control data into a two-sample test of a control group ($i = 0$) vs. a single treatment group ($i = 1$), by setting $k = 1$ in Tarone's (1982a) historical control score statistic (10.5).

10.7*. Set $\lambda = \delta = 0$ in Tarone's (1982a) historical control score statistic (10.5). Show that the resulting quantity for T is identical to the numerator of the Cochran–Armitage trend statistic (6.8). (*Hint*: start by examining (10.5) in the limit as $\delta \to 0$; you may wish to use l'Hôpital's rule (Edwards and Penney, 1990, Section 11.1) for assistance in resolving the limit.)

10.8*. Suppose it is assumed in Tarone's (1982a) historical control score statistic (10.5) that *no* excess study-to-study variability exists in the historical control populations. Thus we set $\delta = 0$. Show that the resulting score statistic, T, is essentially the numerator of the simple Cochran–Armitage trend statistic (6.8), but where the historical control data are pooled together with the current control data.

10.9. Exercise 6.21 gives data on tumorigenicity in male rats, in particular, numbers of organ sites (per animal) exhibiting tumors after exposure to the chemical solvent stabilizer 1,2-epoxybutane. Refer back to that exercise and assume the Poisson distribution is a valid sampling model for these data. To incorporate potential historical data in a test for trend with these data, suppose frequencies for number of tumors among 1975 historical control animals are given as follows:

Number of tumors:	0	1	2	3	4	5
Number of control animals:	121	293	436	455	327	192

Number of tumors:	6	7	8	9	10
Number of control animals:	104	28	17	2	0

Use Tarone's statistic (10.5) to incorporate this control information, and test for increasing trend in these data. Set $\alpha = 0.01$.

11

Survival-data analysis

An important outcome measured in ecological and environmental studies involves data on time to some event. Outcomes such as survival times of organisms or failure times of biological systems after exposure to an environmental insult generally exhibit statistical characteristics for which standard ANOVA-type analyses are inappropriate, even after some transformation of the observed data. Determining a correct statistical model and the corresponding data analysis for time-to-event data requires special methodology, depending on the particular environmental response under study. For instance:

- Cold shock may lead to the death of an exposed insect. Pre-exposing the insect to oxygen-deprived conditions may result in better survival when the insect is then exposed to cold. How do we quantify survival?
- One population of fish lives in an old farm pond. A second population of fish lives in a reservoir that experiences heating by an adjacent power-generating plant. Do these two populations of fish exhibit the same pattern of survival when they have been exposed to mercury? Would genotype play a role in the survival pattern?
- Laboratory animals exposed to increasing doses of an environmental toxin may exhibit differential mortality. How do we assess an increasing trend in mortality over dose?

Exposures to environmental agents can alter the lifespan of an organism, even, in the extreme case, to where the entire life distribution changes. In this chapter, we introduce some basic methods for describing and estimating features of lifetime distributions, testing the equivalence of two or more lifetime distributions, and modeling the effects of predictor variables on these distributions. We will use the terms 'failure' or 'death' equivalently, to indicate an event occurring as part of the lifetime or time-to-event study.

As might be expected, statistical methods for survival data are often complex, and the analyses presented below represent a further step up in level of presentation. Introductory readers may wish to limit their initial readings to the early portions of Sections 11.2, 11.3, and 11.4 before continuing through the more advanced material.

11.1 LIFETIME DISTRIBUTIONS

The random variable observed in a survival-time or failure-time experiment is the lifetime (or, technically, the 'time on test'), T, of the experimental unit under study. This quantity can assume only positive values: $T > 0$.

When studying lifetimes, T, the fundamental measure of survival is the probability that the organism survives at least a specified amount of time. In a generic sense, T represents the time until the occurrence of some event of interest, and we use it to measure the probability that the event does not occur until after some specific time t. This quantity, the **survival probability** or **survivor function**, is denoted as $S(t) = P(T > t)$. If a continuous random variable is considered, the survivor function is simply $S(t) = 1 - F(t)$, where, as in Section 1.1.3, $F(t) = P(T \leq t)$ is the c.d.f. associated with the random variable T. Related to the survivor function is the **hazard function**, $h(t)$, which is the **instantaneous failure rate** of the random variable T. Mathematically, $h(t)$ is defined as

$$h(t) = \lim_{\Delta \to 0} \frac{P[t \leq T < t + \Delta \mid T \geq t]}{\Delta} = \frac{f(t)}{S(t)}$$

where $f(t)$ is the p.d.f. associated with the survival variable T. Hazard functions and survival probability functions are alternate ways of describing a lifetime distribution; they are related via integration of the hazard function,

$$S(t) = \exp\left\{-\int_0^t h(u)\,du\right\},$$

or via differentiation of the survivor function,

$$h(t) = -\frac{\partial \log[S(t)]}{\partial t}.$$

It is helpful to place these functions in context: think of $S(t)$ as the probability that a random individual from a certain population will 'survive' past t years (or whatever time units are under study), and $h(t)$ as the age-specific death rate when an individual from that population is t years old.

Estimation of and inferences on $S(t)$ or $h(t)$ could be accomplished using straightforward methods of statistical inference if it were not for one problem often encountered in survival studies: not all the experimental units die before the study concludes. To avoid this concern, some studies are designed to include a follow-up period in which data are provided on experimental units surviving past the end of the study period. Via such follow-up, the organisms can provide additional lifetime data.

Even so, some survival time/failure time observations can be lost or left unobserved at the completion of the study, resulting in data that provide no explicit information on their actual survival/failure times. In these situations, all that is known is that the observed lifetimes exceed some limiting value, the study's ending time. This phenomenon is known as **censoring**. If we know that an observation exhibits a survival/failure time that exceeds a limiting value, but is otherwise unknown, we say that observation is **right-censored**. Similarly, a **left-censored** observation exhibits a survival or failure time that transpired previous to the study's beginning. Combinations of both left and right censoring can occur in time-to-event data. If all the censoring times are fixed, such as in a study that ends at a fixed date or time, right censoring is then also called **time censoring**, or **Type I censoring**. (Type I *left* censoring is also possible, for example, if data are chemical concentrations that cannot be detected below a certain lower **detection limit**; see the introduction by Adams (1992).) Censoring can also be **random**; it can occur according to some biological or environmental process possibly related to the treatment under study, as with treatment-related toxicity in a carcinogenesis assay discussed in Section 6.3.2, or perhaps be unassociated with it, as with the incidental tumors in Example 8.7.

The most common form of censoring in environmental toxicology is fixed right censoring, and we will limit our attention to this scenario. As we see below, the presence of both observed and right-censored failure times in the data complicates hypothesis testing and parameter estimation in survival analysis. Computer routines and packages such as SAS or S-PLUS can assuage some of these concerns, although these details often are hidden from investigators performing survival analysis via computer. In this chapter, we will illustrate direct (where possible) and computer calculation of statistical quantities useful in survival-data analyses.

11.2 ESTIMATING THE SURVIVOR FUNCTION

11.2.1 Nonparametric estimation: The product-limit estimator

In its simplest form, the estimated survivor function at some time t is the proportion of experimental units still alive at this time if no censoring occurs. In the presence of censoring, not all survival times are observed, and we require special statistical methods for estimating $S(t)$. The most common method is called the Kaplan–Meier or **product-limit (PL) estimator** (Kaplan and Meier, 1958). To define the PL estimator, we introduce the following notation. Let $t_1, t_2, \ldots, t_M$ $(M \leq N)$ be the M distinct observed failure times from a set of N organisms or subjects. Thus no more than $N - M$ of the observations are right-censored. Order these observations from smallest to largest, and denote the ordered set as $t_{(1)} < t_{(2)} < \cdots < t_{(M)}$. (These are referred to as the **order statistics** from a random sample.) That is, the ith largest event time is denoted as $t_{(i)}$. Further, assume that d_i organisms die at time $t_{(i)}$. Let n_i be the number of organisms alive and uncensored immediately before time $t_{(i)}$, and assume that, if present, any censoring is independent of the underlying cause of death. Then, the PL estimator of the survivor function is

$$\hat{S}_{\mathrm{PL}}(t) = \prod_{\{i:t_{(i)}<t\}} \left\{1 - \frac{d_i}{n_i}\right\}, \tag{11.1}$$

where the product is taken over all times of death $t_{(i)}$ that are less than or equal to t, $i = 1, \ldots, M$. Intuitively, $1 - (d_i/n_i)$ should represent an estimate of the probability of surviving past time $t_{(i)}$ conditional on being alive after time $t_{(i-1)}$. Thus, the PL estimator may be viewed as a product of conditional survival probabilities. Notice that we assume no underlying parametric function or form for the true survival probability. Because of this, $\hat{S}_{\mathrm{PL}}(t)$ is viewed as a **nonparametric estimator** of $S(t)$.

Example 11.1 Acute toxicity data

Suppose an investigator is studying the toxic response of newborn laboratory rodents to acute exposure of heavy metals. Preliminary study of $N = 10$ independent animals after dietary exposure to cadmium at 2 mg/kg body weight yields $M = 5$ distinct death times (in days): $t_{(1)} = 1$, $t_{(2)} = 3$, $t_{(3)} = 4$, $t_{(4)} = 6$, $t_{(5)} = 8$. Two observed deaths occur at time $t = 4$ and one observed death occurs at each other time. Thus $d_1 = d_2 = d_4 = d_5 = 1$, and $d_3 = 2$. Further, four of the animals develop unrelated post-natal health problems, and are censored at times $t = 2, 4, 5, 9$, respectively.

The PL estimator calculations from (11.1) proceed for these data as given in Table 11.1. Note in the table that since all animals are alive and uncensored at the start of the study ($t = 0$), it is reasonable to take $\hat{S}_{\mathrm{PL}}(0) = 1$. (Indeed, setting the PL estimate to 1.0 at time $t = 0$ is a generally accepted default with uncensored or right-censored data.)

Table 11.1 PL calculations for acute toxicity data, using (11.1)

i	t	d_i	n_i	$\hat{S}_{PL}(t)$
–	$t < 1$	0	10	1
1	$1 \le t < 3$	1	10	$\left(1-\frac{1}{10}\right) = \frac{9}{10}$
2	$3 \le t < 4$	1	8	$\left(1-\frac{1}{10}\right)\left(1-\frac{1}{8}\right) = \frac{63}{80}$
3	$4 \le t < 6$	2	7	$\left(1-\frac{1}{10}\right)\left(1-\frac{1}{8}\right)\left(1-\frac{2}{7}\right) = \frac{9}{16}$
4	$6 \le t < 8$	1	3	$\left(1-\frac{1}{10}\right)\left(1-\frac{1}{8}\right)\left(1-\frac{2}{7}\right)\left(1-\frac{1}{3}\right) = \frac{3}{8}$
5	$8 \le t$	1	2	$\left(1-\frac{1}{10}\right)\left(1-\frac{1}{8}\right)\left(1-\frac{2}{7}\right)\left(1-\frac{1}{3}\right)\left(1-\frac{1}{2}\right) = \frac{3}{16}$

In Table 11.1, the only values of t at which $\hat{S}_{\mathrm{PL}}(t)$ changes are those $t_{(i)}$s where a death time is observed. By construction, $\hat{S}_{\mathrm{PL}}(t)$ remains constant between observed death times. This is reasonable, since in the absence of any explicit, functional model for $S(t)$, there is no information in the data regarding change in survival at other time points. Table 11.1 illustrates this step-function effect, where each interval of time starting at an observed death time and continuing up to (but not including) the next largest death time yields a constant value for $\hat{S}_{\mathrm{PL}}(t)$.

Censoring affects the PL estimator in (11.1) by changing the number at risk of dying in the next interval of time. At the start, all $n_1 = 10$ of the acute toxicity observations were available to 'fail' prior to time $t_{(1)} = 1$. After one observation failed at time $t_{(1)}$ and one observation was censored at time $t_{(2)} = 3$, however, only $n_2 = 10 - 2 = 8$ observations were candidates for failure at time $t_{(2)}$. The calculations for $\hat{S}_{\mathrm{PL}}(t)$ over $3 \le t < 4$ reflect this effect in Table 11.1.

The estimates in Table 11.1 can be calculated via computer, using, for example, the S-PLUS `survfit` function or SAS PROC LIFETEST. The results of applying the S-PLUS function to the acute toxicity data in Table 11.1 are displayed in Fig. 11.1. There, the expression

```
survfit(Surv(ttt,ttt.status)~1,...)
```

describes an S-PLUS function for estimating the survivor function. The `survfit` function calculates the PL estimator for a survival object created by `Surv(ttt,ttt.status)` for only one group, as denoted by the `1` on the right of the tilde. The failure and censoring times are stored in the object `ttt`; the vector `ttt.status` contains indicators of whether the

```
ttt<-c(1,3,4,4,6,8,2,4,5,9)
ttt.status<-c(1,1,1,1,1,1,0,0,0,0)
sfit<-survfit(Surv(ttt,ttt.status)~1,
       type="kaplan-meier",error="greenwood")
summary(sfit)
```

```
Call: survfit(formula = Surv(ttt, ttt.status) ~ 1)

 time n.risk n.event survival std.err lower 95% CI
    1     10       1    0.900  0.0949       0.7320
    3      8       1    0.788  0.1340       0.5641
    4      7       2    0.562  0.1651       0.3165
    6      3       1    0.375  0.1885       0.1400
    8      2       1    0.188  0.1627       0.0342
```

Fig. 11.1 S-PLUS `survfit` code and output for product-limit estimates with data in Table 11.1.

```
* SAS code to fit toxicity data from Table 11.1;
options linesize = 70;

data actox;
  input stime ievent @@;
  censflag = 1-ievent;
  cards;
1 1  2 0  3 1  4 1  4 1  4 0  5 0  6 1  8 1  9 0

proc lifetest;
  time stime*censflag(1);
```

Fig. 11.2 SAS LIFETEST code to find product-limit estimates for data in Table 11.1.

corresponding element `ttt` is a censoring time (`ttt.status` = 0) or an observed failure time (`ttt.status` = 1).

The SAS PROC LIFETEST code appears in Fig. 11.2. In the code, the `time` statement is required: it is used to indicate the variable containing the survival-time information, here `stime`. The asterisk (`*`) is optional, but if included it must precede a second variable that provides censoring information. In Fig. 11.2, this is the censoring variable `censflag`. The parenthetical detail after the censoring variable indicates to SAS the numerical values that correspond to right censorship. For example, in Fig. 11.2, `stime*censflag(1)` indicates that the variable `stime` contains the values of the survival or censoring time and that if `censflag`

```
                The SAS System
             The LIFETEST Procedure

        Product-Limit Survival Estimates
                              Survival
                              Standard    Number     Number
STIME      Survival   Failure   Error     Failed      Left
0.000       1.0000         0         0        0        10
1.000       0.9000    0.1000    0.0949        1         9
2.000*           .         .         .        1         8
3.000       0.7875    0.2125    0.1340        2         7
4.000            .         .         .        3         6
4.000       0.5625    0.4375    0.1651        4         5
4.000*           .         .         .        4         4
5.000*           .         .         .        4         3
6.000       0.3750    0.6250    0.1885        5         2
8.000       0.1875    0.8125    0.1627        6         1
9.000*           .         .         .        6         0
                 * Censored Observation
```

Fig. 11.3 SAS LIFETEST output (edited) of product-limit estimates for data in Example 11.1.

is 1, the value is right-censored and not a true observed survival time. The resulting output (edited for presentation) appears in Fig. 11.3.

Notice in Table 11.1 or Figs 11.1 and 11.3 that the survival probability estimates never reach zero: for $t \geq 8$, the survival probability is estimated at 3/16, or 18.75%. This is because the last observation is censored, rather than a true death time. (Technically, the survival estimate is 0.1875 at $t = 8$, and is actually undefined over $t > 8$. With no better estimate of survival available from the data, however, we estimate survival past $t = 8$ as *no more than* 18.75%.) If the last observation had been an observed death, the PL estimate at $t = 8$ would drop to zero, and would equal zero for all $t > 8$.

A useful and common way to represent survival probabilities is to graph $\hat{S}_{PL}(t)$ as a function of t. The plot is a step function, with changes occurring only at observed event times (death, failure, etc.). The S-PLUS function `plot` used with a survival object can plot PL estimates as part of its analyses. Alternatively, although not provided in our display, in PROC LIFETEST the `plots` option provides various survival graphics; for example, invoke PROC LIFETEST via

```
proc lifetest plots=(s,ls,lls);
```

to plot estimates of the survivor function (`s`), the log-survivor function (times –1) (`ls`), and the log-log survivor function (`lls`).

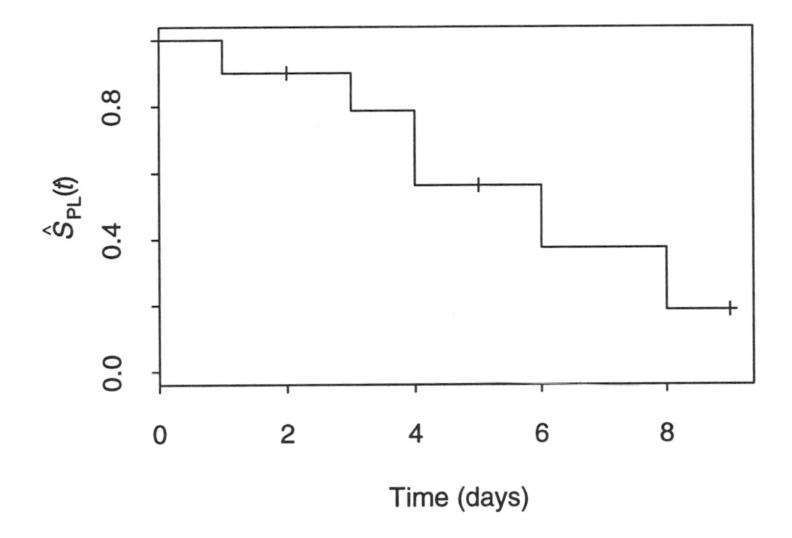

Fig. 11.4 Step-function form of nonparametric PL estimator for acute toxicity data from Table 11.1. Crosses (+) indicate right-censored observations (the cross at $t = 4$ is suppressed since this corresponds also to an event time).

Figure 11.4 displays the step-function effect for the data in Table 11.1. Notice the step-down nature of the plot, and the way the steps occur at only the observed $t_{(i)}$s. This is a common feature of any PL plot. It may seem unusual, since we assume generally that the underlying survivor function is some form of smooth curve. If the number of observed failure times were to increase, however, the staircase features of the PL plot would diminish, and the graph would approach a smooth curve. □

We can use the PL estimator in (11.1) to perform inferences on the underlying survival curve. Fundamental to any such effort is calculation of the standard errors $se[\hat{S}_{PL}(t)]$. For survival probabilities estimated using (11.1), a common method for calculating $se[\hat{S}_{PL}(t_i)]$ at the observed failure times t_i is due to Greenwood (1926). **Greenwood's formula** gives standard errors in terms of their squares, the estimated variances of $\hat{S}_{PL}(t_i)$:

$$\mathrm{Var}[\hat{S}_{PL}(t_i)] = \hat{S}^2_{PL}(t_i)\sum_{j=1}^{i} \frac{q_j}{(n_j)(p_j)}, \tag{11.2}$$

Table 11.2 Greenwood's formula (11.2) calculations for data from Table 11.1

i	–	1	2	3	4	5
t_i	$t<1$	$1\le t<3$	$3\le t<4$	$4\le t<6$	$6\le t<8$	$8\le t$
d_i	0	1	1	2	1	1
n_i	10	10	8	7	3	2
$q_i = d_i/n_i$	0.000	0.100	0.125	0.286	0.333	0.500
$p_i = 1 - q_i$	1.000	0.900	0.875	0.714	0.667	0.500
$\sum_{j=1}^{i} \frac{q_j}{n_j p_j}$	0.000	0.011	0.029	0.086	0.253	0.753
$\hat{S}_{PL}(t_i)$	1.000	0.9000	0.7875	0.5625	0.3750	0.1875
$\text{Var}[\hat{S}_{PL}(t_i)]$	1.000	0.0089	0.0180	0.0273	0.0356	0.0266
$se[\hat{S}_{PL}(t_i)]$	1.000	0.0944	0.1342	0.1651	0.1886	0.1631

where $q_j = d_j/n_j$ and $p_j = 1 - q_j$. From this, $se[\hat{S}_{PL}(t_i)]$ is taken as the square root of (11.2). Table 11.2 illustrates use of Greenwood's formula on the acute toxicity data from Example 11.1.

Once standard errors are computed, a confidence interval on the survival probability at time t_i may be calculated. An obvious choice is to assume large-sample normality for the PL estimator and calculate the Wald interval

$$\hat{S}_{PL}(t_i) \pm \mathfrak{z}_{\alpha/2}\, se[\hat{S}_{PL}(t_i)], \tag{11.3}$$

As in previous chapters, $\mathfrak{z}_{\alpha/2}$ is the upper-($\alpha/2$) critical point from the standard normal distribution. In large samples – we recommend $N \ge 50$ – the interval in (11.3) should cover the true value of $S(t_i)$ with approximate $1 - \alpha$ confidence.

Example 11.2 Confidence limits on survival in mice after exposure to 1,2-epoxybutane

In a laboratory experiment to study toxic effects of exposure to the chemical solvent stabilizer 1,2-epoxybutane, the US National Toxicology Program (NTP) exposed 50 male mice to 50 ppm of the chemical via inhalation for 104 weeks (NTP, 1988); a similar experiment with male rats was described in Exercise 4.20. The mortality experiences of the male mice were summarized via their survival times over the 104-week exposure period (in weeks): 5 mice died prior to the study's end at weeks 3, 89, 93, 98, and 100, and 45 mice were sacrificed at the end of the study (right-censored). Thus the PL estimator of $S(t)$ is straightforward to calculate (Exercise 11.9):

$$\hat{S}_{PL}(t) = \begin{cases} 1.0 & \text{if } 0 \le t < 3 \\ 0.98 & \text{if } 3 \le t < 89 \\ 0.96 & \text{if } 89 \le t < 93 \\ 0.94 & \text{if } 93 \le t < 98 \\ 0.92 & \text{if } 98 \le t < 100 \\ 0.90 & \text{if } 100 \le t\ . \end{cases}$$

If we were interested, for example, in the survival probability at the last death time (t = 100), we see $\hat{S}_{PL}(100) = 0.90$. For a single set of (pointwise) confidence limits on this survival estimate, $N = 50$ here is large enough to validate use of the standard large-sample confidence limits from (11.3). We find $se[\hat{S}_{PL}(100)] = \sqrt{0.2059} = 0.0424$, giving the 95% limits $\hat{S}_{PL}(t_i) \pm z_{\alpha/2} se[\hat{S}_{PL}(t_i)] = 0.9 \pm (1.96)(0.0424) = 0.9 \pm 0.0831$, or $0.8196 < S(100) < 0.9831$. □

One concern with use of the Wald confidence limits in (11.3) is that there is no guarantee they will satisfy the natural restriction $0 \le S(t) \le 1$. To correct this, a slightly more complex set of confidence limits is available, based on the same sort of quadratic construction that led to the confidence limits in equations (2.13), (4.22), and (4.28) (Wilson, 1927): solve a quadratic equation that relates the square of $\{\hat{S}_{PL}(t_i) - S(t_i)\}/se[\hat{S}_{PL}(t_i)]$ with the large-sample critical point $(z_{\alpha/2})^2$, and take as limits the resulting (two-value) solution. Applied to censored time-to-event data, the method was proposed by Rothman (1978), and leads to the following confidence limits on the survival probability at time t_i, $S(t_i)$:

$$\frac{\hat{S}_{PL}(t_i) + C_\alpha^2 \pm C_\alpha\left(2\hat{S}_{PL}(t_i)\left\{1 - \hat{S}_{PL}(t_i)\right\} + C_\alpha^2\right)^{1/2}}{1 + 2C_\alpha^2}, \tag{11.4}$$

where $C_\alpha^2 = (2\nu)^{-1} z_{\alpha/2}^2$, and ν is the **effective sample size**: $\nu = \hat{S}_{PL}(t_i)\{1 - \hat{S}_{PL}(t_i)\}/\text{Var}[\hat{S}_{PL}(t_i)]$. The denominator for ν is the Greenwood variance estimator (11.2). The limits in (11.4) are referred to as **Rothman–Wilson limits** for any $S(t_i)$. For moderate to large samples, where at least $N \ge 10$ but preferably $N \ge 25$, the limits cover the true value of $S(t_i)$ with approximate $1 - \alpha$ confidence (Anderson *et al.*, 1982).

It is important to recognize that both the Wald limits and the Rothman–Wilson limits for $S(t_i)$ are constructed for application to only a single survival probability at a specific failure time, t_i. We call these **pointwise confidence limits**, since they apply only at the single point t_i. For application to all of the observed event times, some adjustment for multiplicity is required. Although a number of multiple-comparison adjustments are possible, we will note only the two simplest: a Bonferroni correction to the critical point based on (5.2), and a Šidák correction similar to that described in Section 5.2. The Bonferroni correction is particularly simple (Afifi *et al.*, 1986): adjust the critical point via $C_\alpha^2 = (2\nu)^{-1}\mathfrak{z}^2_{\alpha/2K}$ to account for the construction of K simultaneous confidence intervals over the K distinct event times, $t_1, \ldots, t_K$. The Šidák correction is slightly more complex: assuming convergence to large-sample normality of $\hat{S}_{PL}(t)$, adjust the critical point to $C_{\tilde{\alpha}}^2 = (2\nu)^{-1}\mathfrak{z}^2_{\tilde{\alpha}/2}$, where $\tilde{\alpha}$ is the Šidák-adjusted value $\tilde{\alpha} = 1 - (1-\alpha)^{1/K}$. In large samples, both constructions provide approximate, simultaneous $1 - \alpha$ confidence, with the Šidák intervals slightly narrower than the Bonferroni intervals.

Example 11.1 (continued) Acute toxicity data
Return to the acute toxicity data in Table 11.1. In the table, the PL estimator of $S(t)$ was calculated as

$$\hat{S}_{PL}(t) = \begin{cases} 1.00 & \text{if } 0 \le t < 1 \\ 0.900 & \text{if } 1 \le t < 3 \\ 0.788 & \text{if } 3 \le t < 4 \\ 0.563 & \text{if } 4 \le t < 6 \\ 0.375 & \text{if } 6 \le t < 8 \\ 0.188 & \text{if } 8 \le t\,. \end{cases}$$

For the survival probability at the first death time ($t = 1$), the estimate is $\hat{S}_{PL}(1) = 0.90$. For a single set of (pointwise) confidence limits on $S(1)$, $N = 10$ here is low enough to require use of the Rothman–Wilson method in (11.4). We have $se[\hat{S}_{PL}(1)] = 0.0944$, so the effective sample size is $\nu = (0.9)(0.1)/(0.0944)^2 = 10.0995$. From these, and at 95% confidence, we find $C_{0.05}^2 = (2\nu)^{-1}\mathfrak{z}^2_{0.025} = (1.96)^2/20.1989 = 0.1902$. This gives the limits

$$\frac{0.9 + 0.1902 \pm (0.4361)\sqrt{(2)(0.9)(0.1) + 0.1902}}{1 + (2)(0.1902)},$$

that is, $(1.0902 \pm 0.2653)/1.3804$, or $0.598 < S(1) < 0.982$.

Of course, these limits hold with only pointwise 95% confidence. If there is interest in simultaneous 95% limits on $S(t)$ at, say, the $K = 2$ points $t = 1$ *and* $t = 3$, a Bonferroni adjustment changes the critical point to $C^2 = \mathfrak{z}^2_{0.0125}/2\nu = (2.241)^2/20.1989 = 0.2487$. At $t = 1$ this gives simultaneous limits

$$\frac{0.9 + 0.2487 \pm (0.4987)\sqrt{(2)(0.9)(0.1) + 0.2487}}{1 + (2)(0.2487)},$$

while at $t = 3$

$$\frac{0.788 + 0.2487 \pm (0.4987)\sqrt{(2)(0.788)(0.212) + 0.2487}}{1 + (2)(0.2487)}.$$

Simplified, we find $0.549 < S(1) < 0.985$ and $0.438 < S(3) < 0.947$, where both intervals hold with simultaneous 95% confidence. Notice that the interval for $S(1)$ has widened, reflecting the additional coverage provided by the simultaneous construction. □

It is also possible to construct distribution-free **simultaneous confidence bands** on $S(t)$, using the information in the PL estimator. Many forms of confidence band are possible, each possessing the feature that the bands cover the true value of $S(t)$ with $1 - \alpha$ simultaneous confidence over all values of t in some pertinent interval. (We describe the nature of the interval in greater detail below.) The confidence bands are based upon a core measure of variability from the Greenwood variance estimator (11.2):

$$\hat{\sigma}^2(t) = N \sum_{\{j: t_j < t\}} \frac{q_j}{n_j p_j},$$

where the notation under the summation symbol indicates summation over all values j such that $t_j < t$. From this, simultaneous bands on $S(t)$ for an interval of values for t are constructed by replacing the critical point $\mathfrak{z}_{\alpha/2}$ in (11.3) with an upward-adjusted value, $\mathscr{E}_\alpha \geq \mathfrak{z}_{\alpha/2}$, that incorporates the simultaneity (Nair, 1984). Notice that this approach yields bands whose widths are constant multiples of $se[\hat{S}_{\mathrm{PL}}(t)]$: $\hat{S}_{\mathrm{PL}}(t_i) \pm \mathscr{E}_\alpha se[\hat{S}_{\mathrm{PL}}(t_i)]$. This gives, in a sense, equal precision at all points t for which the bands are valid, and the construction is often referred to as an **equal-precision confidence band** for $S(t)$.

To ensure the distribution-free character of the equal-precision confidence statement using the adjusted critical point $\mathscr{E}_\alpha$, the bands are constructed to be valid only over an interval of values for t. This interval must be chosen to satisfy certain statistical requirements, and, as it turns out, the adjusted critical point is a function of this range. Specifics for the construction extend beyond the scope of this chapter – see Nair (1984) for greater detail – and we summarize only the necessary elements here. Begin with the function $K(t) = \hat{\sigma}(t)/\{1 + \hat{\sigma}(t)\}$, which reflects the degree of censoring in the observed data: if no censoring is evidenced, $K(t) = 1 - \hat{S}_{\mathrm{PL}}(t)$. We use $K(t)$ to isolate the values of t for which the confidence bands are valid. For user-specified values a and b, find all values of t such that

$$a \leq K(t) \leq b \,. \tag{11.5}$$

Then, the equal-precision construction will yield simultaneous confidence bands for $S(t)$ over the interval defined by (11.5). Since the values of a and b affect the location of the band, however, they also affect the critical point $\mathscr{E}_\alpha$. We acknowledge this dependence by denoting the critical point as $\mathscr{E}_{\alpha,a,b}$. Nair (1984, Table 2) and Weston and Meeker (1991) provide selected values of $\mathscr{E}_{\alpha,a,b}$ for the simple case $a = 1 - b$ at $a = 0.10$ and $a = 0.05$. (The former value is recommended for use in practice.) It is also possible to approximate $\mathscr{E}_{\alpha,a,b}$ numerically, by solving for $\mathscr{E}$ in the non-linear equation

$$\mathscr{E} \exp\left\{-\frac{\mathscr{E}^2}{2}\right\} \log\left\{\frac{(1-a)\,b}{a\,(1-b)}\right\} - \alpha\sqrt{2\pi} = 0\,.$$

The resulting simultaneous confidence bands have the form $\hat{S}_{\mathrm{PL}}(t) \pm \mathscr{E}_{\alpha,a,b}\, se[\hat{S}_{\mathrm{PL}}(t_i)]$, for all t such that (11.5) holds. Unfortunately, these bands suffer the same difficulty evidenced with (11.3): their upper or lower limits may violate the natural restriction $0 \leq S(t) \leq 1$. To correct this, it is recommended that the bands be transformed via some function that ensures containment of the simultaneous limits within the unit interval. We recommend the logit function, $\mathrm{logit}(s) = \log\{s/(1-s)\}$. This transforms $\hat{S}_{\mathrm{PL}}(t)$ to $\mathrm{logit}[\hat{S}_{\mathrm{PL}}(t)]$, modifies the variance of the associated equal-precision limits due to the transformation – that is, $\mathrm{Var}[\mathrm{logit}(\hat{S}_{\mathrm{PL}})] = se[\hat{S}_{\mathrm{PL}}(t_i)]^2/\{\hat{S}_{\mathrm{PL}}(1-\hat{S}_{\mathrm{PL}})\}$ – and then reverses the calculations to report confidence bands on the original, survival probability scale. This leads to distribution-free, simultaneous $1-\alpha$ confidence limits of the form

$$\frac{1}{1 + \exp\left\{-\log\left(\dfrac{\hat{S}_{\mathrm{PL}}(t)}{1-\hat{S}_{\mathrm{PL}}(t)}\right) \pm \mathscr{E}_{\alpha,a,b}\dfrac{se\left[\hat{S}_{\mathrm{PL}}(t)\right]}{\left\{\hat{S}_{\mathrm{PL}}(t)\left[1-\hat{S}_{\mathrm{PL}}(t)\right]\right\}^{1/2}}\right\}}$$

for all t satisfying (11.5). After back-transforming the logits, the confidence limits are now restricted to the unit interval, as intended. For sample sizes as low as 25 (and when censoring is not extreme), the logit transform has been shown to exhibit stable simultaneous coverage properties (Weston and Meeker, 1991).

Transformations for $S(t)$ other than the logit also are possible; for example, the arc sine/square root transform has been shown also to exhibit stable simultaneous coverage properties (Afifi *et al.*, 1986; Borgan and Liestøl, 1990). In fact, the equal-precision construction can be shown to reside within a larger family of confidence bands for $S(t)$, the relatives of which extend over a variety of forms; see Gulati and Padgett (1996), Hollander *et al.* (1997), and the references therein.

11.2.2 Parametric estimation

An alternative to use of the PL estimator of the survivor function is to fit a fully specified probability distribution to the data. When the specification is correct, statistical estimates of the survival probabilities will be more accurate than those based on the PL estimator. (If the specification is incorrect, the estimates can be badly biased.) Common models for survival data include the Weibull distribution (Weibull, 1951), including its special case, the exponential distribution, the lognormal distribution, and possibly the gamma distribution. (The gamma model is less common, since it is somewhat more difficult to implement (Lawless, 1982, Section 5.1), and also since it lacks robustness to model misspecification in selected applications (Sharma and Rana, 1991).) The c.d.f. associated with any of these distributional families, $F(t)$, is modeled as the complement of the survivor function, $F(t) = 1 - S(t)$, as noted in Section 11.1.

When applying parametric functions to model the survivor function, the question arises: What is the correct choice, or even a reasonable choice, for $F(t)$? The simplest way of evaluating this question is via a graphical display: plot some transformation of estimated survival versus some transformation of time. If this pattern is roughly linear, then the proposed model is reasonably consistent with the data. If not, then some other parent distribution for $F(t)$ may be appropriate. The choice of transformation for $S(t)$ and t is determined by the c.d.f. that is considered for $F(t)$. We illustrate these methods with the Weibull model in the next section.

11.2.3 The Weibull and exponential model

Suppose we believe a Weibull distribution is valid for a set of time-to-event data. From Section 1.3.3, the Weibull c.d.f. is $F(t) = 1 - \exp\{-\alpha t^{\gamma}\}$, which corresponds to the survivor function $S(t) = \exp\{-\alpha t^{\gamma}\}$. The corresponding hazard function is $h(t) = \alpha\gamma t^{\gamma-1}$ ($\alpha > 0$, $\gamma > 0$).

The two Weibull parameters α and γ allow for a variety of flexible distributional forms: α acts as a scale parameter, while γ acts as a shape parameter that determines the form and skew of the distribution. γ is viewed also as a **weak link parameter**, in the following sense: assume that the organism or system under study is comprised of γ compartments or components, and that failure in any one compartment leads to organism-wide failure. If, in this **multi-compartment/single-failure model**, it is also assumed that the failure rate (or **hazard rate**) in each compartment is a constant, α, then the Weibull model with parameters α and γ will fit the failure-time distribution (Nelson, 1982, Section 2.4).

For the Weibull model, transformation via logarithms simplifies $S(t)$, and helps in checking for model adequacy. Since $\log\{S(t)\} = -\alpha t^{\gamma}$, we see $\log[-\log\{S(t)\}] = \log(\alpha) + \gamma \log(t)$. Thus, if a Weibull distribution is correct, a complementary log-log transformation of $S(t)$ plotted versus $\log(t)$ should appear linear. Typically the PL estimate of $S(t)$ is used in diagnostic plots, so to examine the quality of a Weibull specification for the survival distribution, one plots $\log[-\log\{\hat{S}_{PL}(t_i)\}]$ vs. $\log(t_i)$, and checks if the resulting graph approximates a straight line. (Computer calculation is necessary for estimating the unknown Weibull parameters, although graphical inspection of $\log[-\log\{\hat{S}_{PL}(t_i)\}]$ vs. $\log(t_i)$ may be used to obtain initial values for the estimation process – see the `parameters` statement used in the S-PLUS code from Fig. 11.6, below). Figure 11.5 illustrates this approach for the acute toxicity data from Example 11.1. As can be seen from the figure, except perhaps for the earliest observation at $t_{(1)} = 1$, the Weibull distribution appears to be a reasonable parametric choice for the acute toxicity data.

The Weibull model is related to another model for failure-time data, the **extreme-value distribution**. If T is distributed as Weibull with parameters α and γ, then $\log(T)$ is distributed as a two-parameter extreme-value distribution. The extreme-value parameters are written as λ and δ; in terms of α and γ these are $\delta = 1/\gamma$ and $\lambda = -\gamma^{-1}\log(\alpha)$. This relationship becomes important as we consider the use of **accelerated failure-time models** to fit parametric survivor functions; see Section 11.4.

From Section 1.3, a special case of the Weibull distribution is the exponential distribution, with survivor function $S(t) = \exp\{-\alpha t\}$, which is a Weibull with $\gamma = 1$. Thus, to check for exponential failure rates in a set of time-to-event data, one can employ the Weibull method described above, and check if the complementary log-log vs. log plot exhibits a slope of approximately +1. (Or, equivalently, a plot of $\log\{S(t)\}$ vs. t will appear approximately linear for exponential failure rates.) If so, the exponential model is indicated.

It is also possible to extend the exponential model to incorporate a **guarantee time** before which no failures can occur (Exercise 11.12). Estimation under this **lagged exponential** model is more complex than for the simple one-parameter exponential form discussed here, however, and we refer interested readers to Lawless (1982, Chapter 3). Where censoring is

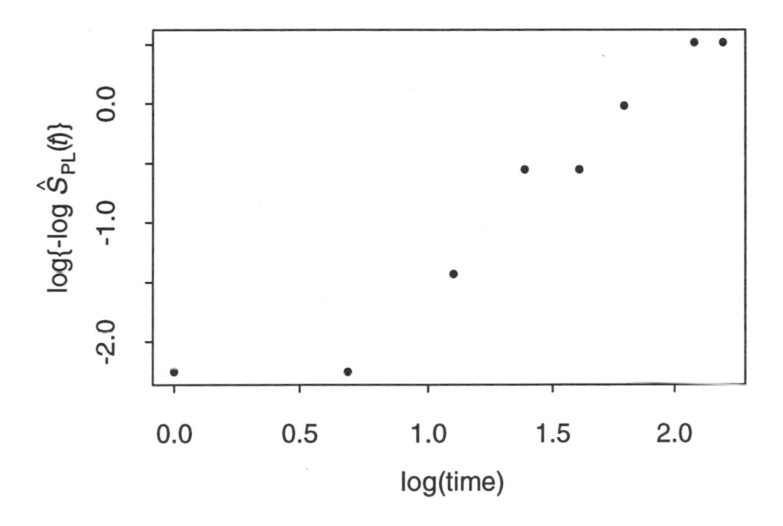

Fig. 11.5 Complementary log-log transformation of $S(t)$ plotted versus $\log(t)$ for acute toxicity data from Table 11.1.

included in the estimation process, the methods require further adjustment; see, for example, Piegorsch (1987b) or Fei and Kong (1994).

It is useful to recall here that the simple one-parameter exponential model exhibits no probabilistic 'memory' of preceding events. That is, the probability of observing a failure between any two time points is dependent only upon the amount of time elapsed between the failures. As in Section 1.3.2, we can see this mathematically by examining the survivor function: the probability of observing a failure after time t is $P[T > t] = \exp\{-\alpha t\}$. The conditional probability of an event occurring after time u, *given that it must occur after time* v $(u > v)$, is $P[T > u \mid T > v]$, which after application of the definition of conditional probability is $P[T > u]/P[T > v]$. Since $P[T > u] = \exp\{-\alpha u\}$, $P[T > u]/P[T > v]$ is $\exp\{-\alpha u + \alpha v\} = e^{-\alpha(u - v)}$. So, the conditional survivor probability depends on u and v only through the length of the time interval, $(u - v)$, and not on the specific values of u and v. In this sense, we say the distribution exhibits a lack of memory of previous events. For the more general Weibull model, however, it can be shown that this memoryless property is not present (Exercise 11.13).

The two parameters of the Weibull model, α and γ, are estimated using maximum likelihood. Recall from Section 2.3.3 that ML estimation requires construction and maximization of the likelihood function. For uncensored data, this is a straightforward calculation. Incorporating censoring in the likelihood, however, adds technical complexities. The following presentation summarizes these calculations, and is intended only for readers interested in acquiring a greater understanding of the procedural details.

As usual, we will work with the log-likelihood function. To express this function under right censoring, we introduce some additional notation. Among the N experimental units, let the total number of observed deaths be $D = \sum_{i=1}^{M} d_i$. We modify our indexing for the observations to distinguish the D units that die from those that are censored. Let u_j be an observed death time if $j = 1, \ldots, D$, and let u_j be a censoring time if $j = D + 1, \ldots, N$. (In effect, we have grouped the data into the first D observed deaths, and into the $N - D$ censoring times. Note, however, that some death times can occur after some censoring times; that is, a value of u_j over $j = 1, \ldots, D$ may nonetheless be larger than a value of u_j from $j = D + 1, \ldots, N$.) In the construction of the likelihood, each observed failure time contributes a survivorship component, $S(u_j)$, and a hazard component, $h(u_j)$, while a censored observation contributes only a survivorship component. For independent observations, the product of these terms forms the likelihood. Thus, the log-likelihood for the parametric survivor models is a sum of the individual log terms

$$\ell = \sum_{j=1}^{N} \log[S(u_j)] + \sum_{j=1}^{D} \log[h(u_j)] \ .$$

For the particular case of a Weibull parent distribution, and assuming that censoring is independent of the underlying cause of death, the log-likelihood under censoring is proportional to

$$\ell(\alpha, \gamma) = D\{\log(\gamma) + \log(\alpha)\} + \sum_{j=1}^{D} (\gamma - 1)\log(u_j) - \alpha \sum_{j=1}^{N} u_j{}^{\gamma}.$$

The ML estimates are found by maximizing $\ell(\alpha, \gamma)$ with respect to α and γ. This yields estimating equations for α and γ of the form

$$\alpha = \frac{D}{\sum_{j=1}^{N} u_j{}^{\gamma}} \ . \tag{11.6}$$

and

$$\frac{D}{\gamma} + \sum_{j=1}^{D} \log(u_j) = \alpha \sum_{j=1}^{N} u_j{}^{\gamma} \log(u_j) \ . \tag{11.7}$$

Solving (11.6) and (11.7) simultaneously yields ML estimates for α and γ.

Approximate standard errors for $\hat{\alpha}$ and $\hat{\gamma}$ may be derived using large-sample likelihood methods for censored data (Lawless, 1982, Appendix E; Nelson, 1982, Section 9.5), which are similar to the large-sample methods discussed in Sections 2.3 and 2.4. For example, to calculate the standard errors, we estimate the elements of the Fisher information matrix as

$$\mathfrak{F}_{\alpha} = D/\hat{\alpha}^2,$$

$$\mathfrak{F}_{\gamma} = \hat{\gamma}^{-2}D + \hat{\alpha}\sum_{j=1}^{N} u_j^{\hat{\gamma}}[\log(u_j)]^2,$$

and

$$\mathfrak{F}_{\alpha\gamma} = \sum_{j=1}^{N} u_j^{\hat{\gamma}}\log(u_j)\ .$$

With these quantities, approximate standard errors for the ML estimators are

$$se[\hat{\alpha}] = \left(\frac{\mathfrak{F}_{\gamma}}{\mathfrak{F}_{\alpha}\mathfrak{F}_{\gamma} - \mathfrak{F}_{\alpha\gamma}^2}\right)^{1/2}$$

and

$$se[\hat{\gamma}] = \left(\frac{\mathfrak{F}_{u}}{\mathfrak{F}_{\alpha}\mathfrak{F}_{\gamma} - \mathfrak{F}_{\alpha\gamma}^2}\right)^{1/2}.$$

If desired, the large-sample covariance of $\hat{\alpha}$ and $\hat{\gamma}$ may be calculated as well; it is $\mathrm{Cov}(\hat{\alpha},\hat{\gamma}) = \mathfrak{F}_{\alpha\gamma}/(\mathfrak{F}_{\alpha}\mathfrak{F}_{\gamma} - \mathfrak{F}_{\alpha\gamma}^2)$.

Approximate $1 - a$ confidence intervals on α and γ may be formed, again using large-sample likelihood methods. The Weibull parameters are assumed positive, however, and to incorporate this restriction into their confidence intervals we refer the *logarithms* of $\hat{\alpha}$ and $\hat{\gamma}$ to large-sample normal distributions (Nelson, 1982, Section 8.5). That is, when $\hat{\alpha}$ is normally distributed with mean α and variance $(se[\hat{\alpha}])^2$, then $\log(\hat{\alpha})$ can be shown to be approximately normally distributed with mean $\log(\alpha)$ and variance $(se[\hat{\alpha}]/\hat{\alpha})^2$. Using this fact to construct a confidence interval on $\log(\alpha)$, we then back-transform the endpoints to generate an interval for α. The result is an approximate $1 - a$ confidence interval for α of the form $\exp\{\log(\hat{\alpha}) \pm \mathfrak{z}_{\alpha/2}se[\hat{\alpha}]/\hat{\alpha}\}$. We operate in the same fashion to find confidence limits on γ. In both cases, the results simplify; for α we find

$$\hat{\alpha}\exp\left\{-\frac{\mathfrak{z}_{a/2}se[\hat{\alpha}]}{\hat{\alpha}}\right\} < \alpha < \hat{\alpha}\exp\left\{\frac{\mathfrak{z}_{a/2}se[\hat{\alpha}]}{\hat{\alpha}}\right\}, \qquad (11.8)$$

and for γ

$$\hat{\gamma}\exp\left\{-\frac{z_{\alpha/2}se[\hat{\gamma}]}{\hat{\gamma}}\right\} < \gamma < \hat{\gamma}\exp\left\{\frac{z_{\alpha/2}se[\hat{\gamma}]}{\hat{\gamma}}\right\}. \tag{11.9}$$

Notice that the intervals in (11.8) and (11.9) are pointwise in nature; that is, they do not adjust for multiplicity, even though they use overlapping data in their computation. Calculations for constructing simultaneous $1-\alpha$ confidence intervals for α and γ are explored in Exercise 11.15.

Example 11.1 (continued) Acute toxicity data
Consider fitting a Weibull model to the acute toxicity data in Table 11.1. ML estimators for α and γ are based on the Weibull estimating equations (11.6) and (11.7), and an S-PLUS program for fitting this model is illustrated in Fig. 11.6. The S-PLUS code employs the function `ms`, which minimizes a user-specified, nonlinear function. The user-specified function here is the *negative* Weibull log-likelihood (`weib.nllh2`). Employed together in Fig. 11.6, the routines yield parameter estimates that minimize the negative log-likelihood. If these quantities minimize the negative of a function, they must also maximize the original function (the log-likelihood), so we take the resulting values as the ML estimators. To calculate standard errors from the estimated variance-covariance matrix, the vector of first derivatives (`gradient` attribute) and the matrix of second derivatives (`hessian` attribute) also are calculated in `weib.nllh2`.

From the figure, we find $\hat{\alpha} = 0.0287$ and $\hat{\gamma} = 1.8644$. The associated survival predictions at a few select time points, along with the corresponding PL estimates, are given in Table 11.3. (The time points presented in the table for estimating $S(t)$ under the Weibull model are chosen from the mid-points of the time intervals in which the PL estimate was constant.) The fit is quite close to the nonparametric PL estimates at these points, supporting our earlier indication of a reasonably good fit for the Weibull model.

In Fig. 11.7, a plot of the estimated Weibull survival curve for these data is superimposed on the previously estimated PL survival curve. The curves

Table 11.3 Weibull model estimates* of $S(t)$ for data from Table 11.1

i	*Interval*	$\hat{S}_{PL}(t)$	t	$\hat{S}(t_i) = exp\{-\hat{\alpha}t^{\hat{\gamma}}\}$
–	$t < 1$	1.000	0.5	0.992
1	$1 \le t < 3$	0.900	2.0	0.901
2	$3 \le t < 4$	0.788	3.5	0.743
3	$4 \le t < 6$	0.563	5.0	0.561
4	$6 \le t < 8$	0.375	7.0	0.339
5	$8 \le t$	0.188	8.0	0.250

* Weibull model ML estimates are $\hat{\alpha} = 0.0287$, $\hat{\gamma} = 1.8644$.

are quite similar, again corroborating the reasonable fit of the Weibull model.

From Fig. 11.6, $se[\hat{\alpha}] = 0.0343$ and $se[\hat{\gamma}] = 0.614$. These lead to approximate 95% confidence intervals based on (11.8) and (11.9) of $0.003 < \alpha < 0.299$ and $0.978 < \gamma < 3.555$, respectively. Notice that the confidence interval for γ contains $\gamma = 1$, so the exponential model may be a valid, simpler form for the distribution of death times with these data. □

In the case of no censoring, $D = N$. Then, the Weibull ML estimates may be used to construct simultaneous $1 - a$ confidence bands on the Weibull survivor function, $\exp\{-\alpha t^{\gamma}\}$, over all $t > 0$. The construction is discussed in detail by Srinivasan and Wharton (1975, Table 3).

A number of computer packages and source codes are available for estimating survival parameters that do not require explicit coding of the log-likelihood (Dixon and Newman, 1991). Many of these include applications for the Weibull model. For example, SAS PROC LIFEREG or the S-PLUS function `survreg` can fit a Weibull model to failure-time data, although the user should be cautioned that both provide parameter estimates in terms of the extreme-value parameters λ (intercept) and δ (scale) and not in terms of α and γ.

If it can be assumed that the simpler, one-parameter exponential model ($\gamma = 1$) is valid for describing the failure times, then the calculations for ML estimation simplify greatly. Recall that the one-parameter exponential survivor function is $S(t) = \exp(-\alpha t)$. Under this model, estimating equation (11.6) for α remains valid, with $\gamma = 1$. The standard error and confidence limits for α simplify, however. For example, an approximate $1 - a$ confidence interval for α is

$$\hat{\alpha}\exp\left\{-\frac{z_{a/2}}{\sqrt{D}}\right\} < \alpha < \hat{\alpha}\exp\left\{\frac{z_{a/2}}{\sqrt{D}}\right\}$$

where D is the number of observed deaths (Nelson, 1982, Section 8.5.2).

11.2.4 The lognormal model

Another common parametric form used for modeling time-to-event data is the lognormal distribution from Section 1.3.4. Recall that a lognormal random variable, U, when transformed via $\log(U)$ becomes a normally distributed random variable. Thus, many connections exist between the log-normal and normal distributions, as we will see in the formulas below. For the sake of brevity, however, we present here only the most basic results for

```
ttt.df<-data.frame(time=ttt,status=ttt.status)
parameters(ttt.df)<-list(alpha=.3,gam.par=1)
# Weibull neg.-log-likelh. with gradient and hessian ...
weib.nllh2<-function(status,t.time,alpha,gam.par) {
   DD<-sum(status)
   t.time.dead<-t.time[status==1]
   value<- -DD*log(alpha) -DD*log(gam.par) -
            (gam.par-1)*sum(log(t.time.dead)) +
            alpha*sum(t.time^gam.par)
   attr(value,"gradient") <- c(
     -DD/alpha + sum(t.time^gam.par),
     -DD/gam.par - sum(log(t.time.dead)) +
        alpha*sum(log(t.time)*t.time^gam.par))
   attr(value,"hessian") <- rbind(
     c(DD/(alpha^2),sum(log(t.time)*t.time^gam.par)),
     c(sum(log(t.time)*t.time^gam.par),
       DD/(gam.par^2) +
       alpha*sum( (log(t.time))^2 * t.time^gam.par )))
   value
}
w.fit2 <- ms(~weib.nllh2(status=ttt.df$status,
              t.time=ttt.df$time,alpha,gam.par),
              start=list(alpha=.1,gam.par=1.0),
              data=ttt.df)
w.fit2
```

```
value: 16.87063
parameters:
      alpha  gam.par
 0.02871909 1.864433
```

```
w.hess<-w.fit2$hessian
w.hess
```

```
           [,1]     [,2]
[1,] 7274.6137  0.00000
[2,]  381.6464 22.67202
```

continued

```
w.hess[1,2]<-w.hess[2,1]   # set up diag. element since
                           # lower diag. is returned
                           # for the hessian

w.vcov<- solve(w.hess)     # estim'd var.-covar. matrix

se.alpha<- sqrt(w.vcov[1,1])
se.gam.par<- sqrt(w.vcov[2,2])
se.alpha
```

```
[1] 0.03429527
```

```
se.gam.par
```

```
[1] 0.6143192
```

Fig. 11.6 S-PLUS code to calculate the (negative) log-likelihood under a Weibull parent distribution and perform ML estimation with the acute toxicity data in Example 11.1.

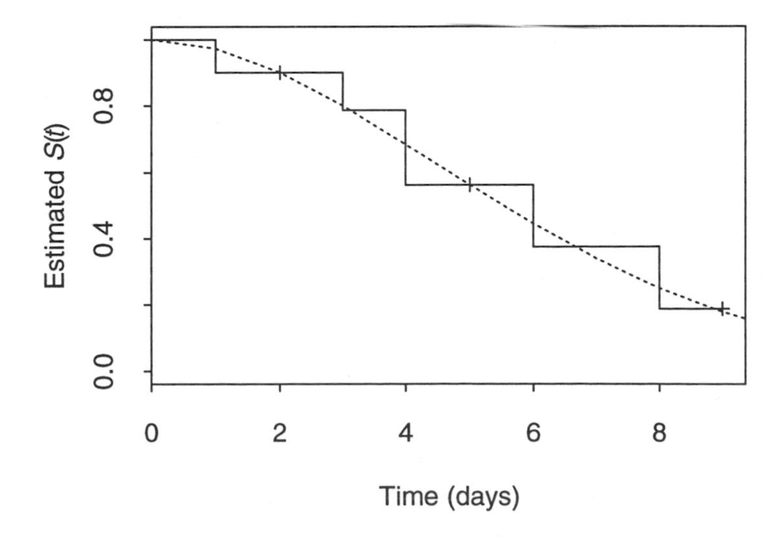

Fig. 11.7 Weibull estimate (---) and nonparametric PL estimator (—) for the acute toxicity data from Table 11.1. Crosses (+) indicate right-censored observations.

time-to-event modeling under the lognormal model. More detail can be found in Nelson (1982, Section 8.2) or Lawless (1982, Section 5.2).

The survivor function for the lognormal model is $S(u) = 1 - \Phi\{[\log(u) - \mu]/\sigma\}$, where μ and σ^2 are the mean and variance, respectively, of $\log(U)$, and $\Phi(\cdot)$ is the standard normal c.d.f. from Section 1.3.4. The hazard function, $h(u) = -\partial\log\{S(u)\}/\partial u$, is found in Exercise 11.2; $h(u)$ can be useful for modeling a nonmonotone hazard rate. To check for lognormal failure rates in a set of time-to-event data, plot $\Phi^{-1}\{1 - S(u)\}$ vs. $\log(u)$. The plot will appear linear if the lognormal assumption is valid.

To estimate the lognormal parameters μ and σ, let u_j be an observed failure time if $j = 1, \ldots, D$, and let u_j be a censoring time if $j = D+1, \ldots, N$. D is the total number of failures. The log-likelihood under the lognormal model is proportional to

$$\ell(\mu,\sigma) = \sum_{j=1}^{D} \log\{\phi(z_j)/\sigma\} + \sum_{j=D+1}^{N} \log\{1 - \Phi(z_j)\},$$

where, for simplicity, $z_j = (u_j - \mu)/\sigma$, and $\phi(z)$ is the standard normal p.d.f.: $\phi(z) = (2\pi)^{-1/2}\exp\{-z^2/2\}$. The ML estimators of μ and σ are found by maximizing $\ell(\mu,\sigma)$ with respect to each parameter. Unfortunately, no closed-form solutions exist for these estimators, and the corresponding estimating equations must be solved iteratively. The estimating equations are, for μ,

$$\sum_{j=D+1}^{N} h(z_j) + D\frac{\bar{u}^* - \mu}{\sigma} = 0, \tag{11.10}$$

and for σ,

$$\sum_{j=D+1}^{N} z_j h(z_j) - D + \frac{D}{\sigma^2}\left\{(s^*)^2 + (\bar{u}^* - \mu)^2\right\} = 0. \tag{11.11}$$

In (11.10) and (11.11), $h(z)$ is the standard normal hazard function, $h(z) = \phi(z)/\{1 - \Phi(z)\}$. Also, $\bar{u}^*$ is the sample mean of the censored observations, $\bar{u}^* = \sum_{j=D+1}^{N} u_j/D$, and $(s^*)^2 = \sum_{j=D+1}^{N}(u_j - \bar{u}^*)^2/D$.

11.2.5* Quantile estimation

When desired, a survivor function estimator allows for estimation of a summary measure of the time-to-event data, the **median event time**. When the event of interest is survival, we often refer to the **median survival time** or **median lethal time**, or LT_{50}. The LT_{50} is defined as that value of t for which the survival reaches 50%. (Notice the similarity to the LD_{50} and

ED_{50} from Exercise 3.16 and Section 7.3.1.) The LT_{50} is used as a summary response measure when comparing multiple observed survival patterns. (As we describe in Sections 11.3 and 11.4, however, there are better, more formal methods for comparing estimated survivor functions.) An estimate of median survival is calculated using the PL estimate from (11.1): take LT_{50} as the lowest value of t at which $\hat{S}_{PL}(t)$ first reaches or drops through 0.5. If the curve never reaches 0.5, then set LT_{50} to ∞, or, equivalently, view it as undefined.

If a parametric model is assumed for the survival times, then to estimate LT_{50} find a value of t such that $\hat{S}(t)$ equals 50%. For example, under a Weibull model, $\hat{S}(t) = \exp\{-\hat{\alpha}t^{\hat{\gamma}}\}$. Setting this to 0.5 and solving for t yields the ML estimator $\hat{LT}_{50} = [-\log(0.5)/\hat{\alpha}]^{1/\hat{\gamma}}$.

More generally, any ***Q*th quantile** of a survivor function may be estimated as the time point at which Q% of the experimental units have died (or failed, etc.). The LT_{50} corresponds to $Q = 50$. In similar fashion to the LT_{50}, we estimate any LT_Q as the earliest time t for which the estimated survival reaches or drops through Q%.

For inferences on LT_{50}, the standard error, $se[\hat{LT}_{50}]$, involves complicated nonlinear combinations of the parameter estimates, particularly under parametric assumptions such as a Weibull distribution for the failure times. These technical details are beyond the scope of this chapter. We refer the reader to Collett (1994, Appendix B) for details on standard error calculations for estimated quantiles, or to Anderson *et al.* (1982) for discussions on confidence intervals for survivor curve quantiles. Computer calculation of $1 - \alpha$ confidence limits on any LT_Q is possible, however, using packages such as SAS or S-PLUS. This is illustrated in the next example.

Example 11.1 (continued) Acute toxicity data

Continuing with the acute toxicity data from Table 11.1, the PL results assign to the interval $4 \le t < 6$ the estimated survival probability $\hat{S}_{PL}(t) = 0.563$, while the interval $6 \le t < 8$ has $\hat{S}_{PL}(t) = 0.375$. Thus, over the interval $6 \le t < 8$, $\hat{S}_{PL}(t)$ drops through 0.50. As such, we estimate LT_{50} as the lowest value in this interval: $\hat{LT}_{50} = 6$.

If survival time is modeled as Weibull, with $S(t) = \exp\{-\alpha t^{\gamma}\}$, we find $\hat{LT}_{50} = [-\log(0.5)/\hat{\alpha}]^{1/\hat{\gamma}} = [0.693/0.0287]^{1/1.8644} = 5.516$. Notice the proximity of the Weibull estimate to the PL estimate.

Many computer programs give confidence intervals for the quantiles of the $\log(T)$ distribution, under both parametric and nonparametric assumptions on the parent distribution. Figure 11.8 depicts use of SAS PROC LIFETEST and PROC LIFEREG to generate nonparametric and parametric estimates, respectively, of any LT_Q. (The parametric calculations in PROC LIFEREG

```
* SAS code to fit toxicity data from Table 11.1;
data actox;
  input stime ievent @@;
  censflag=1-ievent;
  cards;
  1 1  2 0  3 1  4 1  4 1  4 0  5 0  6 1  8 1  9 0
proc lifetest;
 time stime*censflag(1);
proc lifereg;
model stime*censflag(1)=/distribution=Weibull;
output out=wout quantiles=.1 .5 .9 std=std p=predtime;
data wout1; set wout;
 ltime=log(predtime);
 stde=std/predtime;
 uclw= exp(ltime+1.96*stde);     * upper limit on LT(Q);
 lclw= exp(ltime-1.96*stde);     * lower limit on LT(Q);
proc print;
```

Fig. 11.8 SAS PROC LIFETEST and PROC LIFEREG code for calculating event-time quantiles for acute toxicity data in Example 11.1.

for LT_{10}, LT_{50}, and LT_{90} are based upon a Weibull model.)

Applied to the data in Table 11.1, PROC LIFETEST from Fig. 11.8 gives a nonparametric estimate for the LT_{50} of 6, which is similar to that reported above based on nonparametric inversion of $\hat{S}_{PL}(t)$. Alternatively, using PROC LIFEREG in Fig. 11.8 to calculate a parametric estimate of LT_{50} based on a Weibull model gives an estimated LT_{50} of 5.5156 (as expected), with an approximate 95% confidence interval of $3.566 < LT_{50} < 8.530$. □

11.3 NONPARAMETRIC METHODS FOR COMPARING SURVIVAL CURVES

As noted in the previous section, formal comparison of observed survival patterns between two independent groups may be performed via statistical analysis. In this section, we present a formal statistical test for evaluating the equivalence of two survival curves. (We defer discussion on comparing more than two survival curves to Section 11.4, where we consider regression models for survival-time modeling.)

Perhaps the most common method for testing the equivalence of two independent survivor functions is the **log-rank test** (Mantel, 1966), which is based on information in each separate survival pattern's PL estimator. In this sense, the log-rank test is a nonparametric methodology, and it may be viewed as a generalization of common two-sample nonparametric statistical test procedures. The test is also known as the **Cox–Mantel test**, or sometimes the **Mantel–Haenszel test** for comparing survival curves, although these appellations are used to describe more general methods as well. In order to avoid additional confusion, we refer to the procedure simply as the log-rank test.

The data layout for the nonparametric log-rank test is given in Table 11.4. In the table, $t_{(1)}, \ldots, t_{(M)}$ are the ordered death times pooled over the two groups, d_{ij} is the number of deaths occurring at time $t_{(j)}$ in group i, and n_{ij} is the number of experimental units at risk just before time $t_{(j)}$, $i = 1,2$; $j = 1, \ldots, M$. (The '+' sign in the subscript represents summation over that index. For example, the total number of observed deaths at the second death time is $d_{+2} = d_{12} + d_{22}$.) This schematic is often referred to as a **life table**, since it corresponds to the structure used in actuarial studies of lifetime data (Lawless, 1982, Section 2.2); life tables similar to Table 11.4 – many authors also include the estimated survival probabilities – date back to Halley (1693).

Table 11.4 Data schematic for testing equivalence of survival patterns between two independent groups

Group	*Status*	$t_{(1)}$	$t_{(2)}$	...	$t_{(M)}$	*TOTAL*
$i = 1$	Die (d_{1j})	d_{11}	d_{12}	...	d_{1M}	d_{1+}
	At risk (n_{1j})	n_{11}	n_{12}	...	n_{1M}	n_{1+}
$i = 2$	Die (d_{2j})	d_{21}	d_{22}	...	d_{2M}	d_{2+}
	At risk (n_{1j})	n_{21}	n_{22}	...	n_{2M}	n_{2+}
TOTAL	Die (d_{+j})	d_{+1}	d_{+2}		d_{+M}	d_{++}
	At risk (n_{+j})	n_{+1}	n_{+2}	...	n_{+M}	n_{++}

The log-rank test statistic employs the information in the life table to compare number of units at risk between each group at every observed event time. Under a null hypothesis of survival equivalence between groups, the test postulates that the expected number of deaths at any event time is proportional to the corresponding number at risk. One can envision building a 2×2 sub-table of counts (cf. Section 4.1.7) at each time point $t_{(j)}$, where the row category separates those d_{ij} subjects that failed from those that did not fail (among the n_{ij} subjects at risk), and the column category distinguishes the two groups under study. That is, similar to Table 4.1a:

At time point $t_{(j)}$:	*Group 1*	*Group 2*	*TOTAL*
Failures during time interval	d_{1j}	d_{2j}	d_{+j}
Number not failing	$n_{1j} - d_{1j}$	$n_{2j} - d_{2j}$	$n_{+j} - d_{+j}$
Number at risk at $t = t_{(j)}$	n_{1j}	n_{2j}	n_{+j}

As with any such 2×2 table, the expected number of failures (corresponding to the first row) is $e_{ij} = n_{ij}d_{+j}/n_{+j}$, $i = 1,2$; $j = 1, \ldots, M$.

To construct the log-rank statistic, mimic the χ^2 pattern for goodness-of-fit seen in Section 9.3, and compare the observed numbers of deaths, d_{ij}, with their expectations under survival equivalence, e_{ij}, using the familiar quantities $(d_{ij} - e_{ij})^2/e_{ij}$. Combined over all M death times, the statistic is

$$T^2_{\text{calc}} = \frac{\left\{\sum_{j=1}^{M}(d_{1j} - e_{1j})\right\}^2}{\sum_{j=1}^{M}\frac{n_{1j}\, n_{2j}\, d_{+j}(n_{+j} - d_{+j})}{n^2_{+j}(n_{+j} - 1)}}. \tag{11.12}$$

(Notice that in its final form, T^2_{calc} employs only the expected numbers, e_{1j}, from group 1; the values of e_{2j} are not required in the calculation.) In large samples – at least 10 experimental units in each group – the log-rank statistic T^2_{calc} is referred to a χ^2 distribution with 1 *df*. Thus, an approximate P-value for testing survival equivalence is $P \approx \text{P}[\chi^2(1) \geq T^2_{\text{calc}}]$. SAS PROC LIFETEST performs the log-rank test as one of its standard survival analysis routines.

Example 11.3 Spider departure data
As part of a larger study of arachnid ecology, an experiment was conducted to compare the time elapsed before departure from a perch for two species of spider: crab spiders, *Misumenops oblongus*; and jumping spiders, *Metaphidippus galathea*. The independent groups of spiders were placed on a perch in a 25°C chamber and the time until each spider departed the perch was recorded. The ordered departure times (in seconds) were:

Crab spiders: 21, 30, 31, 45, 55, 58, 73, 81, 105, 107
Jumping spiders: 20, 22, 29, 32, 40, 42, 44, 46, 51, 57

Note that these departure times are all observed (there are no censored observations), and that there are no ties. To calculate the log-rank test statistic (11.12) for these data, they are first tabulated as in the data layout from Table 11.4; this is illustrated in Table 11.5. Next, the expected counts

Table 11.5 Portion of spider departure data in schematic form of Table 11.4

Group	*Status*	$t_{(1)} = 20$	$t_{(2)} = 21$	⋯	$t_{(20)} = 107$	*TOTAL*
Jumping	Depart (d_{1j})	1	0	⋯	0	10
	At risk (n_{1j})	10	9	...	0	82
Crab	Depart (d_{2j})	0	1	⋯	1	10
	At risk (n_{2j})	10	10	...	1	128
TOTAL	Depart (d_{+j})	1	1	⋯	1	20
	At risk (n_{+j})	20	19	...	1	210

Table 11.6 Log-rank statistic (11.12) calculations for spider departure data from Table 11.5

j	d_{1j}	$e_{1j} = n_{1j}\dfrac{d_{+j}}{n_{+j}}$	$d_{1j} - e_{1j}$	$\dfrac{n_{1j}n_{2j}d_{+j}(n_{+j} - d_{+j})}{n^2_{+j}(n_{+j} - 1)}$
1	1	$(10)\left(\frac{1}{20}\right) = 0.50$	0.500	0.250
2	0	$(9)\left(\frac{1}{19}\right) = 0.47$	−0.474	0.249
⋮	⋮	⋮	⋮	⋮
20	0	0	0	0
TOTAL	10	–	4.217	3.347

$e_{ij} = n_{ij}d_{+j}/n_{+j}$ are calculated; see Table 11.6. From these, the value of the test statistic is $T^2_{\text{calc}} = (4.217)^2/3.347 = 5.313$. Referring T^2_{calc} to a $\chi^2(1)$ distribution yields a P-value of $P \approx 0.0212$. At $\alpha = 0.05$, a significant difference in the two groups' departure patterns is indicated.

The spider data are reanalyzed using SAS PROC LIFETEST in Figs 11.9 and 11.10. The use of the `strata` statement in PROC LIFETEST separates the survivor function estimates for the two spider species and then later calculates the log-rank statistic for the groups delineated by the classification variable (`species`) in this statement.

The value of the log-rank statistic is displayed in Fig. 11.10 under `Test of Equality over Strata`; see `Log-Rank`. The results in the figure replicate the calculations illustrated above (to two significant digits). Two other statistics, an LR statistic and a Wilcoxon statistic, are also provided. The likelihood ratio statistic tests for equality of survivor functions assuming an exponential parent distribution. The Wilcoxon test is a distribution-free competitor to the log-rank statistic, but is less appropriate with these data. We discuss the Wilcoxon statistic briefly after this example.

```
* SAS code to fit spider data for Example 11.3;
options linesize = 70;

data spider;
  input species $ etime @@;
  cards;
jump 20  jump 22  jump 29  jump 32  jump 40
jump 42  jump 44  jump 46  jump 51  jump 57
crab 21  crab 30  crab 31  crab 45  crab 55
crab 58  crab 73  crab 81  crab 105 crab 107

proc lifetest;
  strata species;
  time etime;
```

Fig. 11.9 SAS PROC LIFETEST code for nonparametric tests of the equality of survivor functions for spider departure data.

```
                 The SAS System
              The LIFETEST Procedure

Testing Homogeneity of Survival Curves over Strata
               Time Variable ETIME

                 Rank Statistics
        SPECIES        Log-Rank      Wilcoxon
        crab            -4.2169       -46.000
        jump             4.2169        46.000

            Test of Equality over Strata
    Test        Chi-Square     DF    Pr > Chi-Square
    Log-Rank        5.3121      1        0.0212
    Wilcoxon        3.1209      1        0.0773
    -2Log(LR)       1.0436      1        0.3070
```

Fig. 11.10 SAS PROC LIFETEST output (edited) from code in Fig. 11.9.

Also, since there was no censoring with these data, other two-sample tests could be applied here. We explore some of these in Exercise 11.27. □

We should note in Example 11.3 that the number of experimental units used for the analysis was low: 10 per sample. Per-group sample sizes should be at least this large to validate use of the χ^2 approximation for T^2.

With small samples it is possible to construct an exact (conditional) test using the log-rank statistic or its positive square root $z_{calc} = \sqrt{T^2_{calc}}$, based on the hypergeometric distribution associated with the 2×2 sub-table of counts at each time point $t_{(j)}$. This is similar to the approach employed with

Fisher's exact test in Section 4.1.7. The computational aspects of this method are quite complex, however, and are beyond the scope of this chapter. Details of the method are available in Soper and Tonkonoh (1993).

Two additional remarks about nonparametric comparisons of survival curves are in order. First, the log-rank test using (11.12) weights differences between the observed (d_{1j}) and expected (e_{1j}) deaths or events equally. No differentiation is made between deaths occurring at times when a large number are at risk and times when a small number are at risk. When very large disparities exist between the numbers at risk, one can modify (11.12) by weighting the observed–expected differences ($d_{1j} - e_{1j}$) by the numbers at risk (n_{+j}). (Other weights might be considered as well.) This is known as the **Wilcoxon test** generalization of the log-rank statistic (Peto and Peto, 1972). The Wilcoxon test is useful if differences between the two groups' failure times are expected early in time, with smaller differences later. By contrast, the log-rank test is more sensitive when the survival/failure rate in one group is a constant multiple of the rate in the other group. (This is called a **proportional hazards assumption**, and is discussed in Section 11.4.2). The data analyst must make the choice of which method to use *before* viewing the data, based on indications from the scientific literature or from previous studies. Lawless (1982, Section 8.2.4) gives more detail on these (and other) methods for comparing survival curves.

Second, both the log-rank and Wilcoxon tests can be extended to testing the equality of more than two survival curves. These involve multivariate generalizations of the corresponding two-sample tests. We will see, however, that the regression models for survival analysis discussed in the next section may be applied to the problem of comparing multiple survival curves, and hence we will not discuss multi-group extensions in any further detail. Interested readers may refer to Collett (1994), Lawless (1982), Peto and Peto (1972), and the references therein.

11.4 REGRESSION MODELS FOR SURVIVAL DATA

When a set of predictor variables or other forms of covariate are included in a comparison of survival curves, it is natural to consider some form of regression model for the analysis. Two general classes of regression models are common in the analysis of time-to-event data. In both classes, time is modeled as a function of the predictor variables/covariates. The first class is known as the **accelerated failure-time (AFT) model**, where the logarithm of survival time is modeled as a linear function of the covariates, or equivalently, the covariates are modeled to act in a multiplicative fashion on survival time. This is similar to the log-linear models described for Poisson regression in Sections 7.2.2 and 8.4.4. The AFT class represents a fully parametric modeling scheme, in that a parametric probability distribution for survival time, T, must be specified in order to estimate and assess the covariates' effects on log-survival.

The second class of survival regression models is the **proportional hazards (PH) model,** where the hazard function associated with a set of predictor variables is modeled as a multiple of some baseline hazard. The PH models may be fully parametric, with the form of the baseline hazard specified as a complete parametric function, or **semi-parametric**, where the form of the baseline hazard is left unspecified. In both the parametric and semi-parametric forms of the PH model, the covariates are taken to act in a multiplicative fashion on the hazard function. In this section, we describe both of these models in detail.

11.4.1 Accelerated failure-time models

The general AFT model takes a log-linear form,

$$\log(t_j) = \beta_0 + \beta_1 X_{1j} + \cdots + \beta_P X_{Pj} + \sigma\varepsilon_j, \tag{11.13}$$

where t_j is survival time associated with the jth experimental unit, and the linear predictor includes P unknown regression coefficients $[\beta_1 \ldots \beta_P]$ corresponding to P predictor variables $[X_1 \ldots X_P]$ for each unit, a scale parameter σ, and a random error term ε_j. The error term is assigned a pre-specified parent distribution. Typically, this is taken as some standardized form, with zero mean and unit variance or unit scale. The intercept parameter β_0 may be related to a location or scale parameter in this parent distribution, while σ is often related to a scale or shape parameter. Also, in (11.13), β_0 is interpreted as the background log-survival rate, associated with subjects whose predictor variables all equal zero: $X_1 = \cdots = X_P = 0$. This need not be a realistic setting for the covariates; it is simply a useful reference point at which to anchor β_0.

The AFT model implies that the survivor functions for two different groups, say $S_1(t)$ and $S_2(t)$, are related via $S_1(t) = S_2(kt)$, where the value k is viewed as an **acceleration constant**. This is a consequence of the multiplicative effects on survival time the model implies.

We also assume that the scale parameter, σ, is constant, although possibly unknown. It is possible to model σ as a function of β_0, of the covariates, or of both, but we will not discuss this here. The approach is known as **functional dispersion modeling**; see Anderson (1991) for an example.

To estimate the $P + 2$ parameters, $\beta_0, \beta_1, \ldots, \beta_P, \sigma$, we apply maximum likelihood. A positive estimate, $b_k > 0$, of a parameter β_k associated with covariate X_k suggests that increases in the value of X_k are related to increased survival time; negative estimates imply decreased survival time. (As usual, these interpretations assume all other predictor variables, $X_1, X_2, \ldots, X_{k-1}, X_{k+1}, \ldots, X_P$, are held constant.) As seen previously with other parametric models, ML estimation requires construction and maximization of a likelihood function. (Selected aspects of the ML fit are presented here for readers interested in a greater understanding of the technical details. Others may wish to skip forward to the continuation of Example 11.1.) We write the observed failure times as t_j when $j = 1, \ldots, D$, and we take t_j as the observed censoring times when $j = D + 1, \ldots, N$. (As above, D is the total number of observed failures.) For simplicity, let

$$z_j = \frac{\log(t_j) - \{\beta_0 + \beta_1 X_{1j} + \cdots + \beta_P X_{Pj}\}}{\sigma}$$

be the jth standardized log-survival time. For any parent distribution with survivor function $S(t) = 1 - F(t)$ and p.d.f. $f(t) = \partial F(t)/\partial t$, the AFT log-likelihood under censoring is written in terms of the z_js:

$$\ell(\beta_0, \beta_1, \ldots, \beta_P, \sigma) \propto \sum_{j=1}^{D} \log\{f(z_j)/\sigma\} - \sum_{j=D+1}^{N} \log\{S(z_j)\}\,. \quad (11.14)$$

For example, in Section 11.2.3 we saw that the Weibull distribution has $S(t) = \exp\{-\alpha t^{\gamma}\}$, which is a function of two parameters, α and γ. (Recall also that the Weibull p.d.f. is $f(t) = \alpha\gamma t^{\gamma-1}\exp(-\alpha t^{\gamma})$.) These parameters are contained in the AFT model specification: the scale parameter α is related to the intercept parameter β_0 in (11.13), via $\beta_0 = -(1/\gamma)\log(\alpha)$. The shape parameter γ is related to σ via $\sigma = 1/\gamma$. The ML estimates of β_k and σ are found by maximizing (11.14) with respect to each of the unknown parameters.

For most models, the ML estimates, b_k and $\hat{\sigma}$, will not possess closed forms, and hence must be calculated iteratively. The general estimating equations are

$$\sum_{j=1}^{D} \frac{f'(z_j)}{f(z_j)} - \sum_{j=D+1}^{N} \frac{f(z_j)}{S(z_j)} = 0$$

for β_0,

$$\sum_{j=1}^{D} X_{kj} \frac{f'(z_j)}{f(z_j)} - \sum_{j=D+1}^{N} \frac{f(z_j)}{S(z_j)} = 0$$

for β_k ($k = 1, \ldots, P$), and

$$\sum_{j=1}^{D} z_j \frac{f'(z_j)}{f(z_j)} - \sum_{j=D+1}^{N} z_j \frac{f(z_j)}{S(z_j)} = 0$$

for σ. To calculate standard errors of the ML estimators, we employ the estimated Fisher information quantities as described in Sections 2.3.3 and 11.2.3.

The likelihood calculations are extensive, and best accomplished via computer. For example, SAS PROC LIFEREG performs these likelihood calculations for a number of possible, user-specified parent distributions, including the exponential, Weibull, and lognormal models.

Example 11.1 (continued) Acute toxicity data
Returning to the acute toxicity data in Table 11.1, the SAS code in Fig. 11.11 illustrates an AFT model fit under a Weibull specification for $f(t)$.

For the acute toxicity data from Table 11.1, the estimated intercept β_0 (and standard error) from this model fit is 1.904 (0.220), with an estimated scale parameter of $\hat{\sigma} = 0.536$ (0.177). As noted above, the ML estimate of the Weibull parameter γ is the reciprocal of the scale parameter based on fitting a Weibull AFT model. Thus, $\hat{\gamma} = 1/0.536 = 1.8644$. The ML estimate of α in the Weibull model is obtained from the relationship $\alpha = \exp(-\beta_0/\sigma)$, resulting in $\hat{\alpha} = \exp(-1.904/0.536) = 0.028$. ❑

```
* SAS code to fit toxicity data from Table 11.1;
options linesize = 70;

data actox;
  input stime ievent @@;
  censflag=1-ievent;
  cards;
1 1  2 0  3 1  4 1  4 1  4 0  5 0  6 1  8 1  9 0

proc lifereg;
  model stime*censflag(1)=/distribution=Weibull;
```

Fig. 11.11 SAS PROC LIFEREG code for fitting an AFT Weibull model to acute toxicity data in Example 11.1.

As a special case of this AFT regression methodology, consider the problem of comparing $P + 1$ different independent survival curves. This is essentially an ANOVA-type scenario, and is therefore amenable to regression modeling and analysis. To apply the AFT model in the $(P + 1)$-group setting, we define a set of P predictor variables that indicate whether the jth observation is taken from the $(k + 1)$st survival pattern/group. That is, define **indicator variables** $I_k(j)$ of the form

$$I_k(j) = \begin{cases} 1 & \text{if observation } j \text{ is from group } k+1 \\ 0 & \text{otherwise,} \end{cases}$$

$j = 1,\ldots, N$; $k = 1, \ldots, P$. This particular coding employs the first group as a reference. Under it, for example, the parameter β_2 represents how survival in the *third* group differs from the first. Let the kth predictor variable X_k be simply I_k, $k = 1, \ldots, P$. Then, the null hypothesis $H_0{:}\beta_1 = \cdots = \beta_P = 0$ corresponds to no difference in log-survival among the $P + 1$ groups. This may be tested using a likelihood ratio (LR) statistic. Calculate the value of the maximized log-likelihood under the null restriction that all the β_k parameters are zero, and denote this value as $\hat{\ell}^{(0)} = \ell(b_0^{(0)},0, \ldots, 0,\hat{\sigma}^{(0)})$. (The '(0)' superscript indicates constrained maximization under the null hypothesis $H_0{:}\beta_1 = \cdots = \beta_P = 0$.) The LR statistic compares this quantity to the unrestricted maximum of the log-likelihood, $\hat{\ell} = \ell(b_0, b_1, \ldots, b_P, \hat{\sigma})$, via the log ratio of maximized likelihoods. This is equivalent to a difference in maximized logarithms: $G^2_{calc} = -2\{\hat{\ell}^{(0)} - \hat{\ell}\}$. As $N \to \infty$, $G^2_{calc} \stackrel{\cdot}{\sim} \chi^2(P)$, so reject H_0 if $P \approx P[\chi^2(P) \geq G^2_{calc}]$ is less than the pre-selected significance level, α.

As suggested above, it is also possible to use any individual β_k parameter as a measure of the increase (or, if the estimate is negative, the decrease) in log-survival time of group k, relative to group 1. This would be appropriate if, for example, group 1 represents a control or reference condition in some environmental application. The significance of this increase (or decrease) is measured via the Wald statistic $W_k = b_k/se[b_k]$. In large samples, refer W_k to a standard normal distribution. The result is significant if the two-sided P-value, $P \approx 2\{1 - \Phi(\mid W_k \mid)\}$, is less than α. Correct any multiple comparisons among the β_ks via a Bonferroni adjustment, or some other appropriate multiplicity correction (Exercise 11.32).

Example 11.3 (continued) Spider departure data

Applied to the spider departure data, we use the AFT model to explore how departure times can be modeled as a function of species. Our preliminary model for these data incorporates a simple two-sample analog of the ANOVA-type linear predictor $\log(t) = \beta_0 + \beta_1 X_1 + \sigma\varepsilon$, where the indicator variable $X_1 = 1$ for a crab spider and $X_1 = 0$ for a jumping spider. As above, interest centers on comparisons between the two species of spider.

A plot of $\log\{-\log[\hat{S}_{\text{PL}}(t_i)]\}$ versus $\log(t_i)$ for each species exhibits roughly linear patterns (Exercise 11.26), suggesting that the Weibull model is a reasonable choice for the departure times. Hence, assuming a Weibull error distribution for ε (or equivalently, an extreme-value distribution for $\log[t]$), the AFT model is fitted using SAS PROC LIFEREG, via the code given in Fig. 11.12. (The first `model` statement fits the full model with X_1; the second `model` statement fits a model with $\beta_1 = 0$, in order to calculate the log-likelihood $\hat{\ell}^{(0)}$ under H_0.) Parameter estimates and standard errors obtained from the SAS output are presented in Table 11.7. Therein, the value of b_1 indicates that crab spiders' departure times are a factor of $\exp\{b_1\} = e^{0.552} = 1.737$ times as long as for jumping spiders. A Wald test of whether this difference is significant, that is, of $H_0{:}\,\beta_1 = 0$, employs the statistic $W_{\text{calc}} = b_1/se[b_1] = 0.552/0.161 = 3.429$. The associated P-value is $P \approx 2\{1 - \Phi(3.429)\} \approx 0.0006$.

Table 11.7 Weibull model estimates under the AFT model for spider departure data

Parameter	*Effect*	*Estimate*	*Std. error*
β_0	Intercept or scale	$b_0 = 3.719$	0.114
β_1	Species effect	$b_1 = 0.552$	0.161
σ	(Reciprocal) shape	$\hat{\sigma} = 0.357$	0.065

```
* SAS code to fit spider data for Example 11.3;
options linesize = 70;

data spider;
  input species $ etime @@;
  cards;
jump 20  jump 22  jump 29  jump 32  jump 40
jump 42  jump 44  jump 46  jump 51  jump 57
crab 21  crab 30  crab 31  crab 45  crab 55
crab 58  crab 73  crab 81  crab 105 crab 107

 proc lifereg;
  class species;
  model etime = species / distribution=Weibull;
  model etime = / distribution=Weibull;
```

Fig. 11.12 SAS PROC LIFEREG code for fitting AFT model to data in Example 11.3.

For the LR test of H_0, the SAS output gives log-likelihoods of $\hat{\ell} = -11.155$ and $\hat{\ell}^{(0)} = -15.138$. This yields $G^2_{calc} = -2(3.983) = 7.966$. We find $P \approx P[\chi^2(1) \geq 7.966] = 0.0048$. In either case, and similar to (although somewhat stronger than) the log-rank test results for these data, a significant difference is indicated between the two species' departure times.

Under a Weibull parent model, the AFT model can be translated back into an estimated survivor function as follows (Collett, 1994, Section 4.7): under the AFT specification, $\log(t) = \beta_0 + \beta_1 X_1 + \sigma\varepsilon$, the random error term is $\varepsilon = [\log(t) - (\beta_0 + \beta_1 X_1)]/\sigma$. The p.d.f. of this quantity can be shown to equal $f_\varepsilon(\varepsilon) = \exp(\varepsilon - \exp\{\varepsilon\})$; with this, take $V = e^{\varepsilon}$. One can determine using statistical transformation theory that $V \sim \mathcal{E}xp(1)$. Thus to find the original survivor function, $S(t) = P[T > t]$, take $S(t) = P[T > t] = P[\log(T) > \log(t)] = P[\beta_0 + \beta_1 X_1 + \sigma\varepsilon > \log(t)]$, which expands to

$$
\begin{aligned}
S(t) &= P\left[\varepsilon > \frac{\log(t) - (\beta_0 + \beta_1 X_1)}{\sigma}\right] \\
&= P\left[V > \exp\left\{\frac{\log(t) - (\beta_0 + \beta_1 X_1)}{\sigma}\right\}\right] \\
&= \exp\left\{-\exp\left[\frac{\log(t) - (\beta_0 + \beta_1 X_1)}{\sigma}\right]\right\}.
\end{aligned}
$$

Since this is also a function of X_1, we extend the notation and write

$$S(t; X_1) = \exp\left\{-\exp\left[-\left(\frac{\beta_0}{\sigma}\right) - \left(\frac{\beta_1}{\sigma}\right)X_1\right] t^{1/\sigma}\right\}.$$

Thus, in this example the estimated survivor function is

$$\hat{S}(t; X_1) = \exp\{-\exp[-10.417 - 1.546 X_1]\, t^{2.801}\}.$$

For jumping spiders ($X_1 = 0$), this is $\hat{S}(t;0) = \exp\{(-2.992\times10^{-5})t^{2.801}\}$, while for crab spiders ($X_1 = 1$) it is $\hat{S}(t;1) = \exp\{(-6.376\times10^{-6})t^{2.801}\}$. These estimate the proportion of spiders remaining at a given time t.

The S-PLUS `survreg` function also can be used to fit the AFT model. Suppose we have a data frame `spider.df`, defined to include a variable representing the event times, `etime`, and an indicator variable for the crab spider species, `icrab`. Then to fit the AFT model, invoke the code

```
survreg(Surv(etime)~icrab,link="log",
dist="extreme",data=spider.df)
```

□

We close this sub-section by noting that advanced statistical methods exist for employing AFT models in a distribution-free fashion, that is, for AFT regressions where it is not possible to specify a form for the parent distribution. The methods employ rank-based statistics, including the PL estimator from (11.1), to estimate covariate effects and survival probabilities. Their complexity is beyond the scope of this chapter, however, and for more details interested readers may consult works by Tsiatis (1990), Wei (1992), and the references therein.

11.4.2 Proportional hazards models

Another important regression formulation employed to model time-to-event data is the **proportional hazards model** (Cox, 1972). This model is written in terms of the hazard function $h(\cdot)$: $h(t,X_1, \ldots, X_P) = g(X_1, \ldots, X_P) \times h(t,0, \ldots, 0)$, where $g(X_1, \ldots, X_P)$ is a function of the covariates that relates the baseline hazard, $h(t,0, \ldots, 0)$, to the hazard of an organism or subject possessing that set of covariates, $h(t,X_1, \ldots, X_P)$. The most common choice for $g(\cdot)$ is a simple exponential: $g(X_1, \ldots, X_P) = \exp(\beta_1 X_1 + \cdots + \beta_P X_P)$, although more general forms are also possible (Hastie and Tibshirani, 1990). The exponential parameterization ensures that the PH model will generate a positive estimate of $h(t,X_1, \ldots, X_P)$, since $\exp(\cdot)$ is always greater than zero. This feature is a necessary component of any PH model, since hazard functions must by definition be positive quantities.

Notice how the PH formulation relates the ratio of a hazard to its baseline:

$$\frac{h(t,X_1, \ldots, X_P)}{h(t,0, \ldots, 0)} = \frac{\exp\{\beta_1 X_1 + \cdots + \beta_P X_P\}\, h(t,0, \ldots, 0)}{\exp\{\beta_1(0) + \cdots + \beta_P(0)\}\, h(t,0, \ldots, 0)}$$

which simplifies to $\exp(\beta_1 X_1 + \cdots + \beta_P X_P)$. This implies that the hazard at time t is a constant multiple, with respect to t, of the baseline. This motivates the term 'proportional' hazard.

It is possible to check the PH assumption via graphical inspection. For example, let $P = 1$. If the PH assumption is true (and if a Weibull distribution is appropriate for the baseline survivor function), a plot of $\log\{-\log[\hat{S}_{PL}(t_i)]\}$ vs. $\log(t_i)$ will produce roughly parallel lines across various levels of X_1. This effect extends naturally to $P > 1$ predictor variables. Also, more formal tests of the PH assumption are available, for example, in Lin and Wei (1991).

We can re-express the PH property so that the logarithm of this hazard ratio is a linear combination of the predictor variables:

$$\log\left\{\frac{h(t,X_1, \ldots, X_P)}{h(t,0, \ldots, 0)}\right\} = \beta_1 X_1 + \cdots + \beta_P X_P \,. \tag{11.15}$$

We do not include an intercept term, β_0, in (11.15), since any background effect is already incorporated in $h(t,0, \ldots, 0)$. If we write the linear predictor as $\eta_j = \beta_1 X_{1j} + \cdots + \beta_P X_{Pj}$, the PH requirement becomes $h(t_j, X_{1j}, \ldots, X_{Pj}) = \exp\{\eta_j\} h(t_j, 0, \ldots, 0)$, $j = 1, \ldots, N$.

Equation (11.15) sheds light on how the regression coefficients are interpreted in the PH model. The kth coefficient, β_k, is the change in log-hazard with each unit increase in X_k. Equivalently, the hazard ratio is multiplied by $\exp(\beta_k)$ for each unit increase in X_k, while holding $X_1, \ldots, X_{k-1}, X_{k+1}, \ldots, X_P$ constant. Contrast this with the AFT regression specification in (11.13): for the AFT model, the covariates act in a multiplicative fashion on time or in an additive fashion on $\log(t)$. For the PH model, the covariates act in a multiplicative fashion on $h(t)$. The linear predictor η_j then acts as a form of time-independent log-**relative risk**, in the sense that the risk of failure for a given realization of the covariates relative to $X_{kj} = 0$ is $\exp\{\eta_j\}$. Under (11.15), this is functionally *un*related to t.

When fitting fully parametric PH models one might ask whether there is a probability model for time where the AFT and PH models coincide. That is, if we are willing to specify fully the parent distribution of t or of $\log(t)$ via the hazard function $h(t)$, can we find a probability model for (11.13) whose hazards possess the PH property (11.15)? The answer is yes: it can be shown that the Weibull model from Section 11.4.1 exhibits the PH property (11.15) when modeled as an AFT regression. Thus, the Weibull PH model can be fitted using AFT regression methods, as described in Section 11.4.1. Indeed, a form specified often for parametric PH baseline hazards with environmental data is the Weibull (AFT) model. Fitting PH models with other distributional forms is less common in practice.

11.4.3 Distribution-free proportional hazards: The Cox regression model

When the hazard $h(t, X_1, \ldots, X_P)$, and hence the distribution of t, is left unspecified, the PH regression model may still be used to estimate the effects of the predictor variables on the failure time. We refer to this as a distribution-free PH approach, since no parent distribution is specified for t. (Once again, the concepts and notation for fitting PH models are somewhat advanced. Introductory readers may wish to skip ahead to the examples for illustrations of the methods before taking up the details that follow.)

The established, fundamental approach to distribution-free estimation under (11.15) was proposed by Cox (1972), and the method often is called

Cox (PH) regression, or Cox's model. In its basic form, the model assumes that D ordered, distinct failure times are observed as $t_{(1)} < t_{(2)} < \cdots < t_{(D)}$, and that an additional $N - D$ subjects yield censored observations. Denote by $\mathcal{R}(t_{(j)})$ the **risk set** at time $t_{(j)}$, that is, the set of (indices of) subjects who have not failed just before time $t_{(j)}$. Then, the probability of failure for any jth subject in $\mathcal{R}(t_{(j)})$ at any time $t > t_{(j)}$ is the hazard for that jth subject, divided by the sum of hazards for all individuals in $\mathcal{R}(t_{(j)})$: $h(t, X_{1j}, \ldots, X_{Pj}) / \sum_{i \in \mathcal{R}(t_{(j)})} h(t, X_{1i}, \ldots, X_{Pi})$. The notation in the denominator indicates a sum over all values i in the set $\mathcal{R}(t_{(j)})$. Notice that, by definition, j must be an element of $\mathcal{R}(t_{(j)})$ for this construction to be valid. Under the PH assumption (11.15), this ratio simplifies to an expression that contains only the β-parameters:

$$\frac{\exp\{\eta_j\}}{\sum_{i \in \mathcal{R}(t_{(j)})} \exp\{\eta_i\}} \tag{11.16}$$

where, as above, η_j is the linear predictor $\beta_1 X_{1j} + \cdots + \beta_P X_{Pj}$.

Cox's (1972) approach for estimating the β-parameters forms a **partial likelihood** that represents a component of the information in the sample related only to the β_ks. This is the product of the probabilities (11.16) over all distinct failure times $t_{(1)} < t_{(2)} < \cdots < t_{(D)}$, mimicking a more formal likelihood construction. The corresponding log-partial likelihood is then

$$\ell^*(\beta_1, \ldots, \beta_P) = \sum_{j=1}^{D} \left\{ \eta_{(j)} - \log\left[\sum_{i \in \mathcal{R}(t_{(j)})} \exp\{\eta_i\} \right] \right\}; \tag{11.17}$$

(Lawless, 1982, Section 7.2.3). Maximizing (11.17) with respect to each β_k yields maximum partial likelihood estimates b_k. As $N \to \infty$, Wald statistics such as $(b_k - \beta_k)/se[b_k]$ are approximately standard normal (Tsiatis, 1981), allowing for LR and Wald-type tests of the covariate effects. Other approaches, such as estimation and inference on the unspecified hazard function $h(\cdot)$ or on the associated survivor function $S(t)$, are also possible, although we will not consider them here; for those details and other extensions of the PH model see Lawless (1982, Section 7.2.4), Andersen (1991), and the references therein. For simultaneous confidence bands on $S(t)$ under a semi-parametric PH assumption, see Lin *et al.* (1994).

The basic PH model as given above requires that the underlying distribution of t is continuous, so that tied observations are essentially nonexistent. In practice, however, ties do occur, and in some cases they can represent a substantial portion of the data. When a limited number of ties are recorded, the Cox model may be modified to correct for their effects. To do so, suppose that d_j subjects are observed to fail at time $t_{(j)}$. Let $\mathcal{D}_j$ be the set of d_j indices representing failures at time $t_{(j)}$. For these d_j subjects, calculate the sums of their corresponding predictor variables $T_{kj} = \sum_{i\in\mathcal{D}_j} X_{ki}$, creating an aggregate set of predictor values that summarizes the covariate effects at time $t_{(j)}$. Then, the partial likelihood components at each $t_{(j)}$ are written as

$$\exp\{\textstyle\sum_{k=1}^{P}\beta_k T_{kj}\}/(\sum_{i\in\mathcal{R}(t_{(j)})}\exp\{\eta_i\})^{d_j},$$

mimicking (11.16). The corresponding log-partial likelihood, $\ell^*(\beta_1, \ldots, \beta_P)$, is proportional to

$$\sum_{j=1}^{D}\left\{\sum_{k=1}^{P}\beta_k T_{kj} - (d_j)\log\left[\sum_{i\in\mathcal{R}(t_{(j)})}\exp\{\eta_i\}\right]\right\}. \tag{11.18}$$

When $d_j = 1$ for all j, (11.18) collapses to the partial likelihood (11.17).

We minimize (11.18) with respect to each β_k to find the maximum partial likelihood estimators b_k. The corresponding estimating equations at each $k = 1, \ldots, P$ are the various first derivatives of (11.18) set equal to zero:

$$\sum_{j=1}^{D} T_{kj} - d_j\frac{\sum_{i\in\mathcal{R}(t_{(j)})} X_{ki}\exp\{\eta_i\}}{\sum_{i\in\mathcal{R}(t_{(j)})}\exp\{\eta_i\}} = 0 \tag{11.19}$$

Unfortunately, closed-form solutions to (11.19) are not commonly available, and we turn again to iterative computer calculations. Both SAS PROC PHREG and the S-PLUS function `coxph` calculate the b_ks and also report their standard errors, $se[b_k]$. A Wald test of the significance of each β_k is then based on the statistic $W_k = b_k/se[b_k]$: reject $H_0\colon\beta_k = 0$ if $P \approx 2(1 - \Phi(\,|W_k|\,)\} < \alpha$ for some pre-assigned significance level, α.

As an alternative to the Wald test of $H_0\colon\beta_k = 0$, one can employ a *partial* likelihood ratio test, which is similar in form to the LR test in Section 11.4.1. Calculate the value of the maximized log-partial likelihood under $\beta_k = 0$, and denote this value by $\hat{\ell}_k^{(0)} = \ell^*(b_0^{(0)}, b_1^{(0)}, \ldots, b_{k-1}^{(0)}, 0, b_{k+1}^{(0)}, \ldots, b_P^{(0)})$. (As above, the '(0)' superscript indicates constrained maximization under the null hypothesis.) The partial LR statistic compares this quantity to the unrestricted maximum of the log-partial likelihood, $\hat{\ell}^* = \ell^*(b_0, b_1, \ldots, b_P)$, via the LR statistic: $G_k^2 = -2\{\hat{\ell}_k^{(0)} - \hat{\ell}^*\}$. As $N\to\infty$, $G_k^2 \,\dot{\sim}\, \chi^2(1)$, and the approximate P-value for testing H_0 is $P \approx P[\chi^2(1) \geq G_k^2]$. Reject H_0 if $P < \alpha$. This can be extended in a straightforward manner to test that any subset of the P parameters is zero, including the global null hypothesis $H_0: \beta_1 = \cdots = \beta_P = 0$. SAS PROC PHREG reports the partial LR statistic for the global H_0 as part of its standard output.

A competitor to both the Wald and LR statistics for testing covariate effects is the score test from Section 2.4.3. Recall that the score test is based on the individual derivatives of ℓ^* with respect to the parameters β_k. For the Cox PH model, score tests on the β_ks often require fewer computations than the Wald or LR statistics, yet are generally as or more stable (Lee *et al.*, 1983). The score test is available in SAS PROC PHREG.

An important case where the score test has a slightly simplified form is the $(P+1)$-group setting, where as in Section 11.4.1 the P covariates are used as indicator variables that identify the independent groups. To test the homogeneity of the associated $P + 1$ survival patterns, the null hypothesis corresponds to the global null hypothesis $H_0: \beta_1 = \cdots = \beta_P = 0$. Then, the score test of homogeneity in survival among all groups is based on the individual scores under H_0:

$$U_k = \left.\frac{\partial \ell^*}{\partial \beta_k}\right|_{\beta_k=0} = \sum_{j=1}^{D} \left\{ d_{kj} - d_{+j}\frac{n_{kj}}{n_{+j}} \right\}$$

$(k = 1, \ldots, P)$, where we extend our notation to the life-table form found in Table 11.4: d_{kj} is the number of observed failures (out of d_{+j}) in group k at time $t_{(j)}$, and n_{kj} is the total number of subjects at risk (out of n_{+j}) in group k at time $t_{(j)}$, $j = 1, \ldots, D$. The $(P + 1)$-group life table schematic for this construction is presented in Table 11.8.

Table 11.8 Data schematic for testing homogeneity in survival patterns over $P + 1$ independent groups

Group	*Status*	$t_{(1)}$	$t_{(2)}$	$\cdots$	$t_{(D)}$	*TOTAL*
$k = 1$	Die (d_{1j})	d_{11}	d_{12}	$\cdots$	d_{1D}	d_{1+}
	At risk (n_{1j})	n_{11}	n_{12}	$\cdots$	n_{1D}	n_{1+}
$\vdots$		$\vdots$	$\vdots$		$\vdots$	$\vdots$
$k = P+1$	Die ($d_{P+1,j}$)	$d_{P+1,1}$	$d_{P+1,2}$	$\cdots$	$d_{P+1,D}$	$d_{P+1,+}$
	At risk ($n_{P+1,j}$)	$n_{P+1,1}$	$n_{P+1,2}$	$\cdots$	$n_{P+1,D}$	$n_{P+1,+}$
TOTAL	Die (d_{+j})	d_{+1}	d_{+2}		d_{+D}	d_{++}
	At risk (n_{+j})	n_{+1}	n_{+2}	$\cdots$	n_{+D}	n_{++}

If we define the quantity I_{kl} as equal to 1 if $k = l$ and zero otherwise, the corresponding Fisher information elements under $H_0: \beta_1 = \cdots = \beta_P = 0$ are

$$\mathfrak{F}_{kl}^{(0)} = \sum_{j=1}^{D} d_{+j} \frac{n_{kj}}{n_{+j}} \left\{ I_{kl} - \frac{n_{lj}}{n_{+j}} \right\}$$

$(k, l = 1, \ldots, P)$. These are the estimated variances and covariances of the U_k; for example, $\mathrm{Cov}(U_k, U_l) = \mathfrak{F}_{kl}^{(0)}$. We collect these information terms together into the Fisher information matrix $\mathbf{F}_0$, with kth row and lth column element denoted by $\mathfrak{F}_{kl}^{(0)}$. Inverting $\mathbf{F}_0$ yields the variance-covariance matrix $\mathbf{V}_0 = \mathbf{F}_0^{-1}$, whose elements are denoted by $\mathfrak{F}_{(0)}^{kl}$. The score statistic for testing survival homogeneity is formed from these latter values:

$$T_{\text{calc}}^2 = \sum_{k=1}^{P} U_k^2 \mathfrak{F}_{(0)}^{kk} + 2 \sum_{k=1}^{P-1} \sum_{l=k+1}^{P} U_k U_l \mathfrak{F}_{(0)}^{kl} . \qquad (11.20)$$

Under H_0, $T_{\text{calc}}^2 \mathrel{\dot\sim} \chi^2(P)$; the P-value is $\mathrm{P}[\chi^2(P) \geq T_{\text{calc}}^2]$. When $P = 1$, so that only two groups' survival patterns are being compared, T_{calc}^2 is similar to the log-rank statistic from Section 11.3, and coincides exactly with it if no ties are observed.

Example 11.3 (continued) Spider departure data

For the spider departure time data, we have $P + 1 = 2$ groups, and a semi-parametric Cox (PH) regression fit to the data once again involves a simple ANOVA-type indicator variable: $X_1 = 1$ for crab spiders and $X_1 = 0$ for jumping spiders. The PH model from (11.15) is then

$$\log\{h(t_j, X_{1j})/h(t_j, 0)\} = \beta X_{1j},$$

where $h(t_j, 0)$ is the hazard of departure in the baseline group (jumping spiders) and $h(t_j, 1)$ is the hazard of departure for crab spiders. Thus we are comparing the departure 'hazard' for crab spiders relative to that for jumping spiders. (Had we reversed the definition of X_1, we would instead be comparing jumping spiders relative to crab spiders.)

SAS PROC PHREG is invoked in a similar manner to PROC LIFEREG; see Fig. 11.13. Applied to the spider departure data, this produces an estimate for β of $b = -1.246$. The estimated relative risk is $\exp(-1.246) = 0.288$, which suggests that the rate of departure is about 29% lower in crab spiders relative to jumping spiders.

The score test of $H_0: \beta = 0$ for the spider departure data yields $T_{\text{calc}}^2 = 5.312$ with 1 df $(P \approx 0.021)$. As expected, this result is identical to the log-rank test result calculated in Section 11.3. The LR test of H_0 in the spider example yields a similar inference: $G_{\text{calc}}^2 = 5.182$ with 1 df $(P \approx 0.023)$.

So does the Wald test of H_0, formed by the ratio of the parameter estimate divided by its standard error. The estimate, $b = -1.246$, has a standard error of 0.570, leading to a Wald statistic of $W_{calc} = 2.188$ ($P \approx 0.029$).

Had we applied S-PLUS to fit the PH model to these data using the `coxph` function, similar inferences would have been achieved. ❑

```
* SAS code to fit spider data for Example 11.3;
options linesize = 70;

data spider;
  input species $ etime @@;
  if species='jump' then x1=0;
  if species='crab' then x1=1;
  cards;
jump 20   jump 22   jump 29   jump 32   jump 40
jump 42   jump 44   jump 46   jump 51   jump 57
crab 21   crab 30   crab 31   crab 45   crab 55
crab 58   crab 73   crab 81   crab 105  crab 107

 proc phreg;
  model etime = x1;
```

Fig. 11.13 SAS PROC LIFEREG code for fitting a Cox regression model to spider departure data in Example 11.3.

11.4.4* Trend tests using the Cox regression model

A special application of the semi-parametric Cox (PH) model is as a test for trend in some ordered covariate, such as a dose of an environmental toxin. Described by Tarone (1975), the method in effect decomposes the log-rank statistic into two components: one that measures linear trend over a single ordered covariate, and a second that measures the departure from linearity of that trend. We will direct attention to the former issue: that of detecting a trend over an ordered covariate.

Suppose we record a single covariate, X_1, over $L + 1$ levels, where the levels $c_0 < c_1 < \cdots < c_L$ represent ordered values, such as dose of an environmental toxin. Typically, we assume the lowest dose is actually a concurrent control, with $c_0 = 0$, although this is not required. To test for a dose effect we could apply the $(L + 1)$-group score statistic from (11.20) to the $L + 1$ groups indexed by c_k. A more powerful test, however, employs the quantitative information in the c_ks. That is, apply the Cox (PH) partial likelihood described in Section 11.4.3, with a single covariate, X_1. Assume X_1 is related to survival via the PH assumption, so that $h(t_j, c_k) = \exp\{\beta c_k\} h(t_j, 0)$ for the jth subject exposed to the kth dose ($j = 1, \ldots, D$; $k = 0, \ldots, L$). Under this model formulation, a test of $H_0: \beta = 0$ is a test for trend over $c_0 < c_1 < \cdots < c_L$. Increasing trend occurs when $\beta > 0$, decreasing trend when $\beta < 0$.

Equations (11.18) and (11.19) give the partial likelihood and estimating equation for β, respectively. Once computed, the maximum partial likelihood estimate of β may be used in LR tests to assess the significance of H_0. For this simple case, however, Tarone (1975) showed that closed-form equations can be derived for the score statistic of H_0, based on a data schematic that is essentially the life table from Table 11.8 modified slightly to incorporate the dose information (Table 11.9). In the table, notice that, as in Tables 11.4 and 11.8, d_{kj} is the number of failures at time $t_{(j)}$ at dose c_k, while n_{kj} is the number of subjects at risk at dose c_k just before time $t_{(j)}$.

Table 11.9 Data schematic for testing trend in survival patterns over $L + 1$ levels of an ordered covariate, X

Level of X	*Status*	$t_{(1)}$	$t_{(2)}$	$\cdots$	$t_{(D)}$	*TOTAL*
$X = c_0$	Die (d_{0j})	d_{01}	d_{02}	$\cdots$	d_{0D}	d_{0+}
	At risk (n_{0j})	n_{01}	n_{02}	$\cdots$	n_{0D}	n_{0+}
$\vdots$		$\vdots$	$\vdots$		$\vdots$	$\vdots$
$X = c_L$	Die (d_{Lj})	d_{L1}	d_{L2}	$\cdots$	d_{LD}	d_{L+}
	At risk (n_{Lj})	n_{L1}	n_{L2}	$\cdots$	n_{LD}	n_{L+}
TOTAL	Die (d_{+j})	d_{+1}	d_{+2}	$\cdots$	d_{+D}	d_{++}
	At risk (n_{+j})	n_{+1}	n_{+2}	$\cdots$	n_{+D}	n_{++}

Using the notation from Table 11.9, the log-partial likelihood is

$$\ell^*(\beta) \propto \sum_{j=1}^{D} \left\{ \beta \sum_{k=0}^{L} d_{kj} c_k - \log\left[\sum_{k=0}^{L} n_{kj} \exp\{\beta c_k\} \right] \right\}. \tag{11.21}$$

To derive the score statistic for testing $H_0: \beta = 0$ from (11.21), first find the log-partial likelihood derivative $\partial\ell^*/\partial\beta$ and evaluate it at $\beta = 0$ to get the score, U_1. This is

$$U_1 = \sum_{j=1}^{D} \sum_{k=0}^{L} c_k \left\{ d_{kj} - d_{+j} \frac{n_{kj}}{n_{+j}} \right\}.$$

A large-sample variance estimate of U_1 is the Fisher information for β, which is derived as $-\partial\ell^*/\partial\beta^2$, evaluated at $\beta = 0$:

$$\mathfrak{F}_{11}^{(0)} = \sum_{j=1}^{D} \frac{d_{+j}(n_{+j} - d_{+j})}{(n_{+j} - 1)} \left\{ \sum_{k=0}^{L} c_k^2 \frac{n_{kj}}{n_{+j}} - \left[\sum_{k=0}^{L} c_k \frac{n_{kj}}{n_{+j}} \right]^2 \right\}.$$

With these values, calculate the score statistic as

$$T_{\text{calc}}^2 = \frac{U_1^2}{\mathfrak{F}_{11}^{(0)}} \tag{11.22}$$

and in large samples refer T_{calc}^2 to $\chi^2(1)$. Reject $H_0: \beta = 0$ when the approximate P-value, $P \approx P[\chi^2(1) \geq T_{\text{calc}}^2]$, is less than α. Since Tarone's score statistic for testing trend can be derived from the life-table data schematic in Table 11.9, it is often referred to as a **life-table trend test**.

Example 11.4 Lethal tumors

An interesting and important application of Tarone's life-table trend test is in environmental carcinogenicity testing. For example, in laboratory studies with rodents undertaken by the US National Toxicology Program (NTP), some chemical carcinogens are found to induce tumors that lead to the death of the animals in a rapid fashion. Such tumors often are described as **instantaneously lethal**. Of course, 'instantaneous' here is not intended to suggest that the tumor develops, affects, harms, and then kills the animal in the blink of an eye! Rather, it represents a rapid and complete occurrence as measured over the time-scale of study. For example, in a classical NTP experiment animals are checked daily for any deaths, but any animals discovered to have died are recorded as dying only during that week of the study. That is, dead animals are identified (and removed) daily, but recorded weekly. (The NTP has since moved to identifying and reporting its event times on a daily basis.) Hence, a quickly developing tumor that develops on, say a Monday, and then affects the animal's systems so profoundly that the animal dies on Friday, would be considered instantaneous from this perspective. Instantaneous lethality represents one end of a spectrum of possible tumor lethalities, with incidental tumor lethality defining the other extreme (as discussed in Example 8.7).

The issue is an important one in long-term studies of carcinogenicity, such as the 105-week studies the NTP undertakes. Tumor incidence generally varies with the age of the animal, so that animals dying early in the study exhibit different incidence and susceptibility patterns than those dying later. Hence, intercurrent mortality among the animals complicates the analysis of tumor incidence, requiring some adjustment for different survival patterns seen among differently treated groups of animals (Dinse, 1994). These **survival-adjusted analyses** are common in long-term toxicity studies

(Peto *et al.*, 1980). The simplest such adjustment is to record the time to tumor and also the age at which the animal dies or is sacrificed (right-censored in the latter case), and incorporate these variables into a regression analysis. Unfortunately, in most carcinogenicity experiments, the tumors are **occult**, that is to say, they develop and grow inside an animal and are identified histopathologically only after the animal dies or is sacrificed at the end of the study. Data analysis of the actual time-course of the cancer's progression is not possible in this case, unless additional assumptions are made about the **context of observation** in which the tumor was recorded (Peto *et al.*, 1980; Archer and Ryan, 1989; McKnight and Wahrendorf, 1992), additional animals are sacrificed at interim times throughout the study (McKnight and Crowley, 1984; Dinse, 1988), and/or complex parametric or semi-parametric models are employed to represent the tumor incidence (Portier and Dinse, 1987; Lindsey and Ryan, 1993).

For instantaneously lethal tumors, however, it is possible to test for increasing carcinogenicity over dose or exposure in a fairly simple manner, while still adjusting for survival differences among the animals. Since the tumor is thought to kill the animal essentially at the same time it develops (under the time-scale of study), the time to death and time to tumor are equivalent, and we can employ the various life-table methods from this chapter to study the time-to-tumor data.

In particular, if the study design calls for a series of increasing doses of the carcinogen, we can apply Tarone's score statistic in (11.22) to test for trend in the time-to-tumor response. The analysis proceeds using the schematic in Table 11.9, where d_{kj} is the number of events – that is, instantaneous tumors – observed at time $t_{(j)}$ at dose c_k, while n_{kj} is the number of animals still alive and assumed tumor-free at dose c_k just before time $t_{(j)}$. Note that an animal dying during the study or sacrificed at the end of the study without a tumor would be considered a right-censored observation. (A word of warning is needed. If this method is applied to tumors that are not instantaneously lethal and also treatment-related toxicity is present, then incorrect inferences may result. In particular, this method when applied to nonlethal tumors tends to reject a true H_0 at rates much greater than the specified Type I error rate (Bailer and Portier, 1988).)

To illustrate, consider the NTP (1993b) study of environmental carcinogenesis in male B6C3F$_1$ mice brought on by exposure to *p*-nitroaniline. This study considered murine carcinogenesis after gavage exposure to the chemical over a 105-week period. Here, we will consider the specific issue of assessing an exposure-related increase in vascular system hemangiosarcomas. These murine tumors are some of the most lethal studied by the

NTP, leading to death in animals with the tumor approximately 12 times faster, on average, than animals without the tumor (Portier *et al.*, 1986). Hence, analyses based on life tables are reasonable when performing survival-adjusted analyses of hemangiosarcoma tumorigenicity.

The vehicle control response in this study showed four of 50 animals (8%) with the tumor. $L = 3$ additional dose groups were reported; their hemangiosarcoma response rates (and exposures, in mg/kg body weight) were 1/50 (2%) at 3 mg/kg, 3/50 (6%) at 30 mg/kg, and 8/50 (16%) at 100 mg/kg. The weekly event data appear in life-table format in Table 11.10.

Table 11.10 *p*-nitroaniline data for occurrence of hemangiosarcoma

Dose (mg/kg)	*Tumor status*	*Time (weeks)* 1	57	81	94	96	103	104	105	*TOTAL*
0	Tumors (d_{0j})	0	0	0	0	1	1	2	0	4
	At risk (n_{0j})	50	49	46	43	43	41	39	34	345
3	Tumors (d_{1j})	0	0	0	0	0	0	0	1	1
	At risk (n_{1j})	50	50	46	41	38	33	33	32	323
30	Tumors (d_{2j})	0	0	0	0	0	0	1	2	3
	At risk (n_{2j})	50	49	46	41	41	37	37	35	336
100	Tumors (d_{3j})	0	0	1	1	0	1	0	5	8
	At risk (n_{3j})	50	49	48	45	44	43	42	39	360
TOTAL	Tumors (d_{+j})	0	0	1	1	1	2	3	8	16
	At risk (n_{+j})	200	197	186	170	166	154	151	140	1364

Applying Tarone's life-table test to the data in Table 11.10 yields $U_1 = 322.649$ and $\mathfrak{F}_{11}^{(0)} = 26\,561.536$. From these, we calculate the Tarone statistic as $T^2_{\text{calc}} = (322.649)^2/26561.535 = 3.919$. The corresponding approximate P-value is $\text{P}[\chi^2(1) \geq 3.919] = 0.048$, which although significant at $\alpha = 0.05$, is not a strong an indication of a significant trend.

For these data, the spread of the dose levels is approximately geometric, that is, each dose is a degree of magnitude higher than its predecessor. This will weigh the higher doses more heavily in the analysis – a form of increased **leverage**; see Neter *et al.* (1996, Section 9.3). Typically, this effect is not intended, and some sort of correction is appropriate to make the dose scale more equi-spaced. A common strategy is to use the natural logarithms of dose, rather than the actual doses. For zero-dose controls we cannot take a logarithm, however, so we approximate a reasonable value for the control level via consecutive-dose average spacing from (6.9): if the original dose levels are, say, w_k ($k = 0, \ldots, L$), with $w_0 = 0$, then take $c_k = \log_{10}(w_k)$ for $k \neq 0$, and calculate

$$c_0 = \log_{10}\{w_1\} - \frac{\log_{10}\{w_L\} - \log_{10}\{w_1\}}{L - 1} .$$

For the doses in Table 11.10, we find $c_1 = 0.4771$, $c_2 = 1.4771$, $c_3 = 2.0$, and then $c_0 = -0.2843$. If we apply Tarone's life-table test to the data in Table 11.10 using these log-doses, we achieve an interesting result: $U_1 = 4.524$ and $\mathfrak{F}_{11}^{(0)} = 12.511$. From these, we calculate Tarone's T^2-statistic as $T^2_{calc} = (4.524)^2/12.511 = 1.636$. The corresponding approximate P-value is $P[\chi^2(1) \geq 1.636] = 0.20$, which is no longer significant.

These marginal results, coupled with the facts that the experiment yielded no other significant neoplastic effects, and that no tumors of any sort were significantly increased in an associated female mouse study, led the NTP to conclude that this chemical exhibited only equivocal evidence of carcinogenicity. ❑

It is important to note that the semi-parametric PH methods from Section 11.4 are only applicable when the numbers of tied observations are low. When many ties are present in the data, more extensive modifications are necessary for the Cox model calculations. The basic approach groups the observations into a set of contiguous time intervals that partition the time line of interest. Thus, ties are grouped into their corresponding time intervals, and then viewed as equivalent to any other observation in that interval. The effect is to create a compartmentalized analysis similar in form to the life-table data schematic in Table 11.8, except that each grouped observation carries with it a set of covariate values, $X_1, \ldots, X_L$. Often, the resulting test statistics are quite similar to the log-rank and score-test T^2-statistics seen above. For further details and examples, see the discussion in Lawless (1982, Section 7.3).

11.5 SUMMARY

In this chapter, basic concepts and models for the analysis of time-to-event/survival data are introduced. It is highlighted that the presence of censored or unobserved events must be accounted for properly in any time-to-event analysis. Both nonparametric and parametric techniques are described for estimating the survivor function and related quantiles. Testing for the equivalence of survivor functions also is discussed. Finally, the incorporation of covariates into the analysis of survivor functions is presented, using either accelerated failure-time or proportional hazards regression models.

Although the language of survival studies used in this chapter is developed primarily in terms of biological outcomes, it is emphasized that survival-data analysis is not restricted to the analysis of biological failure times. Any time-to-event data may be analyzed using the methods introduced here.

It is noted also that many questions and issues regarding analysis of time-to-event data are beyond the scope of this chapter's presentation. Interested readers are encouraged to pursue advanced methods for time-to-event data, as found in specific texts on the analysis of survival data (Lawless, 1982; Nelson, 1982; Collett, 1994). In particular, further illustrations on toxicity-related survival methods, along with SAS computer strategies, can be found in Dixon and Newman (1991).

EXERCISES

(Exercises with asterisks, e.g. 11.2*, typically are more advanced, or require greater knowledge and understanding of probability distributions and/or calculus.)

11.1. Write out the parametric survivor function, $S(t)$ corresponding to the following distribution functions from Section 1.3:

- a. Exponential with parameter $\beta > 0$.
- b. Gamma with parameters $\alpha = 1$ and $\beta > 0$.
- c. Gamma with parameters $\alpha = 2$ and $\beta > 0$.
- d. Lognormal with parameters μ (arbitrary) and $\sigma = 1$.
- e. Lognormal with parameters μ (arbitrary) and $\sigma > 0$.

11.2*. Find the parametric hazard function associated with each of the distributions in Exercise 11.1.

11.3. Can you envision a *sensible* setting from your own field of study in which censored time-to-event data occur? Is it right-censored, left-censored, or both? Explain.

11.4*. For a continuous random variable, T, show that the definition of the hazard function as a limit of a ratio approaches $f(t)/S(t)$. (*Hint*: use the definition of a derivative and recall the relationship between a c.d.f., $F(\cdot)$, and a p.d.f., $f(\cdot)$, for continuous random variables.)

11.5. To study the mortality of insects after exposure to extreme cold ('cold shock'), a preliminary experiment was conducted in a laboratory to measure survival times of grasshoppers after a single shock exposure to extreme cold (4°C). This produced the following survival times (in minutes) among $N = 7$ insects: 3.6, 7.1, 11.2, 12.5, 31.4, 39.5, 46.1.

a. Compute the product-limit survival curve estimate for these data, and plot your estimate as a function of survival time.

b. Calculate the standard error of the product-limit estimates from part (a) at each time point, using Greenwood's formula.

c. Use the standard errors from part (b) to calculate a pointwise confidence interval on the survival at the last observed time point using the Wald limits from (11.3).

d. Use the standard errors from part (b) to calculate a pointwise confidence interval on the survival at the last observed time point using the Rothman–Wilson limits from (11.4). Compare your results to the limits from part (c).

11.6. In a study of fish after exposure to heavy metals in their water supply, $N = 9$ fish exhibited survival times (in hours) of 1.2, 1.3, 3.9, 5.9, 6.5, 7.4, 7.6, 9.2, 16.6, after acute mercury exposure.

a. Compute the product-limit survival curve estimate for these data and plot your estimate as a function of survival time.

b. Calculate the standard error of the product-limit estimates from part (a) at each time point, using Greenwood's formula.

c. Repeat part (a) assuming that times 7.6 and 16.6 correspond to right-censored observations.

11.7. In a laboratory experiment to study carcinogenicity of the solvent stabilizer 1,2-epoxybutane, the US National Toxicology Program (NTP) exposed 50 male rats to 200 ppm of the chemical, via inhalation, for 105 weeks (NTP, 1988); see Exercise 4.20. The mortality experiences of the male rats were summarized via their survival times over the 105-week exposure period (in weeks): 32 rats died prior to the study's end, at weeks 50, 73, 76, 79, 81, 81, 83, 85, 85, 85, 87, 89, 90, 91, 91, 94, 94, 95, 95, 97, 98, 99, 100, 100, 100, 100, 102, 102, 103, 103, 104, 104, and 18 rats were sacrificed at the end of the study (right-censored).

a. Compute the product-limit survival curve estimate for these data and plot your estimate as a function of survival time.

b. Calculate the standard error of the PL estimates from part (a) at each survival time, using Greenwood's formula.

c. Use the standard errors from part (b) to calculate pointwise confidence limits on the survival at each observed time point using the Wald limits from (11.3).

d. How would you adjust the limits computed in part (c) for multiplicity over the many different time points?

11.8. In a laboratory experiment to study carcinogenicity of the solvent stabilizer 1,2-epoxybutane, the US National Toxicology Program (NTP) exposed 50 female rats to 200 ppm of the chemical, via inhalation, for 105 weeks (NTP, 1988), as in Exercise 11.7. The mortality experiences of the female rats were summarized via their survival times over the 105-week exposure period (in weeks): 29 rats died prior to the study's end, at weeks 1, 32, 49, 66, 68, 71, 71, 73, 81, 83, 85, 85, 85, 85, 87, 89, 91, 91, 91, 91, 91, 92, 92, 95, 95, 99, 99, 100, 102, and 21 rats were sacrificed at the end of the study (right-censored).

   a. Compute the nonparametric product-limit survival curve estimate for these data.
   b. Plot the estimate you calculated in part (a) as a function of survival time (in weeks).

11.9. Return to the male mice survival data in Example 11.2.

   a. Verify the calculations of the nonparametric product-limit survival curve estimate for these data. Plot the PL estimate of the curve as a function of survival time (in weeks).
   b. Calculate a Rothman–Wilson 95% confidence interval for the survival at the last death time, $t_5 = 100$. Compare this to the large-sample interval reported in the example.

11.10. Is the equal-precision confidence band from Section 11.2.1 an equal-*width* band? Explain.

11.11*. Recognize that if T is Weibull with parameters α and γ, then $U = \log(T)$ has the extreme-value distribution with parameters $1/\gamma$ and $-\log(\alpha)/\gamma$. From this, find the survivor function, $S(u)$, for an extreme-value random variable.

11.12*. The **shifted** or **lagged exponential** distribution is used to model exponential failure times when a **guarantee time**, $v > 0$, is included, below which the experimental units cannot fail. The distribution has p.d.f. $f(t \mid \beta, v) = (1/\beta)\exp\{-(t - v)/\beta\}$ over $t > v$ (and for $\beta > 0$). (This is also called a **two-parameter exponential** distribution.) Find:

   a. The survivor function $S(t)$ for this distribution.
   b. The hazard function $h(t)$ for this distribution.

11.13*. Show that the memoryless property possessed by the exponential distribution is not shared by the more general Weibull distribution. That is, show that if T is distributed as Weibull with parameters α and $\gamma \neq 1$, $P[T > t \mid T > u]$ is not simply a function of $(t - u)$.

11.14. Give a nonparametric estimate of the median survival time, LT_{50}, for the data in:

a. Exercise 11.5.

b. Exercise 11.6.

c. Exercises 11.7 and 11.8. Notice that these two data sets represent survival of different sexes of rat after exposure to the same environmental toxin. Do the two LT_{50}s you calculated appear comparable?

11.15. Assume that the survival-time data in Exercise 11.5 are distributed according to a Weibull distribution.

a. Find the ML estimates of the Weibull parameters α and γ, using (11.6) and (11.7). (To use (11.7), select a limited series of values for γ and see if the equation can be satisfied by any of these values. If not, narrow your search by choosing a finer grid of values near the value of γ that came closest to satisfying the equation. We recommend limiting your search with these data to the range $0.5 < \gamma < 1.5$.)

b. Calculate the standard errors of your estimators from part (a).

c. Using the standard errors calculated in part (b), compute 90% confidence intervals for α and γ, using (11.8) and (11.9). Based on the interval for γ, is it plausible to infer that the data are in fact exponentially distributed (at the 10% significance level)?

d. Note that the intervals computed in part (c) are pointwise in nature; that is, they do not adjust for multiplicity, even though they use overlapping data in their computation. How would you adjust these two confidence intervals for multiplicity? Do so, and compare your results to those in part (c).

11.16. Assume that the survival-time data in Exercise 11.5 are distributed according to an exponential distribution with single parameter $\alpha > 0$. Find the maximum likelihood estimate of α and construct a 90% confidence interval for α based on this value.

11.17. Assume that the survival time data in Exercise 11.6 are distributed according to a Weibull distribution.

a. Find the ML estimates of the Weibull parameters α and γ, using (11.6) and (11.7). (To use (11.7), select a limited series of values for γ and see if the equation can be satisfied by any of these values. If not, narrow your search by choosing a finer grid of values near the value of γ that came closest to satisfying the equation. We recommend limiting your search with these data to the range $0.5 < \gamma < 3.5$.)

b. Calculate the standard errors for your estimators of α and γ from part (a).

c. Using the standard errors from part (b), compute 95% confidence intervals for α and γ, using (11.8) and (11.9).

11.18. Assume the survival data in Exercise 11.5 are distributed according to a Weibull distribution. Using the results from Exercise 11.15a, give a parametric estimate of the median survival time, LT_{50}, for these data. How does this compare with the nonparametric estimate from Exercise 11.14a?

11.19. Assume the survival data in Exercise 11.6 are distributed according to a Weibull distribution. Using the results from Exercise 11.17a, give a parametric estimate of the median survival time, LT_{50}, for these data. How does this compare with the nonparametric estimate from Exercise 11.14b?

11.20. For survival-time data assumed distributed according to a Weibull distribution with parameters α and γ, what are the general parametric forms for estimates of the time to 90% survival (LT_{90}) and to 10% survival (LT_{10})?

11.21. For survival time data assumed exponentially distributed with parameter α, what is the general parametric form for an estimate of the median survival time LT_{50}?

11.22. For survival time data assumed exponentially distributed with parameter α, what are the general parametric forms for estimates of the time to 90% survival (LT_{90}) and to 10% survival (LT_{10})?

11.23. (Dixon and Newman, 1991) In a study to compare the effect of dissolved oxygen (O_2) concentrations on survival in speckled trout, fish were exposed to one of two different O_2 concentrations, 0.94 and 1.16 mg/l. The survival times (in minutes) were:

at 0.94 mg/l: 20, 22, 24, 26, 26, 29, 29, 31, 34, 34, 41
at 1.16 mg/l: 30, 30, 35, 35, 40, 45, 50, 58, 62, 70, 100.

Perform a log-rank test for these data to assess whether any significant difference exists in the survival patterns. Set $\alpha = 0.05$.

11.24. Refer back to the data in Exercises 11.7 and 11.8. Notice that these two data sets represent survival of different sexes of rat after exposure to the same environmental toxin. To assess whether the two survival patterns are similar, compute a log-rank test for the collected male and female data. Set $\alpha = 0.05$.

11.25. Associated with the experiment from Exercise 11.5, a second, independent set of $N = 7$ grasshoppers were exposed to a different temperature (10°C) to study their mortality patterns after cold shock. This second group exhibited survival times (in minutes) of 2.2, 7.5, 10.9, 22.6, 29.5, 38.5, 44.0. Using an accelerated failure-time regression model under a Weibull error assumption, test whether any significant difference is evidenced between the survival patterns in this 10°C group and the 4°C group from Exercise 11.5. Set $\alpha = 0.10$.

11.26. Verify the comment in Example 11.3 that with the spider departure data, a plot of $\log\{-\log[\hat{S}_{PL}(t_i)]\}$ versus $\log(t_i)$ for each species exhibits roughly linear patterns (and hence use of a Weibull AFT model is reasonable).

11.27. Return to the spider departure data from Example 11.3. Since there is no censoring with these data, we could test for differences between the two species using two-sample methods from previous chapters. For instance:

a. Assume that the distribution of the departure times is unknown, and apply a distribution-free rank-sum test to compare the two species. Set $\alpha = 0.05$. Do your results differ from those achieved in Example 11.3?

b. Since the departure times are positive, a gamma distribution may be an appropriate parent distribution for the data. Calculate the sample mean and sample standard deviations of the two species' data sets. Are the sample coefficients of variation (Section 8.4.2) roughly similar? If so, assume that the distribution of the departure times is gamma, and apply a gamma GLiM with a log link to model the two-group effect. Use your GLiM to compare the two species. Set $\alpha = 0.05$. Do your results differ from those seen using the log-rank test in Example 11.3?

11.28. Return to the spider departure data from Example 11.3. Suppose the experiment was originally designed to cease data collection after 90 seconds. Thus the final two departures times for *M. oblongus* would not have been observed, but would instead have been right-censored at 90 seconds.

a. Assume this is the case with these data and perform a reanalysis using the methods in Section 11.3. Set $\alpha = 0.10$. Do your results differ from those seen in the example?

b. Repeat your reanalysis in part (a), but now assume the right censoring for both groups occurred at 60 seconds. Now how do the results change?

11.29. Repeat Exercise 11.28, but with an accelerated failure-time regression model under a Weibull error assumption, as in Section 11.4.

11.30. The two-sample data from Exercise 11.23 are part of a larger study of O_2 mortality in speckled trout (Dixon and Newman, 1991). In particular, a number of different exposure levels for O_2 were applied, and at any level of O_2 exposure (in mg/l), the data were right-censored at 5000 minutes if any fish lived that long. Including the original data from Exercise 11.23, a larger set of these data is ('+' indicates a right-censored observation):

O_2 *conc.*	*Survival times (min)*
0.77	17, 18, 20, 20, 21, 21, 22, 22, 23
0.94	20, 22, 24, 26, 26, 29, 29, 31, 34, 34, 41
1.16	30, 30, 35, 35, 40, 45, 50, 58, 62, 70, 100
1.55	165, 165, 195, 270, 270, 440, 440, 735, 865, 1400
1.77	240, 675, 995, 2080, 5000^+, 5000^+, 5000^+, 5000^+, 5000^+, 5000^+

Use an accelerated failure-time regression model under a Weibull error assumption to test whether any significant difference is evidenced among the five exposure levels, via:

a. four 0/1 indicator variables that distinguish among the five groups in an ANOVA-type analysis (as in Section 11.4.1), using the lowest concentration (0.77 mg/l) as the reference group.

b. a single regression variable that equals O_2 concentration (that is, test if O_2 concentration affects survival in the AFT model).

11.31. In Exercise 11.30b, compute a 99% confidence interval on the regression coefficient for O_2 exposure. What interpretation do you place on the values from this interval?

11.32. Explain precisely how you would adjust a series of multiple tests among $P > 1$ β_k-coefficients in an AFT regression model using (a) a Bonferroni adjustment, or (b) a Šidák adjustment.

11.33. Apply your methods from Exercise 11.32 to the data in Exercise 11.30a. Set the O_2 concentration at 0.77 mg/l as the 'group 1' reference group.

11.34. Return to the trout survival data in Exercise 11.23. Compute the product-limit survival curve estimate for each level of O_2 concentration. Use X_1 = concentration and plot $\log\{-\log[\hat{S}_{PL}(t_i)]\}$ vs. $\log(t_i)$ over the two levels of X_1. Comment on whether these data appear to satisfy the proportional hazards assumption.

11.35. Extend your results in Exercise 11.34 to the larger set of trout survival data in Exercise 11.30.

11.36*. Show that a Weibull AFT model exhibits the proportional hazards property.

11.37. Use a semi-parametric Cox proportional hazards model to compare survival between males and females in the 1,2-epoxybutane data, from Exercises 11.7 (males) and 11.8 (females). In particular, calculate both the Wald and score test statistics. Set $\alpha = 0.05$. Compare your results with those from Exercise 11.24.

11.38. In an experiment similar to that discussed in Example 11.4, the US National Toxicology Program studied environmental carcinogenesis in female B6C3F$_1$ mice brought on by gavage exposure to benzene (NTP, 1986). Among the many tumors seen in this study, particular interest was directed to malignant lymphomas, which are thought to be strongly lethal in these mice. The response in the vehicle control group showed 15 of 49 animals (31%) with the cancer. $L = 2$ additional dose groups were studied; their lymphoma response rates and exposures, in mg/kg body weight, were 24/45 (53%) at 25 mg/kg, and 24/50 (48%) at 50 mg/kg. (An additional dose group at 100 mg/kg showed 20/49 lymphomas, but we will not include it in our analysis.) Summarized in life-table format into quarterly event data – and placing right-censored observations in a 'terminal sacrifice' category – the data are:

Dose (mg/kg)	*Tumor status*	*Time (quarters)* 3	5	6	7	8	*terminal*	TOTAL
0	Tumors (d_{0j})	0	1	0	1	3	10	15
	At risk (n_{0j})	49	46	45	43	39	30	252
25	Tumors (d_{1j})	0	0	1	3	5	15	24
	At risk (n_{1j})	45	44	44	42	35	26	236
50	Tumors (d_{2j})	0	0	2	5	6	11	24
	At risk (n_{2j})	50	50	50	46	39	24	259
TOTAL	Tumors (d_{+j})	0	1	3	9	14	36	63
	At risk (n_{+j})	144	140	139	131	113	80	747

(Quarters not listed indicate that no new events occurred in that time period.)

a. For these data, perform a simple pairwise analysis on each possible control–dose combination to compare survival patterns under a proportional hazards assumption. That is, compute two test statistics, one comparing control survival with the 25 mg/kg group's survival, and the other comparing control survival with the 50 mg/kg group's survival. Use the methods illustrated in Section 11.3.

b. Repeat the control-vs.-dose analysis in part (a), but now for each of the two control-vs.-dose comparisons, use the Tarone T^2-statistic from (11.22), with only $L = 1$ 'dose' group per comparison.

c. Return to part (a) and adjust your two control-vs.-dose inferences for multiplicity. Use a Bonferroni correction, and operate at a 5% familywise significance level.

d. Apply Tarone's life-table test to these data and assess whether there is any trend in survival over benzene dose. Use Tarone's T^2-statistic from (11.22), and set $\alpha = 0.05$. Does your analysis agree with the Bonferroni-adjusted results from part (c)? Which method do you think is more powerful?

11.39. Use a semi-parametric Cox proportional hazards model to compare survival between O_2 concentrations in the trout survival data from Exercise 11.23. Use a score test, and set $\alpha = 0.01$.

11.40. Employ a Cox proportional hazards model to test for trend in survival across O_2 concentrations in the trout survival data from Exercise 11.30. Use the Tarone life-table test from (11.22), and set $\alpha = 0.01$.

References

Abramowitz, M. and Stegun, I.A. (1972) *Handbook of Mathematical Functions with Formulas, Graphs, and Mathematical Tables*, 10th edn, Wiley Interscience, New York.

Adams, M.J. (1992) Errors and detection limits, in *Methods of Environmental Data Analysis* (ed C.N. Hewitt), Elsevier Applied Science, Amsterdam, pp. 181–212.

Afifi, A.A., Elashoff, R.M. and Lee, J.J. (1986) Simultaneous non-parametric confidence intervals for survival probabilities from censored data. *Statistics in Medicine*, **5**, 653–62.

Ager, D.D. and Haynes, R.H. (1990) Analysis of interactions between mutagens, I. Heat and ultraviolet light in *Saccharomyces cerevisiae*. *Mutation Research*, **232**, 313–26.

Agresti, A. (1990) *Categorical Data Analysis*, John Wiley, New York.

Agresti, A., Wackerly, D. and Boyett, J.M. (1979) Exact conditional tests for cross-classifications: Approximations of attained significance levels. *Psychometrika*, **44**, 75–83.

Aitchison, J. and Silvey, S.D. (1958) Maximum-likelihood estimation of parameters subject to restraints. *Annals of Mathematical Statistics*, **29**, 813–28.

Ames, B.N., McCann, J. and Yamasaki, E. (1975) Methods for detecting carcinogens and mutagens with the *Salmonella*/mammalian microsome mutagenicity test. *Mutation Research*, **31**, 347–64.

Andersen, P.K. (1991) Survival analysis 1982–1991: the second decade of the proportional hazards regression model. *Statistics in Medicine*, **10**, 1931–44.

Anderson, J.R., Bernstein, L. and Pike, M.C. (1982) Approximate confidence intervals for probabilities of survival and quantiles in life-table analysis. *Biometrics*, **38**, 407–16.

Anderson, K.M. (1991) A nonproportional hazards Weibull accelerated failure time regression model. *Biometrics*, **47**, 281–8.

Anscombe, F.J. (1948) The transformation of Poisson, binomial and negative binomial data. *Biometrika*, **35**, 246–54.

Antonello, J.M., Clark, R.L. and Heyse, J.F. (1993) Application of the Tukey trend test procedure to assess developmental and reproductive toxicity. I. Measurement data. *Fundamental and Applied Toxicology*, **21**, 52–8.

Aranda-Ordaz, F.J. (1981) On two families of transformations to additivity for binary response data. *Biometrika*, **68**, 357–63.

Archer, L.E. and Ryan, L.M. (1989) Accounting for misclassification in the cause-of-death test for carcinogenicity. *Journal of the American Statistical Association*, **84**, 787–91.

Armitage, P. (1955) Tests for linear trends in proportions and frequencies. *Biometrics*, **11**, 375–86.

Armstrong, B.G. (1992) Confidence intervals for arithmetic means of lognormally distributed exposures. *AIHA Journal*, **53**, 481–5.

Au, W.W., Heo, M.-Y. and Chiewchanwit, T. (1994) Toxicological interactions between nickel and radiation on chromosome damage and repair. *Environmental Health Perspectives*, **102**, Suppl. 9, 73–8.

Ayer, M., Brunk, H.D., Ewing, G.M., Reid, W.T. and Silverman, E. (1955) An empirical distribution function for sampling with incomplete information. *Annals of Mathematical Statistics*, **26**, 641–7.

Babu, G.J. and Rao, C.R. (1993) Bootstrap methodology, in *Handbook of Statistics Volume 9: Computational Statistics* (ed C.R. Rao), North-Holland/Elsevier, Amsterdam, pp. 627–60.

Babu, G.J. and Singh, K. (1983) Inference on means using the bootstrap. *Annals of Statistics*, **11**, 999–1003.

Bailer, A.J. (1989) Testing variance equality with randomized tests. *Journal of Statistical Computation and Simulation*, **31**, 1–8.

Bailer, A.J. and Oris, J.T. (1993) Modeling reproductive toxicity in *Ceriodaphnia* tests. *Environmental Toxicology and Chemistry*, **12**, 787–91.

Bailer, A.J. and Oris, J.T. (1994) Assessing toxicity of pollutants in aquatic systems, in *Case Studies in Biometry* (eds N. Lange, L. Ryan, L. Billard, D. Brillinger, L. Conquest and J. Greenhouse), John Wiley, New York, pp. 25–40.

Bailer, A.J. and Oris, J.T. (1996) Implications of defining test acceptability in terms of control-group survival in two-group survival studies. *Environmental Toxicology and Chemistry*, **15**, 1242–4.

Bailer, A.J. and Piegorsch, W.W. (1990) Estimating integrals using quadrature methods with an application in pharmacokinetics. *Biometrics*, **46**, 1201–11.

Bailer, A.J. and Portier, C.J. (1988) Effects of treatment-induced mortality and tumor-induced mortality on tests for carcinogenicity in small samples. *Biometrics*, **44**, 417–31.

Bailer, A.J. and Smith, R.J. (1994) Estimating upper confidence limits for extra risk in quantal multistage models. *Risk Analysis*, **14**, 1001–10.

Barndorff-Nielsen, O.E. (1988) *Parametric Statistical Models and Likelihood*, Springer-Verlag, New York.

Bartlett, M.S. (1935) Contingency table interaction. *Journal of the Royal Statistical Society, supplement*, **2**, 248–52.

Bartlett, M.S. (1937) Properties of sufficiency and statistical tests. *Proceedings of the Royal Society, series A*, **160**, 268–82.

Bates, D.M. and Chambers, J.M. (1992) Nonlinear models, in *Statistical Models in S* (eds J.M. Chambers and T.J. Hastie), Wadsworth Brooks/Cole, Pacific Grove, CA, pp. 421–54.

Beal, S.L. (1987) Asymptotic confidence intervals for the difference between two binomial parameters for use with small samples. *Biometrics*, **43**, 941–50.

Bechhofer, R.E. and Dunnett, C.W. (1988) Percentage points of the multivariate Student t distribution, in *Selected Tables in Mathematical Statistics* **11** (eds R.E. Odeh and J.M. Davenport), American Mathematical Society, Providence, RI, pp. 1–371.

Becka, M., Bolt, H.M. and Urfer, W. (1992) Statistical analysis of toxicokinetic data by nonlinear regression (Example – Inhalation pharmacokinetics of propylene). *Archives of Toxicology*, **66**, 450–3.

Becka, M., Bolt, H.M. and Urfer, W. (1993) Statistical evaluation of toxicokinetic data. *Environmetrics*, **4**, 311–22.

Behrens, W.U. (1929) Ein Beitrag zur Fehlerberechnung bei wenigen Beobachtungen. *Landwirtschaftliche Jahrbücher*, **68**, 807–37.

Berger, R.L. (1996) More powerful tests from confidence interval *p* values. *American Statistician*, **50**, 314–18.

Berger, R.L. and Boos, D.D. (1994) *P* values maximized over a confidence set for the nuisance parameter. *Journal of the American Statistical Association*, **89**, 1012–16.

Bernardo, J.M. (1976) Algorithm AS103. The digamma function. *Applied Statistics*, **25**, 315–17.

Bieler, G.S. and Williams, R.L. (1993) Ratio estimates, the delta method, and quantal response tests for increased carcinogenicity. *Biometrics*, **49**, 793–801.

Bieler, G.S. and Williams, R.L. (1995) Cluster sampling techniques in quantal response teratology and developmental toxicology studies. *Biometrics*, **51**, 764–76.

Bishop, Y.M.M., Fienberg, S.E. and Holland, P.W. (1975) *Discrete Multivariate Analysis: Theory and Practice*, MIT Press, Cambridge, MA.

Bliss, C.I. (1935) The calculation of dosage-mortality curves. *Annals of Applied Biology*, **22**, 134–67.

Blot, N.J. and Day, N.E. (1979) Synergism and interaction: are they equivalent? *American Journal of Epidemiology*, **110**, 99–100.

Bock, J. and Toutenburg, H. (1991) Sample size determination in clinical research, In *Handbook of Statistics Volume 8: Statistical Methods in Biological and Medical Sciences* (eds C.R. Rao and R. Chakraborty), North-Holland, Amsterdam, pp. 515–38.

Bond, J.A., Csanady, G.A., Gargas, M.L., Guengerich, F.P., Leavens, T., Medinsky, M.A. and Recio, L. (1994) 1,3-butadiene: Linking metabolism, dosimetry, and mutation induction. *Environmental Health Perspectives*, **102**, Suppl. 9, 87–94.

Boos, D.D. (1992) On generalized score tests. *American Statistician*, **46**, 327–33.

Borgan, Ø. and Liestøl, K. (1990) A note on confidence intervals and bands for the survival function based on transformations. *Scandinavian Journal of Statistics*, **17**, 35–41.

Box, G.E.P. and Cox, D.R. (1964) An analysis of transformations (with discussion). *Journal of the Royal Statistical Society, series B*, **26**, 211–52.

Box, G.E.P., Jenkins, G.M. and Reinsel, G. (1994) *Time Series Analysis: Forecasting and Control*, 2nd edn, Prentice Hall, Englewood Cliffs, NJ.

Boyd, M.N. (1982) Examples of testing against ordered alternatives in the analysis of mutagenicity data. *Mutation Research*, **97**, 147–53.

Bradley, R.A. and Srivastava, S.S. (1979) Correlation in polynomial regression. *Annals of Statistics*, **33**, 11–14.

Breslow, N.E. (1990) Tests of hypothesis in overdispersed Poisson regression and other quasi-likelihood models. *Journal of the American Statistical Association*, **85**, 565–71.

Breslow, N.E. and Day, N.E. (1980) *Statistical Methods in Cancer Research. I. The Analysis of Case-Control Studies*, IARC Scientific Publications, Lyon, France.

Broekhoven, L.H. and Nestmann, E.R. (1991) Statistical analysis of the Salmonella mutagenicity assay, in *Statistics in Toxicology* (eds D. Krewski and C. Franklin), Gordon and Breach, New York, pp. 205–64.

Burridge, J. and Sebastiani, P. (1994) *D*-optimal designs for generalised linear models with variance proportional to the square of the mean. *Biometrika*, **81**, 295–304.

Butterworth, E.W. and Quiring, D.T. (1994) Genotype and environment interact to influence acceptability and suitability of white spruce for a specialist herbivore, *Zeiraphera canadensis*. *Ecological Entomology*, **19**, 230–44.

Calabrese, E.J. and Baldwin, L.A. (1993) Possible examples of chemical hormesis in a previously published study. *Journal of Applied Toxicology*, **13**, 169–72.

Cariello, N.F., Piegorsch, W.W., Adams, W.T. and Skopek, T.R. (1994) Computer program for the analysis of mutational spectra: application to *p53* mutations. *Carcinogenesis*, **15**, 2281–5.

Carr, G.J. and Chi, E.M. (1992) Analysis of variance for repeated measures data: A generalized estimating equations approach. *Statistics in Medicine*, **11**, 1033–40.

Carr, G.J. and Gorelick, N.J. (1995) Statistical design and analysis of mutation studies in transgenic mice. *Environmental and Molecular Mutagenesis*, **25**, 246–55.

Carroll, R.J. and Ruppert, D. (1988) *Transformation and Weighting in Regression*, Chapman & Hall, New York.

Casagrande, J.T., Pike, M.C. and Smith, P.G. (1978) An improved approximate formula for calculating sample sizes for comparing two binomial proportions. *Biometrics*, **34**, 483–6.

Casella, G. (1986) Refining binomial confidence intervals. *Canadian Journal of Statistics*, **14**, 113–29.

Casella, G. and Berger, R.L. (1990) *Statistical Inference*, Duxbury Press, Belmont, CA.

Casella, G. and Robert, C. (1989) Refining Poisson confidence intervals. *Canadian Journal of Statistics*, **17**, 45–57.

Caswell, H. (1989) *Matrix Population Models: Construction, Analysis, and Interpretation*, Sinauer Associates, Sunderland, MA.

Chakraborti, S. and Gibbons, J.D. (1991) One-sided nonparametric comparison of treatments with a standard in the one-way layout. *Journal of Quality Technology*, **23**, 102–6.

Chakraborti, S. and Gibbons, J.D. (1993) Nonparametric comparison of treatments with a standard in the one-way layout: Some design considerations. *Communications in Statistics – Theory and Methods*, **22**, 1–14.

Chambers, J.M. (1992a) Data for models, in *Statistical Models in S* (eds J.M. Chambers and T.J. Hastie), Wadsworth Brooks/Cole, Pacific Grove, CA, pp. 45–94.

Chambers, J.M. (1992b) Linear models, in *Statistical Models in S* (eds J.M. Chambers and T.J. Hastie), Wadsworth Brooks/Cole, Pacific Grove, CA, pp. 95–144.

Chambers, J.M., Freeny, A.E. and Heiberger, R.M. (1992) Analysis of variance; designed experiments, in *Statistical Models in S* (eds J.M. Chambers and T.J. Hastie), Wadsworth Brooks/Cole, Pacific Grove, CA, pp. 145–93.

Chambers, J.M. and Hastie, T.J. (1992a) Statistical models, in *Statistical Models in S* (eds J.M. Chambers and T.J. Hastie), Wadsworth Brooks/Cole, Pacific Grove, CA, pp. 13–44.

Chambers, J.M. and Hastie, T.J. (eds) (1992b) *Statistical Models in S*, Wadsworth Brooks/Cole, Pacific Grove, CA.

Chanter, D.O. (1984) Curtailed sigmoid dose-response models for fungicide experiments. *Applied Statistics*, **11**, 2–11.

Chapman, P.M., Caldwell, R.S. and Chapman, P.F. (1996) A warning: NOECs are inappropriate for regulatory use. *Environmental Toxicology and Chemistry*, **15**, 77–9.

Chatfield, C. (1989) *The Analysis of Time Series: An Introduction*, 4th edn, Chapman & Hall, New York.

Chen, Y.I. and Wolfe, D.A. (1990) A study of distribution-free tests for umbrella alternatives. *Biometrical Journal*, **32**, 47–57.

Cheng, K.F. and Wu, J.F. (1994) Testing goodness of fit for a parametric family of link functions. *Journal of the American Statistical Association*, **89**, 657–64.

Cherian, M.G. (1994) The significance of the nuclear and cytoplasmic localization of metallothionein in human liver and tumor cells. *Environmental Health Perspectives*, **102**, Suppl. 3, 130–5.

Christensen, R. (1991) *Linear Models for Multivariate, Time Series and Spatial Data*, Springer-Verlag, New York.

Clark, S.J. and Perry, J.N. (1989) Estimation of the negative binomial parameter κ by maximum quasi-likelihood. *Biometrics*, **45**, 309–16.

Cochran, W.G. (1954) Some methods for strengthening the common χ^2 tests. *Biometrics*, **10**, 417–51.

Cochran, W.G. and Cox, G.M. (1957) *Experimental Designs*, 2nd edn, John Wiley, New York.

Collett, D. (1994) *Modelling Survival Data in Medical Research*, Chapman & Hall, London.

Collings, B.J. and Margolin, B.H. (1985) Testing goodness of fit for the Poisson assumption when observations are not identically distributed. *Journal of the American Statistical Association*, **80**, 411–18.

Collings, B.J., Margolin, B.H. and Oehlert, G.W. (1981) Analyses for binomial data, with applications to the fluctuation test for mutagenicity. *Biometrics*, **37**, 775–94.

Cordeiro, G.M. (1983) Improved likelihood ratio statistics for generalized linear models. *Journal of the Royal Statistical Society, series B*, **45**, 404–13.

Cordeiro, G.M. and McCullagh, P. (1991) Bias correction in generalized linear models. *Journal of the Royal Statistical Society, series B*, **53**, 629–43.

Cordeiro, G.M., De Paula Ferrari, S.L. and Paula, G.A. (1993) Improved score tests for generalized linear models. *Journal of the Royal Statistical Society, series B*, **55**, 661–74.

Cordeiro, G.M., Botter, D.A. and Ferrari, S.L.d.P. (1994a) Nonnull asymptotic distributions of three classic criteria in generalised linear models. *Biometrika*, **81**,

709–20.

Cordeiro, G.M., Paula, G.A. and Botter, D.A. (1994b) Improved likelihood ratio tests for dispersion models. *International Statistical Review*, **62**, 257–74.

Cothern, C.R. and Ross, N.P. (eds) (1994) *Environmental Statistics, Assessment, and Forecasting*, Lewis Publishers, Boca Raton, FL.

Cox, C. (1987) Threshold dose-response models in toxicology. *Biometrics*, **43**, 511–23.

Cox, D.R. (1958) The regression analysis of binary sequences (with discussion). *Journal of the Royal Statistical Society, series B*, **20**, 215–42.

Cox, D.R. (1972) Regression models and life tables. *Journal of the Royal Statistical Society, series B*, **34**, 187–220.

Cox, D.R. (1988) Some aspects of conditional and asymptotic inference: A review. *Sankhyā, series A*, **50**, 314–37.

Cressie, N.A.C. (1993) *Statistics for Spatial Data*, John Wiley, New York.

Cressie, N.A.C. and Read, T.R.C. (1989) Pearson's X^2 and the loglikelihood ratio statistic G^2: A comparative review. *International Statistical Review*, **57**, 19–43.

D'Agostino, R., Chase, W. and Belanger, A. (1988) The appropriateness of some common procedures for testing equality of two independent binomial proportions. *American Statistician*, **42**, 198–202.

Dahiya, R.C. and Guttman, I. (1982) Shortest confidence and prediction intervals for the log-normal. *Canadian Journal of Statistics*, **10**, 277–91.

Darroch, J.N. and Borkent, M. (1994) Synergism, attributable risk and interaction for two binary exposure variables. *Biometrika*, **81**, 259–70.

Davidian, M. (1990) Estimation of variance functions in assays with possibly unequal replication and nonnormal data. *Biometrika*, **77**, 43–54.

Dean, C.B. (1992) Testing for overdispersion in Poisson and binomial regression models. *Journal of the American Statistical Association*, **87**, 451–7.

Dimitrov, B.D. (1994) Types of chromosomal aberrations induced by seed aging in *Crepis capillaris*. *Environmental and Molecular Mutagenesis*, **23**, 318–22.

Dinse, G.E. (1988) Simple parametric analysis of animal tumorigenicity data. *Journal of the American Statistical Association*, **83**, 638–49.

Dinse, G.E. (1993) Evaluating constraints that allow survival-adjusted incidence analysis in single-sacrifice studies. *Biometrics*, **49**, 399–407.

Dinse, G.E. (1994) A comparison of tumor incidence analyses applicable in single-sacrifice animal experiments. *Statistics in Medicine*, **13**, 689–708.

Dinse, G.E. and Haseman, J.K. (1986) Logistic regression analysis of incidental-tumor data from animal carcinogenicity experiments. *Fundamental and Applied Toxicology*, **6**, 44–52.

Dinse, G.E. and Lagakos, S.W. (1983) Regression analysis of tumour prevalence data. *Applied Statistics*, **32**, 236–48.

Dixon, P.M. and Newman, M.C. (1991) Analyzing toxicity data using statistical models for time-to-death: an introduction, in *Metal Ecotoxicology, Concepts and Applications* (eds M.C. Newman and A.W. McIntosh), Lewis Publishers, Chelsea, MI, pp. 207–42.

Dixon, W.J. and Mood, A.M. (1948) A method for obtaining and analyzing sensitivity data. *Journal of the American Statistical Association*, **43**, 109–26.

Dobo, K.L. and Eastmond, D.A. (1994) Role of oxygen radicals in the chromosomal loss and breakage induced by the quinone-forming compounds, hydroquinone and *tert*-butylhydroquinone. *Environmental and Molecular Mutagenesis*, **24**, 293–300.

Dobson, A.J. (1990) *An Introduction to Generalized Linear Models*, Chapman & Hall, London.

Draper, D., Hodges, J.S., Mallows, C.L. and Pregibon, D. (1993) Exchangeability and data analysis (with discussion). *Journal of the Royal Statistical Society, series A*, **156**, 9–37.

Draper, N.R. and Smith, H. (1981) *Applied Regression Analysis*, 2nd edn, John Wiley, New York.

Duffy, D.E. (1990) On continuity-corrected residuals in logistic regression. *Biometrika*, **77**, 287–94.

Dulout, F.N. and Natarajan, A.T. (1987) A simple and reliable *in vitro* test system for the analysis of induced aneuploidy as well as other cytogenetic end-points using Chinese hamster cells. *Mutagenesis*, **2**, 121–6.

Dunnett, C.W. (1955) A multiple comparison procedure for comparing several treatments with a control. *Journal of the American Statistical Association*, **50**, 1096–121.

Dunnett, C.W. (1964) New tables for multiple comparisons with a control. *Biometrics*, **20**, 482–92.

Dunnett, C.W. (1980) Pairwise multiple comparisons in the unequal variance case. *Journal of the American Statistical Association*, **75**, 796–800.

Dunnett, C.W. (1989) Multivariate normal probability integrals with product correlation structure. *Applied Statistics*, **38**, 564–79.

Dunnett, C.W. and Sobel, M. (1954) A bivariate generalization of Student's *t*-distribution, with tables for certain special cases. *Biometrika*, **41**, 154–69.

Dunnett, C.W. and Tamhane, A.C. (1993) Power comparisons of some step-up multiple test procedures. *Statistics and Probability Letters*, **16**, 55–8.

Durham, S.D. and Flournoy, N. (1995) Up-and-down designs I: Stationary treatment distributions, in *Adaptive Designs* (eds N. Flournoy and W.F. Rosenberger), Institute of Mathematical Statistics, Hayward, CA, pp. 139–57.

Eastwood, B.J. (1993) Semi-parametric estimates of dose-response mutagenic potency of environmental chemicals and an assessment of inter- and intra-laboratory variation using a short-term bioassay. *Canadian Journal of Statistics*, **21**, 436–48.

Edler, L. (1992) Statistical methods for short-term tests in genetic toxicology: The first fifteen years. *Mutation Research*, **277**, 11–33.

Edwards, C.H., Jr and Penney, D.E. (1990) *Calculus and Analytic Geometry*, 3rd edn, Prentice Hall, Englewood Cliffs, NJ.

Efron, B. (1986) Double exponential families and their use in generalized linear regression. *Journal of the American Statistical Association*, **81**, 709–21.

Efron, B. and Tibshirani, R. (1993) *Introduction to the Bootstrap*, Chapman & Hall, New York.

Facemire, C.F., Gross, T.S. and Guillette, L.J.J. (1995) Reproductive impairment in the Florida panther: Nature or nurture? *Environmental Health Perspectives*, **103**, Suppl. 4, 79–86.

Favor, J., Sund, M., Neuhäuser-Klaus, A. and Ehling, U.H. (1990) A dose-response analysis of ethylnitrosourea-induced recessive specific-locus mutations in treated spermatogonia of the mouse. *Mutation Research*, **231**, 47–54.

Fears, T.R., Benichou, J. and Gail, M.H. (1996) A reminder of the fallibility of the Wald statistic. *American Statistician*, **50**, 226–7.

Federer, W.T. (1976) Sampling, blocking, and model considerations for the *r*-row by *c*-column experiment design. *Biometrische Zeitschrift*, **18**, 595–607.

Fei, H. and Kong, F. (1994) Interval estimates for one- and two-parameter exponential distributions under multiple type II censoring. *Communications in Statistics – Theory and Methods*, **23**, 1717–33.

Fieller, E.C. (1940) The biological standardization of insulin. *Journal of the Royal Statistical Society, series B*, **7**, 1–53.

Finette, B.A., Sullivan, L.M., O'Neill, J.P., Nicklas, J.A., Vacek, P.M. and Albertini, R.J. (1994) Determination of *hprt* mutant frequencies in T-lymphocytes from a healthy pediatric population: statistical comparison between newborn, children and adult mutant frequencies, cloning efficiency and age. *Mutation Research*, **308**, 223–32.

Finkelstein, D.M. and Ryan, L.M. (1987) Estimating carcinogenic potency from a rodent tumorigenicity experiment. *Applied Statistics*, **36**, 121–33.

Finney, D.J. (1952) *Statistical Method in Biological Assay*, Chas. Griffin & Co., London.

Finney, D.J. (1971) *Probit Analysis*, 3rd edn, Cambridge University Press, Cambridge.

Fisher, N.I. (1983) Graphical methods in nonparametric statistics: A review and annotated bibliography. *International Statistical Review*, **51**, 25–58.

Fisher, R.A. (1912) On an absolute criterion for fitting frequency curves. *Messenger of Mathematics*, **41**, 155–60.

Fisher, R.A. (1922) On the mathematical foundations of theoretical statistics. *Philosophical Transactions of the Royal Society*, **222**, 309–68.

Fisher, R.A. (1935a) *The Design of Experiments*, Oliver & Boyd, New York.

Fisher, R.A. (1935b) The logic of inductive inference (with discussion). *Journal of the Royal Statistical Society, series A*, **98**, 39–82.

Fisher, R.A. (1939) The comparison of samples with possibly unequal variances. *Annals of Eugenics*, **9**, 174–80.

Fisher, R.A. (1950) The significance of deviations from expectation in a Poisson series. *Biometrics*, **6**, 17–24.

Fisher, R.A. and Yates, F. (1957) *Statistical Tables*, 5th edn, Oliver & Boyd, Edinburgh.

Fisher, R.A., Thornton, H.G. and MacKenzie, W.A. (1922) The accuracy of the plating method of estimating the density of bacterial populations. *Journal of Applied Biology*, **9**, 325–59.

Fleiss, J.L. (1981) *Statistical Methods for Rates and Proportions*, 2nd edn, John Wiley, New York.

Francis, B., Green, M. and Payne, C. (eds) (1993) *The GLIM System – Release 4 – Manual*, Numerical Algorithms Group, Oxford.

Freeman, M.F. and Tukey, J.W. (1950) Transformations related to the angular and the square root. *Annals of Mathematical Statistics*, **21**, 607–11.

Frome, E.L., Smith, M.H., Littlefield, L.G., Neubert, R.L. and Colyer, S.P. (1996) Statistical methods for the blood beryllium lymphocyte proliferation test. *Environmental Health Perspectives*, **104**, Suppl. 5, 957–68.

Fu, Y.-X. and Arnold, J. (1992) A table of exact sample sizes for use with Fisher exact test for 2×2 tables. *Biometrics*, **48**, 1103–12.

Fung, K.Y., Krewski, D., Rao, J.N.K. and Scott, A.J. (1994) Tests for trend in developmental toxicity experiments with correlated binary data. *Risk Analysis*, **14**, 639–48.

Gallant, A.R. and Fuller, W.A. (1973) Fitting segmented polynomial regression models whose join points have to be estimated. *Journal of the American Statistical Association*, **68**, 144–7.

Galloway, S.M., Bloom, A.D., Resnick, M., Margolin, B.H., Nakamura, F., Archer, P. and Zeiger, E. (1985) Development of a standardized protocol for in vitro cytogenetic testing with Chinese hamster ovary cells: Comparisons of results for 22 compounds in two laboratories. *Environmental Mutagenesis*, **7**, 1–51.

Galloway, S.M., Berry, P.K., Nichols, W.W., Wolman, S.R., Soper, K.A., Stolley, P.D. and Archer, P. (1986) Chromosome aberrations in individuals occupationally exposed to ethylene oxide, and in a large control population. *Mutation Research*, **170**, 55–74.

Gart, J.J. and Thomas, D.C. (1982) The performance of three approximate confidence limit methods for the odds ratio. *American Journal of Epidemiology*, **115**, 453–70.

Gart, J.J., Chu, K.C. and Tarone, R.E. (1979) Statistical issues in interpretation of chronic bioassay tests for carcinogenicity. *Journal of the National Cancer Institute*, **62**, 957–74.

Garthwaite, P.H. and Buckland, S.T. (1992) Generating Monte Carlo confidence intervals by the Robbins–Monro process. *Applied Statistics*, **41**, 159–71.

Garwood, F. (1936) Fiducial limits for the Poisson distribution. *Biometrika*, **28**, 437–42.

Gaver, D., Draper, D., Goel, P., Greenhouse, J., Hedges, L., Morris, M. and Waternaux, C. (1992) *Combining Information: Statistical Issues and Opportunities for Research*, National Academy Press, Washington, DC.

Gaver, D.P., Jacobs, P.A. and O'Muircheartaigh, I.G. (1990) Regression analysis of hierarchical Poisson-like event rate data: Superpopulation model effects on predictions. *Communications in Statistics – Theory and Methods*, **19**, 3779–97.

Gaylor, D.W., Sheehan, D.M., Young, J.F. and Mattison, D.R. (1988) The threshold dose question in teratogenesis. *Teratology*, **38**, 389–91.

Genizi, A. and Marcus, R. (1994) Some multiple comparison procedures for comparing treatments within several groups. *Communications in Statistics – Theory and Methods*, **23**, 763–79.

George, V.T. and Elston, R.C. (1993) Confidence limits based on the first occurrence of an event. *Statistics in Medicine*, **12**, 685–90.

Gibbons, J.D. (1993) *Nonparametric Statistics: An Introduction*, Sage Publications, Newbury Park, CA.

Godambe, V.P. (ed) (1991) *Estimating Functions*, Oxford University Press, Oxford.

Gombar, V.K., Enslein, K., Hart, J.B., Blake, B.W. and Borgstedt, H.H. (1991)

Estimation of maximum tolerated dose for long-term bioassays from acute lethal dose and structure by QSAR. *Risk Analysis*, **11**, 509–17.

Gompertz, B. (1825) On the nature of the function expressive of the law of human mortality. *Philosophical Transactions of the Royal Society*, **155**, 513–93.

Good, P. (1994) *Permutation Tests. A Practical Guide to Resampling Methods for Testing Hypotheses*, Springer-Verlag, New York.

Goodman, S.N. (1993) *p* values, hypothesis tests, and likelihood: Implications for epidemiology of a neglected historical debate (with discussion). *American Journal of Epidemiology*, **137**, 485–501.

Govindarajulu, Z. (1988) *Statistical Techniques in Bioassay*, Karger, Basel.

Green, A.S., Chandler, G.T. and Piegorsch, W.W. (1996) Life-stage-specific toxicity of sediment-associated chlorpyrifos to a marine, infaunal copepod. *Environmental Toxicology and Chemistry*, **15**, 1182–231.

Green, R.H. (1979) *Sampling Design and Statistical Methods for Environmental Biologists*, John Wiley, New York.

Green, R.H., Boyd, J.M. and Macdonald, J.S. (1993) Relating sets of variables in environmental studies: the sediment quality triad as a paradigm. *Environmetrics*, **4**, 439–57.

Greenwood, M. (1926) *The 'Error of Sampling' of the Survivorship Tables. Reports on Public Health and Medical Subjects, No. 33, Appendix 1*, HM Stationery Office, London.

Greenwood, P.E. and Nikulin, M.S. (1996) *A Guide to Chi-Squared Testing*, John Wiley, New York.

Griffiths, P. and Hill, I.D. (eds) (1985) *Applied Statistics Algorithms*, Ellis Horwood, Chichester.

Guegan, D. and Pham, D.T. (1989) A note on the estimation of the parameters of the diagonal bilinear model by the method of least squares. *Scandinavian Journal of Statistics*, **16**, 129–36.

Gulati, S. and Padgett, W.J. (1996) Families of smooth confidence bands for the survival function under the general random censorship model. *Lifetime Data Analysis*, **2**, 349–62.

Haining, R. (1990) *Spatial Data Analysis in the Social and Environmental Sciences*, Cambridge University Press, Cambridge.

Hakulinen, T. and Dyba, T. (1994) Precision of incidence predictions based on Poisson distributed errors. *Statistics in Medicine*, **13**, 1513–23.

Hallenbeck, W.H. (1993) *Quantitative Risk Assessment for Environmental and Occupational Health*, CRC Press, Boca Raton, FL.

Halley, E. (1693) An estimate of the degrees of the mortality of mankind, drawn from curious tables of the births and funerals at the city of Breslau; with an attempt to ascertain the price of annuities upon lives. *Philosophical Transactions of the Royal Society of London*, **17**, 596–610, 654–6.

Hamada, C., Wada, T. and Sakamoto, Y. (1994) Statistical characterization of negative control data in the Ames Salmonella/microsome test. *Environmental Health Perspectives*, **102**, Suppl. 1, 115–19.

Hamilton, M.A. (1991) Estimation of the typical lethal dose in acute toxicity studies, in *Statistics in Toxicology* (eds D. Krewski and C. Franklin), Gordon and Breach, New York, pp. 61–88.

Hamilton, M.A., Russo, R.C. and Thurston, R.V. (1977) Trimmed Spearman-Karber method for estimating median lethal concentrations in toxicity bioassays. *Environmental Science & Technology*, **11**, 714–19.

Hardy, G.H. (1908) Mendelian proportions in a mixed population. *Science*, **28**, 49–50.

Haseman, J.K. (1978) Exact sample sizes for use with the Fisher–Irwin test for 2×2 tables. *Biometrics*, **34**, 106–9.

Haseman, J.K. (1984) Statistical issues in the design, analysis and interpretation of animal carcinogenicity studies. *Environmental Health Perspectives*, **58**, 385–92.

Haseman, J.K. (1992) Value of historical controls in the interpretation of rodent tumor data. *Drug Information Journal*, **26**, 191–200.

Haseman, J.K. and Hogan, M.D. (1975) Selection of the experimental unit in teratology studies. *Teratology*, **12**, 165–72.

Haseman, J.K. and Lockhart, A. (1994) The relationship between use of the maximum tolerated dose and study sensitivity for detecting rodent carcinogenicity. *Fundamental and Applied Toxicology*, **22**, 382–91.

Haseman, J.K. and Piegorsch, W.W. (1994) Statistical analysis of developmental toxicity data, in *Developmental Toxicology*, 2nd edn., (eds C. Kimmel and J. Buelke-Sam), Raven Press, New York, pp. 349–61.

Haseman, J.K., Huff, J.E., Rao, G.N., Arnold, J.E., Boorman, G.A. and McConnell, E.E. (1985) Neoplasms observed in untreated and corn oil gavage control groups of F344/N rats and (C57BL/6N × C3H/HeN)F_1 mice. *Journal of the National Cancer Institute*, **75**, 975–84.

Haseman, J.K., Winbush, J.S. and O'Donnell, M.W., Jr (1986) Use of dual control groups to estimate false positive rates in laboratory animal carcinogenicity studies. *Fundamental and Applied Toxicology*, **7**, 573–84.

Haseman, J.K., Huff, J.E., Rao, G.N. and Eustis, S.L. (1989) Sources of variability in rodent carcinogenicity studies. *Fundamental and Applied Toxicology*, **12**, 793–804.

Hastie, T. and Tibshirani, R. (1990) Exploring the nature of covariate effects in the proportional hazards model. *Biometrics*, **46**, 1005–16.

Hastie, T.J. and Pregibon, D. (1992) Generalized linear models, in *Statistical Models in S* (eds J.M. Chambers and T.J. Hastie), Wadsworth Brooks/Cole, Pacific Grove, CA, pp. 195–247.

Hauck, W.W. and Donner, A. (1977) Wald's test as applied to hypotheses in logit analysis. *Journal of the American Statistical Association*, **72**, 851–3.

Haynes, R.H., Eckardt, F. and Kunz, B.A. (1984) The DNA damage-repair hypothesis in radiation biology: Comparison with classical hit theory. *British Journal of Cancer*, **49**, Suppl. 6, 81–90.

Hayter, A.J. (1984) A proof of the conjecture that the Tukey-Kramer multiple comparisons procedure is conservative. *Annals of Statistics*, **12**, 61–75.

Hayter, A.J. (1990) A one-sided Studentized range test for testing against a simple ordered alternative. *Journal of the American Statistical Association*, **85**, 778–85.

Hayter, A.J. and Hsu, J.C. (1994) On the relationship between stepwise decision procedures and confidence sets. *Journal of the American Statistical Association*, **89**, 128–36.

Hayter, A.J. and Stone, G. (1991) Distribution free multiple comparisons for monotonically ordered treatment effects. *Australian Journal of Statistics*, **33**, 335–46.

Hedges, L.V. and Olkin, I. (1985) *Statistical Methods for Meta-Analysis*, Academic Press, Orlando, FL.

Hertzberg, R.C. and Miller, M. (1985) A statistical model for species extrapolation using categorical response data. *Toxicology and Industrial Health*, **1**, 43–57.

Hettmansperger, T.P. (1984) *Statistical Inference Based on Ranks*, John Wiley, New York.

Hettmansperger, T.P. and Norton, R.M. (1987) Tests for patterned alternatives in *k*-sample problems. *Journal of the American Statistical Association*, **82**, 292–9.

Hewitt, C.N. (ed) (1992) *Methods of Environmental Data Analysis*, Elsevier Applied Science, Amsterdam.

Hewlett, P.S. and Plackett, R.L. (1959) A unified theory for quantal response to mixtures of drugs: Noninteractive action. *Biometrics*, **15**, 591–610.

Hinkelmann, K. and Kempthorne, O. (1994) *Design and Analysis of Experiments*, John Wiley, New York.

Hochberg, Y. (1988) A sharper Bonferroni procedure for multiple tests of significance. *Biometrika*, **75**, 800–2.

Hochberg, Y. and Tamhane, A.C. (1987) *Multiple Comparison Procedures*, John Wiley, New York.

Hoel, D.G. (1983) Conditional two sample tests with historical controls, in *Contributions to Statistics: Essays in Honor of Norman L. Johnson* (ed P.K. Sen), North-Holland, Amsterdam, pp. 229–36.

Hoel, D.G. (1985) Mathematical dose-response models and their application to risk estimation, in *Methods for Estimating Risk of Chemical Injury: Human and Non-Human Biota and Ecosystems* (eds V.B. Vouk, G.C. Butler, D.G. Hoel and D.B. Peakall), John Wiley, New York, pp. 347–59.

Hoel, D.G. and Yanagawa, T. (1986) Incorporating historical controls in testing for a trend in proportions. *Journal of the American Statistical Association*, **81**, 1095–9.

Hogan, M.D., Kupper, L.L., Most, B.M. and Haseman, J.K. (1978) Alternatives to Rothman's approach for assessing synergism (or antagonism) in cohort studies. *American Journal of Epidemiology*, **108**, 60–7.

Hogg, R.V. and Tanis, E.A. (1997) *Probability and Statistical Inference*, 5th edn, Macmillan, New York.

Hollander, M., Mckeague, I.W. and Yang, J. (1997) Likelihood ratio-based confidence bands for survival functions. *Journal of the American Statistical Association*, **92**, 215–26.

Hollander, M. and Wolfe, D.A. (1973) *Nonparametric Statistical Methods*, John Wiley, New York.

Holschuh, N. (1980) Randomization and design: I, in *R. A. Fisher: An Appreciation* (eds S.E. Feinberg and D.V. Hinkley), Springer-Verlag, New York, pp. 35–45.

Hommel, G. (1988) A stagewise rejective multiple test procedure based on a modified Bonferroni test. *Biometrika*, **75**, 383–6.

Hommel, G. (1989) A comparison of two modified Bonferroni procedures. *Biometrika*, **76**, 624–5.

Hothorn, L. (1989) Robustness study on Williams- and Shirley-procedure, with application in toxicology. *Biometrical Journal*, **31**, 891–903.

Hothorn, L. (1994) Biostatistical analysis of the micronucleus mutagenicity assay based on the assumption of a mixing distribution. *Environmental Health Perspectives*, **102**, Suppl. 1, 121–5.

Hothorn, L. and Lemacher, W.A. (1991) A simple testing procedure "Control versus *k* treatments" for one-sided ordered alternatives, with application in toxicology. *Biometrical Journal*, **33**, 179–89.

Hothorn, L.A. (1997) Modifications of the closure principle for analyzing toxicological studies. *Drug Information Journal*, **31**, 403–12.

Hsu, J. (1996) *Multiple Comparisons*, Chapman & Hall, New York.

Hudson, D.J. (1966) Fitting segmented curves whose join points have to be estimated. *Journal of the American Statistical Association*, **61**, 1097–129.

Huff, J.E. (1989) Multiple-site carcinogenicity of benzene in Fischer 344 rats and $B6C3F_1$ mice. *Environmental Health Perspectives*, **82**, 125–63.

Huff, J.E., Haseman, J.K., McConnell, E.E. and Moore, J.A. (1986) The National Toxicology Program toxicology data evaluation techniques and long-term carcinogenesis studies, in *Safety Evaluation of Drugs and Chemicals* (ed W.E. Lloyd), Hemisphere Publishing, Washington, DC, pp. 411–46.

Hyrenius, H. (1950) Distribution of 'Student'-Fisher's *t* in samples from compound normal functions. *Biometrika*, **37**, 429–42.

Inman, H.F. and Bradley, E.L. (1994) Hypothesis test and confidence interval estimates for the overlap of two normal distributions with equal variances. *Environmetrics*, **5**, 167–90.

Irwin, J.O. (1937) Statistical method applied to biological assays (with discussion). *Journal of the Royal Statistical Society, supplement*, **4**, 1–60.

Jaffe, J.A. (1994) *Mastering the SAS System*, Van Nostrand Reinhold, New York.

Johnson, N.L. and Kotz, S. (1969) *Distributions in Statistics: Discrete Distributions*, Houghton-Mifflin, Boston.

Johnson, N.L. and Kotz, S. (1972) *Distributions in Statistics: Continuous Multivariate Distributions*, John Wiley, New York.

Johnson, N.L., Kotz, S. and Kemp, A.W. (1992) *Univariate Discrete Distributions*, 2nd edn, John Wiley, New York.

Johnson, N.L., Kotz, S. and Balakrishnan, N. (1994) *Continuous Univariate Distributions, Volume 1*, 2nd edn, John Wiley, New York.

Johnson, N.L., Kotz, S. and Balakrishnan, N. (1995) *Continuous Univariate Distributions, Volume 2*, 2nd edn, John Wiley, New York.

Jolicoeur, P. and Ponteir, J. (1989) Population growth and decline: a four-parameter generalization of the logistic curve. *Journal of Theoretical Biology*, **141**, 563–71.

Jonckheere, A.R. (1954) A distribution-free *k*-sample test against ordered alternatives. *Biometrika*, **41**, 133–45.

Jorgensen, M.A. (1994) Tail functions and iterative weights in binary regression. *American Statistician*, **48**, 230–4.

Kaplan, E.L. and Meier, P. (1958) Nonparametric estimation from incomplete observations. *Journal of the American Statistical Association*, **53**, 457–81.

Kärber, G. (1931) Beitrag zur kollektiven Behandlung pharmakologischer Reihenversuche. *Archiv für Experimentelle Pathologie und Pharmakologie*, **162**, 480–7.

Kastenbaum, M.A., Hoel, D.G. and Bowman, K.O. (1970a) Sample size requirements: one-way analysis of variance. *Biometrika*, **57**, 421–30.

Kastenbaum, M.A., Hoel, D.G. and Bowman, K.O. (1970b) Sample size requirements: randomized block designs. *Biometrika*, **57**, 573–7.

Katoh, Y., Maekawa, M. and Sano, Y. (1995) Mutagenicity of *O*-diazoacetyl-L-serine (azaserine) and 6-diazo-5-oxo-L-norleucine (DON) in a soybean test system. *Mutation Research*, **342**, 37–41.

Kendall, M.G. and Stewart, A. (1967) *The Advanced Theory of Statistics, Volume 2. Inference and Relationship*, 2nd edn, Hafner, New York.

Kerckaert, G.A., Brauninger, R., LeBouf, R.A. and Isfort, R.J. (1996) Use of the Syrian hamster embryo cell transformation assay for carcinogenicity prediction of chemicals currently being tested by the National Toxicology Program in rodent bioassays. *Environmental Health Perspectives*, **104**, Suppl. 5, 1075–84.

Keselman, H.J., Keselman, J.C. and Games, P.A. (1991) Maximum familywise Type I error rate: the least significant difference, Newman–Keuls, and other multiple comparison procedures. *Psychological Bulletin*, **110**, 155–61.

Khuri, A.I. (1993) *Advanced Calculus with Applications in Statistics*, John Wiley, New York.

Kimball, A.W. (1951) On dependent tests of significance in the analysis of variance. *Annals of Mathematical Statistics*, **22**, 600–2.

Kingman, A. and Zion, G. (1994) Some power considerations when deciding to use transformations. *Statistics in Medicine*, **13**, 769–83.

Kleinbaum, D.G., Kupper, L.L. and Muller, K.E. (1988) *Applied Regression Analysis and Other Multivariable Methods*, 2nd edn, Duxbury, Belmont, CA.

Kodell, R.L. and West, R.W. (1993) Upper confidence intervals on excess risk for quantitative responses. *Risk Analysis*, **13**, 177–82.

Kodell, R.L., Chen, J.J. and Moore, G.E. (1994) Comparing distributions of time to onset of disease in animal tumorigenicity experiments. *Communications in Statistics – Theory and Methods*, **23**, 959–80.

Kodell, R.L., Pearce, B.A., Turturro, A. and Ahn, H. (1997) An age-adjusted trend test for the tumor incidence rate for single-sacrifice experiments. *Drug Information Journal*, **31**, 471–87.

Krewski, D. and Franklin, C. (eds) (1991) *Statistics in Toxicology*, Gordon and Breach, New York.

Krewski, D., Smythe, R.T., Dewanji, A. and Colin, D. (1988) Statistical tests with historical controls, in *Carcinogenicity: The Design, Analysis, and Interpretation of Long-Term Animal Studies* (eds H.C. Grice and J.L. Ciminera), Springer-Verlag, New York, pp. 23–38.

Krewski, D., Leroux, B.G., Bleuer, S.R. and Broekhoven, L.H. (1993) Modeling the Ames *Salmonella*/microsome assay. *Biometrics*, **49**, 499–510.

Kroer, N., Coffin, R.B. and Jorgensen, N.O.G. (1994) Comparison of microbial trophic interactions in aquatic microcosms designed for the testing of introduced microorganisms. *Environmental Toxicology and Chemistry*, **13**, 247–58.

Kruglikov, I.L., Polig, E. and Jee, W.S.S. (1993) Statistics of hits to bone cell nuclei. *Radiation and Environmental Biophysics*, **32**, 87–98.

Kshirsagar, A.M. and Smith, W.B. (1995) *Growth Curves*, Marcel Dekker, New York.

Kuehl, R.O. (1994) *Statistical Principles of Research Design and Analysis*, Duxbury Press, Belmont, CA.

Lachenbruch, P.A. (1990) A note on goodness-of-link tests. *Journal of Statistical Computation and Simulation*, **36**, 43–5.

Lachenbruch, P.A. (1992) On the sample size for studies based upon McNemar's test. *Statistics in Medicine*, **11**, 1521–5.

Lahvis, G.P., Wells, R.S., Kuehl, D.W., Stewart, J.L., Rhinehart, H.L. and Via, C.S. (1995) Decreased lymphocyte responses in free-ranging bottlenose dolphins (*Tursiops truncatus*) are associated with increased concentrations of PCBs and DDT in peripheral blood. *Environmental Health Perspectives*, **103**, Suppl. 4, 67–72.

Lau, H.-S. and Lau, A.H.-L. (1991) Effective procedures for estimating beta distribution's parameters and their confidence intervals. *Journal of Statistical Computation and Simulation*, **38**, 139–50.

Lawless, J.F. (1982) *Statistical Models and Methods for Lifetime Data*, John Wiley, New York.

Lee, C.I. and Shen, S.Y. (1994) Convergence rates and powers of six power-divergence statistics for testing independence in 2 by 2 contingency tables. *Communications in Statistics – Theory and Methods*, **23**, 2113–26.

Lee, K.L., Harrell, F.E., Jr, Tolley, H.D. and Rosati, R.A. (1983) A comparison of test statistics for assessing the effects of concomitant variables in survival analysis. *Biometrics*, **39**, 341–450.

Leemis, L.M. (1986) Relationships among common univariate distributions. *American Statistician*, **40**, 143–6.

Leemis, L.M. and Trivedi, K.S. (1996) A comparison of approximate interval estimators for the Bernoulli parameter. *American Statistician*, **50**, 63–8.

Lefkopoulou, M. and Ryan, L. (1993) Global tests for multiple binary outcomes. *Biometrics*, **49**, 975–88.

Lefkopoulou, M., Rotnitzky, A. and Ryan, L. (1996) Trend tests for clustered data, in *Statistics in Toxicology* (ed B.J.T. Morgan), Clarendon Press, Oxford, pp. 179–97.

Lehmann, E.L. (1959) *Testing Statistical Hypotheses*, John Wiley, New York.

Lehmann, E.L. (1983) *Theory of Point Estimation*, John Wiley, New York.

Lemeshow, S. and Hosmer, D.W. (1982) A review of goodness of fit statistics for use in the development of logistic regression models. *American Journal of Epidemiology*, **115**, 92–106.

Lerman, P.M. (1980) Fitting segmented regression models by grid search. *Applied Statistics*, **29**, 77–84.

Leroux, B.G. and Krewski, D. (1993) Components of variation in mutagenic potency values based on the Ames *Salmonella* test. *Canadian Journal of Statistics*, **21**, 448–59.

Leroux, B.G., Fung, K.Y., Krewski, D. and Prentice, R.L. (1994) The use of historical control data in testing for trend in counts. *Statistica Sinica*, **4**, 581–602.

Leslie, P.H. (1945) On the use of matrices in certain population mathematics. *Biometrika*, **33**, 183–212.

Levenberg, K. (1944) A method for the solution of certain non-linear problems in least squares. *Quarterly of Applied Mathematics*, **2**, 164–8.

Levy, K.L. (1975) Large-sample many-to-one comparisons involving correlations, proportions or variances. *Psychological Bulletin*, **82**, 177–9.

Lewontin, R.C. and Felsenstein, J. (1965) The robustness of homogeneity tests in 2×*N* tables. *Biometrics*, **21**, 19–33.

Liang, K.Y. and Zeger, S.L. (1986) Longitudinal data analysis using generalized linear models. *Biometrika*, **73**, 13–22.

Lin, D.Y. and Wei, L.J. (1989) The robust inference for the Cox proportional hazards model. *Journal of the American Statistical Association*, **84**, 1074–8.

Lin, D.Y. and Wei, L.J. (1991) Goodness-of-fit tests for the general Cox regression model. *Statistica Sinica*, **1**, 1–17.

Lin, D.Y., Fleming, T.R. and Wei, L.J. (1994) Confidence bands for survival curves under the proportional hazards model. *Biometrika*, **81**, 73–81.

Lin, F.O. and Haseman, J.K. (1976) A modified Jonckheere test against ordered alternatives when ties are present at a single extreme value. *Biometrische Zeitschrift*, **18**, 623–31.

Lindsey, J.C. and Ryan, L.M. (1993) A three-state multiplicative model for rodent tumorigenicity experiments. *Applied Statistics*, **42**, 283–300.

Lindsey, J.C. and Ryan, L.M. (1994) A comparison of continuous- and discrete-time three-state models for rodent tumorigenicity experiments. *Environmental Health Perspectives*, **102**, Suppl. 1, 9–17.

Lockhart, A.-M., Piegorsch, W.W. and Bishop, J.B. (1992) Assessing overdispersion and dose response in the male dominant lethal assay. *Mutation Research*, **272**, 35–58.

Lovell, D.P. (1995) Population genetics of induced mutations. *Environmental and Molecular Mutagenesis*, **25**, Suppl. 26, 65–73.

Lovell, D.P. (1996) Statistical analysis of genetic toxicology test data, in *Statistics in Toxicology* (ed B.J.T. Morgan), Clarendon Press, Oxford, pp. 33–54.

Mack, G.A. and Wolfe, D.A. (1981) *K*-sample rank tests for umbrella alternatives. *Journal of the American Statistical Association*, **76**, 175–81.

Manly, B.F.J. (1991) *Randomization and Monte Carlo Methods in Biology*, Chapman & Hall, London.

Mann, H.B. and Whitney, D.R. (1947) On a test of whether one of two random variables is stochastically larger than the other. *Annals of Mathematical Statistics*, **18**, 50–60.

Mann, N.R. and Fertig, K.W. (1977) Efficient unbiased quantile estimators for moderate-size complete samples from extreme-value and Weibull distributions; Confidence bounds and tolerance and prediction intervals. *Technometrics*, **19**, 87–93.

Mantel, N. (1966) Evaluation of survival data and two new rank order statistics arising in its consideration. *Cancer Chemotherapy Reports*, **50**, 163–70.

Mantel, N. and Greenhouse, S.W. (1968) What is the continuity correction? *American Statistician*, **22**, 27–30.

Marazzi, A.N. (1993) *Algorithms, Routines and S Functions for Robust Statistics. The FORTRAN Library ROBETH with an Interface to S-Plus*, Chapman & Hall, New York.

Marcus, R., Peritz, E. and Gabriel, K.R. (1976) On closed testing procedures with special reference to ordered analysis of variance. *Biometrika*, **63**, 655–60.

Margolin, B.H. (1985) Statistical studies in genetic toxicology: A perspective from the US National Toxicology Program. *Environmental Health Perspectives*, **63**, 187–94.

Margolin, B.H. (1987) The use of multiple control groups in designed experiments. *Statistical Science*, **2**, 308–10.

Margolin, B.H. (1988) Test for trend in proportions, in *Encyclopedia of Statistical Sciences,* **9** (eds S. Kotz, N.L. Johnson and C.B. Read), John Wiley, New York, pp. 334–6.

Margolin, B.H. and Risko, K.J. (1984) The use of historical control data in laboratory studies. *Proceedings of the International Biometric Conference*, **12**, 21–30.

Margolin, B.H., Kaplan, N. and Zeiger, E. (1981) Statistical analysis of the Ames *Salmonella*/microsome test. *Proceedings of the National Academy of Sciences, USA*, **76**, 3779–83.

Margolin, B.H., Collings, B.J. and Mason, J.J. (1983) Statistical analysis and sample-size determinations for mutagenicity experiments with binomial response. *Environmental Mutagenesis*, **5**, 705–16.

Margolin, B.H., Resnick, M.A., Rimpo, J.Y., Archer, P., Galloway, S.M., Bloom, A.D. and Zeiger, E. (1986) Statistical analyses for in vitro cytogenetic assays using Chinese hamster ovary cells. *Environmental Mutagenesis*, **8**, 183–204.

Margolin, B.H., Kim, B.S. and Risko, K.J. (1989) The Ames *Salmonella*/microsome mutagenicity assay: Issues of inference and validation. *Journal of the American Statistical Association*, **84**, 651–61.

Marquardt, D.W. (1963) An algorithm for least-squares estimation of nonlinear parameters. *Journal of the Society for Industrial and Applied Mathematics*, **11**, 431–41.

Maul, A. (1991) Determination of endpoints in acute and chronic toxicity tests by using generalized linear models, in *Statistical Methods in Toxicology* (ed L. Hothorn), Springer-Verlag, Heidelberg, pp. 132–8.

McCullagh, P. and Nelder, J.A. (1989) *Generalized Linear Models*, 2nd edn, Chapman & Hall, London.

McCulloch, C.E. and Casella, G. (1983) Explicit formulas for confidence interval estimation in discrete distributions, Technical Report no. BU-820-M, Biometrics Unit, Cornell University, Ithaca, NY.

McKnight, B. (1991) Survival analysis, in *Statistics in Toxicology* (eds D. Krewski and C. Franklin), Gordon and Breach, New York, pp. 471–98.

McKnight, B. and Crowley, J. (1984) Tests for differences in tumour incidence based on animal carcinogenesis experiments. *Journal of the American Statistical Association*, **79**, 639–48.

McKnight, B. and Wahrendorf, J. (1992) Tumour incidence rate alternatives and the cause-of-death test for carcinogenicity. *Biometrika*, **79**, 131–8.

McMillan, A. and Silvers, A. (1996) Extending the Dewanji and Kalbfleisch model to incorporate marker data (with discussion). *Environmental and Ecological Statistics*, **3**, 189–206.

Mehta, C.R. and Patel, N.R. (1983) A network algorithm for performing Fisher's exact test in $r \times c$ contingency tables. *Journal of the American Statistical Association*, **78**, 427–34.

Mehta, C. and Patel, N. (1991) *StatXact: Statistical Software for Exact Nonparametric Inference – Manual*, Cytel Software Corp., Boston, MA.

Meier, K.L., Bailer, A.J. and Portier, C.J. (1993) A measure of tumorigenic potency incorporating dose-response shape. *Biometrics*, **49**, 917–26.

Michaelis, L. and Menten, M.L. (1913) Die Kinetik der Invertionwirkung. *Biochemische Zeitschrift*, **49**, 333–69.

Miller, R.G. (1981) *Simultaneous Statistical Inference*, 2nd edn, Springer-Verlag, New York.

Miller, R.G. (1986) *Beyond ANOVA, Basics of Applied Statistics*, John Wiley, New York.

Montelone, B.A., Gilbertson, L.A., Nassar, R., Giroux, C. and Malone, R.E. (1992) Analysis of the spectrum of mutations induced by the rad3-102 mutator allele of yeast. *Mutation Research*, **267**, 55–66.

Moore, D.F. (1986) Asymptotic properties of moment estimators for overdispersed counts and proportions. *Biometrika*, **73**, 583–8.

Moore, D.F. and Tsiatis, A. (1991) Robust estimation of the standard error in moment methods for extra-binomial and extra-Poisson variation. *Biometrics*, **47**, 383–401.

Moore, D.S. and McCabe, G.P. (1993) *Introduction to the Practice of Statistics*, W. H. Freeman & Co., New York.

Morgan, B.J.T. (1993) *Analysis of Quantal Response Data*, Chapman & Hall, New York.

Morgan, B.J.T. (ed) (1996) *Statistics in Toxicology*, Oxford University Press, Oxford.

Motimaya, A.M., Subramanya, K.S., Curry, P.T. and Kitchin, R.M. (1994) Lack of induction of micronuclei by azidothymidine (AZT) in vivo in mouse bone marrow cells. *Environmental and Molecular Mutagenesis*, **23**, 74–6.

Muhle, H., Bellman, B. and Pott, F. (1994) Comparative investigations of the biodurability of mineral fibers in the rat lung. *Environmental Health Perspectives*, **102**, Suppl. 5, 163–8.

Muller, K.E. and Benignus, V.A. (1992) Increasing scientific power with statistical power. *Neurotoxicology and Teratology*, **14**, 211–19.

Myers, L.E., Sexton, N.H., Southerland, L.I. and Wolff, T.J. (1981) Regression analysis of Ames test data. *Environmental Mutagenesis*, **3**, 575–86.

Nair, V.N. (1984) Confidence bands for survival functions with censored data: A comparative study. *Technometrics*, **26**, 265–75.

Nakamura, T. (1986) BMDP program for piecewise linear regression. *Computer Methods and Programs in Biomedicine*, **23**, 53–5.

National Toxicology Program (1984) Toxicology and carcinogenesis studies of 1,3-butadiene in $B6C3F_1$ mice, Technical Report no. 288, US Department of Health and Human Services, Public Health Service, Research Triangle Park, NC.

National Toxicology Program (1986) Toxicology and carcinogenesis studies of benzene in F344/N rats and $B6C3F_1$ mice, Technical Report no. 289, US Department of Health and Human Services, Public Health Service, Research Triangle Park, NC.

National Toxicology Program (1987a) Teratologic evaluation of codeine (CAS No. 76-57-3) administered to CD-1 mice on gestational days 6 through 13, Technical Report no. NTP-87-103, US National Toxicology Program, Research Triangle Park, NC.

National Toxicology Program (1987b) Toxicology and carcinogenesis studies of 1,4-dichlorobenzene in F344/N rats and B6C3F_1 mice, Technical Report no. 319, US Department of Health and Human Services, Public Health Service, Research Triangle Park, NC.

National Toxicology Program (1988) Toxicology and carcinogenesis studies of 1,2-epoxybutane in F344/N rats and B6C3F_1 mice, Technical Report no. 329, US Department of Health and Human Services, Public Health Service, Research Triangle Park, NC.

National Toxicology Program (1989a) Developmental toxicity of boric acid (CAS No. 10043-35-3) administered to CD-1 Swiss mice, Technical Report no. NTP-89-250, US National Toxicology Program, Research Triangle Park, NC.

National Toxicology Program (1989b) Toxicology and carcinogenesis studies of nitrofurantoin in F344/N rats and B6C3F_1 mice, Technical Report no. 341, US Department of Health and Human Services, Public Health Service, Research Triangle Park, NC.

National Toxicology Program (1993a) Toxicology and carcinogenesis studies of 1,3-butadiene in B6C3F_1 mice, Technical Report no. 434, US Department of Health and Human Services, Public Health Service, Research Triangle Park, NC.

National Toxicology Program (1993b) Toxicology and carcinogenesis studies of *p*-nitroaniline in B6C3F_1 mice, Technical Report no. 418, US Department of Health and Human Services, Public Health Service, Research Triangle Park, NC.

Nelder, J.A. and Pregibon, D. (1987) An extended quasi-likelihood function. *Biometrika*, **74**, 221–31.

Nelson, P. (1982) *Applied Life Data Analysis*, John Wiley, New York.

Neter, J., Kutner, M.H., Nachtsheim, C.J. and Wasserman, W. (1996) *Applied Linear Statistical Models*, 4th edn, R.D. Irwin, Chicago.

Neyman, J. and Scott, E.L. (1966) On the use of C(α) optimal tests of composite hypotheses. *Bulletin de l'Institut International de Statistique (Calcutta)*, **41**, 477–97.

Oehlert, G.W., Lee, R.J. and Van Orden, D. (1995) Statistical analysis of asbestos fibre counts. *Environmetrics*, **6**, 115–26.

Oris, J.T. and Bailer, A.J. (1993) Statistical analysis of the *Ceriodaphnia* toxicity test: Sample size determination for reproductive effects. *Environmental Toxicology and Chemistry*, **12**, 85–90.

Ott, W.R. (1995) *Environmental Statistics and Data Analysis*, CRC Press, Boca Raton, FL.

Pagano, M. and Trichler, D. (1983) On obtaining permutation distributions in polynomial time. *Journal of the American Statistical Association*, **78**, 435–40.

Passing, H. (1984) Exact simultaneous comparisons with control in $r{\times}c$ contingency tables. *Biometrical Journal*, **26**, 643–54.

Pasternak, H. and Shalev, B.A. (1992) An algorithm to fit the Gompertz function to growth curves. *Computer Applications in the Biosciences*, **8**, 239–42.

Patil, G.P. and Rao, C.R. (eds) (1994) *Handbook of Statistics Volume 12: Envi-*

ronmental Statistics, North-Holland/Elsevier, Amsterdam.

Patterson, D.G., Jr, Todd, G.D., Turner, W.E., Maggio, V., Alexander, L.R. and Needham, L.L. (1994) Levels of non-*ortho*-substituted (coplanar), mono- and di-*ortho*-substituted polychlorinated biphenyls, dibenzo-*p*-dioxins, and dibenzofurans in human serum and adipose tissue. *Environmental Health Perspectives*, **102**, Suppl. 1, 195–204.

Pearson, J.C.G. and Turton, A. (1993) *Statistical Methods in Environmental Health*, Chapman & Hall, New York.

Pearson, K. (1900) On the criterion that a given system of deviations from the probable in the case of a correlated system of variables is such that it can be reasonably supposed to have arisen from random sampling. *Philosophy Magazine*, **50**, 157–72.

Pendergast, J.F., Gange, S.J., Newton, M.A., Lindstrom, M.J., Palta, M. and Fisher, M.R. (1996) A survey of methods for analyzing clustered binary response data. *International Statistical Review*, **64**, 89–118.

Peto, R. and Peto, J. (1972) Asymptotically efficient rank invariant test procedures. *Journal of the Royal Statistical Society, series A*, **135**, 185–207.

Peto, R., Pike, M., Day, N., Gray, R., Lee, P., Parish, S., Peto, J., Pichards, S. and Wahrendorf, J. (1980) Guidelines for simple, sensitive significance tests for carcinogenic effects in long-term animal experiments. *IARC Monographs, Supplement*, **2**, 311–426.

Piegorsch, W.W. (1987a) Discretizing a normal prior for change point estimation in switching regressions. *Biometrical Journal*, **29**, 777–82.

Piegorsch, W.W. (1987b) Performance of likelihood-based interval estimates for two-parameter exponential samples subject to Type I censoring. *Technometrics*, **29**, 41–9.

Piegorsch, W.W. (1990) Maximum likelihood estimation for the negative binomial dispersion parameter. *Biometrics*, **46**, 863–7.

Piegorsch, W.W. (1991) Multiple comparisons for analyzing dichotomous response data. *Biometrics*, **47**, 45–52.

Piegorsch, W.W. (1992) Nonparametric methods to assess non-monotone dose response: Applications to genetic toxicology, in *Order Statistics and Nonparametrics: Theory and Applications* (eds P.K. Sen and I.A. Salama), North-Holland, Amsterdam, pp. 419–30.

Piegorsch, W.W. (1993) Biometrical methods for testing dose effects of environmental stimuli in laboratory studies. *Environmetrics*, **4**, 483–505.

Piegorsch, W.W. (1994) Environmental biometry: Assessing impacts of environmental stimuli via animal and microbial laboratory studies, in *Handbook of Statistics Volume 12: Environmental Statistics* (eds G.P. Patil and C.R. Rao), North-Holland/Elsevier, Amsterdam, pp. 535–59.

Piegorsch, W.W. and Bailer, A.J. (1994) Statistical approaches for analyzing mutational spectra: Some recommendations for categorical data. *Genetics*, **136**, 403–16.

Piegorsch, W.W. and Margolin, B.H. (1989) Quantitative methods for assessing a synergistic or potentiated genotoxic response. *Mutation Research*, **216**, 1–8.

Piegorsch, W.W., Weinberg, C.R. and Haseman, J.K. (1986) Testing for simple independent action between two factors for dichotomous response data.

Biometrics, **42**, 413–19.

Piegorsch, W.W., Weinberg, C.R. and Margolin, B.H. (1988) Exploring simple independent action in multifactor tables of proportions. *Biometrics*, **44**, 595–603.

Piegorsch, W.W., Lockhart, A.-M.C., Margolin, B.H., Tindall, K.R., Gorelick, N.J., Short, J.M., Carr, G.J., Thompson, E.D. and Shelby, M.D. (1994) Sources of variability in data from a *lacI* transgenic mouse mutation assay. *Environmental and Molecular Mutagenesis*, **23**, 17–31.

Plackett, R.L. and Hewlett, P.S. (1952) Quantal responses to mixtures of poisons (with discussion). *Journal of the Royal Statistical Society, series B*, **14**, 141–63.

Pocock, S.J. (1976) The combination of randomized and historical controls in clinical trials. *Journal of Chronic Diseases*, **29**, 175–88.

Portier, C.J. (1989) Quantitative risk assessment, in *Carcinogenicity and Pesticides. Principles, Issues, and Relationships* (eds N.N. Ragsdale and R.E. Menzer), American Chemical Society, Washington, DC, pp. 164–74.

Portier, C.J. and Bailer, A.J. (1989) Testing for increased carcinogenicity using a survival-adjusted quantal response test. *Fundamental and Applied Toxicology*, **12**, 731–7.

Portier, C.J. and Dinse, G.E. (1987) Semiparametric analysis of tumor incidence rates in survival/sacrifice experiments. *Biometrics*, **43**, 107–14.

Portier, C.J. and Hoel, D.G. (1983) Optimal design of the chronic animal bioassay. *Journal of Toxicology and Environmental Health*, **12**, 1–19.

Portier, C.J. and Hoel, D.G. (1987) Issues concerning the estimation of the TD_{50}. *Risk Analysis*, **7**, 437–47.

Portier, C.J., Hodges, J.C. and Hoel, D.G. (1986) Age-specific models of mortality and tumor onset for historical control animals in the National Toxicology Program's carcinogenicity experiments. *Cancer Research*, **46**, 4372–8.

Posten, H.O. (1992) Robustness of the two-sample *t*-test under violations of the homogeneity of variance assumption. II. *Communications in Statistics – Theory and Methods*, **21**, 2169–84.

Pregibon, D. (1980) Goodness of link test for generalized linear models. *Applied Statistics*, **29**, 14–23.

Pregibon, D. (1982) Score tests in GLIM with applications, in *GLIM82: Proceedings of the International Conference on Generalised Linear Models* (ed R. Gilchrist), Springer-Verlag, New York, pp. 87–97.

Pregibon, D. (1985) Link tests, in *Encyclopedia of Statistical Sciences,* **5** (eds S. Kotz, N.L. Johnson and C.B. Read), John Wiley, New York, pp. 82–5.

Prentice, R.L., Smythe, R.T., Krewski, D. and Mason, M. (1992) On the use of historical control data to estimate dose response trends in quantal bioassay. *Biometrics*, **48**, 459–78.

Press, W.H., Teukolsky, S.A., Vettering, W.T. and Flannery, B.P. (1992) *Numerical Recipes in FORTRAN: The Art of Scientific Computing*, 2nd edn, Cambridge University Press, New York.

Przyborowski, J. and Wilenski, H. (1935) Statistical principles of routine work in testing clover for doddler. *Biometrika*, **27**, 273–92.

Putman, D., San, R.H.C., Bigger, A., Levine, B.S. and Jacobson-Kram, D. (1996) Genetic toxicity assessment of HI-6 dichloride. *Environmental and Molecular Mutagenesis*, **27**, 152–61.

Ralston, M. and Jennrich, R.I. (1979) DUD, A derivative-free algorithm for nonlinear least squares. *Technometrics*, **21**, 7–14.

Ramakrishnan, V. and Meeter, D. (1993) Negative binomial cross-tabulations, with applications to abundance data. *Biometrics*, **49**, 195–207.

Rao, C.R. (1947) Large sample tests of statistical hypotheses concerning several parameters with applications to problems of estimation. *Proceedings of the Cambridge Philosophical Society*, **44**, 50–7.

Razzaghi, M. and Kodell, R. (1992) Box–Cox transformation in the analysis of combined effects of mixtures of chemicals. *Environmetrics*, **3**, 319–35.

Read, C.B. (1993) Freeman-Tukey chi-squared goodness-of-fit statistics. *Statistics and Probability Letters*, **18**, 271–8.

Read, T.R.C. and Cressie, N.A.C. (1988) *Goodness-of-Fit Statistics for Discrete Multivariate Data*, Springer-Verlag, New York.

Reitz, R.H., Ramsey, J.C., Andersen, M.E. and Gehring, P.J. (1988) Integration of pharmacokinetics and pathological data in dose selection for chronic bioassays, in *Carcinogenicity: The Design, Analysis, and Interpretation of Long-Term Animal Studies* (eds H.C. Grice and J.L. Ciminera), Springer-Verlag, New York, pp. 56–64.

Risko, K.J. and Margolin, B.H. (1996) Some observations on detecting extra-binomial variability within the beta-binomial model, in *Statistics in Toxicology* (ed B.J.T. Morgan), Clarendon Press, Oxford, pp. 57–65.

Robens, J.F., Piegorsch, W.W. and Schueler, R.L. (1989) Methods of testing for carcinogenicity, in *Principles and Methods of Toxicology*, 2nd edn (ed A.W. Hayes), Raven Press, New York, pp. 251–74.

Robertson, T., Wright, F.T. and Dykstra, R.L. (1988) *Order Restricted Statistical Inference*, John Wiley, New York.

Roff, D.A. and Bentzen, P. (1989) The statistical analysis of mitochondrial DNA polymorphisms: χ^2 and the problem of small samples. *Molecular Biology and Evolution*, **6**, 539–45.

Rom, D.M. (1990) A sequentially rejective test procedure based on a modified Bonferroni inequality. *Biometrika*, **77**, 663–5.

Rom, D.M. (1992) Strengthening some common multiple test procedures for discrete data. *Statistics in Medicine*, **11**, 511–14.

Rom, D.M., Costello, R.J. and Connell, L.T. (1994) On closed test procedures for dose-response analysis. *Statistics in Medicine*, **13**, 1583–96.

Rosenbaum, P.R. (1987) The role of a second control group in an observational study (with discussion). *Statistical Science*, **2**, 292–316.

Rosenkranz, H.S. and Klopman, G. (1993) Structural relationships between mutagenicity, maximum tolerated dose, and carcinogenicity in rodents. *Environmental and Molecular Mutagenesis*, **21**, 193–206.

Rosenthal, R. and Rubin, D.B. (1983) Ensemble-adjusted p values. *Psychological Bulletin*, **94**, 540–1.

Rothman, K.J. (1978) Estimation of confidence limits for the cumulative probability of survival in life table analysis. *Journal of Chronic Diseases*, **31**, 557–60.

Rotnitzky, A. and Jewell, N.P. (1990) Hypothesis testing of regression parameters in semiparametric generalized linear models for cluster correlated data. *Biometrika*, **77**, 485–97.

Russek-Cohen, E. (1994) STATXACT. *Bulletin of the Ecological Society of America*, **75**, 160–1.
Ryan, L.M. (1985) Efficiency of age-adjusted tests in animal carcinogenicity experiments. *Biometrics*, **41**, 525–31.
Ryan, L.M. (1993) Using historical controls in the analysis of developmental toxicity data. *Biometrics*, **49**, 1126–35.
Sahai, H. and Khurshid, A. (1993a) Confidence intervals for the mean of a Poisson distribution: A review. *Biometrical Journal*, **35**, 857–67.
Sahai, H. and Khurshid, A. (1993b) Confidence intervals for the ratio of two Poisson means. *The Mathematical Scientist*, **18**, 43–50.
Sahai, H. and Misra, S.C. (1992) Comparing the means of two Poisson distributions. *The Mathematical Scientist*, **17**, 60–7.
Salsburg, D.S. (1990) *Statistics for Toxicologists*, Marcel Dekker, New York.
Samuel-Cahn, E. (1996) Is the Simes improved Bonferroni procedure conservative? *Biometrika*, **83**, 928–33.
Samuels, M.L., Casella, G. and McCabe, G.P. (1994) Evaluating the efficiency of blocking without assuming compound symmetry. *Journal of Statistical Planning and Inference*, **38**, 237–48.
Santner, T.J. and Duffy, D.E. (1989) *The Statistical Analysis of Discrete Data*, Springer-Verlag, New York.
Santner, T.J. and Yamagami, S. (1993) Invariant small sample confidence intervals for the difference between two success probabilities. *Communications in Statistics – Theory and Methods*, **22**, 33–59.
SAS Institute Inc. (1989) *SAS/STAT® User's Guide, Version 6, Fourth Edition, Volumes 1–2*, SAS Institute Inc., Cary, NC.
SAS Institute Inc. (1993) *SAS® Technical Report P-243, SAS/STAT® Software: The GENMOD Procedure, Release 6.09*, SAS Institute Inc., Cary, NC.
Satterthwaite, F.E. (1946) An approximate distribution of estimates of variance components. *Biometrics*, **2**, 110–14.
Sawyer, C., Peto, R., Bernstein, L. and Pike, M.C. (1984) Calculation of carcinogenic potency from long-term animal carcinogenesis experiments. *Biometrics*, **40**, 27–40.
Scheaffer, R.L., Mendenhall, W. and Ott, L. (1996) *Elementary Survey Sampling*, 5th edn, Wadsworth, Belmont, CA.
Scheiner, S.M. and Gurevitch, J. (eds) (1993) *The Design and Analysis of Ecological Experiments*, Chapman & Hall, New York.
Schlueter, M.A., Guttman, S.I., Oris, J.T. and Bailer, A.J. (1995) Differential effects of allozyme genotypes on survival of juvenile fathead minnows, *Pimephales promelas*, exposed to copper. *Environmental Toxicology and Chemistry*, **14**, 1727–34.
Schoenfeld, D.A. (1986) Confidence bounds for normal means under order restrictions, with application to dose-response curves, toxicology experiments, and low-dose extrapolation. *Journal of the American Statistical Association*, **81**, 186–95.
Schumacher, M. and Schmoor, C. (1991) Statistical analysis of the Ames assay, in *Statistics in Toxicology* (ed L. Hothorn), Springer-Verlag, Heidelberg, pp. 5–19.

Schwetz, B.A. and Harris, M.W. (1993) Developmental toxicology: Status of the field and contribution of the National Toxicology Program. *Environmental Health Perspectives*, **100**, 269–82.

Searle, S.R. (1982) *Matrix Algebra Useful for Statistics*, John Wiley, New York.

Selby, P.B. and Olson, W.H. (1981) Methods and criteria for deciding whether specific-locus mutation-rate data in mice indicate a positive, negative, or inconclusive result. *Mutation Research*, **83**, 403–18.

Sen, P.K. and Krishnaiah, P.R. (1991) Selected tables for nonparametric statistics, in *Handbook of Statistics Volume 4: Nonparametric methods* (eds P.R. Krishnaiah and P.K. Sen), North-Holland/Elsevier, Amsterdam, pp. 937–58.

Severini, T.A. and Staniswallis, J.G. (1994) Quasi-likelihood estimation in semiparametric models. *Journal of the American Statistical Association*, **89**, 501–11.

Shaffer, J.P. (1995) Multiple hypothesis testing. *Annual Review of Psychology*, **46**, 561–84.

Sharma, K.K. and Rana, R.S. (1991) Robustness of sequential gamma life-testing procedures in respect of expected failure times. *Microelectronics and Reliability*, **31**, 1073–6.

Shi, N.-Z. (1988) Rank test statistics for umbrella alternatives. *Communications in Statistics – Theory and Methods*, **17**, 2059–73.

Shirley, E. (1977) A non-parametric equivalent of Williams' test for contrasting increasing dose levels of a treatment. *Biometrics*, **33**, 386–9.

Shiue, W.-K. and Bain, L.J. (1990) Simple approximate inference procedures for the mean of the gamma distribution. *Journal of Statistical Computation and Simulation*, **34**, 67–73.

Sichel, H.S. (1973) On a significance test for two Poisson variables. *Applied Statistics*, **22**, 50–8.

Šidák, Z. (1967) Rectangular confidence regions for the means of multivariate normal distributions. *Journal of the American Statistical Association*, **62**, 626–33.

Simes, R.J. (1986) An improved Bonferroni procedure for multiple tests of significance. *Biometrika*, **73**, 751–4.

Simpson, D.G. and Dallal, G.E. (1989) BUMP: A FORTRAN program for identifying dose-response curves subject to downturns. *Computers and Biomedical Research*, **22**, 36–43.

Simpson, D.G. and Margolin, B.H. (1986) Recursive nonparametric testing for dose-response relationships subject to downturns at high doses. *Biometrika*, **73**, 589–96.

Simpson, D.G. and Margolin, B.H. (1990) Nonparametric testing for dose-response curves subject to downturns: Asymptotic power considerations. *Annals of Statistics*, **18**, 373–90.

Smith, H.F. (1936) The problem of comparing the results of two experiments with unequal errors. *Journal of the Council of Scientific and Industrial Research*, **9**, 211–12.

Snedecor, G.W. and Cochran, W.G. (1980) *Statistical Methods*, 7th edn, Iowa State University Press, Ames.

Snee, R.D. and Irr, J.D. (1984) A procedure for the statistical evaluation of Ames *Salmonella* assay results: comparison of results among 4 laboratories. *Mutation Research*, **128**, 115–25.

Solomon, D.L. (1983) The spatial distribution of cabbage butterfly eggs, in *Life Science Models, Vol. 4* (eds H. Marcus-Roberts and M. Thompson), Springer-Verlag, New York, pp. 350–66.

Soper, K.A. and Tonkonoh, N. (1993) The discrete distribution used for the log-rank statistic test can be inaccurate. *Biometrical Journal*, **35**, 291–8.

Spearman, C. (1908) The method of 'right and wrong cases' ('constant stimuli') without Gauss's formulae. *Journal of Psychology*, **2**, 227–42.

Spector, P. (1994) *An Introduction to S and S-PLUS*, Duxbury Press, Belmont, CA.

Spouge, J.L. (1994) Computation of the gamma, digamma, and trigamma functions. *SIAM Journal on Numerical Analysis*, **31**, 931–44.

Spurrier, J.D. (1993) Distribution-free and asymptotically distribution-free comparisons with a control in blocked experiments, in *Multiple Comparisons, Selection, and Applications in Biometry* (ed F.M. Hoppe), Marcel Dekker, New York, pp. 97–120.

Spurrier, J.D. and Isham, S.P. (1985) Exact simultaneous confidence intervals for pairwise comparisons of three normal means. *Journal of the American Statistical Association*, **80**, 438–42.

Srinivasan, R. and Wharton, R.M. (1975) Confidence bands for the Weibull distribution. *Technometrics*, **17**, 375–80.

StatSci Division of MathSoft Inc. (1995) *S-Plus Guide to Statistical and Mathematical Analysis, Version 3.3*, MathSoft, Inc., Seattle, WA.

Stead, A.G., Hasselblad, V., Creason, J.P. and Claxton, L. (1981) Modeling the Ames test. *Mutation Research*, **85**, 13–27.

Stebbing, A.R.D. (1982) Hormesis – The stimulation of growth by low levels of inhibitors. *Science of the Total Environment*, **22**, 213–34.

Stoline, M.R. (1991) An examination of the lognormal and Box and Cox family of transformations in fitting environmental data. *Environmetrics*, **2**, 85–106.

Stoline, M.R. (1993) Comparison of two medians using a two-sample lognormal model in environmental contexts. *Environmetrics*, **4**, 323–40.

Student (1908) The probable error of a mean. *Biometrika*, **6**, 1–25.

Styer, P.E. (1994) An illustration of the use of generalized linear models to measure long-term trends in the wet deposition of sulfate, Technical Report no. 18, National Institute of Statistical Sciences, Research Triangle Park, NC.

Suissa, S. and Salmi, R. (1989) Unidirectional multiple comparisons of Poisson rates. *Statistics in Medicine*, **8**, 757–64.

Suter, G.W. (ed) (1993) *Ecological Risk Assessment*, Lewis Publishers, Boca Raton, FL.

Suter, G.W. (1996) Abuse of hypothesis testing statistics in ecological risk assessment. *Human and Ecological Risk Assessment*, **2**, 331–47.

Takeshima, K. and Yanagimoto, T. (1978) Threshold values as safe levels and the hockey stick regression method. *Annals of the Institute of Statistical Mathematics*, **25**, 29–40.

Tamhane, A.C. (1986) A survey of literature on quantal response curves with a view towards application of the problem of selecting the curve with the smallest q-quantile (ED100q). *Communications in Statistics – Theory and Methods*, **15**, 2679–718.

Tan, W.-Y. and Zhang, J.H. (1996) Monte Carlo results for the 3-poly test for animal carcinogenicity experiments. *Environmental Health Perspectives*, **104**, 872–7.

Tarone, R.E. (1975) Test for trend in life table analysis. *Biometrika*, **62**, 679–82.

Tarone, R.E. (1979) Testing the goodness of fit of the binomial distribution. *Biometrika*, **66**, 585–90.

Tarone, R.E. (1982a) The use of historical control information in testing for a trend in Poisson means. *Biometrics*, **38**, 457–62.

Tarone, R.E. (1982b) The use of historical control information in testing for a trend in proportions. *Biometrics*, **38**, 215–20.

Tarone, R.E. (1986) Correcting tests for trend in proportions for skewness. *Communications in Statistics – Theory and Methods*, **15**, 317–28.

Tarone, R.E. and Gart, J.J. (1980) On the robustness of combined tests for trends in proportions. *Journal of the American Statistical Association*, **75**, 110–16.

Thompson, S.K. (1992) *Sampling*, John Wiley, New York.

Tishler, A. and Zang, I. (1981) A new maximum likelihood algorithm for piecewise regression. *Journal of the American Statistical Association*, **76**, 980–7.

Trevan, J.W. (1927) The error of determination of toxicity. *Proceedings of the Royal Society, Series B*, **101**, 483–514.

Tsiatis, A.A. (1980) A note on a goodness-of-fit test for the logistic regression without replication. *Biometrika*, **67**, 250–1.

Tsiatis, A.A. (1981) A large sample study of Cox's regression model. *Annals of Statistics*, **9**, 93–108.

Tsiatis, A.A. (1990) Estimating regression parameters using linear rank tests for censored data. *Annals of Statistics*, **18**, 354–72.

Tukey, J.W. (1994) The problem of multiple comparisons, in *The Collected Works of John W. Tukey, Vol. VIII. Multiple Comparisons: 1948–1983* (ed H.I. Braun), Chapman & Hall, New York, pp. 1–300.

Ulm, K.W. (1991) A statistical method for assessing a threshold in epidemiological studies. *Statistics in Medicine*, **10**, 341–9.

Uusipaikka, E. (1985) Exact simultaneous confidence intervals for multiple comparisons of three or four mean values. *Journal of the American Statistical Association*, **80**, 196–201.

Væth, M. (1985) On the use of Wald's test in exponential families. *International Statistical Review*, **53**, 199–214.

Van Beneden, R.J. (1994) Molecular analysis of bivalve tumors: Models for environmental/genetic interactions. *Environmental Health Perspectives*, **102**, Suppl. 12, 81–3.

van der Hoeven, N. (1997) How to measure no effect. Part III: Statistical aspects of NOEC, EC*x* and NEC estimates. *Environmetrics*, **8**, 255–61.

van der Laan, P. and Verdooren, L.R. (1987) Classical analysis of variance methods and nonparametric counterparts. *Biometrical Journal*, **29**, 635–65.

Van Ewijk, P.H. and Hoekstra, J.A. (1993) Calculation of the EC_{50} and its confidence interval when subtoxic stimulus is present. *Ecotoxicology and Environmental Safety*, **25**, 25–32.

Venables, W.N. and Ripley, B.D. (1997) *Modern Applied Statistics with S-Plus*, 2nd edn, Springer-Verlag, New York.

Venzon, D.J. and Moolgavkar, S.H. (1988) A method for computing profile-likelihood-based confidence intervals. *Applied Statistics*, **37**, 87–94.

Vollset, S.E. (1993) Confidence intervals for a binomial proportion. *Statistics in Medicine*, **12**, 809–24.

Wacholder, S. (1986) Binomial regression in GLIM: Estimating risk ratios and risk differences. *American Journal of Epidemiology*, **123**, 174–84.

Wackerly, D.D., Mendenhall, W. and Scheaffer, R.L. (1996) *Mathematical Statistics with Applications*, 5th edn, Duxbury Press, Belmont, CA.

Wahrendorf, J., Zentgraf, R. and Brown, C.C. (1981) Optimal designs for the analysis of interactive effects of two carcinogens or other toxicants. *Biometrics*, **37**, 45–54.

Wald, A. (1943) Tests of statistical hypotheses concerning several parameters when the number of observations is large. *Transactions of the American Mathematical Society*, **54**, 426–82.

Wallenstein, S. and Bodian, C. (1987) Inferences on odds ratios, relative risks, and risk differences based on standard regression programs. *American Journal of Epidemiology*, **126**, 346–55.

Wang, Y.Y. (1971) Probabilities of the Type I errors of the Welch tests for the Behrens–Fisher problem. *Journal of the American Statistical Association*, **66**, 605–8.

Weber, C.I., Peltier, W.H., Norberg-King, T.J., Horning, W.B., Kessler, F.A., Menkedick, J.R., Neiheisel, T.W., Lewis, P.A., Klemm, D., Pickering, Q.H., Robinson, E.L., Lazorchak, J.M., Wymer, L.J. and Freyberg, R.W. (1989) Short-term methods for estimating the chronic toxicity of effluents and receiving waters to freshwater organisms, 2nd edn, Technical Report no. EPA/600/4-89/001A, US Environmental Protection Agency, Cincinnati, OH.

Wedderburn, R.W.M. (1974) Quasi-likelihood functions, generalized linear models, and the Gauss-Newton method. *Biometrika*, **61**, 439–47.

Wei, L.J. (1992) The accelerated failure time model – a useful alternative to the Cox regression model in survival analysis. *Statistics in Medicine*, **11**, 1871–9.

Weibull, W. (1951) A statistical distribution function of wide applicability. *Journal of Applied Mechanics*, **18**, 293–7.

Weinberg, C.R. (1986) Applicability of the simple independent action model to epidemiologic studies involving two factors and a dichotomous outcome. *American Journal of Epidemiology*, **123**, 162–73.

Weinberg, W. (1908) Über den Nachweis der Vererbung beim Menschen. *Jahreshefte des Vereins für vaterländische Naturkunde in Württemberg*, **64**, 369–82.

Weir, B.S. (1990) *Genetic Data Analysis*, Sinauer Associates, Sunderland, MA.

Weissfeld, L.A., St. Laurent, R.T. and Moulton, L.H. (1991) Confidence intervals for the comparison of two Poisson distributions. *Communications in Statistics – Theory and Methods*, **20**, 3071–81.

Welch, B.L. (1938) The significance of the difference between two means when the

population variances are unequal. *Biometrika*, **29**, 350–62.
West, R.W. and Kodell, R.L. (1993) Statistical methods of risk assessment for continuous variables. *Communications in Statistics – Theory and Methods*, **22**, 3363–76.
Westfall, P.W. and Wolfinger, R.D. (1997) Multiple tests with discrete distributions. *American Statistician*, **51**, 3–8.
Weston, S.A. and Meeker, W.Q., Jr (1991) Coverage probabilities of nonparametric simultaneous confidence bands for a survival function. *Journal of Statistical Computation and Simulation*, **38**, 83–97.
White, D.H. and Hoffman, D.J. (1995) Effects of polychlorinated dibenzo-*p*-dioxins and dibenzofurans on nesting wood ducks (*Aix sponsa*) at Bayo Meto, Arkansas. *Environmental Health Perspectives*, **103**, Suppl. 4, 37–9.
Whittaker, S.G., Zimmermann, F.K., Dicus, B., Piegorsch, W.W., Fogel, S. and Resnick, M.A. (1989) Detection of induced chromosome loss in *Saccharomyces cerevisiae* – An interlaboratory study. *Mutation Research*, **224**, 31–78.
Whittemore, A.S. (1983) Transformations to linearity in binary regression. *SIAM Journal of Applied Mathematics*, **43**, 703–10.
Wilcoxon, F. (1945) Individual comparisons by ranking methods. *Biometrics*, **1**, 80–3.
Williams, D.A. (1971) A test for differences between means when several dose levels are compared with a zero dose control. *Biometrics*, **27**, 103–17.
Williams, D.A. (1972) The comparison of several dose levels with a zero dose control. *Biometrics*, **28**, 519–31.
Williams, D.A. (1986) A note on Shirley's nonparametric test for comparing several dose levels with a zero-dose control. *Biometrics*, **42**, 183–6.
Williams, D.A. (1991) The reliability of tests of hypotheses when overdispersed logistic-linear models are fitted by maximum quasi-likelihood. *Biometrical Journal*, **33**, 259–70.
Williams, D.A. (1996) Overdispersion in logistic-linear models, in *Statistics in Toxicology* (ed B.J.T. Morgan), Clarendon Press, Oxford, pp. 75–84.
Williams, P.L. and Portier, C.J. (1992) Explicit solutions for constrained maximum likelihood estimators in survival sacrifice experiments. *Biometrika*, **79**, 717–29.
Wilson, E.B. (1927) Probable inference, the law of succession, and statistical inference. *Journal of the American Statistical Association*, **22**, 209–12.
Winsor, C.P. (1932) The Gompertz curve as a growth curve. *Proceedings of the National Academy of Sciences, USA*, **18**, 1–8.
Witt, K.L., Gulati, D.K., Kaur, P. and Shelby, M.D. (1995) Phenolphthalein: induction of micronucleated erythrocytes in mice. *Mutation Research*, **341**, 151–60.
Woolf, B. (1955) On estimating the relation between blood group and disease. *Annals of Human Genetics*, **19**, 251–3.
Wypij, D. and Santner, T.J. (1990) Interval estimation of the marginal probability of success for the beta-binomial distribution. *Journal of Statistical Computation and Simulation*, **35**, 169–85.
Yanagawa, T., Kikuchi, Y. and Brown, K.G. (1994) Statistical issues on the no-observed-adverse-effect level in categorical response. *Environmental Health Perspectives*, **102**, Suppl. 1, 95–104.

Yanagimoto, T. and Yamamoto, E. (1979) Estimation of safe doses: Critical review of the hockey stick regression model. *Environmental Health Perspectives*, **32**, 193–9.

Yates, F. (1934) Contingency tables involving small numbers and the χ^2 test. *Journal of the Royal Statistical Society, supplement*, **1**, 217–35.

Yates, F. (1948) The analysis of contingency tables based on quantitative characters. *Biometrika*, **35**, 176–81.

Yates, F. (1984) Tests of significance for 2×2 contingency tables (with discussion). *Journal of the Royal Statistical Society, series A*, **147**, 426–63.

Zeger, S.L. and Liang, K.Y. (1992) An overview of methods for the analysis of longitudinal data. *Statistics in Medicine*, **11**, 1825–39.

Zeiger, E., Risko, K.J. and Margolin, B.H. (1985) Strategies to reduce the cost of mutagenicity screening with the *Salmonella* assay. *Environmental Mutagenesis*, **7**, 901–11.

Zelterman, D. (1987) Goodness-of-fit tests for large sparse multinomial distributions. *Journal of the American Statistical Association*, **82**, 624–9.

Zimmermann, F.K. and Mohr, A. (1992) Formaldehyde, glyoxal, urethane, methyl carbamate, 2,3-butanedione, 2,3-hexanedione, ethyl acrylate, dibromoacetonitrile and 2-hydroxypropionitrile induce chromosome loss in *Saccharomyces Cerevisiae*. *Mutation Research*, **270**, 151–66.

Subject Index